ऋग्वेदः

THE MAGNETO-PLASMA COSMOLOGY OF THE ṚGVEDA

AN ALTERNATIVE TO THE BIG BANG THEORY

SECOND EDITION (FIRST PUBLIC EDITION)

Y. SOHN

THE MAGNETO-PLASMA COSMOLOGY OF THE ṚGVEDA: AN ALTERNATIVE TO THE BIG BANG THEORY

AUTHOR: Youngsinn Sohn

Independently published by the author in the United States
Ellicott City, MD
Printed on acid-free paper

Manuscript print (for copyright registration and private distribution), March & May 2020
Second Edition (First Public Edition), March 2021

ISBN 978-1-7346730-1-2 Paperback

The idea that the Ṛgveda is about the magneto-plasma cosmology may sound fantastical if one adheres to the Darwinian ideology and the views of conventional human history. And yet reasoned interpretations and translations of the hymns of the Ṛgveda, along with other ancient texts, especially the ancient alchemical texts, and insightful examinations of the planetary archaeological evidences in our solar system will inevitably lead one to the conclusion that, in remote antiquity, frontier outposts of very advanced civilisations were established on the planets in our solar system and that this knowledge of cosmic creation must have originated from these advanced civilisations. Moreover, it has been known among learned pundits of India that the Ṛgveda is about the cosmic creation. This known fact that the Ṛgveda is about the cosmic creation is ignored by most of the Vedic students in the groves of academe. And the revered Veda became an irrelevant, distant memory "in the land of its birth". This book is the fruits of the author's efforts to find the true meaning of the hymns of the Ṛgveda and the knowledge they reveal.

Three hundred twenty eight select stanzas from the ten maṇḍalas of the Ṛgveda and six stanzas from the Kaṭha Upaniṣad, Siddhāntaśiromaṇi, and Sūryasiddhānta are translated and interpreted from the original Vedic and Sanskrit texts by the author to draw a complete picture of the magneto-plasma cosmology revealed in the Ṛgveda. The knowledge of the magneto-plasma cosmology revealed in the Ṛgveda is the knowledge of the cosmic creation, the great work of Nature, the celestial Alchemy. Cosmic ionisation and cosmic plasma discharge and release mechanisms, presented in this book for the first time, are based on what the Ṛgveda reveals.

SUBJECTS: Ṛgveda, cosmic creation, magneto-plasma cosmology of the Ṛgveda, cosmic ionisation and cosmic plasma discharge and release mechanisms, Vedic symbolism associated with cosmic creation

This work is dedicated
to Vivekānanda (विवेकानन्दः), to J. J. Goodwin,
to the Brāhmaṇas who have guarded and preserved the Veda for millennia,

And
to Students of the Veda,
to Hannes Alfvén, to Anthony L. Peratt.

Contents

List of Illustrations xii
Acknowledgements xv
Preface xxi
A Note to the Second Edition xxvii
Dear Readers xxix
Notes xxxi

TECHNICAL NOTES xxxi
SANSKRIT TRANSLITERATION xxxiv
ABBREVIATIONS AND SYMBOLS xxxvi

I. The Ṛgveda and Its Interpretation 1

1.1 THE ṚGVEDA AND THE VEDIC LITERATURE 1
1.2 THE VEDĀṄGAS, THE SIX LIMBS OF THE VEDA 3
1.3 THE DATE OF THE ṚGVEDA 4
1.4 PROBLEMS ASSOCIATED WITH TRADITIONAL INTERPRETATIONS OF THE ṚGVEDA 6
1.5 A NEW INTERPRETATION OF THE ṚGVEDA 10
1.6 THE TEXT OF THE ṚGVEDA 12

II. The Heavenly Structures Revealed in the Ṛgveda 15

2.1 THE SHINING ONES, THE DEVAS AND THE ṚṢIS 15

2.2 THE THREEFOLD HEAVENLY SPHERE, THE ALTAR OF COSMIC YAJÑA 18
2.2.1 The Upper Heaven, the Innerfield, and the Lower Heaven 18
2.2.2 The Cosmic Yajña, the Origin of the World and Life 20
2.2.3 Make Me Immortal in That Luminous Threefold Heaven 22

2.3 THE HEIGHT OF THE HEAVENLY VAULT 23
2.3.1 The Height of the Heavenly Vault Revealed in the Sūryasiddhānta 24
2.3.2 The Height of the Heavenly Vault by the Siddhāntaśiromaṇi 26

2.4 THE TWOFOLD HEAVENLY VAULT AND ITS SUBSTRUCTURES 28
2.4.1 The Great Heavens of Two, the Upper and the Lower Heavens 28
2.4.2 The Twofold Heavenly Vault Observed by Chandra and Hubble 29
2.4.3 The Soma Pond in the Vault of the Heaven 31
2.4.4 Vimāna, the Chariot of the Devas 33
2.4.5 The Upper Heaven Revolves Westward, the Lower Heaven Eastward 34
2.4.6 The Heaven, the Red Bull 36
2.4.7 The Structures of the Heaven, the Complete Picture 37

2.5 THE SOMA POND, THE WOMB OF COSMIC CREATION 39
2.5.1 The Soma Pond Observed by Chandra 39
2.5.2 The Soma Pond, the Golden Womb 42
2.5.3 Samudra, the Ocean of Celestial Waters 44
2.5.4 Rodasī, the Pair of Mātā and Duhitā, the Pair of the Stone and the Soma Pond 45
2.5.5 The Radiant Udder of Pṛśni 46
2.5.6 The Moon with Beauteous Wings Glides in the Heaven 47
2.5.7 Sūrya, the Best of Lights 48
2.5.8 Sūrya, the Single Eye, the All-Seeing Eye 51
2.5.9 Sūrya, the Eye of Mitra and Varuṇa 53
2.5.10 Sūrya Takes Hold of the Wondrous Weapon 54
2.5.11 Sūrya, the Refulgent Chariot 55
2.5.12 Uṣā, the Dawn 55
2.5.13 The Celestial Boat Equipped with Oars 58
2.5.14 The Heads of Sorceresses 59
2.5.15 An Owl, an Owlet, and the Sorcerer 59
2.5.16 The Mighty Thunderbolt of Indra 60
2.5.17 Rudra 60
2.5.18 The Maruts Arrive in Lightning Laden Chariots with Spears 63
2.5.19 The Maruts Make Streaming Currents and Flashing Thunderbolts 65
2.5.20 The Maruts Move Swiftly Like Vātas and Are Effulgent Like Tongues of Fires 66
2.5.21 The Maruts Are Spirited Like Serpents 66
2.5.22 The Maruts, the Eternal Flames 67
2.5.23 Triṣṭup, a Libation for the Maruts 68

2.5.24 Vāyu and Vāta 68
2.5.25 Indra Slays the Serpent of the Heaven 71
2.5.26 Indra Delights in Soma Drinking 72
2.5.27 Indra, the Maker of Light and Form 74
2.5.28 Saṃvatsarakālacakraṃ, the Soma Pond and the Wheel of Cosmic Ionisation 75
2.5.29 The Thousand-Pillared Abode of Kings Varuṇa and Mitra 77

2.6 THE SACRIFICIAL POSTS 77
2.6.1 The Sacrificial Post, the Luminous Pillar of the Heaven 77
2.6.2 A Hundred or a Thousand Sacrificial Posts 80
2.6.3 The Remaining Stump of the Destroyed Sacrificial Post 83

2.7 THE STONE, THE CELESTIAL MACHINE OF IONISATION AND COSMIC PLASMA DISCHARGE AND RELEASE 84
2.7.1 The Stone and the Keystone 84
2.7.2 The Two Stones of Trita 86
2.7.3 The Two Stones at the Base of the Soma Pond 87
2.7.4 The Stones Are Impelled by Savitā 89
2.7.5 The Stones Sing and Make Rumbling Sounds 90
2.7.6 The Two Arms and Ten Fingers of the Stone 95
2.7.7 Tvaṣṭā's Ten Maidens 96
2.7.8 The Stone Wheel 97
2.7.9 The Two Stones, Mortar and Pestle 97
2.7.10 Like a Real Comb 101
2.7.11 Kṣetrapati, the Lord of the Field 104
2.7.12 The Adhvaryus Make Sweet Soma Mead 107

2.8 PṚTHIVĪ, THE MILKY WAY 108
2.8.1 The Milky Way Has Four Major Spiral Arms 108
2.8.2 Different Regions of the Milky Way 110
2.8.3 Five Regions 111
2.8.4 Six Divine Expanses 111
2.8.5 Seven Domains 112
2.8.6 Eight Regions 113
2.8.7 A Hymn to Pṛthivī 114

2.9 AŚVATTHA, THE INVERTED TREE, THE VEDIC TREE OF LIFE 115

III. The Divine Architects 119

3.1 TVAṢṬĀ, MAGNETISM IN GENERAL AND THE MAGNETIC FIELD IN PARTICULAR 119
3.1.1 Tvaṣṭā, the First Born 119

3.1.2 The Shaper of All Forms 120
3.1.3 The Possessor of the World 121
3.1.4 Tvaṣṭā Holds the Goblets That Serve the Drink of the Devas 121
3.1.5 Tvaṣṭā, the Divine Architect, the All-Seeing One with Two Winged Arms 122

3.2 The Ṛbhus, the Magnetic Forces 126
3.2.1 The Ṛbhus Revive Their Old Parents and Impel the Yajña 127
3.2.2 The Ṛbhus Guard the Cow and Make the Four Celestial Bowls 128
3.2.3 The Skilful Architects Repose at Agni's Rite of Soma Reception 130
3.2.4 The Ṛbhus Make Two Bay Steeds for Indra 132

IV. Celestial Ionisation and Other Plasma Phenomena 135

4.1 Celestial Ionisation 135
4.1.1 Praise of the Ploughing Wheel 135
4.1.2 The Gamester's Reproach of the Ploughing Wheel That Shatters the Ionic Bond 138
4.1.3 Cosmic Plasma Discharge 140

4.2 Celestial Waters, the Cosmic Plasmas 144
4.2.1 The Adhvaryus Press Out the Celestial Waters 144
4.2.2 Celestial Waters Rise Up and Are Released Day after Day 146
4.2.3 Sūrya Spins the Celestial Waters with His Beams of Light 147
4.2.4 Varuṇa, Mitra, Aryamā: Cosmic Plasma Discharge Modes 149
4.2.5 Varuṇa, the Supreme Monarch 150
4.2.6 Within Varuṇa the Three Heavens Are Placed 151
4.2.7 Mitra and Varuṇa Sustain the Cosmic Sacrifice Through the Cosmic Sacrifice 152
4.2.8 Mitra Beholds the Tillers of the Field with an Unwinking Eye 153

4.3 Aditi, the Double Headed 154
4.3.1 Aditi, the Mother of Kings Mitra and Varuṇa 155
4.3.2 Aditi, the Milch Cow 156
4.3.3 Aditi Unbinds the Sin of the Ionic Bond 156
4.3.4 Aditi, Daughter of Dakṣa 157
4.3.5 Aditi and Diti 159
4.3.6 The Ādityas, the Sons of Aditi 160

4.4 Savitā 161
4.4.1 Savitā, Tvaṣṭā, Viśvarūpa 161
4.4.2 By Savitā's Propulsion Celestial Waters Flow 162
4.4.3 Savitā Impels Our Hymns 163
4.4.4 Three Times a Day Savitā Brings Celestial Waters 163

4.5 VṚTRA, THE SERPENT OF THE HEAVEN, THE PLASMA SHEATH 164
4.5.1 Dāsa and Ārya, the Two Opposite Charges of the Double Layer 164
4.5.2 Let Not My Thread Be Severed Before Time 165

4.6 VARUṆA'S NOOSE, THE SIN OF THE IONIC BOND 166
4.6.1 The Bonds Not Wrought of Ropes 166
4.6.2 Remove the Noose from Us O Varuṇa 167

4.7 VĀK, THE SPEECH OF COSMIC PLASMAS 168
4.7.1 Vāk, the Queen of the Devas 168
4.7.2 A Hymn of Vāk 169

4.8 THE WATCHERS 173
4.8.1 Varuṇa's Watchers 173
4.8.2 Agni's Watchers 174
4.8.3 Soma's Watchers 175

4.9 MĀYĀ 176
4.9.1 Māyā, the Magic Force of Varuṇa 176
4.9.2 Māyā Drives the Flying Horse 177

4.10 ĀTMĀ 178
4.10.1 Ātmā, the Soul of Cosmic Yajña 178
4.10.2 Have I Indeed Drunk Soma? 179

4.11 MANA, THE COSMIC MIND 182

V. In the Beginning 187

5.1 A HYMN OF THE CREATION 187
5.2 WEAVING THE NEBULOUS FABRIC WITH SĀMA HYMNS 191

VI. The Events of Cosmic Plasma Release 197

6.1 AŚVINĀ, THE TWIN JETS 198
6.1.1 Aśvinā Transport the Plasma Sheaths Filled with Soma Mead 199
6.1.2 The Spectacle of Cosmic Eruption, the Dappled Cows Are Released 201
6.1.3 The Mighty Aśvinā Perform Wondrous Deeds 205
6.1.4 The Coursers of Aśvinā Rush Over the Blazing Summit of the Heaven 205
6.1.5 Aśvinā Bring the Two Streams of Light 207
6.1.6 The Two Circumambulatory Paths of Aśvinā 207

6.2 THE CELESTIAL RIVERS 208
6.2.1 Sindhu Flows Upwards Roaring Like a Charging Bull 209
6.2.2 Sindhu Leads the Two Wings of Army Like a Warrior King 210
6.2.3 Seek My Hymn O Gaṅgā, Yamunā, Sarasvatī, Śutudrī, and Paruṣṇī 211
6.2.4 Sindhu, Most Rapid of the Rapids, Traverses the Heavenly Spheres 212
6.2.5 Exalted by Soma Mead Sindhu Grows Bright 214
6.2.6 Mighty Gaṅgā Falls Upon Mt. Meru 214

VII. Agni and Soma 217

7.1 AGNI, PROTON 217
7.1.1 Agni, an Embryonic Particle of Celestial Waters and All Beings 217
7.1.2 Agni Comes Out of the Stone as Celestial Waters 218
7.1.3 The Three Forms of Agni 219
7.1.4 Ekata, Dvita, Trita: Protium, Deuterium, Tritium 219
7.1.4.1 Ekata 219
7.1.4.2 Dvita Brings Slayed Offerings 220
7.1.4.3 Trita Blasts and Whets Like a Blower in the Smelting Furnace 222
7.1.5 Agni's Path: Black, White, Yellow, Red, Purple, and Glorious 226
7.1.6 Frogs Capture One Another Based on Their Speech Patterns 226
7.1.7 The Bird That Is but One, Singers with Hymns Shape It into Many Forms 231

7.2 SOMA, ELECTRON 232
7.2.1 Soma, an Embryonic Particle of Celestial Waters 232
7.2.2 The Thunderbolt of Indra 233
7.2.3 Soma Spins a Triple Filament 233
7.2.4 Soma, the Brahmā of the Devas 234
7.2.5 The Bestower of Light Skilled in a Thousand Arts 236
7.2.6 Soma, the Lord of the Mind 237
7.2.7 Soma, the Lord of Speech 237
7.2.8 Flow Indu, the Frog is Eager for Water 238

VIII. The Cosmology of the Ṛgveda: Summary 241

8.1 WHAT THE ṚGVEDA REVEALS 241
8.1.1 The Universe, the Infinite Ocean of Celestial Hydrogen 242
8.1.2 A Single Source and a Single Cause 242
8.1.3 The Creation and Life are the Magneto-Plasma Phenomena 242
8.1.4 The Twofold Heavenly Vault and Its Substructures are the Magneto-Plasma Structures 243
8.1.5 The Stone, the Celestial Ioniser, the Cold Fire 244

8.1.6 Magnetic Field and Magnetic Forces 244
8.1.7 Māyā 245
8.1.8 Cosmic Yajñas, the Sacred Cosmic Sacrifices 245
8.1.9 The Rains of Cosmic Plasmas and the Watchers 246
8.1.10 The Soma Pond and Plasma Blooms 247
8.1.11 The God Particles 248
8.1.12 By the Hymns of the Singers, Agni is Made into Many Forms 249
8.1.13 The Milky Way Has Four Major Arms and Is Sustained by Cosmic Yajñas 249

8.2 COSMIC PLASMA DISCHARGE AND RELEASE MECHANISMS 250
8.3 HYMNS OF BLESSING 251
8.3.1 The Devas Embody Their Wish and Adorn Their Creations 251
8.3.2 Powerful Be the Cosmic Yajña 252

Epilogue 255

Appendix: Stanzas Translated 259
1 STANZAS LISTED BY SECTIONS IN THIS BOOK 261
2 STANZAS LISTED BY THE ṚGVEDA MAṆḌALAS 269

Bibliography 273
Index 281
ABOUT THE AUTHOR 295

List of Illustrations

<Fig. 1> "A Planetary Nebula Gallery" X-Ray & Optical Images of Planetary Nebulas. 16
<Fig. 2> Open Cluster NGC 290 "Gemstones in the Southern Sky". 16
<Fig. 3> Image of Fermi Bubbles released by NASA. 17
<Fig. 4> The Threefold Heavenly Sphere and the Height of the Heaven. 20
<Fig. 5> The Twofold Heavenly Vault Mapped by the Chandra X-Ray Observatory and the Hubble Space Telescope. 31
<Fig. 6> The Structural Component of the Heaven, the Creator's Marvellous Machine, Revealed in the Ṛgveda. 38
<Fig. 7> The Soma Pond Mapped by Scientists of the Chandra X-Ray Observatory. 41
<Fig. 8> The Stone, the Celestial Ioniser, Mapped by Chandra and Hubble. 85
<Fig. 9> A Noble Comb at Bradford Industrial Museum. 101
<Fig. 10> "Annotated Roadmap to the Milky Way (artist's conception)". 110
<Fig. 11> Holy Fig Tree (अश्वत्थः aśvatthaḥ), *Ficus religiosa*, the Inverted Tree, the Vedic Tree of Life. 116
<Fig. 12> "Artist's Illustration of Galaxy with Jets from a Supermassive Black Hole". 198
<Fig. 13> The Creator's Marvellous Machine and Mechanisms of Cosmic Ionisation and Cosmic Plasma Discharge and Release. 250
<Fig. 14> "*Shiva as the Lord of Dance*, India (Tamil Nadu), c. 950–1000, Los Angeles County Museum of Art, anonymous gift". 258

Acknowledgements

All of the works acknowledged here made this new reading and interpretation of the Ṛgveda much easier and bearable for the author. The works of earlier Vedic scholars were of valuable assistance to this project. Their lifelong commitments to the Vedic studies were truly inspirational.

In his book *Alberuni's India* (AD1030), Alberuni explains that Hindus believed "the Veda, together with all the rites of their religion and country, had been obliterated in the last Dvāpara-yuga … until it was renewed by Vyāsa, the son of Parāśara". Vyāsa is also called Vedavyāsa (वेदव्यासः). The word Vyāsa (व्यासः) means an "arranger" or a "compiler". A Brāhmaṇa (ब्राह्मणः), or several Brāhmaṇas, who contributed to compiling the remnants of the Veda after the obliteration were probably given the name Vedavyāsa. Alberuni also noted that, not long before his time, circa AD1030, Vasukra (वसुक्रः), a famous and determined Brāhmaṇa of Kashmir, had committed himself to the task of writing down the Veda. Vasukra "was afraid that the Veda might be forgotten and entirely vanish out of the memories of men" since he observed the diminishing characters of men who no longer "care much for virtue, nor even for duty".[1] Considering Alberuni's statement and that the oldest extant Vedic manuscripts discovered in Nepal date to circa AD1040[2], one can reasonably conclude that the Veda was probably first recorded in writing by Vasukra around AD800–900.

---

1 Edward C Sachau (ed.), *Alberuni's India: An Account of the Religion, Philosophy, Literature, Geography, Chronology, Astronomy, Customs, Laws and Astrology of India about A.D. 1030*, Vols I & II (Bound in One), reprint, Delhi, Low Price Publications, Vol. I, 2011, pp.126-127.

2 Michael Witzel, 'The Development of the Vedic Canon and its Schools: The Social and Political Milieu', in Michael Witzel (ed.), *Inside the Texts, Beyond the Texts*, Cambridge, MA, Harvard University, 1997, p.259. Harvard Oriental Series, Opera Minora 2.

Before the Veda was recorded in writing by Vasukra, it had been orally transmitted by the Brāhmaṇas who guarded and preserved it for many millennia. To prevent the corruption of the Ṛgveda, many different recitation methods were devised. Saṃhitāpāṭha (संहितापाठः) and Padapāṭha (पदपाठः) are the first two of these recitation methods. Saṃhitāpāṭha is the recitation of the text with phonetic combination (संधिः saṃdhiḥ). Padapāṭha is the recitation of the text, word by word, without phonetic combination. The rest of the recitation methods, which are much more complicated, involve arranging the text using different methods to secure it from possible corruption.[3] Due to these extraordinary measures of preserving the Ṛgveda by the Brāhmaṇas of India, we have the unadulterated text of the Ṛgveda handed down to us from remote antiquity.

In the late 1800s, Max Müller recompiled and published the text of the Ṛgveda using the old manuscripts handed down to us. It had been a little more than a millennium since it was first written down by Vasukra, one audacious Kashmiri Brāhmaṇa, and after several millennia since Vyāsa had recompiled the Veda after the obliteration in the last Dvāpara Yugaṃ (द्वापरयुगं). When Max Müller completed the recompilation and publication of the Ṛgveda text, Ādi Brahma Samāj wrote a congratulatory note to him:

> The Committee of the Ādi Brahma Samāj beg to offer you their hearty congratulations on the completion of the gigantic task which has occupied you for the last quarter of a century. By publishing the Rig-Veda at a time when Vedic learning has by some sad fatality become almost extinct in the land of its birth, you have conferred a boon upon us Hindus, for which we cannot but be eternally grateful.[4]

Without Vasukra's commitment, Müller's latest compilation and publication of the Ṛgveda text, and "western scholarship burrowing into the text", the old manuscripts could have been lost and forgotten. Even if the old manuscripts survived the ravages of time, the text and knowledge revealed within may still have remained out of reach from the students of the Veda.

Though nearly all of the native commentaries on the Ṛgveda were composed after the loss of the original meanings of the Ṛgveda, and subsequent interpretations by Western scholars followed these native commentators' approach of reading the Ṛgveda through the lens of 'sacrificial rituals', it does not diminish the significance of their works and

---

[3] For the recitation methods of the Veda, refer to Pierre-Sylvain Filliozat, 'Ancient Sanskrit Mathematics: An Oral Tradition and a Written Literature', in Karine Chemla (ed.), *History of Science, History of Text*, Dordrecht, Netherlands, Springer, 2004, pp.138–140. Boston Studies in the Philosophy of Science Vol. 238.

[4] F. Max Müller, *India: What Can It Teach Us?* A Course of Lectures Delivered before the University of Cambridge, new edition, London, Longmans, Green, And Co., 1892, p.143.

contributions to establishing the foundation of the Vedic studies and preserving the Vedic traditions.

Niruktaṃ (निरुक्तं), a commentary on the Nighaṇṭu (निघण्टुः), compiled by Yāska (यास्कः c. 500BC) and extensive commentary on the Ṛgveda composed by Sāyaṇa (सायणः died AD1387) help one see how native scholars perceived and understood the Ṛgveda. H. H. Wilson's six volumes (1854-1888) and Ralph Griffith's two volumes (1896, 1897) are the two complete sets of English translations of the ten maṇḍalas of the Ṛgveda. Despite all the mistranslated stanzas and his refusal to translate certain stanzas based on moral ground, still in Griffith's, one finds more grammatically acceptable translations. H. H. Wilson's translation, which was based on Sāyaṇa's commentary, unfortunately, is much less intelligible than Griffith's.

Select stanzas of the Ṛgveda were translated by Max Müller (1891), A. A. Macdonell (1917), Aurobindo (Sri Aurobindo Ashram Trust 1995, 1996), Karen Thomson and Jonathan Slocum (*Ancient Sanskrit online* 2006), and many others. Along with these partial translations, *The Cosmology of the Rigveda: an Essay* by H. W. Wallis (1887), *Vedic Mythology* by A. A. Macdonell (1897), and *Sparks from the Vedic Fire: A New Approach to the Vedic Symbolism* by Vasudeva S. Agrawala (1962) all provide unique and varying perspectives that native and Western scholars had on the Ṛgveda.

For Vedic grammar, *The Roots, Verb-Forms and Primary Derivatives of the Sanskrit Language* (1885) and *Sanskrit Grammar* (1888) by William D. Whitney, and *Vedic Grammar* (1910) and *A Vedic Grammar for Students* (1916) by A. A. Macdonell offered indispensable assistance. For Vedic metre, *Vedic Metre In Its Historical Development* by Vernon Arnold (1905) and *Rig Veda: A Metrically Restored Text with an Introduction and Notes* by Van Nooten and Gary B. Holland (1994) were excellent references. And *A Vedic Concordance* (1906) by Maurice Bloomfield was a convenient tool.

A Sanskrit-English Dictionary by Monier-Williams (1899) is the last comprehensive compilation of the definitions of the Vedic vocabulary handed down to the generations of native Vedic pundits and Sanskrit lexicographers. Its etymologically arranged entries of the Vedic vocabulary make it particularly useful. Often a long list of definitions provided for a Vedic term in Monier-Williams, which sound irrelevant on the face of it, can actually offer critical clues for interpretation and translation of the ancient Vedic hymns. I could not have started this project without *A Sanskrit-English Dictionary* by Monier-Williams (1899) and *A Vedic Grammar for Students* (1916) by A. A. Macdonell.

I've encountered a number of Indian people who wanted to stay away from Monier-Williams, due to his intention behind compiling his Sanskrit-English dictionary, and rely on Amarakoṣa (अमरकोषः amarakoṣaḥ)[5] for Vedic studies. However, Amarakoṣa lacks the comprehensive definitions provided in Monier-Williams, which are critical for interpreting and translating the hymns of the Ṛgveda. In the Preface of his dictionary, Monier-Williams, as the second Boden Chair at the time, stated,[6]

> [I]ts Founder, Colonel Boden, stated most explicitly in his will (dated August 15, 1811) that the special object of his munificent bequest was to promote the translation of the Scriptures into Sanskrit, so as 'to enable his countrymen to proceed in the conversion of the natives of India to the Christian Religion'. ... ... I have made it the chief aim of my professional life to provide facilities for the translation of our sacred Scriptures into Sanskrit, and for the promotion of a better knowledge of the religions and customs of India, as the best key to a knowledge of the religious needs of our great Eastern Dependency. My very first public lecture delivered after my election in 1860 was on 'The Study of Sanskṛit in Relation to Missionary Work in India' (published in 1861). (Quotation marks and the words within the parentheses in the original.)

Nevertheless, in spite of his intention, his dictionary is one of the most valuable assistants available to students of Vedic studies. And one should acknowledge that there were contributions made by hundreds of native Vedic pundits and Sanskrit lexicographers to projects such as Monier-Williams's dictionary. If it were not for Monier-Williams and Macdonell, a good amount of Vedic grammar and the definitions of the Vedic vocabulary still lingering in the minds of Vedic pundits at the time could have been lost forever. That dreadful thought that we could have lost all the knowledge of the Vedic language a little more than a century ago jolts one out of one's sleep at night.

Nicholas Sutton's well organised courses offered at The Oxford Centre for Hindu Studies, which led this author to the Ṛgveda, were helpful in many aspects for reviewing the sophisticated Indian systems of thoughts as an organically grown whole, branched out from the same root. Learning and participating in several of his courses, especially *The Vedas and the Upaniṣads*, *Mahābhārata and Rāmāyaṇa*, and *Purāṇas*, was a delightful experience. His enthusiasm was contagious.

The digital Sanskrit text of Ṛgveda Saṃhitā (ऋग्वेदसंहिता) was provided by Peter Freund at the Maharishi International University (MIU).

---

[5] H.T. Colebrooke, *Kosha or Dictionary of the Sanskrit Language by Umura Singha with an English Interpretation and Annotations*, Calcutta, Haragobinda Rakshit, 1891.
[6] Monier Monier-Williams, *A Sanskrit-English Dictionary: Etymologically and Philologically Arranged with Special Reference to Cognate Indo-European Languages*, new edition, Oxford, The Clarendon Press, 1899, pp. ix-x.

Peter Freund, Jim Funderburk, Gary B. Holland, Karen Makino, Krishna Mukherjee, Karen Thomson, Ian Tresman, and Avinash L. Varna answered my questions graciously. Their answers helped and clarified much more than they could have imagined.

The works of Kristian Birkeland, Irving Langmuir, Hannes Alfvén, and Anthony L. Peratt in the fields of plasma cosmology and astrophysics; Nikola Tesla[7] and Albert Roy Davis and Walter C. Rawls, Jr. in electricity and magnetism; and Robert O. Becker and Gary Selden in bio-electromagnetism provide necessary knowledge of these fields. These works are grounded in observed facts and reality, not pseudo reality or artefacts of mathematical models. The lectures of Vivekānanda (विवेकानन्दः), a truly orthodox and maverick scholar of Indian systems of thoughts, have been a perennial source of encouragement for this project. Vivekānanda left many clues on the nature of the knowledge revealed in the Ṛgveda in his lectures. The clarity of logic and thoughts and the earnest efforts of these scholars have been an inspiration.

Anthony L. Peratt's plasmauniverse.info and Ian Tresman's www.plasma-universe.com provide resources for understanding the nature of cosmic plasmas. Here, one reads about the cosmic plasma phenomena and events, but not "clouds of hot gas".

The Chandra and Hubble images of celestial objects were awe-inspiring, and the technical information about the images provided by NASA science teams was informative. Many hauntingly beautiful images from the Hubble Space Telescope and the Chandra X-Ray Observatory were examined and reexamined. One could easily lose oneself in these celestial images, spending hours probing into the depths of the cosmos.

My gratitude also is due to my late parents, my son Tommy Yoon, and my brothers and sisters and their families for their support throughout this lasting journey of learning. Especially without my son Tommy's support, this project would never have been completed. Tommy's support also enabled me to overcome a series of serious physical challenges I have endured while I was working on this project. I owe them all my heart-filled thank you for their unconditional faith and support.

---

[7] See Thomas C. Martin, *The Inventions, Researches, and Writings of Nikola Tesla*, 2nd ed., New York, The Electrical Engineer, 1894; Nikola Tesla Museum, *Nikola Tesla: Colorado Springs Notes: 1899-1900*, Beograd, Yugoslavia, Nolit, 1978. These publications show that Nikola Tesla understood electric phenomena better than any other scholars in the field of electricity and magnetism before and after him. As Tesla's work is closely examined, one recognises that electricity is much more than the flow of electric charges or of electromagnetic energy, that the better term to describe what Tesla revealed and demonstrated is magneto-plasma-electric phenomena, and that these phenomena are in fact closely associated with what the Ṛgveda reveals.

Preface

The Veda (वेद:) has been revered as the sacred and eternal knowledge of divine origin. It is considered the oldest book in existence by Max Müller.[8] Vivekānanda (विवेकानन्द:), in one of his lectures, stated, "In and through the Vedas, the whole creation has come. All that is called knowledge is in the Vedas. Every word is sacred and eternal, eternal as the created man, without beginning and without end. As it were, the whole of the Creator's mind is in this book. That was the light in which they [Indians] held the Vedas".[9]

After reading the ancient Indian philosophical systems and classical Sanskrit literatures for several years, an attempt to trace the origin of the vast and sophisticated Indian systems of thoughts inevitably led to the Ṛgveda. The Ṛgveda Saṃhitā (ऋग्वेदसंहिता), commonly called the Ṛgveda (ऋग्वेद:), is considered to be the most sacred and eternal knowledge of cosmic creation revealed to men. According to the existing interpretations and translations, however, the Ṛgveda could hardly be regarded as the sacred and eternal knowledge. It was impossible to make sense of those interpretations and translations. It was decided that, to truly understand the meaning of these ancient hymns, it was necessary to read the Ṛgveda in its original Vedic language. This new interpretation of the Ṛgveda started as the author's own reading of the Ṛgveda in order to uncover its original meaning and the knowledge it reveals.

---

8 Max Müller, *A History of Ancient Sanskrit Literature*, 2nd ed., revised, London, Williams and Norgate, 1860, p. 557.

9 *Speeches and Writings of Swami Vivekananda: A Comprehensive Collection with Four Portraits*, 3rd ed., Madras, G.A. Natesan & Co., 1899, p. 295. The quotation is from the book published while Vivekānanda was alive, not from the heavily edited versions published after his death in 1902. The later edited versions lack Vivekānanda's original intention embedded in his spontaneous lectures delivered in colloquial language that enlightened the audience.

When the Ṛgveda was read in the Vedic language, with the help of the Padapāṭha (पदपाठः) text, the parallel between the characteristic attributes of the Vedic Devas (देवाः devāḥ pl.) and the events of celestial Yajña (यज्ञः ritual, sacrifice) described in the Ṛgveda and the observations of celestial objects and cosmic plasma phenomena made by NASA space probes and space telescopes became undeniably clear. Modern scientific observations support the Vedic model of cosmic creation; none of the observations made by NASA contradicts or comes as a surprise to the Vedic model. Supposedly "puzzling" and "incongruous" text started to reveal its meaning, and most importantly, I could make sense of the ancient text. I gazed in awe at a streak of light appearing through the cracked opening of the ancient vault of knowledge. And this was the start of a strenuous, yet illuminating journey.

Though its language is highly metaphorical and poetic, the Ṛgveda is not about religion, nor is it about the history and social customs of Āryans. To understand the Ṛgveda, one has to set the vantage point in intergalactic space, for the subject of the Ṛgveda is the knowledge of cosmic creation (सृष्टिविद्या sṛṣṭividyā). The magneto-plasma cosmology of the Ṛgveda presented here is based on interpretations of select stanzas and hymns, which are considered to be representative of celestial objects and cosmic plasma phenomena associated with cosmic creation, from the ten maṇḍalas (मण्डलानि or मण्डला pl.) of the Ṛgveda. While interpreting the ancient Vedic text, I was careful not to force fit what is presented in the Ṛgveda into the theories and concepts of the standard model of modern astrophysics and to remain true to the text of the Ṛgveda.

As one reads the Ṛgveda, it becomes clear that it contains many knowledge gaps. This is not unexpected since the Veda and the country of India, according to the Hindu tradition, had been obliterated in the last Dvāpara Yugaṃ (द्वापरयुगं). The Veda, as it has been handed down to us, is the remnant that was recompiled by Vyāsa (व्यासः) after its obliteration. Vivekānanda tells us that a major portion of the Veda had been lost, and only a minor portion survived and was handed down to us.[10] However, a considerable section has been preserved enabling us to piece together the knowledge revealed in the extant text of the Ṛgveda: the origin, cause, and process of cosmic creation.

For cross referencing and filling in knowledge gaps in the Ṛgveda, the Atharvaveda (अथर्ववेदः), the Śatapathabrāhmaṇaṃ (शतपथब्राह्मणं), the select Upaniṣads (उपनिषदः pl.), and the Bṛhaddevatā (बृहद्देवता) have been consulted. In addition, chapters of the two astronomical treatises that contain the cosmogony and cosmography of the Milky Way

---

[10] ibid., p.406; Advaita Ashrama, *The Complete Works of Vivekananda*, Vol. 3, 20th print, Mayavati, India, 2007, p.435.

were also consulted: Bhūgolādhyāya of the Sūryasiddhānta (भूगोलाध्यायः, सूर्यसिद्धान्तः) and Bhuvanakośa of the Siddhāntaśiromaṇi (भुवनकोशः, सिद्धान्तशिरोमणिः).

Celestial objects and cosmic plasma phenomena observed by NASA space probes and space telescopes correspond well to what the Ṛgveda reveals. On the other hand, these observed celestial objects and cosmic plasma phenomena are incompatible and inconsistent with the big bang theory. According to the big bang theory, the whole universe is created out of nothing at the moment of the initial singularity, the big bang, and the universe itself is continuously expanding and cooling to the present day. The big bang theory is based on exotic mathematical artefacts, such as 'singularity', 'black hole', 'dark matter', 'dark energy', 'anti-matter', and so on instead of observed facts and reality. The big bang theory does not provide answers about what the conditions were before the big bang. Also not explained is how the initial singularity could arise, which must be a prerequisite for the event of the big bang.[11]

In the Vedic model of cosmic creation, neither the concept of 'singularity' nor the concept of 'black hole' is required. Neither 'dark energy' nor 'dark matter' is necessary. The Vedic model, presented in the Ṛgveda and backed by modern scientific observations made by NASA, is persuasive and elegant; and it just makes sense.

According to the Ṛgveda, the prima materia, the original matter of the universe, is aṃbha (अंभः), hydrogen. Agni (अग्निः proton) and Soma (सोमः electron) are the two embryonic particles of celestial Waters (आपः āpaḥ pl.), that is, cosmic plasmas. The process of cosmic ionisation and cosmic plasma discharge and release is called Yajña (यज्ञः), the sacred cosmic sacrifice. The subtle yet mighty cosmic force (महिमा m. sg. of महिमन्), which is emanated from the vast celestial ocean of aṃbha, sets the cosmic creation in motion. Tvaṣṭā (त्वष्टा) and the Ṛbhus (ऋभवः ṛbhavaḥ pl.) build the twofold Heavenly Vault and its substructures, the Creator's marvellous machine, the machine that initiates the celestial ionisation of hydrogen, discharges and releases cosmic plasmas, and weaves the nebulous fabric of the Milky Way. Tvaṣṭā, magnetism in general and magnetic field in particular, and the Ṛbhus (pl.), the magnetic forces, are the divine architects.

The universe, the infinite ocean of celestial hydrogen, is in a steady state with no beginning and no end. The universe is not to be created, nor to be destroyed; it is eternal. (Infinity is a frightening concept to the finite mind!) What is created and has evolved are the Worlds (भुवनानि bhuvanāni pl.), the galactic systems that create and harbour life within

---

[11] CERN, *Physics*, [website] https://home.cern/science/physics (accessed 30 January 2020).

the universe. Each galactic system consists of the three spheres of the Heaven, two Heavenly Vaults, two pairs of the Stones and the Soma Ponds, and a galaxy. The celestial ioniser that produces cosmic plasmas is called the Stone (ग्रावा grāvā) in the Ṛgveda (see Figure 6).

We shall see how the event of cosmic creation was set in motion through the eyes of the singers of hymns in the sections that follow. The celestial beings involved in cosmic Yajñas are called the Ṛṣis (ऋषयः ṛṣayaḥ pl.), Devas (देवाः devāḥ pl.), Devīs (देव्यः devyaḥ pl.), an infant (शिशुः śiśuḥ), a king (राजा rājā), supreme monarch (संराट् saṃrāṭ), deceased forefathers (मनुष्याः manuṣyāḥ pl.), man (नरः naraḥ sg.), men (नराः narāḥ pl.), heroes (वीराः vīrāḥ pl.), women (नार्यः nāryaḥ pl.), young maidens (युवतयः yuvatayaḥ pl., योषणः yoṣaṇaḥ f. nom. pl. of योषन्), and so on depending on their roles and tasks at hand. Freshly separated charges, Agnis (protons) and Somas (electrons), which are discharged from the Stone, the celestial ioniser, are called infants or youths.

What are revealed in the Ṛgveda include (i) what the condition was before the process of cosmic creation was set in motion, (ii) how the twofold Heavenly Vault, the marvellous Creator's machine, and its structural components are built, (iii) how the cosmic ionisation and plasma discharge and release processes, the breaking up of the ionic bonds of celestial hydrogen and the discharging and releasing of cosmic plasmas, are initiated by the twofold Heavenly Vault and the Stone, the celestial ioniser, and (iv) how the nebulous fabric of the Milky Way is woven and sustained. The twofold Heavenly Vault (नाकः nākaḥ, व्योम vyoma) consists of the Upper Vault and the Lower Vault of the Heaven, which are arranged in the shape of the Ḍamaru (डमरुः), the sacred drum of Śiva (शिवः), or in the shape of an hourglass (see Figures 4, 5, and 6). These Heavenly structures are sustained by the unbroken chain of cosmic Yajñas, the events of celestial ionisation and cosmic plasma discharge and release.

According to the Ṛgveda, the mighty cosmic force, which sets the cosmic creation in motion, the subsequently built Heavenly Vault, and the event of cosmic Yajña are all born of a single source and of a single cause. The celestial aṃbha (अंभः aṃbhaḥ), hydrogen, is the single source; the mighty cosmic force emanated from celestial aṃbha, the single cause. Following the mighty cosmic force, the desire (कामः kāmaḥ), which was the primaeval flow of cosmic Mind (मनः manaḥ), subsequently arises. Enchanting hymns of praise for and supplication to the celestial beings involved in cosmic Yajñas will have direct impact on the cosmic events and on the sustenance of our galactic system, for our mind (मनः manaḥ) and the subsequently risen cosmic Mind (मनः manaḥ) are one and the same, and our soul (आत्मा ātmā) and the Soul (आत्मा ātmā) of the cosmic Yajña are one and the same. This is the basic tenet of the Vedic yajñas and recitations and chantings of the hymns of praise and supplication. The original Vedic yajñas are the ceremonial enactments of cosmic Yajñas.

The Ṛgveda is concerned with the knowledge of cosmic creation. In the extant text of the Ṛgveda, direct discourses on the subjects of religion and social customs are not found. However, this does not imply that there is no connection between the cosmic law and natural philosophy revealed in the Ṛgveda and the spiritual and philosophical principles of the Indian systems of thoughts. The cosmic law, the law of Nature, that operates on the cosmic plane also operates on the planetary plane and on our individual soul and physical body. As above, so below.

In fact, the rest of Vedic literature that include the remaining three Vedas (Yajurveda, Sāmaveda, Atharvaveda), the Brāhmaṇas, the Āraṇyakas, and the Upaniṣads, which are appended to each branch of the four Vedas, and most of the subsequently developed Indian philosophical systems are rooted in the knowledge of cosmic creation and natural philosophy revealed in the Ṛgveda. Smṛtis (स्मृतयः pl.) are the works derived from the Vedas or the auxiliary arms of the Vedas. Purāṇas (पुराणा pl.) are the mythified accounts of cosmic Yajñas and the events associated with the creation of the World, the Bhuvanaṃ (भुवनं). The Bhuvanaṃ refers to the manifested phenomenal World, the whole of the galactic system that includes the threefold Heavenly Spheres, the two Vaults of the Heaven, two pairs of the Stones and the Soma Pond, and the Milky Way (see Figure 6).

To the students of cosmology and astrophysics, the magneto-plasma cosmology of the Ṛgveda will provide an alternative frame for coherent analyses and interpretations of the celestial objects and cosmic plasma phenomena observed by NASA space probes and telescopes and a better understanding of the history of the universe and the galactic systems that have evolved within the universe. The magneto-plasma cosmology of the Ṛgveda is a realistic and delightful alternative to the big bang theory, albeit written in poetic and highly symbolic and metaphorical language. We need to open our minds and expand our horizons to embrace the reality as it is and do away with the big bang theory that projects pseudo reality. Nature does not operate in a convoluted manner; our unenlightened minds do.

Students of alchemy, Gnosticism, Hermeticism, and classical myth will find that the Ṛgveda provides a key to understanding the origin and philosophical heritage of these fields, and to unwrapping the occult symbolism cloaked in myths and superstitions.

Compared to the fragmentary gnostic and hermetic texts,[12] the extant text of the ten maṇḍalas of the Ṛgveda handed down to us is relatively extensive and comprised of systematic and unadulterated knowledge. The Ṛgveda reveals the knowledge of cosmic creation, the magnum opus, the great work of Nature, the celestial Alchemy. Though the Vedic symbolism is not extensively discussed in this book, for the purpose of the book is to present the cosmology of the Ṛgveda, readers who have knowledge and insights in these fields will see that the symbolism and myths associated with cosmic creation revealed in the Ṛgveda run throughout these different but related fields and across the different cultural and religious boundaries, and that the true meanings of the symbolism and myths originally associated with cosmic creation have been lost and corrupted and cloaked in secrecy and superstitions.

This book began as the author's hermeneutic reference for a coherent interpretation and translation of all ten maṇḍalas of the Ṛgveda. Later, it was decided that the remarkable knowledge revealed is worth sharing with students of the Veda, astrophysicists, and others who might be interested in knowing what is revealed in the Ṛgveda and about the true heritage of mankind. Cosmic ionisation and cosmic plasma discharge and release mechanisms, presented in this book for the first time, are based on what the Ṛgveda reveals. It is the robust and complete cosmic creation model. Knowledge of cosmic creation, recorded in the ancient Vedic language and concealed by veils of symbolism and poetic language, is deciphered and presented in this book.

With his sacred drum the Ḍamaru (डमरुः), the marvellous Creator's machine, Śiva (शिवः) dances to produce and discharge the sacred cosmic lights and fires, the cosmic plasmas, the secret fires of cosmic creation. It is the dance of eternity, for the cosmic creation is "eternal as the created man".

May the powerful and enchanting hymns of the Ṛgveda be heard again.

Y. Sohn
March 2020

---

[12] See for example, Brian P. Copenhaver, *Hermetica: The Greek Corpus Hermeticum and the Latin Asclepius in a New English Translation with Notes and Introduction*, Cambridge, Cambridge University Press, 1992; Marvin Meyer (ed.), *The Nag Hammadi Scriptures: the Revised and Updated Translation of Sacred Gnostic Texts Complete in One Volume*, New York, HarperOne, 2007; Willis Barnstone and Marvin Meyer (eds.), *The Gnostic Bible: Gnostic Texts of Mystical Wisdom from the Ancient and Medieval Worlds*, Boston, Shambhala, 2009.

A Note to the Second Edition

It has been about a year since the manuscript of this book was printed for copyright registration and private distribution. It was sent out to a select few, mostly who knew about this book project and a few others in Great Britain, India, and the United States.

The grammatical analyses of all three hundred twenty seven Ṛgveda stanzas and six other stanzas are reexamined, typos and technical errors are corrected, some section titles are adjusted, and certain events and concepts are clarified. Most parts of the Index are reorganised and recompiled. And one more stanza of the Ṛgveda, RV x.66.12, has been added to the section Hymns of Blessing. Now, the fully revised second edition, which is the first public edition, is being offered to the students of the Veda and to those who seek the Vedic knowledge of cosmic creation and the true origin and heritage of mankind.

Y. Sohn
March 2021

Dear Readers

The idea that the Ṛgveda is about the magneto-plasma cosmology may sound fantastical. How could the early humans have known about the magneto-plasma cosmology and the mechanisms of cosmic ionisation and cosmic plasma discharge and release? And yet reasoned interpretations and translations of the Ṛgveda, along with other ancient texts, especially the ancient alchemical texts, and insightful examinations of Earth's unusual and 'controversial' archaeological findings and archaeological objects imaged on our Moon and on Mars and Mars's moons by NASA space probes, will inevitably lead one to the conclusion that, in remote antiquity, frontier outposts of very advanced civilisations were established on the planets in our solar system, including the planet Earth, and that this knowledge of cosmic creation must have originated from these advanced civilisations. Moreover, it has been known among learned pundits of India that the Ṛgveda is about the cosmic creation. This known fact that the Ṛgveda is about cosmic creation is ignored by most of the Vedic students in the groves of academe.

If one's mind is hermetically sealed with the idea of "hot gas" and the big bang theory, the Darwinian ideology, and the views of conventional human history, one will not be able to discern all the evidences presented before one's very eyes, review and analyse them to comprehend reality, and appreciate the knowledge revealed in the Ṛgveda.

Though this book presents the knowledge of cosmic creation and the cosmic plasma phenomena revealed in the Ṛgveda, even without the basics of Vedic language and a general knowledge of cosmic plasma phenomena, the reader can still plough through the book, as a farmer ploughs his field to prepare for planting seeds, and take delight in the glorious hymns of the Ṛgveda and the knowledge they reveal. As you read the wondrous celestial plays of the Ṛgveda, you will find yourself soaring into the Heavenly Spheres, standing face to face with the Creator.

The contents are laid out so as to build readers' understanding of the Heavenly structures and the cosmic plasma phenomena revealed in the Ṛgveda. So, it is recommended that readers start from the beginning and follow through the book.

What is presented in this book is the knowledge ancient and medieval priests, scholars, mystics, oligarchs, and royals all coveted; the knowledge without which no other field of knowledge is complete.

Note that, though keys for deciphering the layers of knowledge hidden in symbolic language are provided in translation and in grammatical analysis of each stanza, most of the translated stanzas are not fully deciphered and are still in the symbolic language as they are in the original text.

For students of cosmology and the plasma universe, this book is written in general language and does not necessarily follow the idioms of the standard model of modern physics. For example, the term 'cosmic plasma' is used instead of 'hot gas', for what it really is. The term 'plasma sheath' is preferred over 'double layer', 'plasma light discharge' is preferred over 'plasma glow discharge', and 'plasma fire discharge' is preferred over 'plasma arc discharge' to better convey what is described in the Ṛgveda.

Often, you will find there are no Vedic word equivalents of the standard model of modern physics. To illustrate, the term equivalent of Grāvā (ग्रावा), the Stone, which is the celestial ioniser and discharger of cosmic plasmas, does not exist in modern astrophysics. Neither the term equivalent of māyā (माया), the magic force of Varuṇa (वरुण:), nor the term equivalent of Indra (इंद्र:), the slayer of Vṛtra, nor the term equivalent of Vimāna (विमान:), the chariot of the Devas, exists in the standard model of modern physics.

As we shall see, māyā refers to the magical force of Varuṇa, the cosmic plasmas. Vṛtra, the serpent of the Heaven with its tail in its mouth, known as ouroboros or ophis in Gnostic and alchemical traditions of ancient Greece and Rome, refers to the plasma sheath, the double layer of two opposite charges, the shoreline of the Soma Pond, or the Ocean of celestial Waters, within which discharged cosmic plasmas grow, bloom, and mature (see Figure 7). Vimāna, the chariot of the Devas, refers to the Soma Pond, the abode of the Devas, the city of the gods.

Notes

TECHNICAL NOTES

(i) Conventionally, when a Sanskrit noun is cited and transliterated, the root form of the noun (प्रातिपदिकं) is used. This convention is not followed in this book. Instead, the nominative case of the noun is chosen. For example, nominative cases such as bhuvanaṃ (भुवनं n. nom. sg.), Tvaṣṭā (त्वष्टा m. nom. sg.), Savitā (सविता m. nom. sg.), and Grāvā (ग्रावा m. nom. sg.) are used instead of using the root forms of these nouns bhuvana (भुवन), Tvaṣṭṛ (त्वष्टृ), Savitṛ (सवितृ), and Grāvan (ग्रावन्). For the name of the river Gaṃgā (गंगा), the word liṃgaṃ (लिंगं), and the names of Nakṣatras of the Heavenly Vault, such as Maṃda (मंदः) and Caṃdra (चंद्रः), the prevalently used phonetic transliterations Gaṅgā, liṅgaṃ, Manda, Candra, and so on are adopted.

For 'Saṃskṛtaṃ (संस्कृतं)', which was the lingua franca of the elite in ancient and medieval India, the prevalently used anglicised form of the word 'Sanskrit' is used. Sanskrit, though derived from the Vedic language and written in the same Devanāgarī (देवनागरी) alphabet as the Vedic language, is the language that follows Pāṇini's grammar. The Vedic language differs from Sanskrit through a much greater antiquity and a great wealth of grammatical forms as Winternitz (1927) points out. And the general lexis of the Vedic language is quite different from Sanskrit and much more refined.

(ii) For the dual word Aśvinā (अश्विना m. nom. du.), it was used as it appears in the Vedic text instead of Aśvins (अश्विन्s), for Aśvinā (du.) is the twins. For another dual word māye (माये f. nom. du. of माया), it was translated as 'two māyās' or 'two divisions of māyā'. In the extant text of the Ṛgveda, the word māyā (माया) is mostly used in the singular, and to indicate that there are two different divisions of māyā, the word māyābheda (मायाभेदः māyābhedaḥ), the two divisions of māyā, was used instead of the dual word māye.

(iii) For citing plural words, such as Ṛṣis (ऋषयः ṛṣayaḥ pl.), except in the dual case mentioned above, the anglicised transliteration scheme was followed by adding 's' to the root form. The Vedic plural form and its original pronunciation are presented within the parentheses as shown above. Ṛṣis is the anglicised plural of Ṛṣayaḥ (ऋषयः pl.).

(iv) For cited Vedic and Sanskrit words, the nasal was written as anusvāra (अनुस्वारः anusvāraḥ) 'ṃ' as in sūktaṃ (सूक्तं) and aṃtarikṣaṃ (अंतरिक्षं) regardless of its position, as it is in the traditional Ṛgveda manuscripts. For the Vedic words that end with visarga (विसर्गः visargaḥ) 'ḥ', the ḥ was dropped when transliterated. However, when a word with visarga 'ḥ' was cited within the parentheses as a reference, the ḥ was kept to show the original form of the Vedic word as in Ṛṣi (ऋषिः ṛṣiḥ).

(v) The Vedic accents, udātta (उदात्तः), anudātta (अनुदात्तः), and svarita (स्वरितः), were indicated in the Vedic text. However, when transliterated, the Vedic accents were omitted. Udātta indicates the acute raised accent and is unmarked in the Vedic text. Anudātta indicates the grave low pitch accent and is marked by a lower horizontal bar. Svarita indicates the mixed tone of moving accent and is marked by an upright vertical bar above a syllable as in the following stanza (RV i.20.6).

उ॒त त्यं च॑म॒सं नवं॒ त्वष्टु॑र्दे॒वस्य॒ निष्कृ॑तं । अक॑र्त च॒तुरः॑ पुनः॑ ॥१.२०.६॥

"When an independent Svarita immediately precedes an Udātta, it is accompanied by the sign of the numeral १ (1) if the vowel is short, and by ३ (3) if it is long, the figure being marked with both the Svarita and the Anudātta".[13] However, on the numerals, due to the limitations of the softwares used to type the manuscript of this book, the svarita and anudātta marks could not be correctly placed. The marks of accents on the numerals slid as in अ॒प्स्व१॒॑तरा (i.105.1). While svarita and the anudātta can be placed right above and below the Devanāgarī alphabets, for some reason, they could not be correctly placed on the numerals.

(vi) For the Ṛgveda stanza number, the Roman numeral indicates the number of the book followed by the number of the hymn and the stanza. For example, 'RV i.20.6' denotes Ṛgveda book (मण्डलं maṇḍalaṃ) 1, hymn (सूक्तं sūktaṃ) 20, stanza (ऋक् ṛk) 6.

---

[13] A. A. Macdonell, *A Vedic Grammar for Students*, Oxford, The Clarendon Press, 1916, p.450.

(vii) A brief grammatical analysis was provided for each stanza translated. It may not necessarily be a complete one; however, the grammatical analysis will prove valuable for students of the Veda since it shows how each stanza is interpreted and translated.

(viii) All the quoted words, phrases, and passages and the cited English definitions of the Vedic words from *A Sanskrit-English Dictionary* by Monier-Williams (1899) are enveloped by "double quotation marks". To place an emphasis on a certain word, 'single quotation marks' are used. To be consistent with the English definitions of Vedic Sanskrit words provided by Monier-Williams's (1899), British English spelling is adopted. Also, in this book, the British punctuation style is adopted with the exceptions of the usage of double and single quotation marks as explained above.

(ix) To distinguish the Gregorian calendar eras, the terms 'anno domini' (AD) and 'before Christ' (BC) are used in this book.

(x) Lastly, for the web sites' URLs cited in this book, if the URL is not found, please do online searches using the key word(s) of the reference and the source. Information on the web is fluid; it can be removed or moved, and the URLs can be changed. And online content can be edited without formal notice.

SANSKRIT TRANSLITERATION

For the Vedic Sanskrit transliteration, the International Alphabet of Sanskrit Transliteration (IAST) scheme was used. The following tables show the IAST conventions.*

Vowels

Vowels (स्वराः)			
Devanāgarī (देवनागरी)	Transcription	Category	
अ (अ)	a A	Guttural	Simple Vowels
आ (आ)	ā Ā		
इ	i I	Palatal	
ई	ī Ī		
उ	u U	Labial	
ऊ	ū Ū		
ऋ	ṛ Ṛ	Cerebral	
ॠ	ṝ Ṝ		
ऌ	ḷ Ḷ	Dental	
ॡ	ḹ Ḹ		
ए	e E	Guttural + Palatal	Diphthongs
ऐ	ai Ai		
ओ (ओ)	o O	Guttural + Labial	
औ (औ)	au Au		
◌ं	ṃ Ṃ	Anusvāra (अनुस्वारः)	
◌ः	ḥ Ḥ	Visarga (विसर्गः)	
ऽ	' (apostrophe)	Avagraha (अवग्रहः) - In general, it is used to indicate a lost 'अ a' sound. - In the Padapāṭha text of the Ṛgveda, it is used to indicate the separation of the components of a compound or the separation of the stem, suffixes, and terminations.	

*Tables are constructed based on Madhav M. Deshpande, *A Sanskrit Primer*, Ann Arbor, Michigan, The University of Michigan Center for South and Southeast Asian Studies, 2003. Michigan Papers on South and Southeast Asia no. 47. For detailed information on the Sanskrit writing system, the alphabet, and the IAST transliteration conventions, refer to Deshpande (2003).

Consonants

Consonants (व्यञ्जनाः) – Devanāgarī and Transcription with Inherent Vowel 'अ (a)' Added					
Velar (Guttural)	Palatals	Cerebrals	Dentals	Labials	Category
क ka Ka	च ca Ca	ट ṭa Ṭa	त ta Ta	प pa Pa	Voiceless*
ख kha Kha	छ cha Cha	ठ ṭha Ṭha	थ tha Tha	फ pha Pha	
ग ga Ga	ज ja Ja	ड ḍa Ḍa	द da Da	ब ba Ba	Voiced
घ gha Gha	झ jha Jha	ढ ḍha Ḍha	ध dha Dha	भ bha Bha	
ङ ṅa Ṅa	ञ ña Ña	ण (ण) ṇa Ṇa	न na Na	म ma Ma	
-	य ya Ya	र ra Ra	ल la La	व va Va	Semivowels
-	श śa Śa	ष ṣa Ṣa	स sa Sa	-	Sibilants
ह ha Ha	-	-	-	-	Aspirate

* Voiceless consonants are pronounced without the vibration of the vocal cords.

ABBREVIATIONS AND SYMBOLS

RV	Ṛgveda (ऋग्वेदः)
B.Yajur	Black Yajurveda (कृष्णयजुर्वेदः, Taittirīya Saṃhitā तैत्तिरीय संहिता)
ŚB	Śatapathabrāhmaṇaṃ (शतपथब्राह्मणं)
Monier-Williams	*A Sanskrit-English Dictionary* compiled by Monier-Williams,1899.

ऽ Avagraha (अवग्रहः)	See the table of the IAST Conventions, Vowels section.
√	The sign in front of a word, as in √दृश्, indicates that the word is the root (धातुः) of a verb.

Other Abbreviations

Noun/Pronoun/Adjective/Participle-Related	Verb/Participle-Related*
m. masculine f. feminine n. neuter mf. masculine or feminine mn. masculine or neuter mfn. masculine or feminine or neuter sg. singular du. dual pl. plural adj. adjective ind. indeclinable pron. pronoun nom. nominative acc. accusative inst. instrumental dat. dative abl. ablative gen. genitive loc. locative voc. vocative	cl. class of verb P or 'active' active voice Ā or 'middle' middle voice pt. participle pre. present tense aor. aorist tense (expresses the immediate past) ipf. imperfect tense (the past tense) prf. perfect tense (expresses the condition attained as the result of a preceding action; can be translated by the past, by the present, by the present perfect) fut. future tense cau. causative mood des. desiderative mood (expresses desire) inj. injunctive mood (expresses an action irrespective of tense or mood; is translated based on the context) int. intensive mood (frequentative; conveys intensification or frequent repetition of the action) ipv. imperative mood opt. optative mood (expresses wish or possibility) sub. subjunctive mood (expresses the will of the speaker)

*Refer to Macdonell, A. A., *A Vedic Grammar for Students*, Oxford, The Clarendon Press, 1916.

I. The Ṛgveda and Its Interpretation

1.1 THE ṚGVEDA AND THE VEDIC LITERATURE

The word Veda (वेदः) means "knowledge, true or sacred knowledge". In general, the 'Veda' in the singular encompasses all branches (शाखाः śākhāḥ, pl.) of the Veda. However, the 'Vedas' in the plural is often used to denote the two or more branches of the Veda. Ṛgveda (ऋग्वेदः), Atharvaveda (अथर्ववेदः), Sāmaveda (सामवेदः), and Yajurveda (यजुर्वेदः) are the four main branches of the Veda. Each branch of the Veda consists of three different classes of literary works: Saṃhitā (संहिता), Brāhmaṇaṃ (ब्राह्मणं), and Āraṇyakaṃ (आरण्यकं).[14]

Saṃhitās are the collections of hymns of praise for and supplication to the Devas, the celestial beings, the deified celestial objects and phenomena. Brāhmaṇas are "prose texts which contain theological discussions, especially observations on the sacrifice" and "the significance of the individual sacrificial rites and ceremonies". Āraṇyakas ("forest texts") and Upaniṣads ("secret doctrines"), which are partially included in or appended to the Brāhmaṇas, contain "the meditations of forests-hermits". In the Āraṇyakas and Upaniṣads, "a good deal of the oldest Indian philosophy" is contained.[15]

Ṛgveda has its own Saṃhitā, the Ṛgveda Saṃhitā (ऋग्वेदसंहिता). Yajurveda has two recensions of Saṃhitā, that is, the Saṃhitā of the Kṛṣṇa Yajurveda (कृष्णयजुर्वेदसंहिता) and the Saṃhitā of the Śukla Yajurveda (शुक्लयजुर्वेदसंहिता). The two remaining Saṃhitās are the Sāmaveda Saṃhitā (सामवेदसंहिता) and Atharvaveda Saṃhitā (अथर्ववेदसंहिता). These

---

[14] M. Winternitz, *A History of Indian Literature*, S. Ketkar (trans.), revised by the author, Calcutta, The University of Calcutta, Vol. I, 1927, pp. 53-54.
[15] ibid., pp. 53.

Saṃhitās are simply called the Ṛgveda, Black Yajurveda, White Yajurveda, Sāmaveda, and Atharvaveda, respectively. The Ṛgveda is the oldest and the most important and probably the only original of the four Vedas. The Ṛgveda Saṃhitā (ऋग्वेदसंहिता) consists of a total of 1028 hymns divided into ten maṇḍalas.[16] Each hymn consists of somewhere between one to fifty-eight stanzas, averaging about ten stanzas in each hymn.

Atharvaveda is "the Veda of Atharvan (अथर्वन्)". A significant portion of the Atharvaveda Saṃhitā has been taken literally from the Ṛgveda. Traditionally Atharvaveda was believed to be the knowledge of the magic spells.[17] However, Atharvā (अथर्वा m. nom. sg. अथर्वन्) is the name of the priest who institutes the fire sacrifice and offers Soma and prayers. In the Ṛgveda, the one who institutes the fire sacrifice is the fire of the Stone, the celestial ioniser, that breaks the atomic structures of hydrogen to produce cosmic plasmas. Thus, the Atharvaveda must be 'the knowledge of the Stone, the celestial ioniser', and it is not about the magic spells against diseases or demons.

The Sāmaveda includes the songs for the Udgātā (उद्गाता), the priest who chants the hymns of Sāmaveda. The verses and stanzas of the Sāmaveda are largely taken from the Ṛgveda. When a stanza is sung by the Udgātā, it is said that a particular melody (सामन् sāman) is "sung upon a particular stanza". Vedic theologians said the melody is originated out of the stanza. The stanza (ऋक्) is therefore called a 'womb (योनिः Yoni)', from which the melody is born.[18] It is not the stanza that is set to melody. Instead the melody is born of the stanza. Please keep that in mind as you read through this book.

The two recensions of Yajurveda include the mantras and sacrificial formulas (यजुः sg., यजूंषि pl.) for the Adhvaryu priests, which are uttered during the sacrificial rites, and the description and theological discussions of sacrificial rites for the priests and theologians.[19] As we shall see, the Adhvaryu priests refer to the firestones of the Stone, the celestial ioniser. Note, the word 'firestones', used in this book, does not mean the stones used for lining furnaces or ovens. Here, the firestones refer to the fiery firestones of the Stone, which are laid in a circular formation, that make up each of the two discs of the Stone, the celestial ioniser (see Figure 8 and section 2.7).

The Vedic literature is much more extensive than the skeletal summary provided here. For a comprehensive list and discussions on the Vedic literature, refer to Winternitz (1927).

---

16 ibid., pp. 54-57.
17 ibid., pp. 119-121.
18 ibid., pp. 165-167.
19 ibid., pp. 169-171.

1.2 THE VEDĀṄGAS, THE SIX LIMBS OF THE VEDA

There are the six Vedāṅgas (वेदाङ्गानि vedāṅgāni pl.), or the six 'limbs' of the Veda. The six limbs are the auxiliary sciences of the Veda and these auxiliary sciences were the subjects of the study, which had to be learned in the Vedic schools in order to understand the Vedic texts.[20] The six Vedāṅgas include Śikṣā (शिक्षा), Chanda (छंदः), Vyākaraṇam (व्याकरणं), Niruktaṃ (निरुक्तं), Jyotiṣaṃ (ज्योतिषं), and Kalpa (कल्पः). Śikṣā is the instruction of correct pronunciation, intonation, and phonetic combination of words in the Veda, especially of the Saṃhitā texts. Niruktaṃ is etymology and etymological interpretations of Vedic vocabulary. Vyākaraṇaṃ is grammar and grammatical analysis.

It appears that the original knowledge of the physical science fields of Kalpa, Chanda, and Jyotiṣaṃ was lost and these science fields were deviated from being the auxiliary sciences of the Veda. For example, Kalpa is considered to be the instructions and rules for conducting sacrificial rituals; Chanda, the study of the metres of the Vedic hymns. Jyotiṣaṃ is viewed as astronomy and a manual for determining the days and hours for the Vedic sacrifices and religious events.[21]

However, when one considers that the Ṛgveda is the treatise on the cosmic creation and the celestial ionisation and cosmic plasma discharge and release, Kalpa in fact must have been the study of the law of plasma physics. Chanda must have been the study of the patterns of Speeches (वाचः vācaḥ pl.), also called hymns or prayers, the patterns of oscillations generated by cosmic plasmas across a broad frequency band of the electromagnetic spectrum (RV x.130.4-6). These Speeches are numerically coded as metres of the stanzas in the Ṛgveda. Jyotis (ज्योतिस्), the root form from which the word Jyotiṣaṃ is derived, means celestial light, the light that appears in the three Spheres of the Heaven (RV ii.17.4). Celestial lights are cosmic plasmas. The Vedāṅga Jyotiṣaṃ therefore must have been the science of cosmic plasma light and fire discharge phenomena.

Unfortunately, none of the original Vedāṅga texts of these science fields survived.[22] If unadulterated ancient texts that contain the original Vedic knowledge in Vedāṅga science fields still exist somewhere, they have not yet been located.

---

[20] ibid., p. 268.

[21] Monier Monier-Williams, *A Sanskrit-English Dictionary*, op. cit., Under each subject headword.

[22] M. Winternitz, *A History of Indian Literature*, op. cit., pp. 288-289.

1.3 THE DATE OF THE ṚGVEDA

Since no chronological data of the ancient Vedic literature exist, establishing the date of the Veda is difficult. As stated in Winternitz (1927), the estimated age of the Ṛgveda varies from about 1000–3000BC and even up to 6000BC.[23] Müller (1919) argues that it may be brave to postulate the date of the Vedic hymns around 2000BC or even 5000BC as a minimum.[24] These estimates are mostly based on the supposed migration of Āryans into India and the astronomical information revealed in the Ṛgveda. For detailed discussions on the basis of different estimates of the age of the Ṛgveda, refer to Müller (1919) and Winternitz (1927). Note, however, the Ṛgveda is not about the astronomy nor history and social customs of Āryan as earlier Vedic scholars assumed it to be. In the Ṛgveda, Ārya (आर्यः), which is translated as 'Āryan' by Griffith and others, is the positive charge of the double layer of a cosmic plasma sheath, not the race which immigrated from central Asia into Northern and Central India. The negative charge of the double layer is called Dāsa (दासः). For Dāsa and Ārya, see section 4.5.1.

It will be equally, if not more, difficult to establish when the Veda was first revealed to men and for how long it had been orally transmitted before its obliteration in the last Dvāpara Yugaṃ. The reader, who is interested in knowing the history of the human race and what the real age of knowledge revealed in the Ṛgveda could be, is encouraged to read *Forbidden Archeology: the Hidden History of the Human Race* by Cremo and Thompson (1993).[25]

So, instead of engaging in this difficult and almost impossible task of establishing the real age of the Ṛgveda, an attempt was made to estimate the period of last Dvāpara Yugaṃ, during which the obliteration of the Veda took place. According to Sri Yukteswar's (1894) calculation of our Solar Yuga cycle (not Mahā Yuga cycle), the last descending 4,800-year Satya Yugaṃ (सत्ययुगं), also called Kṛta Yugaṃ (कृतयुगं), started in 11501BC and ended in 6701BC followed by the last descending 3,600-year Tretā Yugaṃ (त्रेतायुगं 6701–3101BC) and the 2,400-year Dvāpara Yugaṃ (द्वापरयुगं 3101–701BC). In 701BC the world transitioned into the last descending 1,200-year Kali Yugaṃ (कलियुगं), which ended in AD499. We passed 1,200 years of the last ascending Kali Yugaṃ in AD1699.[26]

---

[23] ibid., pp. 293, 296.

[24] F. Max Müller, *The Six Systems of Indian Philosophy*, new impression, London, Longmans, Green and Co., 1919, p. 34.

[25] Michael A. Cremo and Richard L. Thompson, *Forbidden Archeology: The Hidden History of the Human Race*, revised 1st edition, Los Angeles, Bhaktivedanta Book Publishing, Inc., 1998.

[26] Sri Yukteswar Giri, *The Holy Science*, 8th edition, Los Angeles, Self-Realization Fellowship, 1990, pp. 7-20

Now in 2021 we are 322 years into the 2,400 years of the ascending Dvāpara Yugaṃ (AD1699-4099), which is the beginning of the successive ages of awakening "bringing a rapid development in man's knowledge", which will be followed by the Yugas of awakening: the ascending 3,600-year Tretā Yugaṃ (AD4099–7699) and the ascending 4,800-year Satya Yugaṃ (AD7699–12499) completing both the descending 12,000-year and the ascending 12,000-year Yuga cycles. Sri Yukteswar clarifies that this complete 24,000-year Yuga cycle is based on the movements of our sun.

The sun "takes some star for its dual and revolves around it in about 24,000 years of our earth". When the sun passes the nearest point and moves towards the farthest point from the grand centre (विष्णुनाभी viṣṇunābhī or विष्णुनाभिः viṣṇunābhiḥ), the centre of the Milky Way, in its revolution around its dual, our solar system enters a descending 12,000-year Yuga cycle: the descending 4,800-year Satya Yugaṃ → the descending 3,600-year Tretā Yugaṃ → the descending 2,400-year Dvāpara Yugaṃ → the descending 1,200-year Kali Yugaṃ. During this descending Yuga cycle, "the mental virtue of man" gradually reduces to its lowest. Each Yugaṃ has its morning twilight in the beginning and its evening dusk near the end. The morning twilight and evening dusk are the transition from and into another Yugaṃ and it is the period of gradual mutation. The 4,800 years of Satya Yugaṃ is equal to (400-year morning twilight) + (4,000 years) + (400-year evening dusk). The 3,600 years of Tretā Yugaṃ is equal to (300-year morning twilight) + (3,000 years) + (300-year evening dusk), and so on.

When the sun, during its revolution around its dual, passes the farthest point and moves towards the nearest point to the grand centre, our solar system enters an ascending 12,000-year Yuga cycle: the ascending 1,200-year Kali Yugaṃ → the ascending 2,400-year Dvāpara Yugaṃ → the ascending 3,600-year Tretā Yugaṃ → the ascending 4,800-year Satya Yugaṃ. During this ascending Yuga cycle, "the mental virtue of man" gradually awakens to its highest.

Based on Sri Yukteswar's calculation, we can estimate that the obliteration of the Veda took place between 3101BC and 701BC during the last descending Dvāpara Yugaṃ. Accordingly, it can be estimated that the Ṛgveda had been orally transmitted for at least about 2,700 years and up to about 5,100 years without corruption since the Vedas were recompiled by Vyāsa after its obliteration during the last descending Dvāpara Yugaṃ. And the written text of the Ṛgveda has been in existence since Vasukra (वसुक्रः) committed to put the Veda in writing probably around AD800–900.

According to Hindus, the date of the Veda can never be fixed and has never been fixed. Hindus believed the Veda was the knowledge of God. The Veda, being the knowledge of

God, does not owe its authority to anybody, it is itself the authority. The Veda has existed throughout time just as the creation has been eternal without beginning and without end; so is the knowledge of God without beginning and without end.[27]

1.4 PROBLEMS ASSOCIATED WITH TRADITIONAL INTERPRETATIONS OF THE ṚGVEDA

Though the Ṛgveda has been revered as the sacred and eternal knowledge revealed to men and recognised as the most ancient literary texts in the history of mankind, the meanings of the hymns have remained undeciphered. In section i.15 of Niruktaṃ (निरुक्तं), Yāska (यास्कः c. 500BC) cites an earlier commentator named Kautsa (कौत्सः) who asserted that Vedic hymns were meaningless.[28] Yāska lists Kautsa's examples of the meaningless and contradictory Vedic hymns to which he then provides his counter arguments. As shown by Yāska's quotation, and as indicated by Winternitz (1927, Vol 1: 69-70) and Macdonell (1917: xxx), it is apparent that the original meaning of the Veda had been long lost by the time of Kautsa. Otherwise Kautsa would not have argued that the revered text was meaningless. As noted, the celebrated commentary of Sāyaṇa (सायणः died AD1387) on the Ṛgveda was composed long after the original meaning of the Veda had been lost.

In 1858 Theodor Benfey, a German philologist and scholar of Sanskrit, stated,

> [A]nyone who has carefully studied the Indian interpretations knows that absolutely no continuous tradition between the composition of the Vedas and their explanation by Indian scholars can be assumed; that on the contrary, there must have been a long, uninterrupted break in tradition between the genuine poetic remains of Vedic antiquity and their interpretations.[29]

Based on the accounts of Yāska and Alberuni and the German scholar, Theodor Benfey, it seems reasonable to assume that the obliteration of the Veda in the last Dvāpara Yugaṃ (द्वापरयुगं 3101–701BC) contributed to "a long, uninterrupted break in tradition" and probably the loss of the original meaning of the Vedas. There must have been a

---

[27] *From Colombo to Almora: Being a Record of Swami Vivekānanda's Return to India after His Mission to the West*, 2nd ed., Madras, Brahmavadin Press, 1904, p. 24.

[28] Lakshman Sarup, *The Nighaṇṭu and the Nirukta of Śrī Yāskācārya, The Oldest Indian Treatise on Etymology, Philology, and Semantics*, reprint, Delhi, Motilal Banarsidass, 2009, Part II, pp.15-16.

[29] Theodor Benfey, *Göttingische gelehrte Anzeigen unter der Aufsicht der Königl. Gesellschaft der Wissenschaften*, 1858, cited in Karen Thomson, 'A Still Undeciphered...' 2009a, op. cit., pp. 2-3.

catastrophic event, very likely planetary in nature, during the last descending Dvāpara Yugaṃ that caused the obliteration of the Veda and the religious rites of India.

Most interpretations of the Ṛgveda by western scholars adopt the view of 'a ritual interpretation' in the context of native tradition. E. B. Cowell (1866), in his Preface in volume 4 of H. H. Wilson's *Ṛig-Veda-Sanhitā* published posthumously, stated,

> Sāyaṇa stands to the Veda as Eustathius to the Homeric poems; Professor Wilson's work enables the English reader to know what Hindus themselves supposed the *Ṛig-Veda* to mean. It is easy to depreciate native commentators, but it is not so easy to supersede them; and until we have more materials for comparison and study, the arbitrary guesses which are often indulged in by continental scholars seem to me but the conjectures of the *intellectus sibi permissus*, which only impede the progress towards a true system of interpretation in philological as well as physical science.[30] (Italics in the original)

Cowell's statement probably best represents the earnest view of some Western scholars engaged in the Vedic studies in early days. Nevertheless, by the late nineteenth-century, some European scholars already "reached the conclusion that, when it comes to understanding the poems of the *Rigveda*, native tradition was entirely misleading".[31] Rudolf Roth pointed out that "the authors of the commentaries might throw useful light on later theological works, when it came to the songs of the most ancient poets they were 'untaugliche Führer' 'unfit guides'".[32]

As noted by these scholars, earlier commentaries on the Ṛgveda by Yāska (यास्कः circa 500BC), Sāyaṇa (सायणः died AD1387), and others do not provide much needed help, for it is apparent that these commentaries were composed long after the original meaning of the Ṛgveda had been lost. Müller (1891) quoted what Professor von Roth said: "'No one who knows anything about the Veda', they say, 'would think of attempting a translation of it at present. A translation of Rig-veda is a task for the next century'". Müller then added his own opinion: "If by translation we mean a complete, satisfactory, and final translation of the whole of the Rig-veda, I should feel inclined to go even further than Professor von Roth. Not only shall we have to wait till the next century for such a work, but I doubt whether we shall ever obtain it".[33]

---

30 H.H. Wilson, *Ṛig-Veda Sanhitā. A Collection of Ancient Hindu Hymns Constituting the Fifth Ashṭaka or Book of the Rig-Veda; the Oldest Authority for the Religious and Social Institutions of the Hindus*, E.B. Cowell (ed.), London, N. Trübner and Co., 1866, p.vi.

31 Otto Böhtlingk and Rudolf Roth, *Sanskrit-Wörterbuch*, St. Petersburg, Kaiserliche Akademie der Wissenschaften, 1855-75, cited in Karen Thomson, 'A Still Undeciphered Text...' 2009a, op. cit., p. 2.

32 Karen Thomson, 'A Still Undeciphered Text...' 2009a, op. cit., p. 2.

33 F. Max Müller (trans.), *Vedic Hymns Part I: Hymns to the Maruts, Rudra, Vāyu, and Vāta*, Oxford, The Clarendon Press, 1891, p. xi. The Sacred Books of the East Series Vol. XXXII.

In *A History of Indian Philosophy* by Dasgupta (1957), the Veda and Vedic literature are "slightly" touched upon as he stated.[34] He does not shed much light on the Veda and on the Vedic language. He regards the difficulty of understanding technical terms used in the works of the vast Indian philosophical systems as the main issue in interpreting and understanding the Indian systems of thoughts.[35] Winternitz (1927), in his *History of Indian Literature*, provides some insight into the language of the Ṛgveda. Winternitz (1927) suggested that the language of the Ṛgveda is "Ancient Indian" or "Vedic" in a narrow sense, and that "Ancient High Indian" is perhaps the best name for the language of the Ṛgveda. He distinguishes the "Ancient High Indian" language of the Ṛgveda from the 'classical Sanskrit' that follows the grammar of Pāṇini (पाणिनिः).

> The Vedic language hardly differs at all from Sanskrit in its phonetics, but only through a much greater antiquity, and especially through a greater wealth of grammatical forms. Thus for instance, Ancient High Indian has a subjunctive which is missing in Sanskrit; it has a dozen different infinitive-endings, of which but one single one remains in Sanskrit. The aorists, very largely represented in the Vedic language, disappear in the Sanskrit more and more. Also the case and personal endings are still much more perfect in the oldest language than in the later Sanskrit. A later phase of Ancient High Indian appears already in the hymns of the tenth Book of the Ṛgveda and in some parts of the Atharvaveda, and the collections of the Yajurveda.[36]

Thomson provides an extensive review on the current state of Vedic study and problems involved in the interpretation of the text of the Ṛgveda. Thomson (2009b) states about the current state of the Ṛgveda study in an article titled *A Still Undeciphered Text, continued: the reply to my critics*,

> In the *Rigveda* we have a body of highly structured poetry, ..., that predates the work of Homer by a thousand years. And yet nobody, neither in "the land of its birth"..., nor anywhere else in the Indo-European speaking scholarly community, is applying scientific method to its interpretation. ... I do not know of any university department in which research into its interpretation could currently belong. This extraordinary anthology should, in my view, be a respected part of our Indo-European heritage. For the moment, most of its artistry and craft lies hidden from us.[37] (Italics in the original.)

About the language of the Ṛgveda and classical Sanskrit, Thomson (2004) states in her article '*Sacred Mysteries: Why the Rigveda has resisted decipherment*' that the language

---

[34] Surendranath Dasgupta, *A History of Indian Philosophy*, reprint, London, Cambridge University Press, Vol. I, 1957, p. ix.

[35] ibid., pp. viii, 1-3.

[36] M. Winternitz, *A History of Indian Literature*, Vol. I, 1927, op. cit., pp. 41-42.

[37] Karen Thomson, 'A Still Undeciphered Text, continued: the reply to my critics', *The Journal of Indo-European Studies*, Vol. 37, No. 1 & 2, Spring/Summer 2009b, pp. 78-79.

of the Ṛgveda had a musical accent like ancient Greek and "is different from Classical Sanskrit as the language of *Beowulf* is from modern English".[38]

Thomson (2009a) indicates that tenacious influence of native tradition; misinterpretation of key words of the poems that firmly hold grip on the Vedic scholars; scholarly speculation that hymns of the Ṛgveda are deliberate riddles, mischievously designed to prevent those outside the cabal from being able to penetrate to their meaning; and the lack of scientific approach are the main reasons that the text of the Ṛgveda still remain undeciphered.[39]

Wendy Doniger O'Flaherty (1981), in *The Rig Veda: An Anthology*, argues that "The riddles in the *Rig Veda* are particularly maddening because many of them are *Looking Glass* riddles (Why is a raven like a writing desk?): they do not have, nor are they meant to have, answers".[40] Jamison, in her book *The Ravenous Hyenas and the Wounded Sun: Myth and Ritual in Ancient India* published in 1991, suggests that we must learn to cultivate the divine taste of loving the obscure: "As the Brāhmaṇas tell us so often, 'the gods love the obscure', and in investigating Vedic matters we must learn to cultivate at least that divine taste".[41]

For traditional scholars in the "land of its birth", Thomson (2010) states, "And as far as traditional scholars in India are concerned, there is no debate: these poems should simply not be an object of study. What the *Rigveda* 'means' is of no concern, it is the tradition deriving from them that matters". And to traditional scholars of India, "Western scholarship burrowing into the text is, all said and done, an annoyance".[42]

For the latest translation of the Ṛgveda into English by Jamison and Brereton (2014),[43] Thomson (2016) provides a forthright review:

---

38 Karen Thomson, 'Sacred Mysteries: Why the *Rigveda* Has Resisted Decipherment', *Times Literary Supplement*, 26 March 2004, pp.14-15.

39 Karen Thomson, 'A Still Undeciphered Text ...,' 2009a, op. cit., pp. 1-4,11-13.

40 Wendy Doniger O'Flaherty, *The Rig Veda: An Anthology*, Harmondsworth, Penguin Books, 1981, p. 16.

41 Stephanie Jamison, *The Ravenous Hyenas and the Wounded Sun. Myth and ritual in Ancient India*, Ithaca, Cornell University Press,1981, cited in Karen Thomson, 'A Still Undeciphered Text ...,' 2009a op. cit., p.6.

42 Karen Thomson, 'The Plight of the Rigveda in the Twenty-First Century', *The Journal of Indo-European Studies*, Vol. 38, No. 3 & 4, Fall/Winter 2010, p.423.

43 Stephanie W. Jamison and Joel P. Brereton (trans.), *The Rigveda: The Earliest Religious Poetry of India*, Vols. I-III, New York, Oxford University Press, 2014.

> Within its soberly academic trio of hardback volumes, however, seethes an incoherent mix of mumbo-jumbo and misplaced obscenity, most of it apparently meaningless. ... ... With the requirement for making sense removed, *a vogue la galère* atmosphere pervades the whole.[44] (Italics in the original)

Thomson (2016) argued that, in translating the Ṛgveda, "we need to begin with a different assumption: that the poems are as meaningful as their complex grammar, consistent language and word formation, and highly sophisticated metre would suggest".[45] In investigating Vedic matters, we must learn to cultivate our faculty of discernment and erudition, for the Vedic Devas love 'the kindled', not 'the obscure'. In translating the Ṛgveda, as Thomson stated, we must start with understanding that these hymns are "meaningful". The hymns of the Ṛgveda are not "maddening" riddles, for they are the systematic exposition of the Vedic cosmology, though written in symbolic and beautiful poetic language. In translating the Ṛgveda, we must not torture the text or disregard the grammatical rules of the Vedic language.

When asked by Premānanda what will be the good of studying the Vedas, Vivekānanda answered, "It will kill out superstitions".[46] Indeed, scholarly superstitions need to be 'killed' so that we can study and learn the knowledge the Vedas reveal.

1.5 A New Interpretation of the Ṛgveda

The new interpretation and translation presented in this book is based on the author's reading of enchanting hymns and stanzas chosen from the ten maṇḍalas of the Ṛgveda. The hymns of the Ṛgveda take a highly metrical form and adopt skilful metaphors, personification, deification, and prosopopoeia. The Heavenly structures and the celestial plasma phenomena are described or are spoken of in anthropomorphic language by the poets of the Ṛgveda. In many stanzas, unspecified persons or dead (read ionised) forefathers are represented by the poets or speak through the poets. To decipher the meanings of these hymns, one has to set one's vantage point in intergalactic space and wade through thickets of metaphors and prosopopoeia with a general knowledge of cosmic plasma discharge and release phenomena. Only then will one be able to uncover the meanings of these ancient hymns and make sense of this ancient text.

---

[44] Karen Thomson, 'Speak for itself: How the long history of guesswork and commentary on a unique corpus of poetry has rendered it incomprehensible', *Times Literary Supplement*, 8 January 2016, p.3,4.
[45] ibid., p4.
[46] Advaita Ashrama, *The Life of Swami Vivekananda by his Eastern and Western Disciples*, 6th edition, 3rd reprint, Mayavati, India, 1998, Vol. 2, p.654.

The hymns of the Ṛgveda and other Sanskrit texts cited in this book are interpreted and translated directly from the Ṛgveda and other Sanskrit texts into English by the author. This new interpretation is not a fancy of one's imagination; it is based on the English definitions of the Vedic lexis from *A Sanskrit-English Dictionary* by Monier-Williams (1899) and the observations of the cosmic plasma phenomena and cosmic plasma discharge and release events made by NASA space probes and telescopes. In some instances, when the Monier-Williams's English definitions of a Vedic word were out of context due to the corruption caused by the loss of the original meaning, appropriate definitions were adopted based on the context of the given stanza and the etymology of the word. While each Vedic stanza is in a specific metrical form, the metrical form is ignored in its English translation, for the purpose here is to render it as close and accurate to the original meaning of the stanza.

In translating into English, some archaic and formal forms of words, such as ye, thou, abode, and so on are used for a number of reasons. In Sanskrit, nouns and pronouns exist in singular, dual, and plural forms. Since the dual nouns and pronouns cannot be directly translated into English, using 'thou' and 'ye' was found to be convenient, especially in vocative cases and in differentiating the second person dual and plural forms from the second person singular. For example, without using archaic forms thou and ye, it would become 'O you Agni (sg.)', 'O you Aśvinā (du.)', or 'O you Devas (pl.)', which sounds rude and irritating, while 'O thou Agni', 'O ye Aśvinā', or 'O ye Devas' is much more pleasant to the eye and to the ear. Please note that the hymns of the Ṛgveda are probably much older than most of the Vedic scholars believe them to be. The Ṛgveda is one of the most ancient and important books we have inherited from remote antiquity.

The author retained the original definitions of words in the text rather than changing them to the politically correct ones. For example, the words 'men', 'son', or 'wife or husband' are retained as in the original text and have not been replaced with the terms 'people', 'child', or 'partner' that are considered to be politically correct. The author expects the reader to exercise sound judgement and value our historical and cultural heritage. In the Ṛgveda, some celestial objects and phenomena are represented in masculine or neuter nouns and others are in feminine nouns. In some instances, a celestial object is personified in both genders, as in the case of Sarasvatī (f.) and Sarasvān (m.). Both Sarasvatī and Sarasvān are the personified names of the Soma Pond. Occasionally, certain masculine nouns are declined as if they are feminine. For example, the masculine words Heaven (दिव्) and Agni (अग्नि) are declined as feminine in certain stanzas.

Though the key attributes and identities of the Vedic celestial beings are often translated into English and explained in modern terms of physics and astrophysics when necessary, most of the names of the Vedic celestial beings and events, such as Tvaṣṭā (त्वष्टा), Ṛbhus (ऋभवः ṛbhavaḥ pl.), Agni (अग्निः), Soma (सोमः), Varuṇa (वरुणः), Mitra (मित्रः), Aryamā (अर्यमा), the celestial Yajña (यज्ञः), and so on are not always translated into English.

Translating these anthropomorphised poetic names of the Vedic Devas and celestial events into English as 'magnetism in general and the magnetic field in particular', 'magnetic forces', 'proton or a fiery and dry embryonic particle of celestial Waters', 'electron or a cool and moist embryonic particle of Waters', 'cosmic plasmas in general and plasma dark discharge in particular', 'cosmic plasma light discharge', 'cosmic plasma fire discharge', and 'sacred cosmic ritual or sacrifice' respectively would indeed be awkward. Even though the hymns of the Ṛgveda explain the celestial objects and the properties of cosmic plasmas, the mode of expression is diametrically different from that of the books and articles of modern physics and astrophysics. When appropriate, a word or a phrase was borrowed from Griffith's English translation of the Ṛgveda.[47]

Note that every stanza of the Ṛgveda, if not a praise of or supplication to celestial beings for prosperity and the well-being of the Milky Way and the whole of the galactic system, explains the celestial objects of our galactic system, the events and processes of ionisation and cosmic plasma discharge and release, or the properties of cosmic plasmas.

As interpretation and translation of each of the ten maṇḍalas of the Ṛgveda continue, they will be published in due course. With the translation of each maṇḍalaṃ of the Ṛgveda, it is expected that more information on the properties of cosmic plasmas and the mechanisms of ionisation and cosmic plasma discharge and release will be revealed.

The translation of maṇḍalas II-IX of the Ṛgveda will be done first, then maṇḍalas I and X will follow.

1.6 THE TEXT OF THE ṚGVEDA

The Ṛgveda text compiled by F. Max Müller in volumes I-IV (1890-1892) was used for the interpretation and translation. Müller's compilation includes Saṃhitāpāṭha (संहितापाठः) and Padapāṭha (पदपाठः) texts along with the commentary of Sāyaṇa (सायणः). The digital text of the Ṛgveda was provided by Peter Freund at the Maharishi International University (MIU). The text provided by MIU was written in phonetic spellings. Considering that all the

---

47 Ralph T.H. Griffith, *The Hymns of the Rigveda: Translation with a Popular Commentary*, 2nd ed. (complete in two volumes), Vol. I and Vol. II, Benares, E.J. Lazarus and Co. 1896 and 1897.

traditional Vedic manuscripts adopt etymological spellings, MIU's text was compared to Müller's text, and the Vedic words in phonetic spellings were changed to the etymological spellings to preserve the traditional form of the Ṛgveda text.[48] MIU's digital text contained a few systematic digitisation errors, which were corrected. Please note that the numbers of the Ṛgveda hymns and stanzas follow Müller's Ṛgveda text.

There were minor variances in the Vedic poets' names and the names of the Devas to whom the hymns and stanzas were addressed between the two sources of the text. In such cases, Müller's were chosen. It is unclear whether these variances are due to the variances found in the old manuscripts or caused by the arbitrary adjustments made by the editors of the recently published Ṛgveda text.

Along with the English translation of the select stanzas and hymns, the texts of Saṃhitāpāṭha and Padapāṭha are presented in this book for the convenience of Vedic students and researchers.

---

[48] Gary B. Holland, Professor, University of California, Berkeley, Email response to the author, 10 May 2013. Personal communication.

II. The Heavenly Structures Revealed in the Ṛgveda

2.1 The Shining Ones, the Devas and the Ṛṣis

Once we set our vantage point in intergalactic space, we can embark on the journey of uncovering the 'secret' knowledge the Ṛgveda reveals. In the Ṛgveda, the events of cosmic ionisation and cosmic plasma discharge and release are called the celestial Yajñas (यज्ञाः pl.), the cosmic sacrifices. The cosmic plasma phenomena associated with cosmic sacrifices are personified and deified as celestial beings and collectively called Ṛṣis (ऋषयः ṛṣayaḥ pl.) and Devas (देवाः devāḥ pl.). 'Deva' means 'heavenly or shining one'. Ṛṣi (ऋषिः) is a celestial being specifically called 'a singer of sacred hymns', though all the celestial beings in the ionised state sing. Devas and Ṛṣis constantly sing hymns and make Speeches as they perform their duties for celestial Yajñas.

In Figure 1 are four 'X-Ray & Optical Images of Planetary Nebulas' observed by NASA's Chandra X-Ray Observatory and Hubble Space Telescope.[49] These celestial beings are called planetary nebulas in modern physics terminology. They were named so, by early astronomers who observed these celestial objects, for they appeared to resemble the shape of planets when observed with telescopes used in earlier days.[50] Each glowing, hauntingly beautiful celestial being is often referred to as a celestial jewel or gemstone (रत्नं ratnaṃ RV i.1.1, RV i.35.8) in the Ṛgveda.

---

49 NASA Chandra X-Ray Observatory, Images of Planetary Nebula Gallery, *X-Ray & Optical Images of Planetary Nebulas*, https://www.chandra.harvard.edu/photo/2012/pne/ (accessed 25 June 2019).
50 NASA/ESA Hubble Space Telescope, in the image description of *A Cosmic Garden Sprinkler*, https://www.spacetelescope.org/images/potw1241a/ (accessed June 25, 2019).

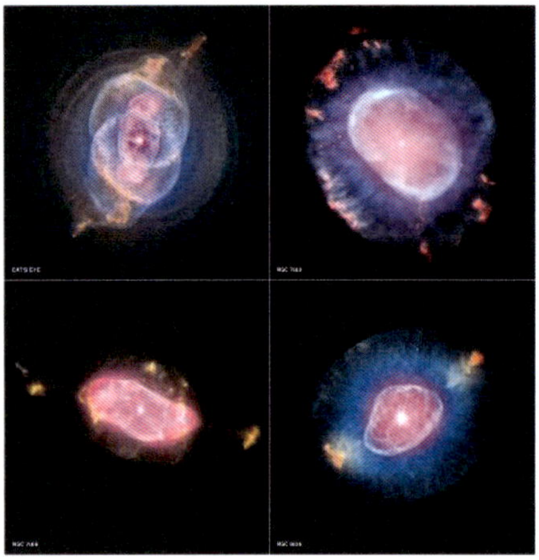

<Figure 1> "A Planetary Nebula Gallery" X-Ray & Optical Images of Planetary Nebulas.
Image credit: X-Ray: NASA/CXC/RIT/J. Kastner et al.; Optical: NASA/STScl
UL: NGC 6543 known as 'Cat's Eye Nebula'
UR: NGC 7662 'Blue Snowball Nebula'
LL: NGC 7009 'Saturn Nebula'
LR: NGC 6826 'The Blinking Eye Nebula' (LR)

Figure 2 is "Gemstones in the Southern Sky", in which celestial gemstones shine brightly in the Open Cluster NGC 290.[51] These celestial gemstones were captured by the Hubble Space Telescope of NASA. These "Gemstones in the Southern Sky" are located about 200,000 light-years away, spanning about 65 light-years across. Like gemstones, the stars of open cluster NGC 290 glitter beautifully. It is apparent that the ancient seers and modern astrophysicists and astronomers share the same sense of perception of luminous celestial beings.

<Figure 2> Open Cluster NGC 290 "Gemstones in the Southern Sky".
Image Credit: NASA/ESA/STScl/E. Olszewski, University of Arizona.

In November 2010, NASA announced that its Fermi Gamma-Ray Space Telescope had revealed a previously unseen structure, which was named Fermi Bubbles, in our Milky Way Galaxy. This structure, which consists of two giant bowls, is anchored upright at the

---

[51] NASA/ESA Hubble Space Telescope, Open Cluster NGC 290 "*Gemstones in the Southern Sky*", https://www.nasa.gov/multimedia/imagegallery/image_feature_550.html (accessed 25 June 2019).

centre of the Milky Way (Figure 3). Julie McEnery, Fermi project scientist at NASA's Goddard Space Flight Center, in her presentation, stated that electrons moving near the speed of light cause gamma-ray emissions and that these fast-moving electrons also power this galactic structure.[52] The edges of the bowls, lighted in blue, are X-Ray emissions. X-Ray emissions in the cosmos are known to indicate the presence of a strong magnetic field, extreme gravity, or explosive forces.[53]

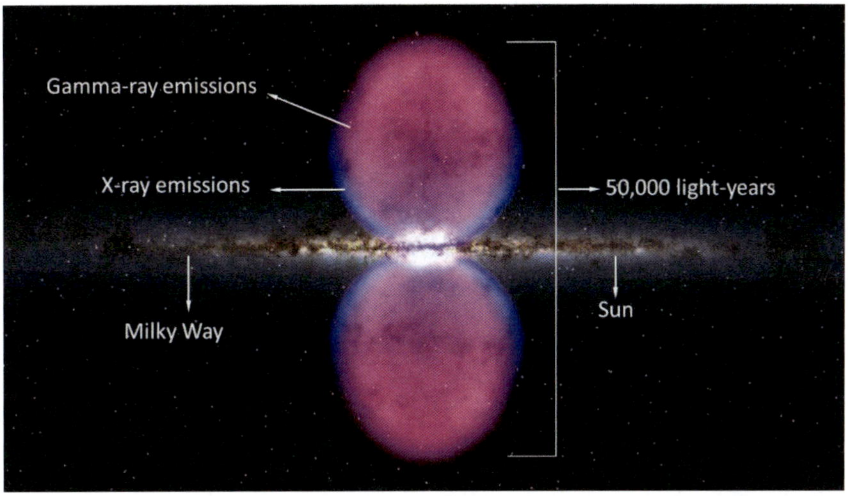

<Figure 3> Image of Fermi Bubbles Released by NASA. Image Credit: NASA's Goddard Space Flight Center. Annotations in the original NASA image. Pointer arrows and annotation size/positions adjusted.

The Fermi Bubbles span 50,000 light-years from top to bottom, which is roughly half of the Milky Way Galaxy's diameter, according to NASA scientists. Subsequent studies also revealed the evidence for gamma-ray Jets[54] and the two smaller, polarised Radio lobes centred on the galactic centre within the Fermi Bubbles.[55] In fact, these newly mapped galactic structure, announced by NASA scientists, agree with the galactic structure presented in the Ṛgveda. According to the Ṛgveda, there are the three spheres of the Heaven (the Upper, the Lower, and the Innerfield) and two Vaults of the Heaven (the Upper and the Lower Vaults). In the Innerfield, or the midheaven, the Milky Way is formed

---

[52] Julie McEnery (Presenter 2), Fermi Project Scientist, 9 November 2010. [website] http://www.nasa.gov/mission_pages/GLAST/ news/new-structure-briefing.html (accessed 21 February 2017).

[53] NASA Chandra X-Ray Observatory, *Discovering the X-Ray Universe: Colossal Clouds of Hot Gas*, http://chandra.harvard.edu/xray_astro/discover.html (accessed 25 June 2019).

[54] Meng Su and Douglas P. Finkbeiner, 2012, 'Evidence for Gama-Ray Jets in the Milky Way', *The Astrophysical Journal* 753:61(13pp), 14 June 2012. doi:10.1088/0004-637X/753/1/61.

[55] Ettore Carretti et al., 'Giant Magnetized outflows from the Centre of the Milky Way', *Nature* 493, 03 January 2013, pp.66-69. doi:10.1038/nature11734.

and sustained. Each of the two Vaults of the Heaven has its own pond, or the Ocean of celestial Waters, within it, which is called the Soma Pond (ह्रदः, सरः, शर्यणावान्) or Samudra (समुद्रः). These Heavenly structures are built and held up by Tvaṣṭā (त्वष्टा), the magnetic field, and the Ṛbhus (ऋभवः ṛbhavaḥ pl.), the magnetic forces.

2.2 THE THREEFOLD HEAVENLY SPHERE, THE ALTAR OF COSMIC YAJÑA

2.2.1 The Upper Heaven, the Innerfield, and the Lower Heaven

The Ṛgveda reveals that Vedi (वेदिः f. nom. sg.), the Altar of cosmic Yajña, that is, the field of cosmic creation, is a threefold sphere consisting of the two bowl-shaped Heavens (द्यावा dyāvā m. nom. du. of दिव् div) and the Innerfield (अंतरिक्षं aṃtarikṣaṃ RV i.73.8). The Innerfield, Antarikṣaṃ, is "the middle of the three spheres or regions of life". The two bowl-shaped Heavens are vertically aligned in the shape of the Ḍamaru (डमरुः), the sacred drum of Śiva (शिवः), or in the shape of an hourglass. The region between the two vertically aligned bowl-shaped Heavens is Antarikṣaṃ, the Innerfield, or the Midheaven (मध्यमः madhyamaḥ), the region where life is created and sustained. (see Figure 4)

There is the Heaven above the Milky Way, the Upper Heaven, and there is the Heaven below the Milky Way, the Lower Heaven. As above, so below. In the Antarikṣaṃ, Pṛthivī (पृथिवी), "the broad and extended One", the Milky Way, is formed and sustained.

The twofold Heaven is called Dhiṣaṇe (धिषणे f. nom. du. RV i.160.1), the two Bowls, for the Heaven is bowl-shaped and there are two Heavens, the Upper and the Lower. Dhiṣaṇe (धिषणे) is the dual case of dhiṣaṇā (धिषणा f. nom. sg.), a goblet or bowl. In the Ṛgveda, the words dhiṣaṇā and camū (चमूः f. nom. sg.) in the singular refer to a bowl-shaped Heaven or a Soma Pond. In their dual forms, dhiṣaṇe (f. nom. du.) and camvā (चम्वा f. nom. du. RV iii.55.20), they refer to the two Heavenly Vaults, the Upper and the Lower Vaults, or to the two Soma Ponds, the upper and the lower Soma Ponds (see Figure 6). When the Soma Pond is used in the plural as ponds, they refer to a hundred or a thousand Soma plants, or plasma blooms, which are also called Soma receptacles or Soma chalices (see Figure 7 and refer to section 2.5.1).

Stanza RV v.60.6 tells us that the threefold Heavenly Sphere consists of the Upper Heaven (उत्तमः uttamaḥ), the Midheaven (मध्यमः madhyamaḥ), and the Lower Heaven (अवमः avamaḥ). The author of the stanza is Śyāvāśva (श्यावाश्वः), a descendant of Atri

(अत्रिः a devourer). Atri, the devourer, is the Stone, the celestial ioniser, as we shall see. It is addressed to the Maruts (मरुतः pl.) and Maruts-Agni (मरुतोवाग्निश्च).

RV v.60.6 author श्यावाश्व आत्रेयः, to मरुतो मरुतोवाग्निश्च, metre त्रिष्टुप् छंदः
यदुत्तमे मरुतो मध्यमे वा यद्वावमे सुभगासो दिवि ष्ठ ।
अतो नो रुद्रा उत वा न्व१स्याग्ने वित्ताद्धविषो यद्यजाम ॥५.६०.६॥
यत् । उत्ऽतमे । मरुतः । मध्यमे । वा । यत् । वा । अवमे । सुऽभगासः । दिवि । स्थ ।
अतः । नः । रुद्राः । उत । वा । नु । अस्य । अग्ने । वित्तात् । हविषः । यत् । यजाम ॥५.६०.६॥
Whether, O blessed Maruts, ye be dwelling in the Upper Heaven, in the Midheaven, or in the Lower Heaven, Thence, O Rudras, O Agni, know that we shall offer this oblation of ours.
(RV v.60.6)

यत् <n. nom. sg. of relative pron. यद् – (with ind. particle वा) whether - or>, उत्तमे <m. loc. sg. of उत्तम – in the Upper Heaven>, मरुतः <m. voc. pl. of मरुत् – O Maruts>, सुभगासः <m. voc. pl. of adj. सुभग – blessed>, मध्यमे <m. loc. sg. of मध्यम – in the Midheaven)>, वा <ind. – or>, यत् वा <whether – or>, अवमे <m. loc. sg. of अवम – in the Lower Heaven>, दिवि <m. loc. sg. of दिव् – in the Heaven>, स्थ <pre. 2nd person pl. of √अस् – ye dwell, exist>;

अतः <ind. – from this, thence>, रुद्राः <m. voc. pl. of रुद्र – O Rudras (sons of Rudras, i.e. the Maruts)>, उत वा <ind. – and also>, अग्ने <m. voc. sg. of अग्नि – O Agni>, नु <ind. – indeed, surely>, नः <m. gen. pl. of अस्मद् – our>, अस्य <m. gen. sg. of pron. इदं – of this>, हविषः <n. gen. sg. of हविस् – of oblation>, वित्तात् <ipv. 2nd person pl. of √विद् – ye know, understand>, यत् <ind. conjunction – that>, यजाम <ipv. 1st person pl. of √यज् – we shall offer>.

The Threefold Heaven is often referred to as Trināka (त्रिनाकः RV ix.113.9), Tridiva (त्रिदिवः RV ix.113.9), Tridhātu Rajasa (त्रिधातुः रजसः RV iii.26.7), Trīṇirajāṃsi (त्रीणिरजांसि RV v.69.1), and Trirocanā (त्रिरोचना RV v.69.1), all indicating that the galactic structure of the Milky Way is a three-tiered structure. Trināka means the 'threefold Heavenly Vault'. Tridiva simply is the 'threefold Heaven'. Tridhātu Rajasa means the 'threefold stratum of the Heavenly Sphere'. Trīṇirajāṃsi is the 'three cultivated Spheres'. Rajāṃsi (रजांसि) is the plural case of 'rajas (रजस्)', which means "coloured or dim space" or "cultivated or ploughed land". So, the Rajāṃsi are the three "coloured", "cultivated or ploughed" spheres prepared by Tvaṣṭā (त्वष्टा) and the Ṛbhus (ऋभवः ṛbhavaḥ pl.) as the field of cosmic Yajña (यज्ञः), the field of cosmic creation. We shall see that these seemingly unrelated definitions for 'rajas' provided by Monier-Williams make sense after all.

Trirocanā is the 'threefold luminous Sphere'. Rocanā (रोचना f. nom. sg.) is a "luminous sphere". The threefold Heaven is called Trirocanā, for all three spheres of the Heaven will be filled with celestial lights and fires once the cosmic Yajña starts. The threefold Heaven, the Upper Heaven, the Innerfield (अंतरिक्षं), or the Midheaven, and the Lower Heaven, is the Altar of cosmic Yajña, on which the marvellous Creator's machine performs cosmic Yajñas, the cosmic sacrifices of celestial ionisation and cosmic plasma discharge and release, three times a day every day. The threefold Heaven is Trimūrti (त्रिमूर्तिः trimūrtiḥ or त्रयीदेहः trayīdehaḥ), the first embodied three.

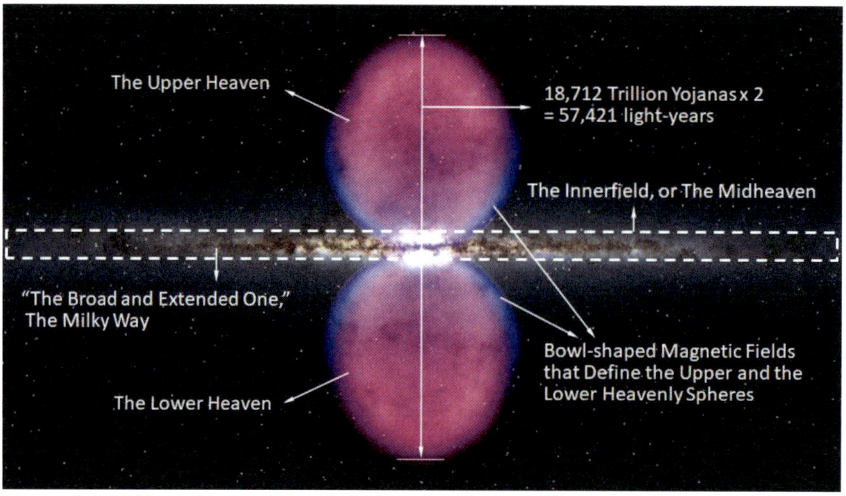

<Figure 4> The Threefold Heavenly Sphere and the Height of the Heaven. Image Credit: NASA's Goddard Space Flight Center. Vedic annotation is added to the original NASA image.

The terms 'the Heaven', 'the Heavenly Vault', and 'the Heavenly sphere' of the Ṛgveda are not to be confused with the 'celestial sphere' of astronomy. In astronomy, the 'celestial sphere' is an arbitrarily projected planet earth-centric sphere. In the Ṛgveda, the Heaven (द्यौः dyauḥ) has a very specific definition. The Heaven refers to the threefold sphere that encompasses the Upper and Lower Vaults of the Heaven, which are defined by the two bowl-shaped magnetic fields, and the Midheaven, or the Innerfield, between the Upper and Lower Heavens. This threefold Heavenly Sphere is shown in Figures 4 and 6.

2.2.2 The Cosmic Yajña, the Origin of the World and Life

According to stanza RV i.164.35, the cosmic Yajña, the cosmic sacrifice, is the origin (नाभिः nābhiḥ) of the World (भुवनं bhuvanaṃ). The World, the Bhuvanaṃ, refers to the whole of the galactic system, which includes the two Heavenly Vaults, the Innerfield, or the Midheaven, the two pairs of the Stones and the Soma Ponds, and the Milky Way (see Figure 6). Through the cosmic Yajña, the World, the Bhuvanaṃ, is sustained. According to the Ṛgveda, the boundary of the Vedi (वेदिः), the Altar of cosmic Yajña, is far beyond the two Vaults of the Heaven and the visible Milky Way. In stanza RV i.164.35, the Stone, the celestial ioniser, is compared to a mighty stallion; the discharge of cosmic plasmas produced by the celestial ioniser, to the seminal stream of that mighty stallion. In RV i.164.35 are the answers to the questions raised in RV i.164.34.

RV i.164.35 author दीर्घतमा औचथ्यः, to विश्वे देवाः, metre त्रिष्टुप् छंदः
इयं वेदिः परो अंतः पृथिव्या अयं यज्ञो भुवनस्य नाभिः ।
अयं सोमो वृष्णो अश्वस्य रेतो ब्रह्मायं वाचः परमं व्योम ॥१.१६४.३५॥

इयं । वेदिः । परः । अंतः । पृथिव्याः । अयं । यज्ञः । भुवनस्य । नाभिः ।
अयं । सोमः । वृष्णः । अश्वस्य । रेतः । ब्रह्मा । अयं । वाचः । परमं । विऽओम ॥१.१६४.३५॥

This Yajña is the origin of the World, this Altar of cosmic Yajña is beyond the limits of the Milky Way. This Soma, the seminal stream of the mighty stallion. The Speeches and the supreme Vault of the Heaven are Brahmā. (RV i.164.35)

इयं <f. nom. sg. of pron. इदं – this>, वेदिः <f. nom. sg. of वेदि – a sacrificial altar, i.e. the altar of cosmic Yajña>, परः <m. nom. sg. of adj. पर – far, remote, ulterior>, अंतः <m. nom. sg. of अंत – an inner part, inside, settlement, limit, boundary>, पृथिव्याः <f. gen. sg. of पृथिवी – of Pṛthivī, i.e. of the Milky Way>, अयं <m. nom. sg. of pron. इदं – this>, यज्ञः <m. nom. sg. of यज्ञ – the cosmic Yajña>, भुवनस्य <n. gen. sg of भुवन – of the World, i.e. the manifested World, the galaxy>, नाभिः <f. nom. sg. of नाभि – navel, centre, origin>;

अयं <m. nom. sg. of pron. इदं – this>, सोमः <m. nom. sg. of सोम – Soma (here, it refers to a draught of electrons)>, वृष्णः <m. gen. sg. of adj. वृषन् – strong, mighty>, अश्वस्य <m. gen. sg of अश्व – of a courser, stallion>, रेतः <n. nom. sg. of रेतस् – a stream, current of rain or water, a seminal stream, i.e. a stream of cosmic plasmas>, ब्रह्मा <m. nom. sg. of ब्रह्मन् – Brahmā, the Creator>, अयं <m. nom. sg. of pron. इदं – this>, वाचः <f. nom. pl. of वाच् – speeches, sounds>, परमं <n. nom. sg. of adj. परम – highest, supreme>, व्योम <n. nom. sg. of व्योमन् – the Vault of the Heaven>.

The term Vyoma (व्योम) used in stanza RV i.164.35 is another name for Nāka (नाकः), the Heavenly Vault. While Nāka implies the shape of the Heaven is a dome or vault, the term Vyoma indicates the functional aspect of the Heaven as an impeller of cosmic Yajña. Vyoman (व्योमन्), the root form of Vyoma, is derived from 'vi-√av (वि-√अव्)' or 'vi-√ve (वि-√वे)'. The vi (वि) is a prefix. The verb √av (√अव्) means "to drive, impel, animate (as a car or horse)". The verb √ve (√वे) means "to weave, interweave, braid". Vyoma is the Heavenly Vault that impels the celestial Yajña and weaves the nebulous fabric of the Milky Way. Note, in this book, the word 'celestial' is used interchangeably with 'cosmic'.

According to RV i.164.35, the Supreme Heavenly Vault (परमं व्योम paramaṃ Vyoma) and the Speeches (वाचः vācaḥ) of cosmic plasmas are Brahmā (ब्रह्मा), the Creator. Speeches, or sounds, or hymns, are the oscillations generated by cosmic plasmas across a broad frequency band of the electromagnetic spectrum. And these Speeches, or sacred hymns, or oscillations, according to the Ṛgveda, play a critical role in cosmic creation and the sustenance of physical reality.

The author of stanza RV i.164.35 is Ṛṣi Dīrghatamā Aucathya (दीर्घतमा औचथ्यः), the descendant of Ucathya (उचथ्यः) who is one of the descendants of the Aṅgirās (अङ्गिरसः aṅgirasaḥ m. nom. pl. of अङ्गिरस्). The Aṅgirās are angels or messengers (compare to Gk. ἄγγαρος ángaros, ἄγγελος ángelos as noted by Monier-Williams); they are celestial lights and fires, the messengers of the Heaven (see Figure 7). The celestial lights and fires are also named the bright arrows, spears, thunderbolt of Indra, Soma plants, golden fleece, and so on. The Aṅgirās are Indra's weapons, the warring angels, who slay Vṛtra, the serpent of the Heaven, in cosmic battles to release the celestial lights and fires that are contained in the Soma Pond.

Dīrghatamā literally means "longest" according to Monier-Williams. Dīrgha means "long, lofty, high" and tama (तम) is "an affix forming the superlative degree of adjectives". Dīrghatamā, the longest One, the lofty stem of a Soma plant, or light-containing plant, or plasma bloom, that grows in the Soma Pond or on Mt. Mūjavān (मूजवान् RV x.34.1) can be tens of thousands of light-years tall. Dīrgha is also the name of the tall reedy Śara (शरः) grass, *Saccharum sara*, which grows in the Soma Pond (RV i.191.3). In the Ṛgveda, grasses, herbs, light-containing plants, and trees refer to Soma plants, the plasma blooms.

The Soma Pond or Mt. Mūjavān is the celestial pond or the mountain, where Soma plants, or light-containing plants, bloom and mature. Mt. Mūjavān is another name for the Soma Pond or Samudra, the Ocean of celestial Waters, that is, the Ocean of cosmic plasmas.

The stanza is addressed to the Viśve Devas (विश्वे देवाः viśve devāḥ pl.). The Viśve Devas are a class of the Devas who are born and grow in the Soma Pond. They are directly associated with Viśvarūpa (विश्वरूपः), the one who is "wearing all forms" that are "many-coloured". He is called Viśvarūpa, for infinite forms that will be born or created are contained in him just as the form of a tree that will grow out of a seed is contained in the seed.

In the Ṛgveda, names of poets are closely associated with the celestial phenomena or events described in a hymn or in a stanza and often provide clues for interpretation of the hymn or the stanza.

2.2.3 Make Me Immortal in That Luminous Threefold Heaven

Stanza RV ix.113.9 is an enchanting supplication of Ṛṣi Kaśyapa to Soma. Kaśyapa pleads with Soma to make him an immortal in the Threefold Heavenly Vault (त्रिनाके trināke, loc.), the Threefold Heaven (त्रिदिवे tridive, loc.), where the luminous spheres are full of lights. Kaśyapa's patronymic is Mārīca (मारीचः), which indicates that he is the descendant of Marīci (मरीचिः m.). Marīci literally means "a particle of light" or "ray of light".

The name Kaśyapa is made up of 'kaśya (कश्य a spirituous liquor) + pa (प the act of drinking)', showing the wont of his Soma drinking. Kaśyapa has "black teeth", which indicates that he is the plasma dark discharge. He is the plasma dark discharge who wishes to be kindled by Soma, streaming electrons, an exhilarating drink. The stanza is addressed to Pavamāna Soma (पवमानः सोमः). Pavamāna means "being purified" or "flowing clear". The ionised and unbound state is described as such. Indu (इंदुः iṃduḥ) is "a drop (especially of Soma)", "a bright drop", "a spark" of light.

RV ix.113.9 author कश्यपः, to पवमानः सोमः, metre पंक्तिः छंदः
यत्रानुकामं चरणं त्रिनाके त्रिदिवे दिवः ।
लोका यत्र ज्योतिष्मंतस्तत्र माममृतं कृधींद्रयिंदो परि स्रव ॥९.११३.९॥
यत्र । अनुऽकामं । चरणं । त्रिऽनाके । त्रिऽदिवे । दिवः ।
लोकाः । यत्र । ज्योतिष्मंतः । तत्र । मां । अमृतं । कृधि । इंद्राय । इंदो इति । परि । स्रव ॥९.११३.९॥
Make me immortal, in that Threefold Heavenly Vault, in that Threefold Heaven, where the pillar of the Heaven is established as desired, There, where the luminous spheres are full of light. Flow O thou Indu, flow richly for Indra. (RV ix.113.9)

यत्र <ind. – in which place, where>, अनुकामं <ind. – as desired>, चरणं <n. nom. sg. of चरण – a pillar, the root (of a tree)>, त्रिनाके <n. loc. sg. of त्रिनाक – in the threefold Heavenly Vault>, त्रिदिवे <n. loc. sg. of त्रिदिव – in the threefold Heaven>, दिवः <m. gen. sg. of दिव् – of the Heaven>;

लोकाः <m. nom. pl. of लोक – the Worlds, spheres>, यत्र <ind. – where>, ज्योतिष्मंतः <m. nom. pl. of adj. ज्योतिष्मत् – brilliant, luminous>, तत्र <ind. – there>, मां <m. acc. sg. of 1st person pron. अस्मद् – me>, अमृतं <m. acc. sg. of अमृत – an immortal, a god>, कृधि <ipv. 2nd person sg. of √कृ – you make>, इंद्राय <m. dat. sg. of इंद्र – for Indra (the slayer of Vṛtra, the serpent of the Heaven, the plasma sheath)>, इंदो <m. voc. sg. of इंदु – O Indu, O Bright Drop of Soma>, परि <ind. – abundantly>, स्रव <ipv. 2nd person sg. of √स्रु – you flow, gush forth>.

2.3 THE HEIGHT OF THE HEAVENLY VAULT

The height of the Heavenly Vault is not found in the extant text of the Ṛgveda. However, we find it in the Sūryasiddhānta (सूर्यसिद्धान्तः) and the Siddhāntaśiromaṇi (सिद्धान्तशिरोमणिः). Prabodhchandra Sengupta, in the Introduction of the reprint of Burgess's translation of the Sūryasiddhānta by Gangooly (1935), states that most of the later Indian astronomical treatises were composed between AD499 and AD1150. Sūryasiddhānta is one of the eighteen Siddhāntas that were considered to be revelations from the gods by the Hindu tradition.[56] According to Sengupta, the Sūryasiddhānta is a composite work that was composed between about AD400 and AD750.[57] Very likely, the Sūryasiddhānta was recompiled after the loss of the original text, just as the Veda was recompiled by Vyāsa after the obliteration in the last Dvāpara Yugaṃ. Siddhāntaśiromaṇi, on the other hand, is known as a work of Bhāskara (भास्करः circa AD1150), also known as Bhāskara II, a celebrated astronomer of India.

According to stanza 90 of Bhūgolādhyāya (भूगोलाध्यायः) of Sūryasiddhānta and stanza 67 of Bhuvanakośa (भुवनकोशः) of Siddhāntaśiromaṇi, the height of the Heavenly Vault, the hemispherical Bowl of the Heaven, is 28,710.8 light-years. The measure of 28,710 light-years falls in the middle of estimates made by NASA scientists and astrophysicists. NASA

---

56 P. Gangooly (ed.), *Translation of the Sūrya-siddhānta: a Textbook of Hindu Astronomy with Notes and an Appendix by Rev. Ebenezer Burgess*, reprint, Calcutta, The University of Calcutta, 1935, p. viii.
57 ibid., p. ix.

estimated it to be 25,000 light-years and Meng Su and Douglas P. Finkbeiner (2012) provided the height of a single Fermi Bubble, which is the Vault of the Heaven according to the Ṛgveda, to be as tall as 10 kiloparsec (kpc),[58] which is about 32,600 light-years.

Both Bhūgolādhyāya of Sūryasiddhānta and Bhuvanakośa of Siddhāntaśiromaṇi provide the cosmography of the Milky Way. It is apparent that some information included in these cosmography chapters are corrupt and mixed with mythified accounts. Fortunately, the figures describing the height of the Heavenly Vault survived without corruption since there is no possibility that Brahmāṇḍa, the 'Egg of Brahmā', could be confused with any other celestial objects in our solar system. In these astronomical treatises, Brahmāṇḍa, the Egg of Brahmā, is the term used for the Heavenly Vault.

The major source of the corruption, it seems, is the confusion between the celestial objects of the galactic structure of the Milky Way and the celestial objects of our solar system. The confusion was probably caused by the loss of the original meanings of the Vedas and the cosmography chapters of these astronomical treatises. In the Ṛgveda, Pṛthivī (पृथिवी) is "the broad and extended One", the Milky Way; Sūrya (सूर्यः) is the fiery, kindled proton, the ion of hydrogen, not the sun in our solar system; and the moon (चंद्रः candraḥ), the bright drop of Soma, the cool and moist electron, not the moon of the planet Earth. Apparently, the planets in our solar system including Earth, our sun, and Earth's moon are named after the original celestial objects of our galactic system.

The rivers Gaṅgā, Yamunā, and Sarasvatī of India are named after the majestic celestial rivers Gaṅgā (गंगा), Yamunā (यमुना), and Sarasvatī (सरस्वती) of the Ṛgveda as we shall see in the later sections. In the Ṛgveda, rivers represent the flowing currents of cosmic plasmas. Considering that the date of the compilation of these two astronomical treatises is at least several millennia later than the date of the recompilation of the Veda by Vyāsa and after the original meaning of the Vedic text had been long lost, the corruptions and confusions introduced into the texts of these two cosmography chapters are not unexpected.

2.3.1 The Height of the Heavenly Vault Revealed in the Sūryasiddhānta

Stanza 90 of Bhūgolādhyāya (भूगोलाध्यायः) of Sūryasiddhānta (सूर्यसिद्धान्तः) states that the height at the centre of the 'revolving Hemispherical Bowl Brahmāṇḍa (ब्रह्माण्डसम्पुटपरिभ्रमणं)' is eighteen quadrillion, seven hundred and twelve trillion, eighty billion, eight hundred and

---

[58] Meng Su and Douglas P. Finkbeiner, 2012, 'Evidence for Gama-Ray Jets in the Milky Way', *Astrophysical Journal* 753:61(13pp), 14 June 2012. doi:10.1088/0004-637X/753/1/61.

sixty-four million (18,712,080,864,000,000) yojanas (योजनानि). The Sanskrit text of the Sūryasiddhānta is from the Maharishi International University.

Stanza 90 भूगोलाध्यायः सूर्यसिद्धान्तः (Bhūgolādhyāya, Sūryasiddhānta)
खव्योमखत्रयखसागरषट्कनागव्योमाष्टशून्ययमरूपनगाष्टचन्द्राः ।
ब्रह्माणडसम्पुटपरिभ्रमणं समन्तादभ्यन्तरे दिनकरस्य करप्रसारः ॥९० भूगोलाध्यायः सूर्यसिद्धान्तः ॥
The total (height) at the centre of Brahmā's Egg, the Revolving Hemispherical Bowl, the maker and discharger of light, is eighteen quadrillion, seven hundred and twelve trillion, eighty billion, eight hundred and sixty-four million yojanas. (90 Bhūgolādhyāya, Sūryasiddhānta)

(<ख (0), व्योम (0), खत्रय (000, three zeros), ख (0), सागर (4), षट्क (6), नाग (8), व्योम (0), अष्ट (8), शून्य (0), यम (2), रूप (1), नग (7), अष्ट (8), चन्द्रा (1) → So the number given is 18,712,080,864,000,000>;

ब्रह्माण्डसम्पुटपरिभ्रमणं <ब्रह्माण्ड n. Brahmā's Egg + सम्पुट m. a hemispherical bowl + परिभ्रमणं n. revolving – Brahmā's Egg, the revolving hemispherical bowl>, समन्तादभ्यन्तरे <समन्तात् ind. wholly, completely, i.e. total (length or height) + अभ्यन्तरे n. loc. in the middle, at the centre – total (height) at the centre>, दिनकरस्य <m. gen. sg. of दिनकर – of the maker of light>, करप्रसारः <कर m. ray of light + प्रसारः m. spreader, discharger – a spreader of light, discharger of light>.)

The first line of the stanza is the total height written in the bhūtasaṃkhyā (भूतसंख्या) system, which is the method of writing numbers in the names of the objects that are associated with particular numerical values. The numbers are written with the last digit first and going backward.[59] Sūryasiddhānta correctly describes the Vault of the Heaven as the maker and discharger of celestial lights, cosmic plasmas.

In the stanza, Brahmāṇḍasampuṭa, the hemispherical bowl-shaped Egg of Brahmā, is the name given for the hemispherical bowl-shaped Vault of the Heaven. The Sanskrit text makes it clear that it refers only to a single Vault of the Heaven, not the two Vaults of the Heaven aligned together in the shape of the Ḍamaru. For the conversion of yojanas to miles, many different conversion units such as 4, 5, or 9 miles were regarded as equal to one yojana by different sources.[60] Others used 7.55 miles and 8 miles per yojana.[61] However, according to Monier-Williams, about 9 miles per yojana is a more accurate measure for one yojana. Accordingly, the conversion unit of 9 miles per yojana was applied. To calculate the number of miles in one light-year, as shown below, the speed of light 186,000 miles per second, 86,400 seconds a day, and 365 days a year are applied.

miles in one light-year = 186,000 x 86,400 x 365 = 5,865,696,000,000 miles
18,712,080,864,000,000 yojanas x 9 miles per yojana = 168,408,727,776,000,000 miles
168,408,727,776,000,000 miles/5,865,696,000,000 miles per light-year = 28,710.78 light-years

---

59 P. Gangooly (ed.), *Translation of the Sūrya-siddhānta,* 1935, op. cit., p. iv.
60 Monier Monier-Williams, *A Sanskrit-English Dictionary*, 1899, op. cit., under the headword 'yojana (योजन)'.
61 Richard L. Thompson, *Vedic Cosmography and Astronomy*, 1996, op. cit., pp. 30, 32.

28,710.78 light-years x 2 = 57,421.56 light-years for the total height of the two vertically aligned Vaults of the Heaven (see Figure 4)

The result of the conversion shows the height for a single hemispherical bowl-shaped Vault of the Heaven to be 28,710.8 light-years; when it is multiplied by 2 to get the total height of the two vertically aligned bowl-shaped Vaults of the Heaven, we obtain the height of 57,421 light-years as compared to the estimated measures of 50,000 light-years by NASA and about 64,000 light-years (10 kpc above and below of the centre of the Milky Way) by Meng Su and Douglas P. Finkbeiner (2012). (see Figures 3 and 4)

Note that, in this stanza, the hemispherical bowl-shaped Brahmāṇḍa is described as revolving (परिभ्रमणं paribhramaṇaṃ). The stanza RV i.164.38 and the stanza 55 of Bhūgolādhyāya of Sūryasiddhānta confirm that the two Vaults of the Heaven revolve in opposite directions of one another.

2.3.2 The Height of the Heavenly Vault by the Siddhāntaśiromaṇi

Stanza 67 of Bhuvanakośa (भुवनकोशः) of Siddhāntaśiromaṇi (सिद्धान्तशिरोमणिः) states that the vertical length of the Hemispherical Cauldron Bowl Brahmāṇḍa, or the Vault of the Heaven, is eighteen quadrillion, seven hundred twelve trillion, sixty-nine billion, two-hundred million (18,712,069,200,000,000) yojanas. The number provided in yojana is slightly different from that of the Sūryasiddhānta. However, when the number in yojana is converted, the same height of 28,710.8 light-years for the hemispherical bowl-shaped Brahmāṇḍa is acquired. The Sanskrit text is from D. Arkasomayaji (2000).[62]

Stanza 67 भुवनकोशः सिद्धान्तशिरोमणिः (Bhuvanakośa, Siddhāntaśiromaṇi)
कोतिघ्नैर्नखनन्दषट्कनखभूभूभृद्भुजङ्गेन्दुभि ज्योतिश्शास्त्रविदो वदन्ति नभसः कक्षामिमां योजनैः ।
तद्ब्रह्माराडकटाहसंपुटतटे केचिज्जगुर्वेष्टनं केचित् प्रोचुरदृश्यदृश्यकगिरिं पौराणिकाः सूरयः
॥६७ भुवनकोशः सिद्धान्तशिरोमणिः ॥
Knowers of astronomical science assert the circumference of the Heaven to be eighteen quadrillion, seven hundred and twelve trillion, sixty-nine billion, two-hundred million yojanas in length. Some have argued this to be the height at the side of Brahmā's Egg, the Hemispherical Cauldron Bowl. Other learned Brāhmaṇas well read in Purāṇas have proclaimed that to be the length of the encircling Mt. Adṛśyadṛśyaka. (67 Bhuvanakośa, Siddhāntaśiromaṇi)

<कोटि (10,000,000), घ्नैः (mn. inst. pl. of घ्न – multiplied by), नख (20), नन्द (9), षट्क (6), नख (20), भू (1), भूभृत् (7), भुजंग (8), अन्दुभि (1) → 18,712,069,200,000,000>, ज्योतिश्शास्त्रविदः <ज्योतिः शास्त्र astronomical science + विदः knowers – knowers of astronomical science>, वदन्ति <pre. 3rd person pl. of √वद् – they say>, नभसः <n. gen. sg. of नभस् – of the

[62] D. Arkasomayaji, *Siddhāntaśiromaṇi of Bhāskarācārya*, 2nd ed., Tirupati, Rashtriya Sanskrit Vidyapeetha, 2000, p.65 (६५). Rashtriya Sanskrit Vidyapeetha Series 67.

Heaven>, कक्षां <f. acc. sg. of कक्ष – circumference>, इमां <f. acc. sg. of इदं –this>, योजनैः <n. inst. pl. of योजन – by yojanas>;

तद्ब्रह्माण्डकटाहसंपुटतटे <तद् that + ब्रह्माण्ड n. Brahmāṇḍa, Brahmā's Egg + कटाह m. a semi-spheroidal cauldron + संपुट m. a hemispherical bowl + तटे m. loc. at the side – at the side of Brahmāṇḍa, Brahmā's Egg, the hemispherical cauldron bowl>, केचित् <के + चिद् – some>, जगुः <prf. 3rd person pl. of √गा or √गै – they have gone towards, spoke, pursued>, वेष्टनं <m. acc. sg. of adj. वेष्टन – encircling>, अदृश्यदृश्यकगिरिं <अदृश्यदृश्यक (= दृश्यादृश्य = लोकालोक) + गिरिं m. acc. sg. of अदृश्यदृश्यकगिरि – Mt. Adṛśyadṛśyaka>, केचित् <others>, पौराणिकाः <m. nom. pl. of पौराणिक – Brāhmaṇas well read in Purāṇas>, सूरयः <m. nom. pl. of सूरि– learned men>, प्रोचुः <prf. 3rd person pl. of √प्रवच् – they have proclaimed>.

18,712,069,200,000,000 yojanas x 9 miles per yojana = 168,408,622,800,000,000 miles
168,408,622,800,000,000 miles/5,865,696,000,000 miles per light-year = 28,710.76 light years

28,710.76 light-years x 2 = 57,421.52 light-years for the total length of the two vertically aligned Vaults of the Heaven

The first part of the stanza (कोतिघ्नैर्नखनन्दषट्कनखभूभूभृद्भुजङ्गेन्दुभि) is the numbers written in the bhūtasaṃkhyā (भूतसंख्या) system.[63] The encircling Adṛśyadṛśyakagiriṃ (अदृश्यदृश्यकगिरिं) literally means the 'encircling Invisible-Visible Mountain', which has the same meaning as Lokāloka (लोकालोकः). It is "a mythical belt or circle of mountains surrounding the outermost of the seven seas and dividing the visible world from the region of darkness". The encircling Adṛśyadṛśyaka Mountain is the outermost spiral Arm of the Milky Way. The 'visible world' is the spiral Arms of the Milky Way; the 'region of darkness', the invisible world, is the dark 'celestial ocean' between the spiral Arms of the Milky Way.

In this stanza, Bhāskara shows the varying opinions about what the given number is for, providing further evidence of the confusion introduced later due to the loss and corruption of the original knowledge of cosmography. According to Bhāskara, the experts well versed in astronomical science said the number given is the circumference of the Heaven; other learned Brāhmaṇas, well read in the Purāṇas, considered it to be the length of the outermost spiral Arm of the Milky Way. Apparently, the ones who argued that this was the height of the side of Brahmā's Egg, the Hemispherical Cauldron Bowl, that is, the Vault of the Heaven, were correct.

Sūryasiddhānta and Siddhāntaśiromaṇi provide the height of the Heavenly Vault, which is remarkably close to the estimated values by NASA scientists and Meng Su and Douglas P. Finkbeiner (2012). Apparently, we have had the knowledge of the Heavenly structures and the height of the Heaven, for at least several millennia, likely for much longer, though no one realised it till now, and NASA confirmed the validity of the Vedic knowledge.

---

63 कोटि (10,000,000) घ्नैः (multiplied by), नख (20)→10,000,000 x 20 = 200,000,000; नन्द (9), षट्क (6), नख (20), भू (1), भूभृत् (7), भुजंग (8), अन्दुभि (1). So, the number given is 18,712,069,200,000,000.

2.4 THE TWOFOLD HEAVENLY VAULT AND ITS SUBSTRUCTURES

2.4.1 The Great Heavens of Two, the Upper and the Lower Heavens

As shown in Figures 4 and 6, the twofold Heavenly Vault consists of the two bowl-shaped Heavens (द्यावा dyāvā du.), the Upper Vault of the Heaven and the Lower Vault of the Heaven, aligned vertically in the shape of the Ḍamaru (डमरुः), the sacred drum of Śiva. The Two Heavens are referred to as Dhiṣaṇe (धिषणे du. RV i.160.1), the Two Bowls. In stanza RV iv.56.5, Dyavī, the Two Heavens, are praised. Here, the Two Heavens are addressed in feminine vocative dual cases, Dyavī (द्यवी O two Heavens) and Śucī (शुची O Radiant Two).

RV iv.56.5 author वामदेवः, to द्यावापृथिव्यौ, metre गायत्री छंदः
प्र वां महि द्यवी अभ्युपस्तुतिं भरामहे । शुची उप प्रशस्तये ॥४.५६.५॥
प्र । वां । महि । द्यवी इति । अभि । उपऽस्तुतिं । भरामहे । शुची इति । उप । प्रऽशस्तये ॥४.५६.५॥
To both of you, O Great Heavens of Two, we offer our song of praise.
O Radiant Two! Glory upon you both. (RV iv.56.5)

प्र <ind. – forward, forth>, वां <m. dat. du. of 2nd person pron. युष्मद् – to both of you>, महि <ind. – great>, द्यवी <f. voc. du. of दिव् – O Heavens of Two>, अभि <ind. – towards, for>, उपस्तुतिं <f. acc. sg. of उपस्तुति – invocation, song of praise>, भरामहे <pre. 1st person pl. of √भृ – we bring, offer, lift up, raise>;
शुची <f. voc. du. of शुचि – O Radiant Two>, उप <ind. – upon>, प्रशस्तये <f. dat. sg. of प्रशस्ति – glory to>.

Dyavī (द्यवी f. voc. du.) takes an unusual form of the feminine declension of div (दिव् the Heaven) probably to rhyme with Śucī (शुची f. du.). Though the entire hymn RV iv.56 is addressed to the Twofold Heaven and Pṛthivī (द्यावापृथिव्यौ), the first and third stanzas praise the Two Heavens and the Milky Way (द्यावापृथिव्यौ = द्यावा Dyāvā, two Heavens + पृथिवी Pṛthivī, the Milky Way), and the fourth stanza praises Rodasī (रोदसी du.). Rodasī is the pair consisting of the Stone and the Soma Pond (see section 2.5.4). The fifth stanza, RV iv.56.5, praises the Two Heavens.

The Two Heavens are also referred to as Rajasī (रजसी du. RV i.160.4), 'two Spheres of Light'. The Upper Sphere is called Raja Uttamaṃ (रजः उत्तमं RV ix.22.5); the Lower Sphere, Raja Uparaṃ (रजः उपरं RV i.62.5). In addition, the Two Heavens are described as 'Two Bowls of Noble Birth' (सुजन्मनी धिषणे sujanmanī dhiṣaṇe RV i.160.1). Often the two Heavenly Spheres are simply referred to as the Upper (परः paraḥ) and the Lower (अवरः avaraḥ) as in stanza RV x.88.17.

In RV x.88.17, it is indicated that the two Soma Ponds, one in the Upper Vault and the other in the Lower Vault of the Heaven, must coordinate to perform the Soma Yajña, the event of releasing cosmic plasmas. To release cosmic plasmas matured in the Soma Pond as cosmic twin Jets, the two Soma Ponds have to 'speak' or 'sing' in concert. The 'feast of drinking Soma (सधमादः)' refers to the event of receiving discharged celestial Waters, that is, cosmic plasmas, from the Stone, the celestial ioniser, into the Soma Pond. Stanza RV x.88.17 is addressed to the Devas Sūrya and Vaiśvānara (सूर्यवैश्वानरौ).

RV x.88.17 author मूर्धवानांगिरसो वामदेव्यो वा, to सूर्यवैश्वानरौ, metre त्रिष्टुप् छंदः
यत्रा॒ वदे॑ते॒ अव॑रः॒ पर॑श्च य॒ज्ञन्योः॑ कत॒रो नौ॒ वि वे॑द ।
आ शे॑कु॒रित्स॑ध॒मादं॑ सखा॑यो॒ नक्षं॑त य॒ज्ञं क इ॒दं वि वो॑चत् ॥१०.८८.१७॥
यत्र॑ । वदे॑ते॒ इति॑ । अव॑रः । पर॑ः । च॒ । य॒ज्ञ॒ऽन्योः॑ । क॒त॒रः । नौ॒ । वि । वे॒द॒ ।
आ । शे॒कुः॒ । इत् । स॒ध॒ऽमादं॑ । सखा॑यः । नक्षं॑त । य॒ज्ञं । कः । इ॒दं । वि । वो॒च॒त् ॥१०.८८.१७॥
Which of us two knows when both should speak together, lower or upper of the two Yajña leaders? The companions arrived and partook in the feast of drinking Soma. Now who should declare this Yajña? (RV x.88.17)

यत्र <ind. – where, on which occasion, when>, वदेते <pre. middle 3rd person du. of √वद् – the two speak, sing>, अवरः <m. nom. sg. of adj. अवर – below, lower>, परः <m. nom. sg. of adj. पर – highest, upper>, च <ind. – and, both>, यज्ञन्योः <(यज्ञ + न्योः) m. gen. du. of यज्ञनि – of two leaders of Yajña>, कतरः <m. nom. sg. of कतर – which of two?>, नौ <m. gen. du. of 1st person pron. अस्मद् – of us two>, वि <ind. perfix to verb वेद (√विद्) to intensify the meaning of the verb – truly, really>, वि-वेद <pre. 3rd person sg. of वि-√विद् – (which of us two) knows, understands>;

आ-शेकुः <prf. 3rd person pl. of आ-√शक् – they (companions) helped, aided, partook>, इत् <(इद्) ind. a particle of affirmation – indeed, surely>, सधामादं <m. acc. sg. of सधमाद – drinking feast (of Soma)>, सखायः <m. nom. pl. of सखि – companions, friends>, नक्षंत <inj. middle 3rd person pl. of √नक्ष् – they (companions) arrived>, यज्ञं <m. acc. sg. of यज्ञ – celestial Yajña>, कः <m. nom. sg. of interrogative pron. क – who?>, इदं <ind. – here, now>, वि-वोचत् <opt. 3rd person sg. of वि-√वच् – (with कः) who should declare, announce>.

2.4.2 The Twofold Heavenly Vault Observed by Chandra and Hubble

The Twofold Heavenly Vault, which is molded into the shape of the Ḍamaru, or of an hourglass, was observed by the Chandra X-Ray Observatory and the Hubble Space Telescope (see Figure 5). Figure 5 shows (1) Hourglass Nebula, (2) Southern Crab Nebula, (3) Hubble 12 Nebula, (4) Twin Jet Nebula, and (5) Pictor A Galaxy. The first four images are from data acquired by the Hubble Wide Field Space Telescope Camera 2 (WFPC2). These four images probably show the Heavenly Vaults at different stages of structural development. Or, each image may show different phenomena mapped by different analytical algorithms or thresholds. The fifth image is the Radio-wave data acquired by Australia Telescope Compact Array (ATCA).[64]

---

[64] NASA Chandra X-Ray Observatory, *Pictor A: Blast from Black Hole in a Galaxy Far, Far Away*, http://chandra.si.edu/photo/2016/pictora/ (accessed 26 June 2019).

In each of the images 1 through 4 of Figure 5, what are mapped are probably the ionised particles, atoms, and molecules aligned along the surface of the bowl-shaped magnetic field of the Heavenly Vault. For the Hourglass Nebula, explanation is given that ionised nitrogen (mapped in red), hydrogen (green), and doubly-ionised oxygen (blue) were mapped.[65] According to the Ṛgveda, the Twofold Heavenly Vault is a magneto-plasma structure sustained by the unbroken chain of cosmic ionisation and plasma discharge and release. The Vault of the Heaven is the seat of the Soma Pond that contains the receptacles of Soma pressed out by the Stone (ग्रावा Grāvā), the celestial ioniser.

The observed twofold Heavenly Vault is described as "a pair of aging stars"[66] associated with the events of "the slow death of Sun-like stars".[67] According to the Ṛgveda, however, the twofold Vault of the Heaven is the Creator's marvellous machine, the wheel of celestial ionisation and cosmic plasma discharge and release. The twin Jets observed[68] emerging from the two Soma Ponds are one of the two mechanisms of cosmic plasma release. The two Heavenly Vaults, molded in the shape of the Ḍamaru, may be revealing the innate topology of the magnetic field structure of Nature, that is, two domains (two vaults) of north and south magnetic polarities defined by the bowl-shaped magnetic fields and the innerfield between the two domains, as revealed in the Ṛgveda and suggested by Albert Roy Davis and Walter C. Rawls, Jr. (1996).[69]

Two radio lobes of Pictor A Galaxy observed by M. Hardcastle et al. (release date February 2016), shown in the fifth image of Figure 5, correspond to the radio lobes observed within the Fermi-Bubbles by Ettore Carretti et al. (2013).[70] These lobes are the two Soma Ponds one located in the Upper Vault of the Heaven and the other in the Lower Vault of the Heaven. At the base of each Soma Pond is the Stone, the celestial ioniser, which is formed by two vertically stacked discs of firestones laid in a circular formation (see Figure 6). Ettore Carretti et al. (2013) reported that they have observed two giant radio lobes emanating from the centre of the Milky Way. According to the Ṛgveda, as we shall see, these "giant magnetized outflows" are cosmic plasmas that are discharged by the Stone, the celestial ioniser, into the Soma Pond, blooming into tall ridge-like structures.

---

65 NASA/ESA Hubble Space Telescope, *The Hourglass Nebula*, http://www.spacetelescope.org/images/opo9607a/ (accessed 26 June, 2019).
66 NASA/ESA Hubble Space Telescope, *Southern Crab Nebula*, http://www.spacetelescope.org/images/opo9932b/ (accessed 26 June 2016).
67 NASA/ESA Hubble Space Telescope, *The Hourglass Nebula*, op. cit.
68 Ettore Carretti et al., 'Giant magnetized outflows...', 2013, op. cit., pp. 66-69.
69 Albert Roy Davis and Walter C. Rawls, Jr., *Magnetism and its Effects on the Living System*, 12th printing, Metairie, Louisian, Acres U.S.A., 1996, pp. 21-24, 56-57.
70 Ettore Carretti et al., 2013, op. cit.

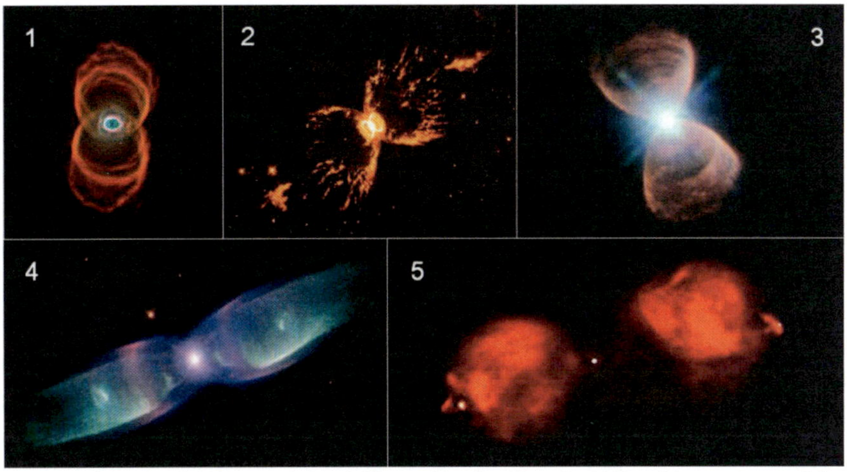

<Figure 5> The Twofold Heavenly Vault Mapped by the Chandra X-Ray Observatory and the Hubble Space Telescope. Images are NOT on the same scale.

1. *Hourglass Nebula* (MyCn 18) by Wide Field and Planetary Camera 2 (WFPC2) of Hubble. Credit: Raghvendra Sahai and John Trauger (JPL), the WFPC2 science team and NASA/ESA. **2**. *Southern Crab Nebula* (IRAS 14085-5112) by Hubble WFPC2. Credit: Romano Corradi, Instituto de Astrofisica de Canarias, Tenerife, Spain; Mario Livio, Space Telescope Science Institute, Baltimore, Md.; Ulisse Munari, Osservatorio Astronomico di Padova-Asiago, Italy; Hugo Schwarz, Nordic Optical Telescope, Canarias, Spain; and NASA/ESA. **3**. *Hubble 12 Nebula* (PN G111.8-02.8 or PN Hb 12) by Hubble WFPC2. Credit: NASA, ESA, Acknowledgement: Josh Barrington. **4**. *Twin Jet Nebula* (M2-9) observations by Hubble WFPC2. Credit: Bruce Balick (University of Washington), Vincent Icke (Leiden University, The Netherlands), Garrelt Mellema (Stockholm University), and NASA/ESA. **5**. *Pictor A Galaxy (Radio)*, by Australia Telescope Compact Array (ATCA). Credit: X-Ray: NASA/CXC/Univ of Hertfordshire /M. Hardcastle et al., Radio: CSIRO/ATNF/ATCA. (Only Radio observation is shown in this image.)

In the Ṛgveda, this ridge-like structure is called a lofty ridge of the Heaven (दिवः बृहत् सानु RV v.59.7) or the ridge of the mountains (सानु गिरीणां RV vi.61.2). The discharged cosmic plasmas blooming into ridge-like structures are also called the Soma plants (सोमाः RV i.135.6, सोमासः RV i.137.2) or light-containing plants (ओषधयः RV x.97.22, RV x.17.14); and they are named 'plasma blooms' in this book (see Figure 7).

2.4.3 The Soma Pond in the Vault of the Heaven

Stanza RV i.164.33 explains that, within each of the two Bowls of the Heaven, aligned back to back (उत्तानयोः चम्वोः loc. du.) with the openings outwards, is an inner womb (योनिः अंतः), the Soma Pond (see section 2.5.2). This Soma Pond is also called Duhitā, the milker (दुहिता duhitā, RV iii.55.12) or the daughter milch Cow, for the Soma Pond is the one that drinks in cosmic plasmas discharged from the Stone, the celestial ioniser (see section 2.5.4). Into this Soma Pond, the Heaven discharges the ionised and freed embryonic particles, Agnis and Somas, of cosmic plasmas. Thus, this stanza simply explains that ionised embryonic particles of cosmic plasmas produced by the Stone, the celestial ioniser, are discharged into the Soma Pond by the Heaven. Stanza RV i.164.33 is one of the grotesquely misinterpreted stanzas of the Ṛgveda.[71]

RV i.164.33 author दीर्घतमा औचथ्यः, to विश्वे देवाः, metre त्रिष्टुप् छंदः
द्यौर्मे पिता जनिता नाभिरत्र बंधुर्मे माता पृथिवी महीयं ।
उत्तानयोश्चम्वो३ योनिरंतरत्रा पिता दुहितुर्गर्भमाधात् ॥१.१६४.३३॥
द्यौः । मे । पिता । जनिता । नाभिः । अत्र । बंधुः । मे । माता । पृथिवी । मही । इयं ।
उत्तानयोः । चम्वोः । योनिः । अंतः । अत्र । पिता । दुहितुः । गर्भं । आ । अधात् ॥१.१६४.३३॥
The Heaven is my Father, my progenitor. This great World, the Milky Way, is my close kin, Mother. In the interior of each of the two Bowls, which are lying back to back with the openings outwards, is the Soma Pond. Here into this Soma Pond, the Heaven discharged the embryonic particles. (RV i.164.33)

द्यौः <m. nom. sg. of दिव् – the Heaven>, मे <m. gen. sg. of 1st person pron. अस्मद् – my>, पिता <m. nom. sg. of पितृ – father>, जनिता <m. nom. sg. of जनितृ – progenitor>, नाभिः <f. nom. sg. of नाभि – a close relation>, अत्र <ind. – here>, बंधुः <m. nom. sg. of बंधु – a kinsman, relative>, मे <m. gen. sg. of 1st person pron. अस्मद् – my>, माता <f. nom. sg. of मातृ – mother>, पृथिवी <f. nom. sg. of पृथिवी – Pṛthivī, "the broad and extended One", i.e. the Milky Way>, मही <f. nom. sg. of मही – the great World>, इयं <f. nom. sg. of इदं – this>;

उत्तानयोः <f. loc. du. of adj. उत्तान – two lying back to back with the opening outwards>, चम्वोः <f. loc. du. of चमू – in the two bowls>, योनिः <m. nom. sg. of योनि – the womb, i.e. the Soma Pond. see section 2.5.2>, अंतः <(अंतर्) ind. – within, in the interior (with loc.)>, अत्र <ind. – here, in this place>, पिता <m. nom. sg. of पितृ – father, i.e. the Heaven>, दुहितुः <f. dat. sg. of दुहितृ – to the extractor or drawer, i.e. to the Soma Pond. see section 2.5.4>, गर्भं <m. acc. sg. of गर्भ – an embryo, i.e. the embryonic particle (ionised, freed Agni, proton, or Soma, electron) of celestial Waters. see sections 7.1.1 & 7.2.1>, आ-अधात् <ipf. 3rd person sg. of आ-√धा – he placed, deposited>.

In the Ṛgveda, the Heaven is often called Father; Pṛthivī (पृथिवी), the Milky Way, Mother. The two Bowls of the Heaven are lying back to back with the mouths of the two Bowls open in outward directions (see Figures 3, 4, 5, 6). The word uttānayoḥ (उत्तानयोः) is the locative dual case of the adjective uttāna (उत्तान), which means "stretched out, lying on the back, upright, turned so that the mouth or opening is uppermost (as a vessel)". For 'camū (चमूः f. nom. sg.)', Monier-Williams indicates it is 'a vessel into which the Soma is poured'. This stanza, however, clearly states that the inner womb (योनिः अंतः), the Soma Pond, is within the two Bowls (उत्तानयोः चम्वोः uttānayoḥ camvoḥ, loc. du.), the Upper and the Lower Vaults of the Heaven. In the context of this stanza, the two Bowls refer to the Upper and the Lower Vaults of the Heaven.

Yoni (योनिः), a masculine word, has various meanings: the womb; place of birth, source, origin, fountain; place of rest, repository, receptacle, seat, abode, lair, nest, stable. In this stanza, duhituḥ (दुहितुः) is the dative singular case of duhitṛ (दुहितृ milker, drawer), that is,

---

[71] For example, see the translation of this stanza in Wendy Doniger O'Flaherty, *The Rig Veda: An Anthology*, Harmondsworth, Penguin Books, 1981, p. 79.

the Soma Pond. Monier-Williams's definition of duhitṛ is "a daughter (the milker or drawing milk from her mother)". The verb √duh (√दुह्), from which duhitṛ (दुहितृ) is derived, means "to milk (a cow or an udder), to extract". Hence, Duhitā (दुहिता) is the one who drinks in milk (celestial plasmas) from mother milch Cow, the Stone, the celestial ioniser, not a real person 'daughter'.

In stanza RV iii.55.12, of the 'two milch Cows (धेनू f. nom. du.)' who produce milk, or cosmic plasmas, one is called Mātā (माता), the producer of milk, that is, the Stone, the celestial ioniser; and the other Duhitā (दुहिता), the drinker of the produced milk or cosmic plasmas, that is, the Soma Pond. In this case, Mātā may be translated as 'mother milch Cow' and Duhitā 'daughter milch Cow'. However, it needs to be understood that Duhitā is the one who drinks in the milk, the Soma Pond, and is not a 'daughter' in the conventional sense (see section 2.5.4). It seems that lack of understanding of these technical terms led to a grotesque mistranslation of this stanza.

2.4.4 Vimāna, the Chariot of the Devas

Vimāna (विमानः) is another name for the Soma Pond. Vimāna (विमानः) means "a car or chariot of the gods" (RV iii.26.7). In the Ṛgveda, the Soma Pond is described as the abode or chariot of the Devas. Definitions provided for Vimāna by Monier-Williams include "any mythical self-moving aerial car", which implies that the meaning of Vimāna can include the Vault of the Heaven as well. Vimāna (विमानः) is a compound word of vi (वि m. "a bird", ind. "in two parts") + māna (मानः m. a building, house, dwelling); it literally means 'a flying abode' of the Devas. If one takes the meaning of vi (वि) as "in two parts", then Vimāna means the 'twofold abode of the Devas'. The poet is Ātmā (आत्मा) and it is addressed to Agni-Ātmā.

RV iii.26.7 author आत्मा, to अग्निरात्मा वा, metre त्रिष्टुप् छंदः
अ॒ग्निर॑स्मि॒ जन्म॑ना जा॒तवे॑दा घृ॒तं मे॒ चक्षु॑र॒मृतं॑ म आ॒सन् ।
अ॒र्कस्त्रि॒धातू॒ रज॑सो वि॒मानो॒ऽज॑स्रो घ॒र्मो ह॒विर॑स्मि॒ नाम॑ ॥३.२६.७॥
अ॒ग्निः । अ॒स्मि॒ । जन्म॑ना । जा॒तऽवे॑दाः । घृ॒तं । मे॒ । चक्षुः॑ । अ॒मृतं॑ । मे॒ । आ॒सन् ।
अ॒र्कः । त्रि॒ऽधातुः॑ । रज॑सः । वि॒ऽमानः॑ । अज॑स्रः । घ॒र्मः । ह॒विः । अ॒स्मि॒ । नाम॑ ॥३.२६.७॥
I, Agni, know, by birth, all beings created and born as my property; the fertilising rain, my nectar, my light. I am the threefold fire of the Heavenly Sphere, the chariot of the Devas, the inexhaustible cauldron. I am also called an offering to fire. (RV iii.26.7)

अग्निः <m. nom. sg. of अग्नि – Agni>, अस्मि <pre.1st person sg. of √अस् – I am>, जन्मना <n. inst. sg. of जन्मन् – by birth>, जातवेदाः <m. nom. sg. of adj. जातवेदस् – "having whatever is born or created as his property", "all-possessor", "knowing all created beings">, घृतं <n. nom. sg. of घृत – clarified butter, fertilising rain ("the fat which drops from heaven"), here it refers to Soma>, मे <m. gen. sg. of 1st person pron. अस्मद् – my>, चक्षुः <n. nom. sg. of चक्षुस् – the eye, light>, अमृतं <n. nom. sg. of अमृत – the nectar, food>, मे <m. gen. sg. of 1st person pron. अस्मद् – my>, आसन् <ipf. 3rd person pl. of √अस् – they are>;

अर्कः <m. nom. sg. of अर्क – fire>, त्रिधातुः <m. nom. sg. of adj. त्रिधातु – consisting of 3 parts, threefold>, रजसः <n. gen. sg. of रजस् – of "coloured or dim space", of the Heavenly Sphere>, विमानः <m. nom. sg. of विमान – Vimāna, a chariot of the Devas>, अजस्रः <m. nom. sg. of adj. अजस्र – not to be obstructed, perpetual, i.e. inexhaustible>, घर्मः <m. nom. sg. of घर्म – a cauldron, boiler (refers to the Stone, the celestial ioniser)>, हविः <n. nom. sg. of हविस् – an offering to fire>, अस्मि <pre.1st person sg. of √अस् – I am>, नाम <ind. – by name, i.e. named, called>.

In stanza RV iii.26.7, Agni, the fiery embryonic particle of cosmic plasmas, declares that he alone is the knower of all beings, which are created as his property, by birth, for all are in his possession and under his control. Soma is his nectar, his food. The fertilising rain, the cosmic plasmas, is his light and his eye. Note the number of Agnis, protons, in the nucleus determines the identity of the chemical element and the number of Somas, electrons. Thus, he claims that he is the knower and possessor of all beings. He also declares himself an offering to the fire. Agni offers himself to the fire of the Stone, the celestial ioniser. After the destruction of his body, that is, freed from the ionic bond, he emerges as freed Agni along with Soma and the blissful individuated soul (आत्मा ātmā). Note also that he claims to be the threefold fire of the Heavenly Sphere: Ekata (एकतः protium), Dvita (द्वितः deuterium), and Trita (त्रितः tritium). See sections 7.1.3 and 7.1.4 for the threefold fire of Agni. What is implied in the stanza is that the different cosmic plasma phenomena and celestial objects described in the stanza are different modifications or manifestations of a single being, that is, Agni, proton.

2.4.5 The Upper Heaven Revolves Westward, the Lower Heaven Eastward

Stanza RV i.164.38 reveals that the Upper Vault and the Lower Vault of the Heaven revolve in opposite directions to one another. In the stanza 'he' refers to the twofold Heavenly Vault; one Vault turns westward and the other Vault eastward. 'Bearing fruit' indicates having the Soma Pond filled full of cosmic plasma blooms, which are the fruits of ionisation. 'She' refers to the World of mortals, Pṛthivī (पृथिवी), the "broad and extended One", the Milky Way. She, the Milky Way, continually spreads out in different directions.

RV i.164.38 author दीर्घतमा औचथ्यः, to विश्वे देवाः, metre त्रिष्टुप् छंदः
अपाङ् प्राङेति स्वधया गृभीतोऽमर्त्यो मर्त्येना सयोनिः ।
ता शश्वंता विषूचीना वियंता न्य१ न्यं चिक्युर्न नि चिक्युरन्यमं ॥१.१६४.३८॥
अपाङ् । प्राङ् । एति । स्वधया । गृभीतः । अमर्त्यः । मर्त्येन । सऽयोनिः ।
ता । शश्वंता । विषूचीना । विऽयंता । नि । अन्यं । चिक्युः । न । नि । चिक्युः । अन्यं ॥१.१६४.३८॥
Bearing fruit, westwards and eastwards turns he by inherent power, he the immortal who has a common origin with the World of mortals. And she spreads out continuously in different directions. People observe the one, and fail to mark the other. (RV i.164.38)

अपाङ् <m. nom. sg. of adj. अपाञ्च् – western (opposed to प्राञ्च्), westwards>, प्राङ् <(=प्राञ्) m. nom. sg. of adj. प्राञ्च् – eastern, eastwards>, एति <pre. 3rd person sg. of √इ – he (the Heaven) goes, walks, runs>, स्वधया <f. inst. sg. of स्वधा – by self-power, by inherent power>, गृभीतः <m. nom. sg. of adj. गृभीत – impregnated, bearing fruit>, अमर्त्यः <m.

nom. sg. of adj. अमर्त्य – imperishable, immortal>, मर्त्येन <m. inst. sg. of मर्त्य – with "the world of mortals">, सयोनिः <m. nom. sg. of adj. सयोनि – having the same womb, having a common origin with>;

ता <f. nom. sg. of pronominal base त (of pron. तद्) – it/she>, शश्वंता <f. inst. sg. of शश्वन्त् – constantly, continually, continuously>, विषूचीना<f. nom. sg. of adj. विषूचीन – in different directions>, वियंता <(वियन्ता) prf. 3rd person sg. of वि-√यं– she (Pṛthivī, the Milky Way) has spread out, she has held apart>. अन्यं <n. acc. sg. of अन्य – अन्य … अन्य the one ... the other>, नि-चिक्युः <prf. 3rd person pl. of नि-√चि – they observe, perceive>, न <ind. – not>, नि-चिक्युः <prf. 3rd person pl. of नि-√चि – they observe, perceive>, अन्यं <see above – the other>.

He, the Heavenly Vault, is mysterious and hidden, and she, the Milky Way, manifested and apparent. So, people see the manifested World, the Milky Way, but fail to mark the hidden, mysterious Heavenly Vault. Note that the twofold Heavenly Vault, which is named Fermi Bubbles by NASA, was unveiled by NASA's Fermi Gamma-Ray Space Telescope in 2010. It took the Gamma-Ray Space Telescope to unveil the "previously unseen" structure. Though one (he, the Heavenly Vault) is immortal and the other (she, the Milky Way) is the World of mortals, both have a common origin. They both are born of a single source, aṃbha (अंभः aṃbhaḥ), celestial hydrogen.

Stanza 55 of Bhūgolādhyāya of Sūryasiddhānta provides more specific information about the rotations of the Upper and the Lower Vault of the Heaven.

Stanza 55 भूगोलाध्यायः सूर्यसिद्धान्तः (Bhūgolādhyāya, Sūryasiddhānta)
सव्यं भ्रमति देवानामपसव्यं सुरद्विषाम् । उपरिष्टाद् भगोलोऽयं व्यक्षे पश्चान्मुखः सदा
॥५५ भूगोलाध्यायः सूर्यसिद्धान्तः ॥
The Vault of the Heaven of the Devas revolves to the left, the Vault of Heaven of the opposite of the Devas to the right. Above the latter, at the terrestrial latitude, it always turns westwards.
(55 Bhūgolādhyāya, Sūryasiddhānta)

सव्यं <m. acc. sg. of adj. सव्य – to the left, i.e. clockwise>, भ्रमति <pre. 3rd person sg. of √भ्रम् – (it/he) revolves>, देवानां <m. gen. pl. of देव – of the Devas>, अपसव्यं <m. acc. sg. of adj. अपसव्य – away from the left, to the right, anticlockwise>, सुरद्विषां <m. gen. pl. of सुरद्विष् (सुर a Deva, divinity + √द्विष् to hate, be hostile, to be a rival) – of the rival or opposite of the Devas>;

उपरिष्टाद् <(उपरिष्टात्) ind. – above>, भगोलः <m. nom. sg. of भगोल – the Vault of the Heaven>, अयं <m. nom. sg. of pron. इदं – this, i.e. the Vault of the Heaven of the opposite of the Devas, the Lower Vault>, व्यक्षे <m. loc. sg. of व्यक्ष – at "having no latitude", at "the equator", at the terrestrial latitude>, पश्चान् <(=पश्चात्) ind. – westwards>, मुखः <m. nom. sg. of adj. मुख – turning, facing, turned towards>, सदा <ind. – always>.

To read this stanza, one has to imagine oneself standing with a compass in front of the twofold Heavenly Vault that is vertically aligned with the Upper Vault of the Devas at the top (see Figure 6). Then we can see that the Upper Vault of the Devas revolves to the left, or clockwise, and the Lower Vault of the Opposite of the Devas revolves to the right, or anticlockwise. And at the terrestrial, or at the zero latitude, that is, at the latitude where the disc of the Milky Way is located, it turns always in a westerly direction. This indicates that the disc of the Milky Way Galaxy always revolves clockwise. 'Of the Devas' denotes 'of the north or of the Upper Vault of the Heaven', and 'Of the Opposite of the Devas' denotes the 'of the south or of the Lower Vault of the Heaven'. The expression 'of the Devas' versus 'of the Opposite of the Devas' is analogous to that of 'Arctic' versus 'Antarctic'.

As you stand in front of the twofold Heavenly Vault with a compas, the north pole of the compass needle will vertically align with the Upper Vault of the Heaven, and the south pole of the compass needle will vertically align with the Lower Vault of the Heaven. What follows then is the Upper Vault of the Heaven is in fact the south magnetic pole and the Lower Vault of the Heaven is the north magnetic pole of the Heaven.[72] Please note the north pole of a compass is attracted by the south magnetic pole of the twofold Heavenly Vault and vice versa.

2.4.6 The Heaven, the Red Bull

The Heaven (द्यौः dyauḥ) is the bull (वृषा v.36.5) or the red bull (उस्रियः वृषभः RV v.58.6). The author of stanza RV v.58.6 is Ṛṣi Śyāvāśva Ātreya (श्यावाश्वः आत्रेयः), 'Dark horse, the descendant of Atri (अत्रिः), the devourer'. The stanza is addressed to the Maruts (pl.). The Maruts are the Soma plants, or plasma blooms, personified.

When the Maruts (pl.) set out with the coursers, the celestial Waters (आपः āpaḥ) are agitated and the trees are shattered to be released as cosmic twin Jets (अश्विना Aśvinā du.). When the cosmic twin Jets shoot upwards, the Heaven, the red bull, thunders the celestial Waters downward as well. This downward release of cosmic plasmas is called the river Gaṅgā (गंगा), which is the celestial river that falls upon Mt. Meru, the Galactic Bar, at the centre of the Milky Way (stanza 37 of Bhuvanakośa, Siddhāntaśiromaṇi).

RV v.58.6 author श्यावाश्व आत्रेयः, to मरुतः, metre त्रिष्टुप् छंदः
यत्प्रायासिष्ट पृषतीभिरश्वैर्वीळुपविभिर्मरुतो रथेभिः ।
क्षोदंत आपो रिणते वनान्यवोस्त्रियो वृषभः क्रंदतु द्यौः ॥५.५८.६॥
यत् । प्र । अयासिष्ट । पृषतीभिः । अश्वैः । वीळुपविऽभिः । मरुतः । रथेभिः ।
क्षोदंते । आपः । रिणते । वनानि । अव । उस्रियः । वृषभः । क्रंदतु । द्यौः ॥५.५८.६॥
When ye set out with the dappled cows and coursers, O Maruts, on your chariots with strong-wrought golden tires, The celestial Waters are agitated, the trees are shattered. Let the Heaven, the red bull, thunder downwards. (RV v.58.6)

यत् <conjunction. – when>, प्र <ind. prefix to verb अयासिष्ट (√या)>, प्र -अयासिष्ट <aor. 2nd person pl. of प्र -√या – ye (Maruts) proceeded, set out>, पृषतीभिः <f. inst. pl. of पृषती – with dappled cows>, अश्वैः <m. inst. pl. of अश्व – with steeds, coursers>, वीळुपविभिः <(=वीडुपविभिः) m. inst. pl. of वीडुपवि – with strong-wrought golden tires (वीळु adj. strong

[72] Cesare Emiliani, *Planet Earth: Cosmology, Geology, and Evolution of Life and Environment*, reprint, New York, Cambridge University Press, 2007, pp.228-229.

+ पवि m. “esp. a golden tire on the chariot of the अश्विन्s and मरुत्s”)>, मरुतः <m. voc. pl. of मरुत् – O Maruts>, रथेभिः <m. inst. pl. of रथ – with chariots, on chariots>;

क्षोदंते <pre. middle 3rd person pl. √क्षुद् – celestial Waters are agitated>, आपः <f. nom. pl. of अप् – celestial Waters, i.e. cosmic plasmas>, रिणते <pre. middle 3rd person pl. of √री – they (trees) are shattered>, वनानि <n. nom. pl. of वन – trees, forests>, अव <ind. preposition – down, downwards>, उस्रियः <m. nom. sg. of adj. उस्रिय – bright red>, वृषभः <m. nom. sg. of वृषभ – the Bull>, क्रंदतु <ipv. 3rd person sg. of √क्रंद् – let the Heaven roar, thunder>, द्यौः <m. nom. sg. of दिव् – the Heaven>.

Dappled cows represent celestial lights, or Agnis, tied to the pillars of plasma blooms, also called trees or sacrificial posts. The coursers represent the forces behind the swiftly flowing currents of cosmic plasmas. Maruts' chariots are the chariots of fire, the flames of cosmic plasma blooms. The Maruts are about to be released as cosmic twin Jets.

2.4.7 The Structures of the Heaven, the Complete Picture

In stanza RV i.20.6, the two Vaults of the Heaven and the two Soma Ponds are described as four sacrificial Vessels or Bowls. Tvaṣṭā (त्वष्टा) first made one and the Ṛbhus (ऋभवः pl.) made it into four Vessels (चतुरः चमसाः). Tvaṣṭā is the magnetic field and the Ṛbhus are the magnetic forces. The author of the stanza is Ṛṣi Medhātithi Kāṇva (मेधातिथिः काण्वः ऋषिः), a son of Kaṇva (spirits) who is an attendant to Soma. The stanza is addressed to the Ṛbhus (pl.).

RV i.20.6 author मेधातिथिः काण्वः, to ऋभवः, metre गायत्री छंदः
उत त्यं चमसं नवं त्वष्टुर्देवस्य निष्कृतं । अकर्त चतुरः पुनः ॥१.२०.६॥
उत । त्यं । चमसं । नवं । त्वष्टुः । देवस्य । निःऽकृतं । अकर्त । चतुरः । पुनरिति ॥१.२०.६॥
The sacrificial Vessel newly made by the Deva Tvaṣṭā,
Ṛbhus fashioned four Vessels out of that one. (RV i.20.6)

उत <ind. – and, also>, त्यं <m. acc. sg. of त्यद् – that (often put after उत in the beginning of a sentence)>, चमसं <m. acc. sg. of चमस – a sacrificial vessel or bowl>, नवं <m. acc. sg. of adj. नव – new>, त्वष्टुः <m. gen. sg. of त्वष्टृ – of Tvaṣṭā>, देवस्य <m. gen. sg. of देव – of the Deva>, निष्कृतं <m. acc. sg. of adj. निष्कृत – prepared, made ready>;

अकर्त <ipf. 3rd person pl. of √कृ – they (Ṛbhus) made>, चतुरः <m. acc. pl. of चतुर् – four>, पुनः <(पुनर्) ind. – once more, moreover>.

Figure 6 shows the complete picture of the twofold Heavenly Vault and its substructures revealed in the Ṛgveda. There are the Upper Heaven and the Lower Heaven. Between the two Heavens is the Innerfield (अंतरिक्षं), the Midheaven. Within each of the Upper and Lower Vaults of the Heaven is the Reedy Soma Pond (शर्यणावान् śaryaṇāvān), or Samudra (समुद्रः samudraḥ), the Ocean of celestial Waters. And the two discs of the Stone (ग्रावाणा grāvāṇā du. 'the two stones' RV ii.39.1) are located at the base of each Soma Pond. Note that the Soma Pond, which is the crown of King Varuṇa, sits upon the Stone, the celestial ioniser, at the base of the Heavenly Vault. These substructures will be discussed in sections 2.5, 2.6, and 2.7 that follow.

In the Milky Way, the galaxy of ours, the twofold Heavenly Vault and the Soma Pond have been observed and mapped by scientists using data acquired by the Chandra X-Ray Observatory and the Hubble Space Telescope as discussed at the beginning of this Chapter. The Stone, the celestial ioniser, of the Milky Way has not been recognised by the scientists yet. However, one can easily see that the two extremely bright spots at the bases of the Upper and Lower Vaults of the Heaven in Figures 3 and 4 are where the Stones, the celestial ionisers, of the Upper and the Lower Vaults are located.

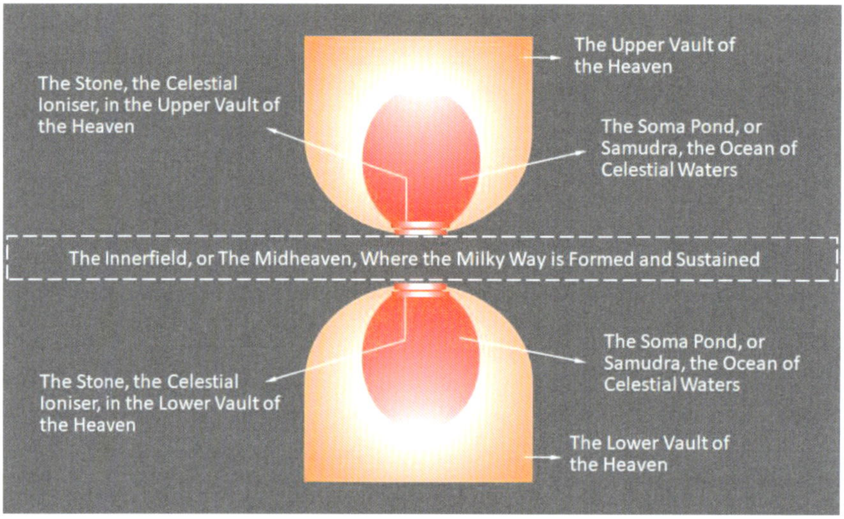

<Figure 6> The Structural Component of the Heaven, the Creator's Marvellous Machine, Revealed in the Ṛgveda. Schematic diagram. Diagram is NOT to scale.

As noted earlier, this magnetic field structure in the shape of the Ḍamaru, the drum of Śiva, seems intrinsic in all matter. As we shall see in Chapter III, according to the Ṛgveda, Tvaṣṭā, the magnetic field, and the Ṛbhus, the magnetic forces, are the divine architects who shape all forms from Agni to the galactic structure of the Milky Way.

2.5 THE SOMA POND, THE WOMB OF COSMIC CREATION

Included in this section are the cosmic plasma events and phenomena that take place in the Soma Pond (ह्रदः hradaḥ or सरः saraḥ), also called Samudra (समुद्रः), the Ocean of celestial Waters. Since the celestial ionisation and cosmic plasma discharge and release events, which are the core of cosmic Yajña, take place in the Soma Pond, in the Ṛgveda, the Soma Pond and its associated cosmic plasma phenomena are described from many different perspectives using a myriad of names, metaphors, and symbolic representations.

2.5.1 The Soma Pond Observed by Chandra

The Soma Pond (सरः RV viii.7.10; ह्रदः RV i.52.7, RV x.142.8) is the womb of cosmic creation, into which the produced cosmic plasmas by the Stone are discharged, and from which the matured cosmic plasmas are released. This womb of cosmic creation is called by many names such as Reedy Soma Pond (शर्यणावान् RV i.84.14), Samudra (समुद्रः RV x.142.8), Rudra (रुद्रः RV ii.33, RV vii.46), Indra's Belly (कुक्षिः RV i.8.7, जठरः RV iii.42.5), Candra (चंद्रः moon RV vi.6.7), Candramā (चंद्रमाः moon RV i.105.1), Sūrya (सूर्यः RV v.63.7), Saṃvatsara (संवत्सरः RV i.164.44, RV vii.103.1), Udder of Pṛśni (पृश्न्याः ऊधः RV ii.34.2), and Lord of the Forest (RV iii.8.6 and 11).

Forest (वनं RV ix.88.5), Mt. Mūjavān (मूजवान् RV x.34.1), Mountain (अद्रिः adriḥ RV i.7.3, पर्वतः parvataḥ RV v.32.2), Celestial Boat equipped with oars (नौः अरित्रपरणी RV x.101.2 nauḥ aritraparaṇī), Sarasvatī (सरस्वती RV vi.52.6), Abode for Cattle (गोष्ठः goṣṭhaḥ RV vi.28.1), and Ram (मेषः meṣaḥ RV i.52.1) are some of the many other names and symbolic representations of the Soma Pond.

The Soma Pond, or Samudra, is further described as the Sacrificial Post or Pillar of the Heaven (यूपः yūpaḥ RV iii.8.1-5), Thousand-Pillared Abode (सदः सहस्रस्थूणं sadaḥ sahasra-sthūṇaṃ) of Kings Varuṇa and Mitra (RV ii.41.5), Goddess of the Forest (अरण्यानी araṇyānī RV x.146.1-6), and Golden Womb (हिरण्यगर्भः hiraṇyagarbhaḥ RV x.121.1).

The Soma Pond, with Soma plants growing and blooming in it, has been observed by the Chandra X-Ray Observatory and mapped by teams of Chandra scientists. Images 1, 2, and 3 of Figure 7 are views from the top. Image 4, probably in a different developmental stage, shows the roaring filamentary structures of cosmic plasmas. In the Ṛgveda, these blooming cosmic plasmas on the magnetic field-aligned filamentary structures are called Soma plants (सोमाः somāḥ pl. RV i.135.6, सोमासः somāsaḥ pl. RV iii.36.3) or light-containing plants (ओषधयः oṣadhayaḥ pl. RV x.97.22). Soma plants, or light-containing plants, refer to the discharged cosmic plasmas blooming in the Soma Pond. Here, 'light' refers to cosmic plasma; thus, Soma plants, or light-containing plants, are the blooming cosmic plasma plants (see Figure 7).

In the Ṛgveda, these Soma plants, or plasma blooms, are called and described as Nakṣatras (नक्षत्रा=नक्षत्राणि constellations RV i.50.2, RV iii.54.19, RV x.68.11), a purifying sieve (पवित्रं pavitraṃ RV ix.96.6), a filament of Soma plant (अंशुः aṃśuḥ RV ix.74.5), beloved sheep's golden fleece (अव्यः वारः प्रियः हरिः RV ix.7.6), and sheep's fleeces (रोमाणि अव्यया RV i.135.6). They are peacock-tailed (मयूरशेप्या adj. RV viii.1.25) and they are peacock's plumes (मयूररोमाणः RV iii.45.1). In RV i.191.3, they are called Śara grass (शरः), Kuśara grass (कुशरः), Darbha grass (दर्भः), Muñja grass (मौंजः=मुञ्जः) and, in RV iv.58.5, golden reed (हिरण्ययः वेतसः). These are the grasses, on which the Devas are asked to sit and drink Soma.

They are also called the pillars in radiant form (वसानाः स्वरवः RV iii.8.9), sacrificial posts (यूपाः yūpāḥ pl. RV iii.8.6-10), the Maruts (मरुतः pl. RV ii.34.2), Uṣās (उषसः pl. RV i.92.1-2), arrows of Rudra (सायकानि RV ii.33.10), glittering spears of the Maruts (भ्राजदृष्टयः RV i.31.1), thunderbolts (वज्रासः RV i.80.8) or hooks (अंकाः aṃkāḥ RV i.162.13) of Indra, stalls of cattle (गोत्रा गवां gotrā gavāṃ RV vi.65.5), rivers (सिंधवः sindhavaḥ RV i.90.6, RV i.93.5, RV vi.52.6), mountains (गिरयः girayaḥ RV i.37.12, पर्वताः parvatāḥ RV i.84.14), and trees or forests (वनानि vanāni RV v.58.6).

What is more, the Soma plants, or plasma blooms, are described as heads of sorceresses (शीर्षा यातुमतीनां śīrṣā yātumatīnāṃ RV i.133.2), horns (शृंगा śṛṃgā RV i.140.6), horns of bulls (शृंगाणि शृंगिणां śṛṃgāṇi śṛṃgiṇāṃ RV iii.8.10). These are the horns of the Altar, the Altar of cosmic creation. They are also described as Soma chalices that contain Soma juice, the celestial ambrosia. These chalices are called by various names, such as kalaśā (कलशाः water pots RV ix.96.23), somadhānā (सोमधानाः Soma receptacles RV iii.36.8), the Soma chalice (ब्राह्मणं n. 'the Soma chalice of Brāhmaṇa' RV ii.36.5), and so on.

As can be seen in Figure 7, a bundle of Soma plants, or plasma blooms, is wrapped in Vṛtra (वृत्रः), also called Ahi (अहिः RV i.32.11), the serpent of the Heaven, the plasma sheath. Vṛtra in the singular refers to the outermost plasma sheath, that is, the shoreline of the Soma Pond, the Womb of cosmic creation. And the Vṛtra is symbolically represented by the serpent with its tail in its mouth. This Vṛtra is known as ouroboros or ophis in Gnostic and alchemical traditions of ancient Greece and Rome. Vṛtras in the plural, the serpents or dragons of the Heaven, refer to individual plasma sheaths of Soma plants or to a fleet of cosmic plasma sheaths sailing through the Heavenly Spheres. These plasma sheaths are the serpent people, the Nāgas (नागाः nāgāḥ pl.), the celestials.

When Vṛtra is viewed in a positive perspective, he is called Rudra (रुद्रः), who gathers and guards the celestial lights, that is, cosmic plasmas, until they become matured enough to be released. On the other hand, when the plasma sheath is viewed in a negative perspective, it is called the enemy of the Devas, for eventually it needs to be destroyed so that the salubrious celestial lights, or cosmic plasmas, can be released to sustain Bhuvanaṃ, the World, the whole of the galactic system.

In the Ṛgveda, in some instances, Vṛtra is called Tvaṣṭā, the magnetic field. In stanza RV iii.48.4, it is said the Indra conquered Tvaṣṭā to release Soma. In stanzas RV x.8.8-9, Vṛtra is called the son of Tvaṣṭā. In these stanzas, it is implied that the plasma sheath is a magnetic field-aligned structure in nature.

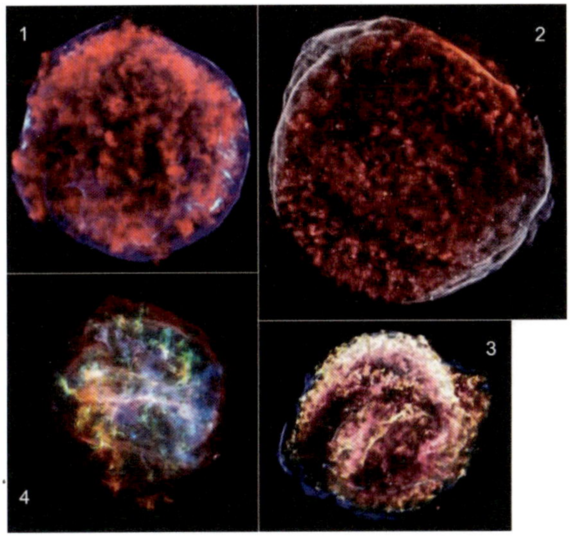

<Figure 7> The Soma Pond Mapped by Scientists of the Chandra X-Ray Observatory. Images are NOT on the same scale.
1. *SN1572 (Tycho Supernova Remnants)*, about 55 light-years across. Image credit: X-Ray: NASA/CXC/ Rutgers/K. Eriksen et al. Optical: DSS.
2. *SN1006 Supernova Remnants*, about 70 light-years across. Image credit: NASA/CXC/ Middlebury College/F. Winkler.
3. *Kepler's Supernova*, about 19-33 light-years across. Image credit: NASA/CXC/SAO/D. Patnaude.
4. *Cassiopeia A Supernova Remnants*, 29 light-years across. Image credit: NASA/CXC/Penn State/S. Park et al.

Vṛtra, in misty purple, white, blue, and red, is shown in the images of Figure 7. It looks as though a piece of delicate silk chiffon is wrapped around a bouquet of wild flowers or a pink rose bud or bright drops of light or reedy plasma blooms. These beautiful celestial bouquets of plasma blooms are about 55, 70, 19-33, and 29 light-years across in size respectively, according to the scientists who mapped these celestial bouquets. In the Ṛgveda, these plasma blooms are called Soma plants (सोमाः), or light-containing plants (ओषधयः). In RV i.64.10, they are called the eternal flames (अनंतशुष्माः anaṃtaśuṣmāḥ). One may compare these Soma plants, or light-containing plants, to the burning bush on the mountain of God, where Moses beheld the angel of the Lord appearing to him in a flame of fire from the midst of the bush, which was burning but not consumed by the flames (Exodus 3:1-3).

2.5.2 The Soma Pond, the Golden Womb

In RV x.121.1, the Soma Pond is called Hiraṇyagarbha (हिरण्यगर्भः), the Golden Womb. The Hiraṇyagarbha holds up and sustains the Heaven and the Milky Way through the unbroken chain of cosmic ionisation and cosmic plasma discharge and release. In the stanza, the Heaven refers to the twofold Heavenly Vault; Pṛthivī (पृथिवी), to the Milky Way. We shall see that the Soma Pond, the Golden Womb, is also called the Pillar of the Heaven or the Sacrificial Post (see section 2.6).

Hymn RV x.121 is addressed to Ka (कः), which literally means 'Who?' The author is Hiraṇyagarbha Prājāpatya (हिरण्यगर्भः प्राजापत्यः), 'Hiraṇyagarbha, the son of Prajāpati'. Monier-Williams's definitions of 'Ka (कः)' include "lord of creatures" and "a divinity presiding over procreation". Stanzas RV x.121.1-4 reveal that 'Ka' is the Soma Pond, the Golden Womb of cosmic creation.

RV x.121.1 author हिरण्यगर्भः प्राजापत्यः, to कः, metre त्रिष्टुप् छंदः
हिरण्यगर्भः समवर्तताग्रे भूतस्य जातः पतिरेक आसीत् ।
स दाधार पृथिवीं द्यामुतेमां कस्मै देवाय हविषा विधेम ॥१०.१२१.१॥
हिरण्यऽगर्भः । सं । अवर्तत । अग्रे । भूतस्य । जातः । पतिः । एकः । आसीत् ।
सः । दाधार । पृथिवीं । द्यां । उत । इमां । कस्मै । देवाय । हविषा । विधेम ॥१०.१२१.१॥
In the beginning arose the Golden Womb, born only Lord of all beings.
He sustains this Milky Way and the Heaven. To which Deva should we offer our oblation?
(RV x.121.1)

हिरण्यगर्भः <m. nom. sg. of हिरण्यगर्भ – Hiraṇyagarbha, the Golden-Womb, i.e. the Soma Pond, the Womb of cosmic creation>, सं-अवर्तत <ipf. 3rd person sg. of सं-√वृत् – it/he took shape, came into being, arose>, अग्रे <ind. – in the beginning>, भूतस्य <n. gen. sg. of भूत – of all beings>, जातः <m. nom. sg. of adj. जात – born, grown, produced>, पतिः <m. nom. sg. of पति – a lord, ruler>, एकः <m. nom. sg. of adj. एक – alone, single, that one only>, आसीत् <ipf. 3rd person sg. of √अस् – there was>;

सः <m. nom. sg. of pron. तद् – he>, दाधार <prf. 3rd person sg. of √धृ – he holds up and maintains>, पृथिवीं <f. acc. sg. of पृथिवी – Pṛthivī, "the broad and extended One", i.e. the Milky Way>, द्यां <m. acc. sg. of दिव् – the Heaven>, उत <ind. – and, even>, इमां <f. acc. sg. of इदं - this>, कस्मै <m. dat. sg. of interrogative pron. क – to whom, for whom, to which>, देवाय <m. dat. sg. of देव – to the Deva>, हविषा <n. inst. sg. of हविस् – with oblation, offering>, विधेम <opt. 1st person pl. of √विध् – we may worship, honour, offer>.

In RV x.121.2, the Golden Womb is described as 'the world of immortality and death'. In the Womb of cosmic creation, the ionic bonds of hydrogen are destroyed by the fire of the Stone, and from this destruction or death, the immortals, the celestial lights, the cosmic plasmas, arise. So, it is the world of immortality and death.

RV x.121.2 author हिरण्यगर्भः प्राजापत्यः, to कः, metre त्रिष्टुप् छंदः
य आत्मदा बलदा यस्य विश्व उपासते प्रशिषं यस्य देवाः ।
यस्य छायामृतं यस्य मृत्युः कस्मै देवाय हविषा विधेम ॥१०.१२१.२॥
यः । आत्मऽदाः । बलऽदाः । यस्य । विश्वे । उपऽआसते । प्रऽशिषं । यस्य । देवाः ।

यस्य । छाया । अमृतं । यस्य । मृत्युः । कस्मै । देवाय । हविषा । विधेम ॥१०.१२१.२॥
He who bestows soul and imparts vital force, whose precept all the Devas honour,
Whose abode is the world of immortality and death. To which Deva should we offer our oblation? (RV x.121.2)

यः <m. nom. sg. of pron. यद् – he who>, आत्मदाः <m. nom. sg. of adj. आत्मदा – granting life or soul>, बलदाः <m. nom. sg. of adj. बलदा – conferring or imparting vigour, force>, यस्य <m. gen. sg. of pron. यद् – whose>, विश्वे <m. nom. pl. of pron. विश्व – all>, उपासते <pre. 3rd person pl. of √उपास् (उप-√आस्) – they honour, acknowledge>, प्रशिषं <f. acc. sg. of प्रशिस् – order, precept, law>, यस्य <m. gen. sg. of pron. यद् – whose>, देवाः <m. nom. pl. of देव – Devas>;

यस्य <m. gen. sg. of pron. यद् – whose>, छाया <f. nom. sg. of छाया – "a covered place, house">, अमृतं <n. nom. sg. of अमृत – "world of immortality">, यस्य <m. gen. sg. of pron. यद् – whose>, मृत्युः <m. nom. sg. of मृत्यु – death>, कस्मै देवाय हविषा विधेम <see RV. x.121.1 – to which Deva should we offer our oblation?>.

In RV x.121.3, it is stated that the Soma Pond is full of vital force and strewn lights. The bipeds refer to men or the Devas; the quadrupeds refer to cattle and steeds, that is, the celestial lights and streaming plasma currents respectively.

RV x.121.3 author हिरण्यगर्भः प्राजापत्यः, to कः, metre त्रिष्टुप् छंदः
यः प्राणतो निमिषतो महित्वैक इद्राजा जगतो बभूव ।
य ईशे अस्य द्विपदश्चतुष्पदः कस्मै देवाय हविषा विधेम ॥१०.१२१.३॥
यः । प्राणतः । निऽमिषतः । महिऽत्वा । एकः । इत् । राजा । जगतः । बभूव ।
यः । ईशे । अस्य । द्विऽपदः । चतुःऽपदः । कस्मै । देवाय । हविषा । विधेम ॥१०.१२१.३॥
He who is full of vital force and strewn lights, he who is the sole Monarch of this World,
He who possesses these bipeds and quadrupeds. To which Deva should we offer our oblation? (RV x.121.3)

यः <m. nom. sg. of pron. यद् – he who>, प्राणतः <m. nom. sg. of प्राणत – one with full of vital force>, निमिषतः <m. nom. sg. of निमिषत – the one with lightning, twinkling (lights)>, महित्वा <ind. pt. of √मह् – being esteemed highly, praised highly>, एकः <m. nom. sg. of adj. एक – one, that one only, sole>, इत् <(इद्) ind. particle of affirmation – indeed>, राजा <m. nom. sg. of राजन् – king, ruler>, जगतः <n. gen. sg. of जगत् – of this World>, बभूव <prf. 3rd person sg. of √भू – he has become, is>;

यः <m. nom. sg. of pron. यद् – he who>, ईशे <prf. middle 3rd person sg. of √ईश् – he owns, possesses (gen.)>, अस्य <n. gen. sg. of इदं – of this>, द्विपदः <n. gen. sg. of द्विपद् – of bipeds (collectively), i.e. men or mankind (Agnis, the Devas)>, चतुष्पदः <n. gen. sg. of चतुष्पद् – of quadrupeds (collectively), i.e. cattle (celestial lights) and steeds (streaming currents or forces)>, कस्मै देवाय हविषा विधेम <see RV. x.121.1 – to which Deva should we offer our oblation?>.

In RV x.121.4, plasma blooms in the Soma Pond are described as snowy mountains. According to the stanza, the Arms of the Milky Way (प्रदिशः pradiśaḥ pl.) and the upper and lower discs of Stones (बाहू bāhū du. the two arms) are in the possession of Ka (कः). To make a sensible interpretation, the stanza needs to be read along with stanzas RV v.43.4 and RV i.164.42.

RV x.121.4 author हिरण्यगर्भः प्राजापत्यः, to कः, metre त्रिष्टुप् छंदः
यस्येमे हिमवंतो महित्वा यस्य समुद्रं रसया सहाहुः ।

यस्येमाः प्रदिशो यस्य बाहू कस्मै देवाय हविषा विधेम ॥१०.१२१.४॥
यस्य । इमे । हिमऽवंतः । महिऽत्वा । यस्य । समुद्रं । रसया । सह । आहुः ।
यस्य । इमाः । प्रऽदिशः । यस्य । बाहू इति । कस्मै । देवाय । हविषा । विधेम ॥१०.१२१.४॥
Whose snowy mountains are glorified, whose Ocean they have inundated with Soma mead,
His are the Arms of the Milky Way and his are two arms of the Stone. To which Deva should we offer our oblation? (RV x.121.4)

यस्य <m. gen. sg. of pron. यद् – whose>, इमे < m. nom. pl. of pron. इदं – these>, हिमवंतः <m. nom. pl. of हिमवत् – frosty or snowy mountains>, महित्वा <ind. pt. of √मह् – being esteemed highly, being glorified>, यस्य <m. gen. sg. of pron. यद् – whose>, समुद्रं <m. acc. sg of समुद्र – the Ocean of celestial Waters>, रसया <(=रसेन) m. inst. sg. of रस – with sap or juice, water, draught, elixir, i.e. with Soma mead or cosmic plasmas>, सह <ind. – with>, आहुः <prf. 3rd person pl. of √अह् – they have occupied, pervaded>;

यस्य <m. gen. sg. of pron. यद् – whose, of which>, इमाः <f. nom. pl. of इदं – these>, प्रदिशः <f. nom. pl. of प्रदिश् – the pointed-out regions, i.e. the Arms of the Milky Way (see section 2.8.1)>, यस्य <m. gen. sg. of pron. यद् – whose, of which>, बाहू <mf. nom. du. of बाहु – two arms, i.e. the upper and the lower discs of the Soma Press, the Stone, the celestial ioniser (see section 2.7.6)>, कस्मै देवाय हविषा विधेम <see RV. x.121.1 – to which Deva should we offer our oblation?>.

2.5.3 Samudra, the Ocean of Celestial Waters

In RV x.142.7 and 8, the Soma Pond is called Samudra, the Ocean of celestial Waters, where flowery Dūrvā grasses spring up. In these two stanzas, it is explained that, as Agni, proton, a fiery and dry embryonic particle, is discharged into the Samudra, he oscillates and jumps here and there, causing flowery Dūrvā grasses to spring up in the Ocean of celestial Waters.

The author is Stambamitra (स्तम्बमित्रः) and both stanzas are addressed to Agni. The name Stambamitra (स्तम्बः stambaḥ + मित्रः mitraḥ) literally means 'a tuft of kindled plasmas', which refers to a tuft of plasma blooms, or Soma plants. Stamba (स्तम्बः) is "a clump or tuft of grass". Mitra (मित्रः) represents the plasma light discharge, the kindled plasmas, the blooming plasmas, as we shall see in section 4.1.3.

RV x.142.7 author स्तंबमित्रः, to अग्निः, metre अनुष्टुप् छंदः
अपामिदं न्ययनं समुद्रस्य निवेशनं । अन्यं कृणुष्वेतः पंथां तेन याहि वशाँ अनु ॥१०.१४२.७॥
अपां । इदं । निऽअयनं । समुद्रस्य । निऽवेशनं । अन्यं । कृणुष्व । इतः । पंथां । तेन । याहि । वशान् । अनु ॥१०.१४२.७॥
This is the receptacle of the Samudra, home of celestial Waters.
Take one of the paths and set out as you desire. (RV x.142.7)

अपां <f. gen. pl. of अप् – of celestial Waters, i.e. cosmic plasmas>, इदं <n. nom. sg. of pron. इदं – this>, न्ययनं <n. nom. sg. of न्ययन – entrance, gathering-place, receptacle>, समुद्रस्य <m. gen. sg of समुद्र – of the Samudra, i.e. the Ocean of cosmic plasmas, the Soma Pond >, निवेशनं <n. nom. sg. of निवेशन – home, dwelling place>;

अन्यं <m. acc. sg. of pron. अन्य – one, other>, पंथां <m. acc. sg. of पथिन् – a way, path, course>, कृणुष्व <ipv. 2nd person sg. of √कृ – you take>, याहि <ipv. 2nd person sg. of √या – you set out, proceed, advance>, वशान् <m. acc. pl. of वश – wishes, desires>, अनु <ind. – after, according to>.

RV x.142.8 author स्तंबमित्रः, to अग्निः, metre अनुष्टुप् छंदः
आयने ते परायणे दूर्वा रोहंतु पुष्पिणीः । ह्रदाश्च पुंडरीकाणि समुद्रस्य गृहा इमे ॥.१४२.८॥
आऽअयने । ते । पराऽअयने । दूर्वाः । रोहंतु । पुष्पिणीः । ह्रदाः । च । पुंडरीकाणि । समुद्रस्य । गृहाः । इमे
॥१०.१४२.८॥
On your way, hither and yon, let the flowery Dūrvā grasses spring up.
These pools and lotus blooms are the mansions of the Samudra, the Ocean of celestial Waters. (RV x.142.8)

आयने <n. loc. sg. of आयन – coming, approaching>, ते <n. nom. du. of तद् – both>, परायणे <n. loc. sg. of परायण – way of departure>, दूर्वाः <f. acc. pl. of दूर्वा – Dūrvā grasses>, रोहंतु <ipv. 3rd person pl. of √रुह् – let them to ascend, spring up>, पुष्पिणीः <f. acc. pl. of adj. पुष्पिन् – flowering, blossoming>;

ह्रदाः <m. nom. pl. of ह्रद – lakes, pools, ponds>, च <ind. – and, both>, पुंडरीकाणि <n. nom. pl. of पुंडरीक (पुण्डरीक) – lotus flowers>, समुद्रस्य <m. gen. sg. of समुद्र – of the Samudra, the Ocean of cosmic plasmas>, गृहाः <m. nom. pl. of गृह – houses>, इमे <m. nom. pl. of इदं – these>.

Note, in stanza RV x.142.8, pools (ह्रदाः pl.), lotus blooms (पुंडरीकाणि pl.), Dūrvā grasses (दूर्वाः pl.), and mansions (गृहाः pl.) are in the plural, while in RV x.142.7, the receptacle (न्ययनं) and home (निवेशनं) are in the singular. When the words such as, pond, receptacle, mansion, pillar, mountain, and so on are used in the singular, they denote the Soma Pond as a single unit. When these words are used in the plural, they refer to Soma plants, or plasma blooms, that grow in the Soma Pond, which are also called hundreds of ponds or Soma receptacles or bright pillars or mansions or mountains, and so on (see Figure 7).

2.5.4 Rodasī, the Pair of Mātā and Duhitā, the Pair of the Stone and the Soma Pond

According to RV iii.55.12 and 13, Rodasī is the pair of milch Cows: Mātā (माता) and Duhitā (दुहिता). Mātā, the mother milch Cow, is also called Devī Iḷā (इळा). The mother milch Cow is the Stone (ग्रावा grāvā), the celestial ioniser, who yields milk, the cosmic plasmas, by breaking up the ionic bonds of hydrogen. Duhitā, the daughter milch Cow, is the Udder, the Soma Pond, who drinks in all of the discharged milk, the cosmic plasmas. In RV iii.55.13, the mother milch Cow, the Stone, is called the Inexhaustible, for it produces milk, the cosmic plasmas, uninterrupted.

The author of stanzas RV iii.55.12-13 is Prajāpati Vaiśvāmitra Vācya (प्रजापतिः वैश्वामित्रः वाच्यः). Vaiśvāmitra is Prajāpati's patronymic, Vācya, his matronymic. Both stanzas are addressed to Rodasī. As has been explained on a number of occasions, the authors of the Ṛgveda hymns are personified celestial objects or phenomena and not necessarily the actual personages. The authors are always associated with the Deva(s) or Devī(s) to whom the hymns or stanzas are addressed, and often the names of the poets provide a key to interpreting and translating the stanzas.

RV iii.55.12 author प्रजापतिर्वैश्वामित्रो वाच्यो वा, to रोदसी, metre त्रिष्टुप् छंदः
माता च यत्र दुहिता च धेनू सबर्दुघे धापयेते समीची । ऋतस्य ते सदसीळे अंतर्महद्देवानामसुरत्वमेकं ॥३.५५.१२॥
माता । च । यत्र । दुहिता । च । धेनू इति । सबर्दुघे इति सबःऽदुघे । धापयेते इति । समीची इति संऽईची ।
ऋतस्य । ते इति । सदसि । ईळे । अंतः । महत् । देवानां । असुरऽत्वं । एकं ॥३.५५.१२॥
When the two milch Cows, Mātā and Duhitā, together, yield milk and nourish
At the seat of sacrifice, I praise the two. Great is the one and only supreme spirit of the Devas. (RV iii.55.12)

यत्र <ind. – on which occasion, when>, माता <m. nom. sg. of मातृ – a mother>, च <ind. – and>, दुहिता <f. nom. sg. of दुहितृ – "a daughter (the milker or drawing milk from her mother)">, च <ind. – and>, धेनू <f. acc. du. of धेनु – two milch Cows>, सबर्दुघे <f. nom. du. of adj. सबर्दुघ – yielding milk or nectar>, धापयेते <pre. 3rd person du. of √धे – the two nourish>, समीची <f. nom. du. of adj. सम्यञ्च् or समीच् – together, turned towards each other, facing one another>;

ऋतस्य <n. gen. sg. of ऋत – of sacrifice>, सदसि <n. loc. sg. of सदस् – at the seat, in the assembly>, अंतः <(अंतर्) ind. – (with loc.) in, at>, ईळे <(=ईडे) pre. 1st person sg. of √ईड् – I praise>, ते <f. acc. du. of pron. तद् – the two>, महत् <n. nom. sg. of adj. महत् – great, important, eminent>, देवानां <m. gen. pl. of देव – of the Devas>, असुरत्वं <n. nom. sg. of असुरत्व – supreme spirit, divine dignity>, एकं <n. nom sg. of adj. एक – alone, solitary, that one only>.

RV iii.55.13 author प्रजापतिर्वैश्वामित्रो वाच्यो वा, to रोदसी, metre त्रिष्टुप् छंदः
अनयस्या वत्सं रिहती मिमाय कया भुवा नि दधे धेनुरूधः ।
ऋतस्य सा पयसापिन्वतेळा महद्देवानामसुरत्वमेकं ॥३.५५.१३॥
अन्यस्याः । वत्सं । रिहती । मिमाय । कया । भुवा । नि । दधे । धेनुः । ऊधः ।
ऋतस्य । सा । पयसा । अपिन्वत । इळा । महत् । देवानां । असुरऽत्वं । एकं ॥३.५५.१३॥
She bellows praising the Udder of the Inexhaustible. With which sacrificial fire has the milch Cow laid her Udder? This Iḷā has been overflowing with the milk of sacrifice. Great is the one and only supreme spirit of the Devas. (RV iii.55.13).

अन्यस्याः <f. gen. sg. of अन्या – of the inexhaustible>, वत्सं <mn. acc. sg. of वत्स – the breast, udder (refers to the Soma Pond)>, रिहती <(रिहति) pre. 3rd person sg. of √रिह् – she praises>, मिमाय <prf. 3rd person sg. of √मा – she bellows>, कया <ind. or f. inst. sg. of interrogative pron. किं – in what manner? with which?>, भुवा <f. inst. sg. of भू – with "sacrificial fire">, नि <ind. prefix to verb दधे (√धा)>, नि-दधे <prf. 3rd person sg. of नि-√धा – she has put or laid down>, धेनुः <f. nom. sg. of धेनु – a milch cow>, ऊधः <n. acc. sg. of ऊधस् – the udder>;

ऋतस्य <n. gen. sg. of ऋत – of sacrifice>, पयसा <n. inst. sg. of पयस् – with juice, milk, water, i.e. with cosmic plasmas>, अपिन्वत<ipf. 3rd person sg. of √पिन्व् – she has been overflowing>, इळा <(=इडा) f. nom. sg. of इळा – the goddess Iḷā (इळा), also written as Iḍā (इडा)>, महत् देवानां असुरत्वं एकं <see RV iii.55.12 – Great is the one and only supreme spirit of the Devas>.

As noted, the Inexhaustible refers to Mātā, the mother milch Cow, the Stone, the celestial ioniser. The sacrificial fire, with which the milch Cow laid her Udder, the Soma Pond, is Trita (त्रितः), tritium (^3H). Trita is one of the three fires of Agni, proton. The three fires of Agni are Ekata, Dvita, and Trita (see section 7.1.4).

2.5.5 The Radiant Udder of Pṛśni

In stanza RV ii.34.2, the Soma Pond, the Samudra, the Ocean of celestial Waters, is called Pṛśni's radiant Udder. Note that 'Maruts (pl.)' is another name for Soma plants, or plasma blooms. Rudra is father of the Maruts.

RV ii.34.2 author गृत्समदः, to मरुतः, metre जगती छंदः
द्यावो न स्तृभिश्चितयंत खादिनो व्य१ऽभ्रिया न द्युतयंत वृष्टयः ।
रुद्रो यद्वो मरुतो रुक्मवक्षसो वृषाजनि पृश्न्याः शुक्र ऊधनि ॥२.३४.२॥
द्यावः । न । स्तृऽभिः । चितयंत । खादिनः । वि । अभ्रियाः । न । द्युतयंत । वृष्टयः ।
रुद्रः । यत् । वः । मरुतः । रुक्मऽवक्षसः । वृषा । अजनि । पृश्न्याः । शुक्रे । ऊधनि ॥२.३४.२॥
Maruts, adorned with rings, are flashing forth like thunder-born torrents as the Heavens are decked with strewn lights, When mighty Rudra begot you Maruts with golden ornaments on your breasts in Pṛśni's radiant Udder. (RV ii.34.2)

द्यावः <m. nom. pl. of दिव् – the Heavens>, न <ind. – like, as>, स्तृभिः <m. inst. pl. of स्तृ – with strewn lights>, चितयंत <inj. passive cau. 3rd person pl. of √चित् – they (the Heavens) are decked, piled up>, खादिनः <m. nom. pl. of adj. खादिन् – decorated with bracelets or rings>, अभ्रियाः <f. nom. pl. of adj. अभ्रिय – thundercloud-born>, न <ind. – like>, वि <ind. prefix to verb. द्युतयंत (√द्युत्)>, वि-द्युतयंत <inj. passive cau. 3rd person pl. of वि-√द्युत् – they (the Maruts) are flashing forth, shining forth>, वृष्टयः <f. nom. pl. of विष्टि – rains, torrents>;

रुद्रः <m. nom sg. of रुद्र – Rudra>, मरुतः <m. acc. pl. of मरुत् – the Maruts>, रुक्मवक्षसः <m. acc. pl. of adj. रुक्मवक्षस् – golden breasted, having golden ornaments on the breast>, यत् <conjunction – that, when>, वः <m. acc. pl. of 2nd person pron. युष्मद् – you (pl.), the Maruts>, वृषा <m. nom. sg. of adj. वृषन् – mighty, great>, अजनि <aor. 3rd person sg. of √जन् – he generated, begot>,पृश्न्याः <f. gen. sg. of पृश्नि – of Pṛśni (dappled cow)> शुक्रे <n. loc. sg. of adj. शुक्र – bright, resplendent>, ऊधनि <n. loc. sg. of उधन् – in udder, i.e. the Soma Pond>.

In the stanza, the Heavens are in the plural denoting the three spheres of the Heaven: the Upper Heaven, the Midheaven, and the Lower Heaven. Pṛśni is the celestial Cow, the Stone, the celestial ioniser, who produces milk, cosmic plasmas. The author is Gṛtsamada (गृत्समदः) who is probably 'eager for rapturous Soma'. The stanza is addressed to the Maruts.

2.5.6 The Moon with Beauteous Wings Glides in the Heaven

In stanza RV i.105.1, the Soma Pond is described as the Moon (चंद्रमाः candramāḥ) with beauteous wings gliding in the celestial Waters. The flashing thunderbolts refer to the plasma blooms and they are compared to the beauteous wings of the Moon. In the Moon are the abodes of the flashing thunderbolts; and the flashing thunderbolts seek out their abodes in the Moon, the Soma Pond. The author Āptya Trita, tritium, the fire of Agni, declares to Rodasī that the abodes of the thunderbolts in the Moon and the pressed out Soma drinks are his riches (RV i.105.1 and 2). The golden rims refer to the individual sheaths of plasma blooms. Though the stanzas RV i.105.1 and RV i.105.2 presented here are addressed to Viśve Devas (विश्वे देवाः viśve devāḥ, pl.), the stanzas explain the cosmic plasma phenomena that occur in the Soma Pond.

RV i.105.1 author आप्त्यस्त्रितः कुत्सो वा, to विश्वे देवाः, metre पङ्क्तिः छंदः
चंद्रमा अप्स्व१न्तरा सुपर्णो धावते दिवि ।
न वो हिरण्यनेमयः पदं विंदंति विद्युतो वित्तं मे अस्य रोदसी ॥१.१०५.१॥

चंद्रमाः । अप्ऽसु । अंतः । आ । सुऽपर्णः । धावते । दिवि ।
न । वः । हिरण्यऽनेमयः । पदं । विंदंति । विऽद्युतः । वित्तं । मे । अस्य । रोदसी इति ॥१.१०५.१॥
In the Heaven, the moon with beauteous wings glides in the celestial Waters,
As flashing thunderbolts with golden rims seek your abode, the riches of mine, O Rodasī.
(RV i.105.1)

चंद्रमाः <m. nom. sg. of चन्द्रमस् – the moon>, अप्सु <f. loc. pl. अप् – in the celestial Waters>, अंतः <ind. – within, between>, आ <ind. – (with a preceding loc.) in, at, on>, सुपर्णः <m. nom. sg. of adj. सुपर्ण – having beautiful wings>, धावते <pre. 3rd person sg. of √धव् – it (the moon) runs, flows>, दिवि <m. loc. sg. of दिव् – in the Heaven>;

न <ind. – like, as>, वः <m. gen. pl. of युष्मद् – your (pl.)>, हिरण्यनेमयः <f. nom. pl. of adj. हिरण्यनेमि – having golden fellies or wheels or circular rims>, पदं <n. acc. sg. of पद – station, abode, home>, विंदंति <pre. 3rd person pl. √विद् 6P – they find, discover, seek out>, विद्युतः <f. nom. pl. of विद्युत् – lightnings, flashing thunderbolts>, वित्तं <n. nom. sg. of वित्त – wealth, property>, मे <m. gen. sg. of 1st person pron. अस्मद् – mine, my>, अस्य <m. gen. sg. of pron. इदं – of this (emphasis of मे)>, रोदसी <n. voc. du. of रोदस् – O Rodasī (the pair of the Stone and Soma Pond)>.

RV i.105.2 author आप्त्यस्त्रितः कुत्सो वा, to विश्वे देवाः, metre पङ्क्तिः छंदः
अर्थमिद्वा उ अर्थिन आ जाया युवते पतिं । तुंजाते वृष्ण्यं पयः परिदाय रसं दुहे वित्तं मे अस्य रोदसी ॥१.१०५.२॥
अर्थं । इत् । वै । ऊं इति । अर्थिनः । आ । जाया । युवते । पतिं ।
तुंजाते इति । वृष्ण्यं । पयः । परिऽदायं । रसं । दुहे । वित्तं । मे । अस्य । रोदसी इति ॥१.१०५.२॥
Surely one who seeks wealth draws it to one’s self as a wife her husband.
The two strike and I yield mighty Soma juice, bestowing this elixir, the riches of mine, O Rodasī.
(RV i.105.2)

अर्थं <mn. acc. sg. of अर्थ – aim, object, wealth>, इत् <इद् ind. a particle of affirmation – even, just>, वै <ind. a particle of emphasis and affirmation, followed by उ – indeed, truly>, उ <ind. an enclitic copula>, अर्थिनः <m. dat. sg of अर्थिन् – to one who wants or desires>, जाया <f. nom. sg. of जाया – a wife>, आ-युवते <pre. middle 3rd person sg. of आ-√यु 6Ā – he/she procures, draws or pulls towards one's self>, पतिं <m. acc. sg. of पति – a husband>;

तुंजाते <(तुञ्जाते) pre. middle 3rd person du. of √तुज् 7Ā – the two (the upper and lower discs of the Stone) hit, strike>, वृष्ण्यं <n. acc. sg. of adj. वृष्ण्य – manly, vigorous, mighty>, पयः <n. acc. sg. of पयस् – “juice, (esp.) milk, water, rain”, i.e. Soma juice, or celestial Waters, or cosmic plasmas>, परिदाय <ind. pt. of परि-√दा – granting, bestowing>, रसं <m. acc. sg. of रस – sap, nectar, elixir>, दुहे <pre. middle 1st person sg. of √दुह् 2Ā or 6Ā – I, Āptya Trita, give milk (cosmic plasmas), yield any desired object>, वित्तं मे अस्य रोदसी <see RV i.105.1 – the riches of mine, O Rodasī>.

In RV i.105.2 above, ‘the two’ refers to the upper and the lower discs of the Stone. ‘The two’ strike hydrogen atoms, and I, Āptya Trita, tritium, the fire of the Stone, yield Soma juice, the celestial elixir.

2.5.7 Sūrya, the Best of Lights

Sūrya (सूर्यः) of the Ṛgveda is not the sun in our solar system. Sūrya is the celestial light associated with kindled Agnis, the positively charged hydrogen ions. In hymn RV x.170, though only four stanzas, the characteristic attributes of Sūrya are described succinctly. In these stanzas, Sūrya is described as luminous and lofty, the most splendid, the best of all lights (ज्योतिरुत्तमं = ज्योतिः उत्तमं RV x.170.3), the light cast into the receptacle, that is, into the Soma Pond, accompanied by Viśve Devas and Tvaṣṭā, the divine architect (RV x.170.4). Sūrya is the prop of the Heaven. He is also described as the Vṛtra-slayer, the slayer of the Devas' enemies (RV x.170.2). Sūrya spreads out the powerful, imperishable light to see far and wide (RV x.170.3).

RV x.170.1 author विभ्राट् सौर्यः, to सूर्यः, metre जगती छंदः
विभ्राड्बृहत्पिबतु सोम्यं मध्वायुर्दधद्यज्ञपतावविह्रुतं ।
वातजूतो यो अभिरक्षति त्मना प्रजाः पुपोष पुरुधा वि राजति ॥१०.१७०.१॥
विऽभ्राट् । बृहत् । पिबतु । सोम्यं । मधु । आयुः । दधत् । यज्ञऽपतौ । अविऽह्रुतं ।
वातऽजूतः । यः । अभिऽरक्षति । त्मना । प्रऽजाः । पुपोष । पुरुधा । वि । राजति ॥१०.१७०.१॥
Luminous and lofty, may he drink Soma mead and impart the lord of Yajña unbroken vigour.
He who, driven by Vāta, guards the aftergrowths and nourishes them well, shines forth.
(RV x.170.1)

विभ्राट् <m. nom. sg. of adj. विभ्राज् – shining, splendid, luminous>, बृहत् <m. nom. sg. of adj. बृहत् – lofty, tall, mighty>, पिबतु <ipv. 3rd person sg. of √पा – may he drink>, सोम्यं <n. acc. sg. of adj. सोम्य – containing Soma>, मधु <n. acc. sg. of मधु – mead>, आयुः <n. acc. sg. of आयुस् – life, vital power, vigour>, दधत् <sub. 3rd person sg. of √धा – may he bestow, impart>, यज्ञपतौ <m. loc. sg. of यज्ञपति – on the Lord of Yajña>, अविह्रुतं <n. acc. sg. of adj. अविह्रुत – unbent, unbroken, uninterrupted>;

वातजूतः <m. nom. sg. of adj. वातजूत – driven by Vāta>, यः <m. nom. sg. of यद् – he who>, अभिरक्षति <pre. 3rd person sg. of अभि-√रक्ष् – he guards, protects>, प्रजाः <f. acc. pl. of प्रजा – offspring, aftergrowths (refers to Soma plants or plasma blooms)>, त्मना <ind. – yet, really, indeed, certainly>, पुपोष <prf. 3rd person sg. of √पुष् 1P – he causes to thrive, nourishes>, पुरुधा <ind. – frequently, abundantly>, वि-राजति <pre. 3rd person sg. of वि-√राज् – he is illustrious or shines forth>.

In RV x.170.1 above, the aftergrowths refer to the Soma plants, or plasma blooms, that grow back after the previously matured ones were released by Indra. Note that Monier-Williams's definitions of prajā (प्रजाः prajāḥ) include "after-growth (of plants)", which is an accurate description of the regrowing Soma plants, the light-containing plants (ओषधयः oṣadhayaḥ pl.), the plasma blooms, after the release of the previously matured ones.

RV x.170.2 author विभ्राट् सौर्यः, to सूर्यः, metre जगती छंदः
विभ्राड्बृहत्सुभृतं वाजसातमं धर्मन्दिवो धरुणे सत्यमर्पितं ।
अमित्रहा वृत्रहा दस्युहंतमं ज्योतिर्जज्ञे असुरहा सपत्नहा ॥१०.१७०.२॥
विऽभ्राट् । बृहत् । सुऽभृतं । वाजऽसातमं । धर्मन् । दिवः । धरुणे । सत्यं । अर्पितं ।
अमित्रऽहा । वृत्रऽहा । दस्युहन्ऽतमं । ज्योतिः । जज्ञे । असुरऽहा । सपत्नऽहा ॥१०.१७०.२॥
O luminous and lofty prop of the Heaven, cherished, victorious, the satyaṃ cast into the receptacle. He, the light, the foe-slayer, the Vṛtra-slayer, the best slayer of the Devas' enemy, was born, the slayer of Asura, the rivals-slayer. (RV x.170.2)

विभ्राट् <n. voc. sg. of adj. विभ्राज् – luminous, shining>, बृहत् <n. voc. sg. of adj. बृहत् – lofty, tall, great>, सुभृतं <n. nom. sg. of adj. सुभृत – well-borne, well-cherished>, वाजसातमं <n. nom. sg. of adj. वाजसातम – winning wealth, victorious>, धर्मन् <n. voc. sg. of धर्मन् – support, prop>, दिवः <m. gen. sg. of दिव् – of the Heaven>, धरुणे <n. loc. sg. of धरुण – into the stay, prop, receptacle (the Soma Pond)>, सत्यं <n. nom. sg. of सत्य – satyaṃ>, अर्पितं <n. nom. sg. of adj. अर्पितं – thrown, cast into>;

अमित्रहा <m. nom. sg. of अमित्रहन् – a foe-slayer>, वृत्रहा <m. nom. sg. of वृत्रहन् – a Vṛtra-slayer>, दस्युहंतमं <n. nom. sg. of superlative adj. दस्युहंतम – the best slayer of the enemy of the Devas>, जोतिः <n. nom. sg. of ज्योतिस् – light>, जज्ञे <prf. middle 3rd person sg. of √जन् – he was born>, असुरहा <m. nom. sg. of असुरहन् – a slayer of Asura>, सपत्नहा <m. nom. sg. of सपत्नहन् – a rivals-slayer>.

In RV x.170.2, Sūrya is described as the luminous and lofty Heaven's prop, the Pillar of the Heaven. Sūrya is also the slayer of Vṛtra just as Indra and Agni are described as such. Sūrya is the satyaṃ (सत्यं) cast into the receptacle, the Soma Pond, indicating that satyaṃ is the kindled Agnis. In the Ṛgveda, Sūrya and Agni are often called satyaṃ. Since Agnis are the protons who will be the foundation of all manifested reality, they are called satyaṃ, the reality or truth.

The foe, enemy, the rival of the Devas, and Vṛtra all refer to the negative aspects of plasma sheaths, for plasma sheaths prevent the release of celestial lights and fires, cosmic plasmas. Eventually plasma sheaths, the foes of the Devas, need to be destroyed in order to release salubrious cosmic plasmas.

RV x.170.3 author विभ्राट् सौर्यः, to सूर्यः, metre जगती छंदः
इदं श्रेष्ठं ज्योतिषां ज्योतिरुत्तमं विश्वजिद्धनजिदुच्यते बृहत् ।
विश्वभ्राड्भ्राजो महि सूर्यो दृश उरु पप्रथे सह ओजो अच्युतं ॥१०.१७०.३॥
इदं । श्रेष्ठं । ज्योतिषां । ज्योतिः । उत्ऽतमं । विश्वऽजित् । धनऽजित् । उच्यते । बृहत् ।
विश्वऽभ्राट् । भ्राजः । महि । सूर्यः । दृशे । उरु । पप्रथे । सहः । ओजः । अच्युतं ॥१०.१७०.३॥
This light, the best of lights, the most splendid, the winner of riches, delights in thundering hymn. Great Sūrya, shining, all-illuminating, spreads mighty, imperishable light to see far and wide. (RV x.170.3)

इदं <n. nom. sg. of इदं – this>, श्रेष्ठं <n. nom. sg. of adj. श्रेष्ठ – most splendid>, ज्योतिषां <n. gen. pl. of ज्योतिस् – of all lights>, ज्योतिः <n. nom. sg. of ज्योतिस् – light>, उत्तमं <n. nom. sg. of superlative उत्तम – the best, excellent>, विश्वजित् <n. nom sg. of adj. विश्वजित् – all-winning>, धनजित् <n. nom. sg. of adj. धनजित् – acquiring all wealth>, उच्यते <pre. middle 3rd person sg. of √उच् – he takes pleasure in, he delights in>, बृहत् <n. acc. sg. of adj. बृहत् – clear, loud (sounds), i.e. roaring or thundering hymn, Speech>;

विश्वभ्राट् <m. nom. sg. of adj. विश्वभ्राज् – all-illuminating>, भ्राजः <m. nom. sg. of adj. भ्राज – shining, glittering>, महि <ind. – great>, सूर्यः <m. nom. sg. of सूर्य – Sūrya>, दृशे <m. dat. sg. of दृश् – to see, view>, उरु <ind. – widely, far>, पप्रथे <prf. middle 3rd person sg. of √प्रथ् – he spreads>, सहः <n. acc. sg. of adj. सहस् – powerful, mighty>, ओजः <n. acc. sg. of ओजस् – vigour, splendour, light>, अच्युतं <n. acc. sg. of adj. अच्युत – imperishable>.

RV x.170.4 author विभ्राट् सौर्यः, to सूर्यः, metre आस्तारपंक्तिः छंदः
विभ्राजञ्ज्योतिषा स्व१ रगच्छो रोचनं दिवः ।
येनेमा विश्वा भुवनान्याभृता विश्वकर्मणा विश्वदेव्यावता ॥१०.१७०.४॥
विऽभ्राजन् । ज्योतिषा । स्वः । अगच्छः । रोचनं । दिवः ।
येन । इमा । विश्वा । भुवनानि । आऽभृता । विश्वऽकर्मणा । विश्वदेव्यऽवता ॥१०.१७०.४॥
Shining forth with your light, you have come to the luminous sphere of the Heaven,
Accompanied by Viśve Devas and Viśvakarmā, by whom the World and all beings were brought to existence. (RV x.170.4)

विभ्राजन् <m. nom. sg. of pre. active pt. विभ्राजत् (वि-√भ्राज्) – he shining forth>, ज्योतिषा <n. inst. sg. of ज्योतिस् – with light>, स्वः <m. nom sg. of adj. स्व – his own>, अगच्छः <ipf. 2nd person sg. of √गं – you have come>, रोचनं <n. acc. sg. of रोचन – to luminous sphere>, दिवः <m. gen. sg. of दिव् – of the Heaven>;

येन <m. inst. sg. of यद् – by whom>, इमा <n. nom. pl. of इदं – these>, विश्वा <n. nom. pl. of विश्व – all>, भुवनानि <n. nom. pl. of भुवन – the World and beings>, आभृता <n. nom. pl. of adj. आभृत – produced, caused to exist>, विश्वदेव्यवता <m. inst. sg. of adj. विश्वदेव्यावत् – accompanied by Viśve Devas>, विश्वकर्मणा <m. inst. sg. of विश्वकर्मन् – with Viśvakarmā, "all-makers", "the divine creative architect", i.e. Tvaṣṭā>.

RV x.170.4 explains that Sūrya has come to the luminous sphere of the Heaven accompanied by the Viśve Devas and Viśvakarmā (विश्वकर्मा), Tvaṣṭā, the divine architect, who brought the World into existence. Tvaṣṭā represents magnetism in general and magnetic field in particular. Viśve Devas are the Devas who are born and grow in the Soma Pond, such as Agni, Soma, the Maruts, Rudra, and the Dawns.

2.5.8 Sūrya, the Single Eye, the All-Seeing Eye

Sūrya is called the Single Eye (एकं अक्षि ekaṃ akṣi RV ix.9.4). Stanza RV ix.9.4, which is addressed to Soma Pavamāna, states that the Single Eye is glorified by the seven rivers. In the extant text of the Ṛgveda, Agni and Sūrya are called the Eye and, specifically, Sūrya is called the Single Eye (RV ix.9.4) and the All-Seeing Eye (विश्वचक्षाः viśvacakṣāḥ RV i.50.2). But why 'seven' rivers? In RV i.50.8, it is said that 'seven ruddy mares (सप्त हरितः)' drive Sūrya's chariot. Very likely the 'seven ruddy mares' and the 'seven rivers' refer to the seven major divisions of the electromagnetic spectrum: gamma ray, x-ray, ultraviolet, visible, infrared, microwave, and radio.

RV ix.9.4 author असितः काश्यपो देवलो वा, to पवमानः सोमः, metre गायत्री छंदः
स सप्त धीतिभिर्हितो नद्यो अजिन्वदद्रुहः । या एकमक्षि वावृधुः ॥९.९.४॥
सः । सप्त । धीतिऽभिः । हितः । नद्यः । अजिन्वत् । अद्रुहः । याः । एकं । अक्षि । ववृधुः ॥९.९.४॥
Impelled by Speeches he, Soma, animated the seven guileless rivers,
Which have glorified the Single Eye. (RV ix.9.4)

सः <m. nom. sg. of 3rd person pron. तद् – he, Soma>, सप्त <f. acc. pl. of सप्तन् – seven>, धीतिभिः <f. inst. pl. of धीति – by prayers, i.e. by hymns, speeches>, हितः <m. nom. sg. of adj. हित – impelled, urged on>, नद्यः <f. acc. pl. of नदी – roarers, rivers>, अजिन्वत् <ipf. 3rd person sg. of √जिन्व् – he incited, stirred, animated>, अद्रुहः <f. acc. pl. of अद्रुह् – free from malice or treachery, i.e. guileless>;

याः <f. nom. pl. of यद् – who, which (pl.)>, एकं <n. acc. sg. of एक – one, single>, अक्षि <n. acc. sg. of अक्षि – the eye>, ववृधुः <prf. 3rd person pl. of √वृध् – they have magnified, glorified>.

In RV i.50.2, the Nakṣatras refer to the constellations of cosmic plasma blooms. The flaming locks in RV i.50.8 also refer to the blooming plasmas in the Soma pond (see Figure 7). Both stanzas are addressed to Sūrya. In RV i.50.2, Sūra (सूरः) is another name for Sūrya (सूर्यः).

RV i.50.2 author प्रस्कण्वः काण्वः, to सूर्यः, metre गायत्री छंदः
अप त्ये तायवो यथा नक्षत्रा यंत्यक्तुभिः । सूराय विश्वचक्षसे ॥१.५०.२॥
अप । त्ये । तायवः । यथा । नक्षत्रा । यंति । अक्तुऽभिः । सूराय । विश्वऽचक्षसे ॥१.५०.२॥
The Nakṣatras arise with celestial lights, spreading in a continuous stream
For Sūra, the All-Seeing Eye. (RV i.50.2)

अप <ind. – away, away from>, त्ये <m. nom. pl. of त्यद् – those>, तायवः <m. nom. pl. of adj. तायु (derived from √ताय्) – spread in a continuous stream or line>, यथा <ind. – in which manner or way>, नक्षत्रा <n. nom. pl. of नक्षत्र – "in the Vedas Nakṣatras are considered as abodes of the gods, ...", here Nakṣatras refer to the constellations of plasma blooms, which are abodes of the Devas>, यंति <pre. 3rd person pl. of √इ – they flow, arise>, अक्तुभिः <m. inst. pl. of अक्तु – with rays or lights>;
सूराय <m. dat. sg. of सूर – for Sūra (=Sūrya)>, विश्वचक्षसे <n. dat. sg. of विश्वचक्षस् – for the all-seeing eye>.)

RV i.50.8 author प्रस्कण्वः काण्वः, to सूर्यः, metre गायत्री छंदः
सप्त त्वा हरितो रथे वहंति देव सूर्य । शोचिष्केशं विचक्षण ॥१.५०.८॥
सप्त । त्वा । हरितः । रथे । वहंति । देव । सूर्य । शोचिःऽकेशं । विऽचक्षण ॥१.५०.८॥
Seven ruddy mares draw the chariot that carries thee, O Deva Sūrya,
Thou with flaming locks, O radiant One. (RV i.50.8)

सप्त <m. nom. pl. of सप्तन् – seven>, त्वा <m. acc. sg. of 2nd person pron. युष्मद् – you>, हरितः <m. nom. pl. of हरित् – pale yellow or ruddy mares>, रथे <m. loc. sg. of रथ – in the chariot>, वहंति <pre. 3rd person pl. of √वह् – they carry, transport, draw>, देव <m. voc. sg. of देव – O Deva>, सूर्य <m. voc. sg. of सूर्य – O Sūrya>;
शोचिष्केशं <m. acc. sg. of adj. शोचिष्केश – flamed-hair, having flaming locks>, विचक्षण <m. voc. sg. of विचक्षण – O shining or radiant One>.

As noted earlier, Pṛthivī (पृथिवी), which is often described as a circular disc, is the Milky Way, not the planet earth. Sun (sol) is the kindled Agni (proton); Moon (luna) is Soma (electron). In the Vedas and the ancient mythical and alchemical texts, Jupiter (इज्यः), Saturn (मंदः), Mars (क्षितिसुतः), Venus (शुक्रः), Mercury (इंदुजः), Sun (सूर्यः), and Moon (चंद्रः) are the constellations (नक्षत्राणि nakṣatrāṇi pl.) of the Heavenly Vault, not the sun and planets of our solar system. In the Vedas the Nakṣatras are considered as abodes of the gods.[73] Stanzas 29-31 of Bhūgolādhyāya of Sūryasiddhānta explain that these Nakṣatras revolve within the circumference of the Brahmāṇḍa, the Vault of the Heaven.[74]

Jupiter (इज्यः) is the Vault of the Heaven, the Brahmāṇḍa, the honoured teacher of the gods. Saturn (मंदः) is the Soma Pond, the womb of cosmic creation where the discharged cosmic plasmas gather, grow, mature, and bloom. In the Ṛgveda, the Soma Pond, or Rudra, is called a physician (भिषक् RV ix.112.1). The Stone, the celestial ioniser, which produces and discharges cosmic plasmas into the Soma Pond, is described as salubrious

---

[73] See Monier Monier-Williams, *A Sanskrit-English Dictionary*, 1899, op. cit., p. 524, under the headword 'nakṣatra (नक्षत्र)'.
[74] Bāpū Deva Sāstri (ed.), *Translation of the Sūryasiddhānta by Bāpū Deva Sāstri, and of the Siddhānta Śiromani by the late Lancelot Wilkinson*, Calcutta, Baptist Mission Press,1861, p. 79.

(अनमीव adj. RV vii.46.2). Cosmic plasmas, in the Ṛgveda, are compared to healing medicines. Saturn, the Soma Pond, dispenses the salubrious medicines, the cosmic plasmas, once they are matured and ready to be released. Even today, an early cipher for Saturn, Rx, is still written on nearly every prescription issued by physicians.[75] Manda (मंदः maṃdaḥ), one of the many names for Saturn, is very likely derived from the verb '√mand' (√मंद्), which means "to rejoice, be glad or delighted, be drunk or intoxicated", "to shine, be splendid or beautiful". It is mistakenly believed by many that the name of Saturn, Manda (मंदः maṃdaḥ), is derived from the adjective 'manda' (मन्द) which means "slow, tardy, moving slowly". The Saturn, the Soma Pond, is delighted and exhilarated by the draughts of Somas discharged from the Stone; it is splendid and beautiful with cosmic plasmas blooming in it.

Mars (क्षितिसुतः kṣitisutaḥ), the 'destroyer (क्षितिः) of the pressed Soma (सुतः)', represents Indra, the celestial warrior, who slays Vṛtra, the serpent of the Heaven, with his weapon, the thunderbolts or the hooks or the arrows or the spears, to release celestial lights. Venus (शुक्रः) represents plasma blooms, the light-containing plants, the Soma plants. The long, wavy, flaming locks of Venus are a symbolic representation of tall, wavy plasma blooms, which measure tens of thousands of light-years long (see Figure 7).

Mercury (इंदुजः iṃdujaḥ) is the Stone, the Soma Press, the celestial ioniser. Induja (इंदुजः), one of the many names for Mercury, literally means 'father (जः) of the bright drops of Soma (इंदुः)', that is, the producer of the bright drops of Soma, which is the Stone. Sun (सूर्यः sūryaḥ, sol) is the kindled Agni, the kindled proton. Moon (चंद्रः caṃdraḥ, luna) is Soma, electron, or the draught of Soma, the current of electron. Note, in stanza 31 of Bhūgolādhyāya of Sūryasiddhānta, the moon is written in the plural as Indava (इंदवः iṃdavaḥ pl. moons). Moons in the plural denote the bright drops of Soma, the draughts of electron.

2.5.9 Sūrya, the Eye of Mitra and Varuṇa

In RV x.37.1, Sūrya is called the Eye of Mitra and Varuṇa who sees far and wide. Sūrya is also called the light born of the Devas and the son of the Heaven.

RV x.37.1 author अभितपाः सौर्यः, to सूर्यः, metre जगती छंदः
नमो मित्रस्य वरुणस्य चक्षसे महो देवाय तदृतं सपर्यत ।
दूरेदृशे देवजाताय केतवे दिवस्पुत्राय सूर्याय शंसत ॥१०.३७.१॥
नमः । मित्रस्य । वरुणस्य । चक्षसे । महः । देवाय । तत् । ऋतं । सपर्यत ।
दूरेऽदृशे । देवऽजाताय । केतवे । दिवः । पुत्राय । सूर्याय । शंसत ॥१०.३७.१॥

---

75 Dennis William Hauck, *The Complete Idiot's Guide to Alchemy*, New York, Alpha Books (Penguin Group (USA) Inc.), 2008, p. 119.

Glory to the Eye of Mitra and Varuṇa. Offer ye gladly this sacrifice to the Deva.
Sing ye praise upon Sūrya, the son of the Heaven, who sees far and wide, the light born of the Devas. (RV x.37.1)

नमः <ind. – salutation or glory to (with dat.)>, मित्रस्य <m. gen. sg. of मित्र – of Mitra>, वरुणस्य <m. gen. sg. of वरुण – of Varuṇa>, चक्षसे <n. dat. sg. of चक्षस् – to the Eye>, देवाय <m. dat. sg. of देव – to the Deva>, तत् <n. acc. sg. of pron. तद् – that>, ऋतं <n. acc. sg. ऋत – sacrifice>, सपर्यत <ipv. 2nd person pl. of nominal verb √सपर्य – ye offer, dedicate>;

दूरेदृशे <m. dat. sg. of adj. दूरेदृश् – seeing far and wide>, देवजाताय <m. dat. sg. of adj. देवजात – "god-born", i.e. born of the Devas>, केतवे <m. dat. sg. of केतु – to light>, दिवः <m. gen. sg. of दिव् – of the Heaven>, पुत्राय <m. dat. sg. of पुत्र – to a son>, सूर्याय <m. dat. sg. of सूर्य – to Sūrya>, शंसत <ipv. 2nd person pl. of √शंस् – ye recite, praise>.

2.5.10 Sūrya Takes Hold of the Wondrous Weapon

Stanza RV v.63.4, addressed to Mitra-Varuṇa, explains about Sūrya as well. Sūrya, the son of the Heaven, takes hold of the wondrous weapon, the weapon for slaying Vṛtra. The wondrous weapon refers to the plasma blooms, which are also called the thunderbolts of Indra, the bright arrows of Rudra and Agni, the lances of the Maruts, and so on.

RV v.63.4 author अर्चनाना आत्रेयः, to मित्रावरुणौ, metre जगती छंदः
तमभ्रेण वृष्ट्या गूहथो दिवि पर्जन्य द्रप्सा मधुमंत ईरते ॥५.६३.४॥
माया । वां । मित्रावरुणा । दिवि । श्रिता । सूर्यः । ज्योतिः । चरति । चित्रं । आयुधं ।
तं । अभ्रेण । वृष्ट्या । गूहथः । दिवि । पर्जन्य । द्रप्सा । मधुऽमंतः । ईरते ॥५.६३.४॥
Your extraordinary power māyā, O Mitra and Varuṇa, resides in the Heaven; Sūrya, the light, takes hold of the wondrous weapon; and ye keep Sūrya in secret in the Heaven with thunder-cloud and rain. O Parjanya, bright drops full of honey arise. (RV v.63.4)

माया <f. nom. sg. of माया – māyā, extraordinary or supernatural power, sorcery, witchcraft magic>, वां <m. gen du. of 2nd person pron. युष्मद् – of you two>, मित्रावरुणा <m. nom. du. (dual compound) – Mitra-Varuṇa>, दिवि <m. loc. sg. of दिव् – in the Heaven>, श्रिता <f. nom. sg. of adj. श्रित – situated in, contained in>, सूर्यः <m. nom. sg. सूर्य – Sūrya>, ज्योतिः <n. nom. sg. of ज्योतिस् – light, fire>, चरति <pre. 3rd person sg. of √चर् – he undertakes, takes hold>, चित्रं <n. acc. sg. of adj. चित्र – bright, bright-coloured, wonderful>, आयुधं <n. acc. sg. of आयुध – a weapon>;

तं <m. acc. sg. of pron. तद् – him, Sūrya>, अभ्रेण <n. inst. sg. of अभ्र – with thunder-cloud>, वृष्ट्या <f. inst. sg. of वृष्टि – with rain>, गूहथः <pre. 2nd person du. of √गुह् – ye (du.) conceal, hide, keep secret>, दिवि <m. loc. sg. – in the Heaven>, पर्जन्य <m. voc. sg. of पर्जन्य – O Parjanya (Parjanya is the rain personified, i.e. the streaming currents of celestial Waters, cosmic plasmas)>, द्रप्सा <m. nom. pl of द्रप्स – bright drops>, मधुमंतः <m. nom. pl. of adj. मधुमन्त् – possessed of honey, mead>, ईरते <pre. middle 3rd person pl. of √ईर् – they rise, arise from>.

Māyā is the magic power of Mitra and Varuṇa. In the stanza, rain refers to the streaming cosmic plasmas discharged from the Stone, the celestial ioniser. Thunder-cloud refers to plasma blooms; it is the cloud of blooming cosmic plasmas. Mitra and Varuṇa kept Sūrya in secret midst the blooming cosmic plasmas. Parjanya (पर्जन्यः) is the streaming current of cosmic plasmas personified. In another stanza (RV v.83.6), Parjanya is asked to send down the rain of the Heaven. In RV v.83.6, the rain of the Heaven refers to the discharged cosmic plasmas streaming upwards into the Some Pond and the released cosmic plasmas raining down towards the Milky Way from the Soma Pond.

2.5.11 Sūrya, the Refulgent Chariot

In RV v.63.7, Sūrya is described as the refulgent Chariot placed in the Heaven by Mitra and Varuṇa. The World refers to the whole of the galactic system. With their wise decree and the magic power (माया māyā) of supreme Spirit (असुरः), Mitra and Varuṇa guard the divine laws, the laws of magnetism and cosmic plasmas. And through the cosmic Yajña, they govern the whole galactic system and put Sūrya in the Heaven.

The stanza indicates that, through the unbroken chain of cosmic ionisation and cosmic plasma discharge and release, the galactic system is sustained. Mitra and Varuṇa themselves are the Asura, the supreme Spirit.

RV v.63.7 author अर्चनाना आत्रेयः, to मित्रावरुणौ, metre जगती छंदः
धर्मणा मित्रावरुणा विपश्चिता व्रता रक्षेथे असुरस्य मायया ।
ऋतेन विश्वं भुवनं वि राजथः सूर्यमा धत्थो दिवि चित्र्यं रथं ॥५.६३.७॥
धर्मणा । मित्रावरुणा । विपःऽचिता । व्रता । रक्षेथे इति । असुरस्य । मायया ।
ऋतेन । विश्वं । भुवनं । वि । राजथः । सूर्यं । आ । धत्थः । दिवि । चित्र्यं । रथं ॥५.६३.७॥
With your wise decree and the magic power of supreme Spirit, Mitra and Varuṇa, ye guard the divine laws. Through sacrifice ye govern the whole World and put Sūrya, the refulgent Chariot, in the Heaven. (RV v.63.7)

धर्मणा <n. inst. sg. of धर्मन् – with established order, steadfast decree (of a god, esp. of मित्र-वरुण >, मित्रावरुणा <m. voc. du. (a dual compound) – O Mitra-Varuṇa>, विपश्चिता <n. inst. sg. of adj. विपश्चित् – wise, learned>, व्रता <n. acc. pl. of व्रत – ordinances, laws>, रक्षेथे <pre. middle 2nd person du. of √रक्ष् – ye (du.) guard, watch, protect>, असुरस्य <m. gen. sg. of असुर – of Asura (supreme spirit)>, मायया <f. inst. sg. of माया – with māyā, magic power>;

ऋतेन <n. inst. sg. of ऋत – with/through sacrifice>, विश्वं <n. acc. sg. of adj. विश्व – all, whole>, भुवनं <n. acc. sg. of – the World, i.e our galactic system>, वि-राजथः <pre. 2nd person du. of वि-√राज् – ye (du.) govern, rule>, सूर्यं <m. acc. sg. of सूर्य – Sūrya>, आ-धत्थः <pre. 2nd person du. of आ-√धा – ye (du.) put down, place>, दिवि <m. loc. sg. of दिव् – in the Heaven>, चित्र्यं <m. acc. sg. of adj. चित्र्य – brilliant>, रथं <m. acc. sg. of रथ – chariot>.

2.5.12 Uṣā, the Dawn

Uṣā (उषाः, sg.), the Dawn, when addressed in the singular, is the Udder, or the Soma Pond, personified. When addressed in the plural, the Dawns (उषसः uṣasaḥ, pl.) are plasma blooms, or Soma plants, or bright drops of light, personified. Hymn RV i.92, with four stanzas, explains who the Dawns are (RV i.92.1-3) and who the Dawn is (RV i.92.4). The author of the hymn is Gotama (गोतमः) who belongs to the family of Aṅgirā (अङ्गिराः m. nom. sg. of अङ्गिरस्), an angel, the messenger of the Heaven. Gotama literally means 'the excellent or the largest celestial cow, or light'. Rāhūgaṇa (राहूगणः) is Gotama's patronymic, which means 'the tribe or troop (गणः), which seizes the celestial light (राहुः the seizer)', that is, the tribe of Vṛtras, the tribe of plasma sheaths. The plasma sheaths seize celestial lights and confine them; thus called 'the tribe of Seizers'. The hymn is addressed to the Dawn (उषाः uṣāḥ f. nom. sg. of उषस्).

In RV i.92.1, the flame of the Dawns refers to plasma blooms, or Soma plants. The mother Cows, the firestones of the celestial ioniser, cast the herds of ruddy lights. And the Dawns make them appear in the eastern region of the Heavenly Vault. The eastern region is the region from where the discharged celestial lights, or cosmic plasmas, arise from the Stone, the celestial ioniser (see Figure 6). For the Stone, see section 2.7.

RV i.92.1 author गोतमो राहूगणः, to उषाः, metre जगती छंदः
एता उ त्या उषसः केतुमक्रत पूर्वे अर्धे रजसो भानुमंजते ।
निष्कृण्वाना आयुधानीव धृष्णवः प्रति गावोऽरुषीर्यंति मातरः ॥१.९२.१॥
एताः । ऊं इति । त्याः । उषसः । केतुं । अक्रत । पूर्वे । अर्धे । रजसः । भानुं । अंजते ।
निःऽकृण्वानाः । आयुधानिऽइव । धृष्णवः । प्रति । गावः । अरुषीः । यंति । मातरः ॥१.९२.१॥
The Dawns raised their flame and make the light to appear in the eastern region of the Heavenly Sphere. As the bold ones prepare their weapons, the mother Cows cast herds of ruddy lights. (RV i.92.1)

एताः <f. nom. pl. of pron. एतद् – these>, उ <ind. particle implying assent, calling, command>, त्याः <f. nom. pl. of pron. त्यद् – those (used after another demonstrative)>, उषसः <f. nom. pl. of उषस् – Dawns>, केतुं <m. acc. sg. of केतु – flame, torch>, अक्रत <ipf. 3rd person pl. of √कृ – they placed, raised>, पूर्वे <m. loc. sg. of adj. पूर्व – fore, front, eastern>, अर्धे <m. loc. sg. of अर्ध – in the region>, रजसः <n. gen. sg. of रजस् – of the Heavenly Sphere>, भानुं <m. acc. sg. of भानु – light or ray of light>, अंजते <pre. middle 3rd person pl. of √अंज् (√अञ्ज्) – they cause to appear>;

निष्कृण्वानाः <m. nom. pl. of pre. middle pt. of निष्कृण्वान (निष्-√कृ) – they preparing, restoring>, आयुधानि <n. acc. pl. of आयुध – weapons>, इव <ind. – like, as>, धृष्णवः <m. nom. pl. of धृष्णु – bold, courageous ones (such as Agni, Indra, Soma, the Maruts)>, प्रति <ind. prefix to verb यंति (√इ) – on, upon, towards>, गावः <f. acc. pl. of गो (rarely, but can be acc.) – cattle, "the herds of the sky", i.e. the herds of lights>, अरुषीः <f. acc. pl. of adj. अरुष – red, reddish>, प्रति-यंति <pre. 3rd person pl. of प्रति-√इ – they blow, spread>, मातरः <f. nom. pl. of मातृ – "the divine mothers", mother Cows, i.e. the firestones of the celestial ioniser>.

Stanza RV i.92.2 explains that the Dawns fastened the red cows to the beams of light and turned those beams, to which the red cows are fastened, into the pathways of cosmic plasma flow. The beams of light refer to the triply-spun filaments of plasma blooms. These pathways of cosmic plasma flow are called celestial rivers in the Ṛgveda.

RV i.92.2 author गोतमो राहूगणः, to उषाः, metre जगती छंदः
उदपप्तन्नरुणा भानवो वृथा स्वायुजो अरुषीर्गा अयुक्षत ।
अक्रन्नुषासो वयुनानि पूर्वथा रुशंतं भानुमरुषीरशिश्रयुः ॥१.९२.२॥
उत् । अपप्तन् । अरुणाः । भानवः । वृथा । सुऽआयुजः । अरुषीः । गाः । अयुक्षत ।
अक्रन् । उषसः । वयुनानि । पूर्वऽथा । रुशंतं । भानुं । अरुषीः । अशिश्रयुः ॥१.९२.२॥
Ruddy beams of light soared upwards and yoked the herd of well-fastened red cows.
The Dawns fastened the red cows to the bright beams of light and turned the bright beams to the pathways of flow. (RV i.92.2)

उत् <(उद्) ind. particle – up, upwards>, अपप्तन् <aor. 3rd person pl. of √पत् – they soared upwards>, अरुणाः <m. nom. pl. of adj. अरुण – tawny, ruddy>, भनवः <m. nom. pl. of भानु – lights, rays of light>,

वृथा <ind. – at will, at pleasure, easily>, सुआयुजः <f. acc. pl. of adj. सुआयुज् – well-fastened>, अरुषीः <f. acc. pl. of adj. अरुष – red, reddish>, गाः <f. acc. pl. of गो – cows, the herd of cattle>, अयुक्षत <aor. middle. 3rd person pl. of √युज् 7Ā – they harnessed>;

अक्रन् <ipf. 3rd person pl. of √कृ – they (the Dawns) made, turned (with two acc.)>, उषसः <f. nom. pl. of उषस् – the Dawns>, वयुनानि <n. acc. pl. of वयुन – paths (of flow)>, पूर्वथा <ind. – formerly, previously>, रुशंतं <m. acc. sg. of adj. रुशंत् – brilliant, bright>, भानुं <m. acc. sg. of भानु – a ray of light (the triply-spun filament of a plasma bloom)>, अरुषीः <f. acc. pl. of अरुष (f. अरुषी) – reddish cows>, अशिश्रयुः <aor. 3rd person pl. of √श्रि – they caused to lay on, fasten on>.

In RV i.92.3, the Dawns praise the women, yoked together as a team, who bestow beatitude, the supreme blessedness. Discharged cosmic plasmas blooming in the Soma Pond are described as beatitude, the supreme blessedness. The women are the maidens of Tvaṣṭā, that is, the firestones of the Stone. These firestones, or maidens, are yoked together in a circle, forming a Ring of Fire (see Figure 8). Here in the stanza RV i.92.3, the bounteous offerer of Soma sacrifice refers to the Dawn.

RV i.92.3 author गोतमो राहूगणः, to उषाः, metre जगती छंदः
अर्चंति नारीरपसो न विष्टिभिः समानेन योजनेना परावतः ।
इषं वहंतीः सुकृते सुदानवे विश्वेदह यजमानाय सुन्वते ॥१.९२.३॥
अर्चंति । नारीः । अपसः । न । विष्टिऽभिः । समानेन । योजनेन । आ । पराऽवतः ।
इषं । वहंतीः । सुऽकृते । सुऽदानवे । विश्वा । इत् । अह । यजमानाय । सुन्वते ॥१.९२.३॥
They praise the women, yoked as a team, who are skilful in their works of offering beatitude,
Bestowing libation to all the worlds and to the virtuous, bounteous offerer of Soma sacrifice.
(RV i.92.3)

अर्चंति <pre. 3rd person pl. of √अर्च् – they praise>, नारीः <f. acc. pl. of नारी – women (maidens of Tvaṣṭā)>, अपसः <f. acc. pl. of adj. अपस् – skilful in any art>, न <ind. – like, as>, विष्टिभिः <f. inst. pl. of विष्टि – with their services, works>, समानेन <n. inst. sg. of adj. समान – same, one>, योजनेन <n. inst. sg. of योजन – joined or yoked as a team>, आ <ind. adverb – (the degree of skilfulness) fully, really>, परावतः <mn. gen. sg. of adj. परावत् – of offering beatitude>;

इषं <f. acc. sg. of इष् – a draught, libation>, वहंतीः <f. acc. pl. of pre. active pt. वहन्ती (√वह्) – them (the women) procuring, bestowing>, सुकृते <m. dat. sg. of adj. सुकृत् – doing good, virtuous>, सुदानवे <m. dat. sg. of adj. सुदानु – bounteous, munificent>, विश्वा <n. acc. pl. of विश्व – to all the worlds>, इत् <(इद्) ind. particle – indeed, assuredly>, अह <ind. particle of affirmation – surely, certainly>, यजमानाय <m. dat. sg. of adj. यजमान – sacrificing, worshipping, officiating>, सुन्वते <m. dat. sg. of सुन्वत् – to the offerer of Soma sacrifice (here, refers to the Dawn)>.

The Dawn, the Soma Pond, as the morning light fills the udder, by bringing celestial lights to the whole World, drives darkness away (RV i.92.4).

RV i.92.4 author गोतमो राहूगणः, to उषाः, metre जगती छंदः
अधि पेशांसि वपते नृतूरिवापोर्णुते वक्ष उस्रेव बर्जहं ।
ज्योतिर्विश्वस्मै भुवनाय कृण्वती गावो न व्रजं व्यु१◌◌षा आवर्तमः ॥१.९२.४॥
अधि । पेशांसि । वपते । नृतूःऽइव । अप । ऊर्णुते । वक्षः । उस्राऽइव । बर्जहं ।
ज्योतिः । विश्वस्मै । भुवनाय । कृण्वती । गावः । न । व्रजं । वि । उषाः । आवरित्यावः । तमः ॥१.९२.४॥
The Dawn, like a dancer, puts ornaments on her bosom, as the morning light fills the udder.
Bringing light to the whole World, Dawn, you drove the darkness away, as cows to the cowshed.
(RV i.92.4)

अधि <ind. – prefix to verb वपते (√वप्)>, पेशांसि <n. acc. pl. of पेशस् – ornaments, embroidery>, अधि-वपते <pre. middle 3rd person sg. of अधि-√वप् – she (the Dawn) puts on, fastens>, नृतूः <m. nom. sg. of नृतु – a dancer>, इव <ind. – like, in the same manner>, अप <ind. particle – on the outside of (bosom)>, वक्षः <n. acc. sg. of वक्षस् – the bosom, chest>, ऊर्णुते <pre. middle 3rd person sg. of √ऊर्णु – she (the morning light) covers, surrounds>, उस्रा <f. nom. sg. of उस्रा – the morning light (personified as a red cow)>, इव <ind. – like, as>, बर्जहं <m. acc. sg. of बर्जह् – the udder, i.e. the Soma Pond>;

ज्योतिः <n. acc. sg. of ज्योतिस् – light>, विश्वस्मै <n. dat. sg. of adj. विश्व – whole>, भुवनाय <n. dat. sg. of भुवन – to the World, i.e. to our galactic system>, कृण्वती <f. nom. sg. of pre. active pt. कृण्वती (√कृ) – she bringing, placing>, गावः <f. acc. pl. of गो – cattle, cows>, न <ind. – like, as if>, व्रजं <m. acc. sg. of व्रज – to the cowshed>, वि <ind. – away, off>, उषाः <f. nom. sg. of उषस् – Dawn>, आवः <ipf. 2nd person sg. of √अव् – you drove, impelled>, तमः <n. acc. sg. of तमस् – darkness>.

2.5.13 The Celestial Boat Equipped with Oars

In RV x.101.2, the Soma Pond is described as the celestial boat equipped with oars (नौः अरित्रपरणी). In other stanzas, the Soma Pond is also called the well-oared celestial boat (दैवी नौः स्वरित्रा RV x.63.10), hundred-oared boat (शतारित्रा नौः RV i.116.5), animated winged-boat (प्लवः आत्मन्वान् पक्षी RV i.182.5), and the boat of sacrifice (यज्ञिया नौः RV x.44.6). In these stanzas, Soma plants, or plasma blooms, are compared to the oars or the wings of the celestial boat (see Figure 7).

RV x.101.2 is addressed to Viśve Devas (विश्वे देवाः pl.), the priests (ऋत्विजः ṛtvijaḥ pl.) who conduct celestial Yajñas. The author is Budha Saumya (बुधः सौम्यः), 'Sage, the son of Soma'. As noted by Monier-Williams, Budha translates as "a wise or learned man". Saumya is the patronymic of Budha and means "belonging to Soma" or having "the nature and qualities" of Soma. As the son of Soma, Budha also refers to Mecury, the Stone.

RV x.101.2 author बुधः सौम्यः, to विश्वे देवा ऋत्विजो वा, metre त्रिष्टुप् छंदः
मंद्रा कृणुध्वं धिय आ तनुध्वं नावमरित्रपरणीं कृणुध्वं ।
इष्कृणुध्वमायुधारं कृणुध्वं प्रांचं यज्ञं प्र णयता सखायः ॥१०.१०१.२॥
मंद्रा । कृणुध्वं । धियः । आ । तनुध्वं । नावं । अरित्रऽपरणीं । कृणुध्वं ।
इष्कृणुध्वं । आयुधा । अरं । कृणुध्वं । प्रांचं । यज्ञं । प्र । नयत । सखायः ॥१०.१०१.२॥
In deep tone, offer ye hymns and weave. Build ye a boat equipped with oars for transport.
Prepare the weapons, O friends! Bring them to the fore and lead the Yajña forward.
(RV x.101.2)

मंद्रा <mf. inst. sg. of मंद्र – with/in deep, low tone of the voice>, कृणुध्वं <ipv. 2nd person pl. of √कृ – ye make, offer>, धियः <f. acc. pl. of धी – prayers, hymns>, आ-तनुध्वं <ipv. 2nd person pl. of आ-√तन् – ye extend, stretch, spin out, weave>, नावं <f. acc. sg. of नौ – ship, boat>, अरित्रपरणीं <f. acc. sg. of adj. अरित्रपरण – crossing over by means of oars>, कृणुध्वं <ipv. 2nd person pl. of √कृ – ye make, build>;

इष्कृणुध्वं <ipv. 2nd person pl. इष्-√कृ – ye arrange, prepare>, आयुधा <n. acc. pl. – weapons (plasma blooms)>, अरं <ind. adverb – suitably, ready>, कृणुध्वं <ipv. 2nd person pl. of √कृ (with acc. प्रांचं) – ye bring, procure>, प्रांचं <प्राञ्चं m. acc. sg. of adj. प्रांच् – directed forwards, being in front>, यज्ञं <m. acc. sg. of यज्ञ – Yajña>, प्र <ind. – before, forth, forward>, नयत <ipv. 2nd person pl. of √नी – ye lead, conduct>, सखायः <m. voc. pl. – O friends! assistants!>.

2.5.14 The Heads of Sorceresses

In RV i.133.2, plasma blooms, or Soma plants, are described as the heads of sorceresses (शीर्षा यातुमतीनां śīrṣā yātumatīnāṃ). The stanza is addressed to Indra. Indra crushes the heads of sorceresses with his thunderbolts to release celestial lights contained within the blades of Soma plants. In other stanzas, Indra is said to strike the head of Vṛtra, the outermost plasma sheath, the serpent of the Heaven, which defines the shoreline of the Soma Pond, with his thunderbolt (RV i.52.10, RV viii.6.6).

RV i.133.2 author परुच्छेपो दैवोदासिः, to इंद्रः, metre अनुष्टुप् छंदः
अभिव्लग्या चिदद्रिवः शीर्षा यातुमतीनां । छिंधि वटूरिणा पदा महावटूरिणा पदा ॥१.१३३.२॥
अभिऽव्लग्यं । चित् । अद्रिऽवः । शीर्षा । यातुऽमतीनां । छिंधि । वटूरिणा । पदा । महाऽवटूरिणा । पदा ॥१.१३३.२॥
O thou armed with thunderbolts, crushing the sorceresses' heads,
Split them with a big blow, with a mighty big blow. (RV i.133.2)

अभिव्लग्या <ind. pt. of अभि-√व्लग् – catching, seizing, pressing hard>, चित् <(चिद्) ind. particle – even, indeed>, अद्रिवः <m. voc. sg. of अद्रिवत् – O thou armed with stones or thunderbolts>, शीर्षा <(=शीर्षाणि) n. acc. pl. of शीर्षन् – heads>, यातुमतीनां <f. gen. pl. of यातुमती – of sorceresses>;

छिंधि <ipv. 2nd person sg. of √छिद् – you cut off, pierce, split>, वटूरिणा <m. inst. sg. of adj. वटूरिन् – broad, wide>, पदा <m. inst. sg. of पद् – with a fall, fall-down, i.e. with a blow>, महा <(=महेन) n. inst. sg. of adj. मह – mighty>, वटूरिणा पदा <m. inst. sg. of वटूरिन् and पद् – with a big blow>.

2.5.15 An Owl, an Owlet, and the Sorcerer

In RV vii.104.22, the Soma Pond that contains plasma blooms is compared to the sorcerer in the shape an owl or an owlet, to the sorcerer like a hound or a ruddy goose, to the sorcerer like a Garuḍa, the mythical firebird, and to the sorcerer like a vulture. The stanza is a supplication to Indra to slay the wicked sorcerer. The author is Vasiṣṭha (वसिष्ठः), who is 'the owner of the cow of plenty'. The stanza is addressed to Indra.

RV vii.104.22 author वसिष्ठः, to इंद्रः, metre त्रिष्टुप् छंदः
उलूकयातुं शुशुलूकयातुं जहि श्वयातुमुत कोकयातुं ।
सुपर्णयातुमुत गृध्रयातुं दृषदेव प्र मृण रक्ष इंद्र ॥७.१०४.२२॥
उलूकऽयातुं । शुशुलूकऽयातुं । जहि । श्वऽयातुं । उत । कोकऽयातुं ।
सुपर्णऽयातुं । उत । गृध्रऽयातुं । दृषदाऽइव । प्र । मृण । रक्षः । इंद्र ॥७.१०४.२२॥
Slay thou the sorcerer shaped like an owl or an owlet, the sorcerer like a hound or like a ruddy goose, The sorcerer like a Garuḍa, or the sorcerer like a vulture. O Indra, like a millstone, crush thou the wicked sorcerer. (RV vii.104.22)

उलूकयातुं <m. acc. sg. of उलूकयातु – a sorcerer shaped like an owl>, शुशुलूकयातुं <m. acc. sg. of शुशुलूकयातु – a sorcerer shaped like an owlet>, जहि <ipv. 2nd person sg. of √हन् – you slay>, श्वयातुं <m. acc. sg. of श्वयातु – a sorcerer like a dog, hound>, उत <ind. – and, or>, कोकयातुं <m. acc. sg. – a sorcerer like a ruddy goose>;

सुपर्णयातुं <m. acc. sg. of सुपर्णयातुं – a sorcerer like a mythical bird (Garuḍa)>, उत < ind. – and, or>, गृध्रयातुं <m. acc. sg. of गृध्रयातु – a sorcerer like a vulture>, दृषदा <f. nom. sg. of दृषद (=दृषद्) – a millstone, mortar>, इव <ind. – like, in the

same manner as>, प्र-मृण <ipv. 2nd person sg. of प्र-√मृण् – you crush>, रक्षः <n. acc. sg. of रक्षस् – an evil being or demon, i.e. wicked sorcerer>, इंद्र <m. voc. sg. of इंद्र – O Indra>.

2.5.16 The Mighty Thunderbolt of Indra

The bundle of Soma plants, or plasma blooms, shown in Figure 7 is also called the mighty thunderbolt of Indra. Indra uses his thunderbolt to slay Vṛtra (वृत्रः), the plasma sheath. Stanza RV i.52.7 describes how the thunderbolt of Indra was forged by Tvaṣṭā. As the waves of cosmic plasmas and the exhilarating hymns reach Indra and fill the Soma Pond, Tvaṣṭā, the magnetic field, observes it, magnifies the force suitably, and forges Indra's thunderbolt. The stanza is addressed to Indra by Ṛṣi Savya Āṅgirasa (सव्यः आंगिरसः). Savya is "a fire lighted at a person's death", that is, a fire lighted at the moment of ionisation. Thus, Savya is the flame of Agni, the plasma blooms, the eternal flames.

RV i.52.7 author सव्य आंगिरसः, to इंद्रः, metre जगती छंदः
ह्रदं न हि त्वा न्यृषंत्यूर्मयो ब्रह्माणींद्र तव यानि वर्धना ।
त्वष्टा चित्ते युज्यं वावृधे शवस्ततक्ष वज्रमभिभूत्योजसं ॥१.५२.७॥
ह्रदं । न । हि । त्वा । निऽऋषंति । ऊर्मयः । ब्रह्माणि । इंद्र । तव । यानि । वर्धना ।
त्वष्टा । चित् । ते । युज्यं । ववृधे । शवः । ततक्ष । वज्रं । अभिभूतिऽओजसं ॥१.५२.७॥
O Indra, as those exhilarating hymns and the waves of Waters reach you and fill your lake, Tvaṣṭā observes it, strengthens the force suitably, and forges your thunderbolt of overpowering might. (RV i.52.7)

ह्रदं <m. acc. sg. – a lake, pond, i.e. the Soma Pond>, न <ind. – like, as>, हि <ind. particle – for, indeed>, त्वा <m. acc. sg. of 2nd person pron. युष्मद् – you>, न्यृषंति <pre. 3rd person pl. of √न्यृष् – they flow into, fill (acc.)>, ऊर्मयः <m. nom. pl. of ऊर्मि – waves, billows (of celestial Water, i.e. cosmic plasmas)>, ब्रह्माणि <n. nom. pl. of ब्रह्मन् – the sacred hymns>, इंद्र <m. voc. sg. – O Indra>, तव <m. gen. sg. of 2nd person pron. युष्मद् – your>, यानि <n. nom. pl. of pron. यद् – those>, वर्धना <n. nom. pl. of adj. वर्धन – strengthening, animating, exhilarating>;

त्वष्टा (m. nom. sg. of त्वष्टृ – Tvaṣṭā>, चित्ते <pre. middle 3rd person sg. of √चित् – he observes, takes notice of>, ते <m. gen. sg. of युष्मद् – your>, युज्यं <n. acc. sg. of adj. युज्य – equal in rank or power, proper, suitable>, ववृधे <prf. middle 3rd person sg. of √वृध् – he strengthens, augments, increases>, शवः <n. acc. sg. of शवस् – power, force>, ततक्ष <prf. 3rd person sg. of √तक्ष् – he forms, fashions>, वज्रं <m. acc. sg. of वज्र – thunderbolt>, अभिभूत्योजसं <m. acc. sg. of adj. अभिभूत्योजस् – having superior power>.

2.5.17 Rudra

Rudra (रुद्रः) represents the Vṛtra or the Soma Pond as a gatherer and protector of cosmic plasmas. The gathered and blooming cosmic plasmas are called sacrificial grasses or Soma plants. The name of Rudra (रुद्रः), as noted by Monier-Williams, is either derived from √rud (√रुद् to howl, roar) or from √rudh (√रुध् 7P to obstruct, restrain, withhold), which describe Rudra's key attributes. Rudra is filled with cosmic plasmas that sing, roar, and make Speeches constantly; thus, he is a roarer. Rudra, as described in RV vii.46.3 and 4, gathers and protects celestial lights within the confines of his sheath; thus, Rudra represents the positive aspect of Vṛtra, plasma sheath, within which all the bright drops of cosmic plasmas are gathered and protected until cosmic plasmas are matured and ready to be released. He is the protector of the hearth, the Soma Pond, within which the flames of Agni are ablaze. Note, Rudra does not represent Śiva as is commonly believed.

Rudra is the father of the Maruts (RV ii.33.1). He is the tawny Bull (RV ii.33.15) and Tvaṣṭā's Bull in the mansion of the Moon (RV i.84.15). Here the Moon refers to the Soma Pond. Rudra has a strong bow and swiftly flying arrows and is armed with sharp-pointed weapons (RV vii.46.1). Rudra, like Indra, is armed with thunderbolts (RV ii.33.3).

The four stanzas of hymn RV vii.46 describe the representative attributes of Rudra. The author Vasiṣṭha (वसिष्ठः), 'the owner of the cow of plenty', is the Soma Pond personified. Vasiṣṭha's patronymic Maitrāvaruṇi (मैत्रावरुणि) indicates he is a descendant of Mitra and Varuṇa. The hymn is addressed to Rudra.

Stanza RV vii.46.1 explains that Rudra is the owner of his own abode, the Soma Pond, and equipped with sharp-pointed weapons and swift arrows, which refer to plasma blooms, the Soma plants. Bharata is another name for Rudra.

RV vii.46.1 author वसिष्ठः, to रुद्रः, metre जगती छंदः
इमा रुद्राय स्थिरधन्वने गिरः क्षिप्रेषवे देवाय स्वधाव्ने ।
अषाळ्हाय सहमानाय वेधसे तिग्मायुधाय भरता शृणोतु नः ॥७.४६.१॥
इमाः । रुद्राय । स्थिरऽधन्वने । गिरः । क्षिप्रऽइषवे । देवाय । स्वधाऽव्ने ।
अषाळ्हाय । सहमानाय । वेधसे । तिग्मऽआयुधाय । भरत । शृणोतु । नः ॥७.४६.१॥
These songs are for Rudra, the Deva with the strong bow and swift arrows and his own abode, Invincible, O Bharata, victorious, virtuous, with sharp-pointed weapons. May he hear our songs. (RV vii.46.1)

इमाः <f. nom. pl. of pron. इदं – these>, गिरः <f. nom. pl. of गिर् – praises, songs>, रुद्राय <m. dat. sg. of रुद्र – for Rudra>, स्थिरधन्वने <m. dat. sg. of adj. स्थिरधन्वन् – having strong bow>, क्षिप्रेषवे <m. dat. sg. of adj. क्षिप्रेषु – having quick arrows>, देवाय <m. dat. sg. of देव – for the deva (refers to Rudra)>, स्वधात्रे <m. dat. sg. of adj. स्वधावन् – self-positioned, self-powered, having own place or home>;

अषाळ्हाय <(= अषाढाय) m. dat. sg. of adj. अषाढ – invincible>, सहमानाय <m. dat. sg. of adj. सहमान – victorious>, वेधसे <m. dat. sg. of adj. वेधस् – virtuous>, तिग्मायुधाय <m. dat. sg. of adj. तिग्मायुध – having or casting sharp weapons, flames>, भरत <m. voc. sg. of भरत – O Bharata (the name of Rudra)>, शृणोतु <ipv. 3rd person sg. of √श्रु – may he hear>, नः <m. acc. pl. of 1st person pron. अस्मद् – us, i.e. our songs>.

In RV vii.46.2, the Soma Pond is called the country of Avantī (अवंतीः avaṃtī); its inhabitants are plasma blooms, the Soma plants. Dura (दुरः duraḥ), who dispenses medicines, that is, cosmic plasmas, is described as 'salubrious'. Dura (दुरः from √दॄ to break asunder, split open, rend, divide) is the one who breaks the ionic bonds of hydrogen asunder in order to produce cosmic plasmas. Dura is another name for the Stone.

RV vii.46.2 author वसिष्ठः, to रुद्रः, metre जगती छंदः
स हि क्षयेण क्षम्यस्य जन्मनः साम्राज्येन दिव्यस्य चेतति ।
अवन्नवंतीरुप नो दुरश्चरानमीवो रुद्र जासु नो भव ॥७.४६.२॥
सः । हि । क्षयेण । क्षम्यस्य । जन्मनः । सांऽराज्येन । दिव्यस्य । चेतति ।
अवन् । अवंतीः । उप । नः । दुरः । चर । अनमीवः । रुद्र । जासु । नः । भव ॥७.४६.२॥
He, with his supreme power and his abode, cares for celestial people of his land.
Guarding the country of Avantī and its inhabitants, O Rudra, abide thou amid our tribes with our salubrious Dura. (RV vii.46.2)

सः <m. nom. sg. of pron. तद् – he>, हि <ind. particle – surely, indeed>, क्षयेण <m. inst. sg. of क्षय – with abode, dwelling-place, seat>, साम्राज्येन <n. inst. sg. of साम्राज्य – with universal sovereignty, supreme power>, क्षम्यस्य <m. gen. sg. of क्षम्य (=क्षम्) – of "the ground, earth", i.e. of his land or field (the Soma Pond)>, जन्मनः <n. gen. sg. of जन्मन् – creature, being, race, people>, दिव्यस्य <n. gen. sg. of adj. दिव्य – divine, heavenly, celestial>, चेतति <pre. 3rd person sg. of √चित् – he attends to, cares for (with gen.)>;

अवन् <m. nom. sg. of pre. active pt. अवत् (√अव्) – he guarding>, अवंतीः <f. acc. pl. of अवंती (=अवंति) – the country of Avantī (avaṃtī) and its inhabitants, i.e. the Soma Pond (country) and plasma blooms, or Soma plants (its inhabitants)>, उप <ind. – together with>, नः <m. gen. pl. of 1st person pron. अस्मद् – our>, दुरः <m. nom. sg. of दुर (from √दॄ to break asunder, split open, rend, divide) – Dura (who breaks the ionic bonds asunder, i.e. the Stone, the celestial ioniser)>, चर <ipv. 2nd person sg. of √चर् – (with pt.) you continue do, be engage in>, अनमीवः <m. nom. sg. of adj. अनमीव – salubrious, salutary>, रुद्र <m. voc sg. of रुद्र – O Rudra>, जासु <f. loc. pl. of जा – among races, tribes>, नः <m. gen. pl. of अस्मद् – our>, भव <ipv. 2nd person sg. of √भू – you stay, abide>.

In RV vii.46.3, the children and grandchildren refer to the aftergrowths of plasma blooms, or Soma plants, that will continuously spring up after the previously matured ones are severed and released (see section 2.6.3). The gracious One (स्वपिवातः svapivātaḥ) refers to Rudra. In the Stanza, medicines (भेषजा) refer to Soma plants, or plasma blooms. They, the matured plasma blooms, ask Rudra to have his launched arrow pluck them altogether and release them from the Soma Pond, but not to inflict harm on their children and grandchildren who will spring up and grow and bloom until they are fully matured.

RV vii.46.3 author वसिष्ठः, to रुद्रः, metre जगती छंदः
या ते दिद्युदवसृष्टा दिवस्परि क्ष्मया चरति परि सा वृणक्तु नः ।
सहस्रं ते स्वपिवात भेषजा मा नस्तोकेषु तनयेषु रीरिषः ॥७.४६.३॥
या । ते । दिद्युत् । अवऽसृष्टा । दिवः । परि । क्ष्मया । चरति । परि । सा । वृणक्तु । नः ।
सहस्रं । ते । सुऽअपिवात । भेषजा । मा । नः । तोकेषु । तनयेषु । रिरिषः ॥७.४६.३॥
Let your launched arrow, which flies through the field of the Heaven, pluck us altogether.
For you, O gracious One, a thousand Soma plants are gathered. Inflict no harm on our children and grandchildren. (RV vii.46.3)

या <f. nom. sg. of pron. यद् – which>, ते <m. gen. sg. of pron. युष्मद् – your>, दिद्युत् <f. nom. sg. of दिद्युत् – an arrow, missile, thunderbolt>, अवसृष्टा <f. nom. sg. of adj. अवसृष्ट – thrown, launched>, क्ष्मया <f. inst. sg. of क्षम्य – through "the earth", i.e. through the field (the Soma Pond)>, दिवः <m. gen. sg. of दिव् – of the Heaven>, चरति <pre. 3rd person sg. of √चर् – it (arrow) travels through, flies through>, परि <ind. – fully, richly, altogether>, सा <f. nom. sg. of 3rd person pron. तद् – it (arrow)>, वृणक्तु <ipv. 3rd person sg. of √वृज् – let (it) pluck, pull up, gather>, नः <m. acc. pl. of 1st person pron. अस्मद् – us>;

सहस्रं <ind. – a thousand>, ते <m. dat. sg. of 2nd person pron. युष्मद् – for you (Rudra)>, स्वपिवात <m. voc. sg. of स्वपिवात – O Svapivāta, O gracious One (Svapivāta is the name of Rudra, one who understands and means well)>, भेषजा <(=भेषजानि) n. nom. pl. of भेषज – remedies, medicines, i.e. gathered plasma blooms, or Soma plants>, मा <ind. – not>, नः <m. gen. pl. of अस्मद् – our>, तोकेषु <n. loc. pl. of तोक – on children>, तनयेषु <n. loc. pl. of तनय – on grandchildren>, रीरिषः <inj. 2nd person sg. of √रिष् – you harm, destroy>.

In the last stanza, RV vii.46.4, the instigator is the Stone, the celestial ioniser. Note that, in this stanza, a bundle of triply-braided filaments of plasma blooms is compared to the 'net of the instigator' and plasma blooms, or Soma plants, are referred to as sacrificial grasses (see Figure 7). They, the newly sprung aftergrowths, ask Rudra to protect, and not to slay, nor abandon them until they are fully matured.

RV vii.46.4 author वसिष्ठः, to रुद्रः, metre त्रिष्टुप् छंदः
मा नो वधी रुद्र मा परा दा मा ते भूम प्रसितौ हीळितस्य ।
आ नो भज बर्हिषि जीवशंसे यूयं पात स्वस्तिभिः सदा नः ॥७.४६.४॥
मा । नः । वधीः । रुद्र । मा । परा । दाः । मा । ते । भूम । प्रऽसितौ । हीळितस्य ।
आ । नः । भज । बर्हिषि । जीवऽशंसे । यूयं । पात । स्वस्तिऽभिः । सदा । नः ॥७.४६.४॥
Slay us not, nor give us away, O Rudra, nor abandon your wealth gathered in the net of the Instigator. Let us be amidst the life-giving sacrificial grass. Protect us evermore, O ye Devas, with prosperity. (RV vii.46.4)

मा <ind. – not, do not>, नः <m. acc. pl. of अस्मद् – us>, वधीः <sub. 3rd person sg. of √वध् – (with मा) may he not slay>, रुद्र <m. voc. sg. of रुद्र – O Rudra>, मा <ind. – not>, परा <ind. prefix to verb दाः (√दा)>, परा-दाः <sub. middle 3rd person sg. of परा-√दा 3Ā – (with मा) may he not give up>, मा <ind. – not>, ते <m. gen. sg. of 2nd person pron. युष्मद् – your>, भूम <n. acc. sg. of भूमन् – "a being, (pl.) the aggregate of all existing things", "wealth", i.e. gathered blooming plasmas>, प्रसितौ <f. loc. sg. of प्रसिति – in the net>, हीळितस्य <(=हीडितस्य) m. gen. sg. of हीडित – of the instigator or producer (cf. √हीड् to pull, tear; हीड an instigator), i.e. the celestial ioniser that tears the atomic structure apart>;

आ <ind. prefix to verb भज (√भज्)>, नः <m. acc. pl. of अस्मद् – us>, आ-भज <ipv. 2nd person sg. of आ-√भज् – you let or cause to share, partake in>, बर्हिषि <n. loc. sg. of बर्हिस् – among "sacrificial grass", "a bed or layer of Kuśa grass", i.e. among plasma blooms, or Soma plants>, जीवशंसे <n. loc. sg. of adj. जीवशंस – "praised by living beings", vivifying, life-giving>, यूयं <m. voc. pl. of 2nd person pron. युष्मद् – O ye Devas>, पात <ipv. 2nd person pl. of √पा – ye watch, guard, protect>, स्वस्तिभिः <n. inst. pl. of स्वस्ति – with well-beings, prosperity>, सदा <ind. – always, ever>, नः <m. acc. pl. of 1st person pron. अस्मद् – us>.

2.5.18 The Maruts Arrive in Lightning Laden Chariots with Spears

As noted, the Maruts are the Soma plants, or plasma blooms, personified and they are always addressed in the plural. The Maruts produce streaming currents (वाताः pl.) and flashing thunderbolts with their powers (RV i.64.5). They are the roarers (धुनयः pl. RV i.64.5) and the tongues of Agnis (अग्नीनां जिह्वाः RV x 78.3). They are like the exalted horn(s) of oxen (गवां शृंगं उत्तमं RV v.59.3). They impel celestial Waters (RV i.64.5). They possess mysterious forces (सुमायाः RV i.88.1). The Maruts are agitators or shakers (RV i.37.6, RV i.87.3, RV viii.20.5, RV x.78.3) as Indra, Agni, Soma, and Indu are.

The Maruts are well-adorned with bright lances and ruddy flames (RV i.64.8). They wear strong man's rings (वृषखादयः RV i.64.10) or beautiful rings (सुखादयः RV i.87.6). They are the archers (अस्तारः) who hold arrows (RV i.64.10). They are the eternal flames (अनंतशुष्माः) and they are men or heroes (नरः m. nom. pl. of नृ) as Agni and others are. They form a troop or band (गणः RV v.56.1, RV i.64.9; शर्धः RV i.37.1) and are of one mind (RV ii.34.5). As such, they are always addressed in the plural.

They sing beautifully and bring libation to the Devas. And their chariots are lightning laden (RV i.88.1). In RV i.88.1, the lightning laden chariots, spears, and winged horses all refer to Soma plants, or plasma blooms, each representing a different aspect of plasma blooms. The noblest libation is the libation of Soma.

RV i.88.1 author गोतमो राहूगणः, to मरुतः, metre प्रस्तारपंक्तिः छंदः
आ विद्युन्मद्भिर्मरुतः स्वर्कै रथेभिर्यात ऋष्टिमद्भिरश्वपर्णैः । आ वर्षिष्ठया न इषा वयो न पप्तता सुमायाः ॥१.८८.१॥
आ । विद्युन्मत्ऽभिः । मरुतः । सुऽअर्कैः । रथेभिः । यात । ऋष्टिमत्ऽभिः । अश्वऽपर्णैः ।
आ । वर्षिष्ठया । नः । इषा । वयः । न । पप्तत । सुऽमायाः ॥१.८८.१॥
Come hither, Maruts, in your lightning laden chariots, singing delightful songs, with winged horses, armed with spears. Come flying to us like birds with noblest libation, O ye of magic forces. (RV i.88.1)

आ <ind. a prefix to verb यात (√या)>, विद्युन्मद्भिः <m. inst. pl. of adj. विद्युन्मत् – containing lightning>, मरुतः <m. voc. pl. – O Maruts>, स्वर्कैः <m. inst. pl. of adj. स्वर्क – singing beautifully>, रथेभिः <m. inst. pl. of रथ – with chariots>, आ-यात< ipv. 2nd person pl. of आ-√या – ye come near>, ऋष्टिमद्भिः <m. inst. pl. of adj. ऋष्टिमत् – armed with spears>, अश्वपर्णैः <m. inst. pl. of adj. अश्वपर्ण – with horses with wings, with winged horses>;

आ <ind. prefix to verb पप्तत (√पत्)>, वर्षिष्ठया <f. inst. sg. of adj. वर्षिष्ठा – highest, greatest>, नः <m. dat. pl. of अस्मद् – to us>, इषा <f. inst. sg. of इष् – with libation>, वयः <m. nom. pl. of वि – birds>, न <ind. – like>, आ-पप्तत <ipv. 2nd person pl. of आ-√पत् – ye fly towards, come flying>, सुमायाः <f. voc. pl. of सुमाया – O ye of magic forces>.

In RV ii.34.5, the Maruts are asked to come for rapturous Soma mead as swans seeking their nest. Here, the udders of flaming milch cows refer to plasma blooms, or Soma plants.

RV ii.34.5 author गृत्समदः, to मरुतः, metre जगती छंदः
इंधन्वभिर्धेनुभी रप्शदूधभिरध्वस्मभिः पथिभिर्भ्राजदृष्टयः ।
आ हंसासो न स्वसराणि गंतन मधोर्मदाय मरुतः समन्यवः ॥२.३४.५॥
इंधन्वऽभिः । धेनुऽभिः । रप्शदूधऽभिः । अध्वस्मऽभिः । पथिऽभिः । भ्राजत्ऽऋष्टयः ।
आ । हंसासः । न । स्वसराणि । गंतन । मधोः । मदाय । मरुतः । सऽमन्यवः ॥२.३४.५॥
With unveiled flaming milch cows whose udders are filled full, with bright spears, along your paths, Like swans seeking their nests, O Maruts, of one mind, come for an exhilarating drink of Soma mead. (RV ii.34.5)

इंधन्वभिः <m. inst. pl. of adj. इंधन्वन् – flaming>, धेनुभिः <m. inst. pl. of धेनु – with milch cows>, रप्शदूधभिः <m. inst. pl. of adj. रप्शदूधन् – having a full udder>, अध्वस्मभिः <m. inst. pl. of adj. अध्वस्मन् – unveiled>, पथिभिः <m. inst. pl. of पथिन् – along/through paths>, भ्राजदृष्टयः <m. nom. pl. of adj. भ्राजदृष्टि – having bright spears>;

आ <ind. prefix to the verb गंतन (√गं)>, हंसासः <m. nom. pl. of हंस – swans>, न <ind. – as, like>, स्वसराणि <n. acc. pl. of स्वसर – nests>, आ-गंतन <ipv. 2nd person pl. of आ-√गं – ye come>, मधोः <n. gen. sg. of मधु – of Soma mead>, मदाय <m. dat. sg. of मद – for exhilarating or intoxicating drink>, मरुतः <m. voc. pl. of मरुत् – O Maruts>, समन्यवः <m. voc. pl. of adj. समन्यु – having the same mind>.

2.5.19 The Maruts Make Streaming Currents and Flashing Thunderbolts

Stanzas RV i.64.5 and 7 and RV x.78.3 describe characteristic attributes of the Maruts. The Maruts milk celestial udders and cause the field, the Soma Pond, to overflow with milk, or cosmic plasmas (RV i.64.5). In the context of the stanza, bhūmi (भूमि), the ground or field, refers to the Soma Pond, though it could be the Milky Way. The celestial udders in the plural refer to Soma plants, or plasma blooms.

RV i.64.5 author नोधा गौतमः, to मरुतः, metre जगती छंदः
ईशानकृतो धुनयो रिशादसो वातान्विद्युतस्तविषीभिरक्रत ।
दुहंत्यूधर्दिव्यानि धूतयो भूमिं पिन्वंति पयसा परिज्रयः ॥१.६४.५॥
ईशानऽकृतः । धुनयः । रिशादसः । वातान् । विऽद्युतः । तविषीभिः । अक्रत ।
दुहंति । ऊधः । दिव्यानि । धूतयः । भूमिं । पिन्वंति । पयसा । परिऽज्रयः ॥१.६४.५॥
Competent masters, roarers, devourers of the foe, they produced the streaming currents and flashing thunderbolts with their powers. The shakers, running around, milk the celestial udders and cause the field to overflow with milk. (RV i.64.5)

ईशानकृतः <m. nom. pl. of ईशानकृत् – competent masters, lords>, धुनयः <m. nom. pl. of धुनि – roarers>, रिशादसः <m. nom. pl. of रिशादस् – devourers of enemies>, वातान् <m. acc. pl. of वात – streaming currents>, विद्युतः <f. acc. pl. of विद्युत् – lightnings, flashing thunderbolts>, तविषीभिः <f. inst. pl. of तविषी – with powers>, अक्रत <ipf. 3rd person pl. of √कृ – they produced, made>;

दुहंति <pre. 3rd person pl. of √दुह् – they milk, extract>, ऊधः <n. acc. pl. of उधस् = ऊधस् – udders>, दिव्यानि <n. acc. pl. of adj. दिव्य – celestial>, धूतयः <m. nom. pl. of धूति – shakers, agitators>, भूमिं <f. acc. sg. of भूमि – ground, field (here, it refers to the Soma Pond)>, पिन्वंति <pre. 3rd person pl. of √पिन्व् – they cause to swell, overflow>, पयसा <n. inst. sg. of पयस् – with milk, water, i.e. with cosmic plasmas>, परिज्रयः <m. nom. pl. of adj. परिज्रि – running around>.

RV i.64.7 author नोधा गौतमः, to मरुतः, metre जगती छंदः
महिषासो मायिनश्चित्रभानवो गिरयो न स्वतवसो रघुष्यदः ।
मृगा इव हस्तिनः खादथा वना यदारुणीषु तविषीरयुग्ध्वं ॥१.६४.७॥
महिषासः । मायिनः । चित्रऽभानवः । गिरयः । न । स्वऽतवसः । रघुऽस्यदः ।
मृगाःऽइव । हस्तिनः । खादथ । वना । यत् । आरुणीषु । तविषीः । अयुग्ध्वं ॥१.६४.७॥
Mighty, skilled in the art of magic, shining with light, strong like mountains, moving swiftly,
Like the wild beasts with dexterous hands, ye devour the forests when ye assume your powers upon the ruddy mares. (RV i.64.7)

महिषासः <m. nom. pl. of adj. महिष – mighty>, मायिनः <m. nom. pl. of adj. मायिन् – skilled in magic art>, चित्रभानवः <m. nom. pl. of adj. चित्रभानु – shining with light>, गिरयः <m. nom. pl. of गिरि – mountains>, न <ind. – like>, स्वतवसः <m. nom. pl. of adj. स्वतवस् – self-strong, inherently powerful, firmly rooted>, रघुष्यदः <m. nom. pl. of adj. रघुष्यद् – moving swiftly>;

मृगाः <m. nom. pl. of मृग – forest animals, wild beasts>, इव <ind. – like, in the same manner as>, हस्तिनः <m. nom. pl. of adj. हस्तिन् – dexterous with the hands, limbs>, खादथ <pre. 2nd person pl. of √खाद् – ye eat, devour, feed>, वना <n. acc. pl. of वन – forests, trees>, यत् <conjunctive – that, when>, आरुणीषु <f. loc. pl. of आरुणी – on, among the ruddy mares, i.e. among the red flames>, तविषीः <f. acc. pl. of तविषी – powers, strengths>, अयुग्ध्वं <aor. middle 2nd person pl. of √युज् – ye set to work, employ, apply>.

2.5.20 The Maruts Move Swiftly Like Vātas and Are Effulgent Like Tongues of Fires

In stanza x.78.3, the Maruts are likened to Vātas (वातासः pl.), to the tongues of fires, and to the warriors clad in armour. They are also rich in oblations.

RV x.78.3 author स्यूमरश्मि भार्गवः, to मरुतः, metre त्रिष्टुप् छंदः
वातासो न ये धुनयो जिगत्नवोऽग्नीनां न जिह्वा विरोकिणः ।
वर्मण्वंतो न योधाः शिमीवंतः पितॄणां न शंसाः सुरातयः ॥१०.७८.३॥
वातासः । न । ये । धुनयः । जिगत्नवः । अग्नीनां । न । जिह्वाः । विऽरोकिणः ।
वर्मण्ऽवंतः । न । योधाः । शिमीऽवंतः । पितॄणां । न । शंसाः । सुऽरातयः ॥१०.७८.३॥
Roaring, swiftly moving like Vātas, effulgent as tongues of fires,
Mighty are they as warriors clad in armour, rich in oblations like the fathers' invocations.
(RV x.78.3)

वातासः <m. nom. pl. of वात – Vātas>, न <ind. – as, like>, ये <m. nom. pl. of pron. यद् – those who, who>, धुनयः <m. nom. pl. of adj. धुनि – roaring, boisterous>, जिगत्नवः <m. nom. pl. of adj. जिगत्नु – moving swiftly>, न <ind. – as, like>, अग्नीनां <m. gen. pl. of अग्नि – of fires>, जिह्वाः <m. nom. pl. of जिह्व – tongues>, विरोकिणः <m. nom. pl. of adj. विरोकिन् – shining, radiant>;

वर्मण्वंतः <m. nom. pl. of adj. वर्मण्वत् (=वर्मवत्) – having armour or a coat of mail>, न <ind. – like, as>, योधाः <m. nom. pl. of योध – warriors>, शिमीवंतः <m. nom. pl. of adj. शिमीवत् – mighty, strong>, न <ind. – like, as>, पितॄणां <m. gen. pl. of पितृ – of fathers>, शंसाः <m. nom. pl. of शंस – invocations, praise, blessings>, सुरातयः <m. nom. pl. of adj. सुराति – rich in gifts, oblations (cf. राति grace, gift, oblation)>.

In the Ṛgveda, Father (पिता pitā), in the singular, generally refers to the Heaven. Fathers, (पितरः), in the plural, refer to forefathers or deceased ancestors. The forefathers or ancestors are the reborn (read ionised) Agnis and Somas after losing their bodies, that is, after the ionic bonds were broken, through the cosmic ionisation process. Thus, the fathers' invocations allude to the hymns, the Speeches, the oscillations generated by cosmic plasmas discharged earlier.

2.5.21 The Maruts Are Spirited Like Serpents

In stanza RV i.64.8, the Maruts are described as spirited like serpents. With lances and dappled mares the Maruts impel the darkness.

RV i.64.8 author नोधा गौतमः, to मरुतः, metre जगती छंदः
सिंहा इव नानदति प्रचेतसः पिशा इव सुपिशो विश्ववेदसः ।

क्षपो जिन्वंतः पृषतीभिर्ऋष्टिभिः समित्सबाधः शवसाहिमन्यवः ॥१.६४.८॥
सिंहाःऽइव । नानदति । प्रऽचेतसः । पिशाःऽइव । सुऽपिशः । विश्वऽवेदसः ।
क्षपः । जिन्वंतः । पृषतीभिः । ऋष्टिऽभिः । सं । इत् । सऽबाधः । शवसा । अहिऽमन्यवः ॥१.६४.८॥
They roar like lions, wise, all-knowing, beautifully adorned like tawny deer,
With lances and dappled mares, spirited like serpents, together, they impel the darkness with their might. (RV i.64.8)

सिंहाः <m. nom. pl. of सिंह – lions>, इव <ind. – in the same manner as, like>, नानदति <int. 3rd person pl. of √नद् – they thunder, roar>, प्रचेतसः <m. nom. pl. of adj. प्रचेतस् – attentive, mindful, wise>, पिशाः <m. nom. pl. of पिश – deer, tawny or red deer>, इव <ind. – in the same manner as, like>, सुपिशः <m. nom. pl. of adj. सुपिश् – beautifully adorned>, विश्ववेदसः <m. nom. pl. of adj. विश्ववेदस् (=विश्वविद्) – knowing everything, omniscient>;

क्षपः <f. acc. pl. of क्षप् – nights, darkness>, जिन्वंतः <m. nom. pl. of pre. active pt. जिन्वत् (√जिन्व्) – they impel, incite, cause to move quickly>, पृषतीभिः <f. inst. pl. of पृषती – with dappled cows or mares>, ऋष्टिभिः <f. inst. pl. of ऋष्टि – with lances, spears>, सं <ind. – with, together>, इत् <(इद्) ind. – indeed (here, added to सं to express emphasis)>, सबाधः <(सबाधस्) ind. – urgently, eagerly>, शवसा <ind. – with power, might>, अहिमन्यवः <m. nom. pl. of adj. अहिमन्यु – spirited like serpents>.

2.5.22 The Maruts, the Eternal Flames

In RV i.64.10, the Maruts are called the archers who held arrows in their hands. They are also called the eternal flames. The strong man's rings the Maruts are wearing refer to plasma sheaths. According to the Ṛgveda, plasma blooms, or Soma plants, are individual plasma sheaths. The arrows refer to the plasma blooms.

RV i.64.10 author नोधा गौतमः, to मरुतः, metre जगती छंदः
विश्ववेदसो रयिभिः समोकसः संमिश्लासस्तविषीभिर्विरप्शिनः ।
अस्तार इषुं दधिरे गभस्त्योरनंतशुष्मा वृषखादयो नरः ॥१.६४.१०॥
विश्वऽवेदसः । रयिऽभिः । संऽओकसः । संऽमिश्लासः । तविषीभिः । विऽरप्शिनः ।
अस्तारः । इषुं । दधिरे । गभस्त्योः । अनंतऽशुष्माः । वृषऽखादयः । नरः ॥१.६४.१०॥
All-knowing, possessed of wealth, endowed with mighty powers, exuberant,
The archers, eternal flames, heroes, wearing the strong man's rings, held the arrows in their hands. (RV i.64.10)

विश्ववेदसः <m. nom. pl. of adj. विश्ववेदस् (=विश्वविद्) – all-knowing, omniscient>, रयिभिः <m. inst. pl of रयि – with treasures, wealth>, समोकसः <m nom. pl. of adj. समोकस् – furnished with, possessed of (inst.)>, संमिश्लासः <m. nom. pl. of adj. संमिश्ल = संमिश्र – furnished or endowed with>, तविषीभिः <f. inst. pl. of तविषी – with powers, strength>, विरप्शिनः <m. nom. pl. of adj. विरप्शिन् – exuberant, powerful, mighty>;

अस्तारः <m. nom. pl. of अस्तृ – archers>, इषुं <mf. acc. sg. of इषु – a constellation (of plasma blooms), an arrow>, दधिरे <prf. 3rd person pl. of √धा – they put, held>, गभस्त्योः <m. loc. du. of गभस्ति – in the two hands>, अनंतशुष्माः <m. nom. pl. of अनंतशुष्म (अनंत adj. eternal, boundless + शुष्माः m. nom. pl. flames) – eternal flames>, वृषखादयः <m. nom. pl. of adj. वृषखादि (वृष a bull, strong man, the chief of a class + खादि the ring worn on the hands or feet by the Maruts) – having strong man's rings (as the Maruts); the strong man's rings refer to plasma sheaths>, नरः <m. nom. pl. of नृ – heroes>.

2.5.23 Triṣṭup, a Libation for the Maruts

In RV viii.7.1, Triṣṭup, a metre of four pādas of eleven syllables each, is described as a libation for the Maruts. As noted earlier, in the Ṛgveda, the Speeches or hymns, the frequencies of oscillations generated by cosmic plasmas, are numerically coded as metres of the stanzas. In this stanza, it is stated that when the singer pours out the Triṣṭup, the Maruts shine forth.

In the Ṛgveda, all the Devas sing. However, the Stone, the celestial ioniser, is the most prominent singer. Therefore, it can be understood that the singer refers to the Stone, the celestial ioniser, who sings and pours out libations, that is, cosmic plasmas. The mountains in the plural refer to the individual plasma blooms while the mountain in the singular refers to the Soma Pond.

RV viii.7.1 author पुनर्वत्सः काण्वः, to मरुतः, metre गायत्री छंदः
प्र यद्वस्त्रिष्टुभमिषं मरुतो विप्रो अक्षरत् । वि पर्वतेषु राजथ ॥८.७.१॥
प्र । यत् । वः । त्रिऽस्तुभं । इषं । मरुतः । विप्रः । अक्षरत् । वि । पर्वतेषु । राजथ ॥८.७.१॥
O Maruts, when the singer poured out the Triṣṭup as a libation for you,
Ye shine forth amid the mountains. (RV viii.7.1)

प्र <ind. prefix to verb अक्षरत् (√क्षर्)>, यत् <(यद्) conjunction – when>, वः <m. dat. pl. of 2nd person pron. युष्मद् – for you (pl.)>, त्रिष्टुभं <f. acc. sg. of त्रिष्टुभ् – triṣṭup (a metre of 4 pādas of 11 syllables each)>, इषं <f. acc. sg. of इष् – a draught, libation>, मरुतः <m. voc. pl. of मरुत् – O Maruts>, विप्रः <m. nom. sg. of विप्र – a seer, singer (refers to the Stone)>, प्र-अक्षरत् <ipf. 3rd person sg. of प्र-√क्षर् – he (the singer) poured out, gave forth a stream>;

वि <ind. prefix to verb राजथ (√राज्)>, पर्वतेषु <m. loc. pl. of पर्वत – amid mountains, i.e. amid plasma blooms>, वि-राजथ <pre. 2nd person pl. of वि-√राज् – ye shine forth>.

2.5.24 Vāyu and Vāta

The word Vāta (वात), as an adjective, means "blown". For Vāyu (वायुः), 'vā (वा)' means "to bestow anything (acc.) by blowing"; 'yu (यु)', as an adjective, means "going, moving". So it is reasonable to interpret Vatā as ionised particles 'blown' out of the celestial ioniser and Vāyu as the 'blower', the force behind the streaming charged particles, the cosmic plasma currents. Vāyu and Vāta are both 'cosmic wind', the draught of forces and charged cosmic plasma particles. In RV viii.26.21 and 22, Vāyu is called Tvaṣṭā's son-in-law (जामाता), indicating that Vāyu is associated with the impelling force of the magnetic field.

Though Vāyu is connected with Indra and addressed as Indra-Vāyu in many hymns of the Ṛgveda (RV i.2, i.135, iv.46, iv.47, vii.90, vii.91), hymn RV x.168 reveals that Vāyu and Vāta are the phenomena associated with the draughts of Somas and Agnis. The four stanzas of hymn RV x.168, though addressed to Vāyu, have both Vāyu and Vāta appear together. Vāta and Vāyu are treated as different aspects of the same phenomenon.

In RV x.168.1, Vāyu is described as the majestic chariot of Vāta. Note the Stone, the celestial ioniser, is located at the base of the Heavenly Vault (see Figure 6). As Vāyu, the majestic chariot of Vāta, makes its appearance out of the Stone, the celestial ioniser, it touches the Heaven. As it advances, it produces the beams of light, that is, the plasma blooms, or Soma plants, and casts the ionised and charged particles. And the sound of the majestic chariot reverberates. In the stanza, 'reṇu (रेणुः)' is a freed, ionised particle. Note 'reṇu (रेणु)' is derived from √ri (√रि), as noted by Monier-Williams, which means "to release, set free" or "to be shattered or dissolved". So, the reṇu is the ionised and charged particle freed from the atomic structure of hydrogen.

RV x.168.1 author अनिलो वातायनः, to वायुः, metre त्रिष्टुप् छंदः
वातस्य नु महिमानं रथस्य रुजन्नेति स्तनयन्नस्य घोषः।
दिविस्पृग्यात्यरुणानि कृण्वन्नुतो एति पृथिव्या रेणुमस्यन् ॥१०.१६८.१॥
वातस्य । नु । महिमानं । रथस्य । रुजन् । एति । स्तनयन् । अस्य । घोषः ।
दिविऽस्पृक् । याति । अरुणानि । कृण्वन् । उतो इति । एति । पृथिव्या । रेणुं । अस्यन् ॥१०.१६८.१॥
As it breaks open and makes its appearance, the sound of the majestic Vāta's chariot reverberates. Reaching the Heaven, it marches, producing the ruddy beams of light, casting ionised particles with cosmic dust. (RV x.168.1)

वातस्य <m. gen. sg of वात – of Vāta>, नु <ind. – now, so now, now then>, महिमानं <n. nom. sg. of pre. middle pt. महिमान (√मह्) – it (chariot) being mighty, glorious, majestic>, रथस्य <m. gen. sg. of रथ – of the chariot>, रुजन् <m. nom. sg. of pre. active pt. of रुजत् (√रुज्) – it breaks open, dashes>, एति <pre. 3rd person sg. of √इ – it makes its apprearance, arises>, स्तनयन् <m. nom. sg. of pre. active cau. pt. स्तनयत् (√स्तन्) – it resounds, reverberates>, अस्य <m. gen. sg. of pron. इदं– of this>, घोषः <m. nom. sg. of घोष – the sound, roar>;

दिविस्पृक् <m. nom. sg. of adj. दिविस्पृश् – touching the Heaven>, याति<pre. 3rd person sg. of √या – it proceeds, marches, travels>, अरुणानि <n. acc. pl. of अरुण – red, ruddy (beams of light)>, कृण्वन् <m. nom. sg. of pre. active pt. कृण्वत् (√कृ) – it producing>, उतो <(=उत+उ) ind. – and also>, एति <pre. 3rd person sg. of √इ – it goes on with or continues (with a pt.)>, पृथिव्या <f. inst. sg. of पृथिवी – with "earth regarded as one of the elements", i.e. with atoms and molecules, or cosmic dust>, रेणुं <m. acc. sg. of रेणु (from √रि to release, set free) – freed, ionised particles>, अस्यन् <m. nom. sg. of pre. active pt. अस्यत् (√अस् 4P) – it throwing, casting>.

In RV x.168.2, 'they' refers to the ruddy beams of light, the blooming cosmic plasmas. The Vāta's places refer to plasma blooms, or Soma plants. The purpose of the assembly is to prepare for the Soma sacrifice, which is the event of releasing the cosmic plasmas gathered and matured in the Soma Pond.

RV x.168.2 author अनिलो वातायनः, to वायुः, metre त्रिष्टुप् छंदः
सं प्रेरते अनु वातस्य विष्ठा ऐनं गच्छंति समनं न योषाः ।
ताभिः सयुक्सरथं देव ईयतेऽस्य विश्वस्य भुवनस्य राजा ॥१०.१६८.२॥
सं । प्र । ईरते । अनु । वातस्य । विऽस्थाः । आ । एनं । गच्छंति । समनं । न । योषाः ।
ताभिः । सऽयुक् । सऽरथं । देवः । ईयते । अस्य । विश्वस्य । भुवनस्य । राजा ॥१०.१६८.२॥
They arise together to the Vāta's places, and come to this assembly as morning lights.
With them, harnessed to the same chariot, Vāyu, the Deva, the Monarch of the whole World, speeds forth. (RV x.168.2)

सं <ind. – with, together>, प्रेरते <pre. middle 3rd person pl. of √प्रेर् (प्र-√ईर्) – they arise, appear>, अनु <ind. adverb – after, then, thereupon>, वातस्य <m. gen. sg. of वात – of Vāta>, विष्ठाः <f. acc. pl. of विष्ठा – places, stations>, आ <ind. prefix to verb गच्छन्ति (√गं)>, एनं <n. acc. sg. of pronominal base एन – this, that>, आ-गच्छन्ति <pre. 3rd person pl. of आ-√गं – they come>, समनं <n. acc. sg. of समन – to an assembly, festival>, न <ind. – as>, योषाः <f. nom. pl. of योषा – dames, "especially applied to उषस्", the Dawns, i.e. morning lights, first lights>;

ताभिः <f. inst. pl. of pron. तद् – with them>, सयुक् <m. nom. sg. of adj. सयुज् – yoked, joined, harnessed>, सरथं <ind. – on the same chariot>, देवः <m. nom. sg. of देव – the Deva (Vāyu)>, ईयते <pre. middle 3rd person sg. of √इ – he moves, runs quickly>, अस्य <n. gen. sg. of pron. इदं – of this>, विश्वस्य <n. gen. sg. of adj. विश्व – whole, entire>, भुवनस्य <n. gen. sg. – of the World, of the whole galactic system>, राजा <m. nom. sg. of राजन् – Monarch, King>.

Vāyu, the companion of Waters, once released from the Soma Pond, travels on the paths in the Midheaven (RV x.168.3). Imagine a fleet of plasma sheaths, with Vāyu and Vāta on-board, navigating through the Midheaven, energising the Milky Way.

RV x.168.3 author अनिलो वातायनः, to वायुः, metre त्रिष्टुप् छंदः
अंतरिक्षे पथिभिरीयमानो न नि विशते कतमच्चनाहः ।
अपां सखा प्रथमजा ऋतावा क्व स्विज्जातः कुत आ बभूव ॥१०.१६८.३॥
अंतरिक्षे । पथिऽभिः । ईयमानः । न । नि । विशते । कतमत् । चन । अहरिति ।
अपां । सखा । प्रथमऽजाः । ऋतऽवा । क्व । स्वित् । जातः । कुतः । आ । बभूव ॥१०.१६८.३॥
Travelling on the paths in the Midheaven, not even a single day he rests.
Born foremost, a performer of sacred works, the companion of Waters. Where was he born?
From where originated he? (RV x.168.3)

अंतरिक्षे <n. loc. sg. of अंतरिक्ष – in the Innerfield, the Midheaven>, पथिभिः <m. inst. pl. of पथिन् – through/on the paths>, ईयमानः <m. nom. sg. of pre. middle. pt. ईयमान (√इ) – he travelling>, न <ind. – not>, नि-विशते <pre. 3rd person sg. of नि-√विश् – he rests, settles down>, कतमत् <n. nom. sg. of कतम – (with negative and चन) not even one>, चन <ind. – not even>, अहः <n. nom. sg. of अहस् – a day>;

अपां <f. gen. pl. of अप् – of Waters, i.e. of cosmic plasmas>, सखा <(=सखि) f. nom. sg. – a companion>, प्रथमजाः <m. nom. sg. of adj. प्रथमजा – firstborn, born foremost>, ऋतावा <m. nom. sg. of ऋतावन् – a performer of sacred works, faithful, holy>, क्व स्वित् <(क्व स्विद्) ind. – in what place, where>, जातः <m. nom. sg. of adj. जात – born>, कुतः <(कुतस्) ind. – from where>, आ-बभूव <prf. 3rd person sg. of आ-√भू – he has originated>.

Though RV x.168.4 is addresssed to Vāyu, Vāta is praised in the stanza. Vāta is the germ of the World. Here, in RV x.168.4, Vāta is called the ātmā (आत्मा), the soul, of the Devas whereas Soma is Brahmā (ब्रह्मा) of the Devas (RV ix.96.6). Brahmā is the Creator; ātmā, the individuated soul or self.

RV x.168.4 author अनिलो वातायनः, to वायुः, metre त्रिष्टुप् छंदः
आत्मा देवानां भुवनस्य गर्भो यथावशं चरति देव एषः ।
घोषा इदस्य शृण्विरे न रूपं तस्मै वाताय हविषा विधेम ॥१०.१६८.४॥
आत्मा । देवानां । भुवनस्य । गर्भः । यथाऽवशं । चरति । देवः । एषः ।
घोषाः । इत् । अस्य । शृण्विरे । न । रूपं । तस्मै । वाताय । हविषा । विधेम ॥१०.१६८.४॥
The soul of the Devas, the germ of the World, this Deva moves about as his will inclines him.
His roars are heard, as his handsome form is seen. May we worship this Vāta with our oblation.
(RV x.168.4)

आत्मा <m. nom. sg. of आत्मन् – ātmā, the individuated soul, self>, देवानां <m. gen. pl. of देव – of the Devas>, भुवनस्य <n. gen. sg. of भुवन – of the World, i.e. of the galatic system>, गर्भः <m. nom. sg. of गर्भ – embryo, germ>, यथावशं <ind. – according to inclination>, चरति <pre. 3rd person sg. of √चर् – he moves about>, देवः <m. nom. sg. देव – the Deva>, एषः <m. nom. sg. of pron. एतद् – this>;

घोषाः <m. nom. pl. of घोष – roars, sounds, thunders>, इत् <(इद्) ind. particle of affirmation – indeed>, अस्य <m. gen. sg. of इदं – of this, his>, शृण्विरे <pre. middle 3rd person pl. of √श्रु – they are heard>, न <ind. – like, as>, रूपं <n. nom. sg. of रूप – handsome form, splendour; here, the verb दृश्यते (pre. passive 3rd person sg. of √दृश्) is omitted for the governing of metre>, तस्मै <m. dat. sg. of pron. तद् – to this>, वाताय <m. dat. sg. of वात – to/for Vāta>, हविषा <n. inst. sg. of हविस् – with oblation>, विधेम <opt. 1st person pl. of √विध् – may we worship, honour (with dat.)>.

2.5.25 Indra Slays the Serpent of the Heaven

Indra (इंद्रः) is most prominently known for slaying Vṛtra, the Serpent of the Heaven, in order to release the celestial lights, the cosmic plasmas, gathered in the Soma Pond. Indra is described as 'thunderer' or 'wielder of thunderbolt' (वज्री m. nom. sg. of वज्रिन् RV i.7.2, RV vi.20.7). He slays Vṛtra with his thunderbolt (वज्रः RV i.52.7, RV viii.6.6, RV x.113.5; अशनिः RV x.87.4). Indra's thunderbolt is a hundred-limbed (वज्रं शतपर्व RV viii.6.6) or a thousand-spiked (वज्रः सहस्रभृष्टिः RV i.80.12). Indra's weapon is also described as a long hook (दीर्घः अंकुशः RV viii.17.10, RV x.134.6). (A long hook indeed. It is more or less 20,000–25,000 light-years in length.) 'A hundred-limbed' and 'a thousand-spike' are the descriptions of plasma blooms, Soma plants, the blooming cosmic plasmas.

Slaying Vṛtra, the Serpent of the Heaven, and releasing the blooming cosmic plasmas are the heroic deeds of Indra (RV viii.93.1) and of Aśvinā (RV x.61.5). The Soma Pond is Indra's Belly (कुक्षिः RV i.8.7, RV iii.36.8; जठरः RV iii.42.5) and Soma plants, or plasma blooms, are his ruddy beard (हरिश्मशारुः RV x.96.8). Indra has beautiful jaws (सुशिप्र adj. RV v.36.5) or two jaws (शिप्रे f. du. RV v.36.2). Beautiful jaws refer to the firestones of the Stone, the celestial ioniser, and the two jaws refer to the two discs of the Stone.

Indra delights in drinking Soma (RV viii.32.1, 5, 7, 8) or Indu (इंदुः iṃduḥ), the bright drop of Soma. Indra is also called the Bull (वृषभः RV viii.93.1, वृषा RV v.36.5) as Agni and Soma are. Note, in RV v.58.6, the Heavenly Vault is called the Red Bull (उस्रियः वृषभः). Stanzas RV viii.93.1-3, which are addressed to Indra, describes his prominent role in releasing cosmic plasmas by slaying Vṛtra, Ahi (अहिः), the Serpent of the Heaven.

RV viii.93.1 author सुकक्षः, to इंद्रः, metre गायत्री छंदः
उद्घेदभि श्रुतामघं वृषभं नर्यापसं । अस्तारमेषि सूर्य ॥८.९३.१॥
उत् । घ । इत् । अभि । श्रुतऽमघं । वृषभं । नर्यऽअपसं । अस्तारं । एषि । सूर्य ॥८.९३.१॥
To the bull, the possessor of renowned treasures, the performer of heroic deeds,
The thrower of the thunderbolt, O Sūrya, you have come. (RV viii.93.1)

उत् <ind. particle of deliberation>, घ <ind. – (proceeded by the other particle उत्) verily, indeed>, इत् <(इद्) particle of affirmation – just>, अभि <ind. prefix to verb एषि (√इ) – to, towards, on, upon>, श्रुतामघं <m. acc. sg. of adj. श्रुतामघ –

having renowned treasures>, वृषभं <m. acc. sg. of वृषभ – the bull, i.e. Indra>, नर्यापसं <m. acc. sg. of adj. नर्यापस् – performing manly or heroic deeds (releasing cosmic plasma by slaying Vṛtra)>;

अस्तारं <m. acc. sg. of अस्तृ – to the thrower, shooter (of thunderbolt)>, अभि-एषि <pre. 2nd person sg. of अभि-√इ – you come near, reach, obtain>, सूर्य <m. voc. sg. of सूर्य – O Sūrya>.

In stanza RV viii.93.2, the ninety-nine fortresses refer to plasma blooms, or Soma plants. Ahi (अहिः), the serpent of the Heaven, is another name for Vṛtra, the outermost plasma sheath, which confines cosmic plasmas within the Soma Pond.

RV viii.93.2 author सुकक्षः, to इंद्रः, metre गायत्री छंदः
नव यो नवतिं पुरो बिभेद बाह्वोजसा । अहिं च वृत्रहावधीत् ॥८.९३.२॥
नव । यः । नवतिं । पुरः । बिभेद । बाहुऽओजसा । अहिं । च । वृत्रऽहा । अवधीत् ॥८.९३.२॥
He, who broke nine-and-ninety fortresses down with the strength of his arms,
The slayer of Vṛtra, just smote Ahi dead. (RV viii.93.2)

नव <f. acc. pl. of नवन् – nine>, नवतिं <f. acc. sg. of नवति – ninety>, यः <m. nom. sg. of relative pron. यद् – he who>, पुरः <f. acc. pl. of पुर् – ramparts, fortresses>, बिभेद <prf. 3rd person sg. of √भिद् – he broke down>, बाह्वोजसा <n. inst. sg. of बाह्वोजस् – with the strength of arm>;

अहिं <m. acc. sg. of अहि – Ahi, the serpent of the Heaven>, च <ind. – and, more over>, वृत्रहा <m. nom. sg. of वृत्रहन् – a slayer of Vṛtra>, अवधीत् <aor. 3rd person sg. of √वध् – he just killed, slayed>.

In RV viii.93.3, 'horses' refers to the forces behind the currents of cosmic plasmas; 'cattle', to the celestial lights, the cosmic plasmas; and 'grain', to hydrogen atoms.

RV viii.93.3 author सुकक्षः, to इंद्रः, metre गायत्री छंदः
स न इंद्रः शिवः सखाश्वावद्गोमद्यवमत् । उरुधारेव दोहते ॥८.९३.३॥
सः । नः । इंद्रः । शिवः । सखा । अश्वऽवत् । गोऽमत् । यवऽमत् । उरुधाराऽइव । दोहते ॥८.९३.३॥
He, Indra, our auspicious companion, yields mighty streams
Rich in horses, cattle, and grain. (RV viii.93.3)

सः <m. nom. sg. of 3rd person pron. तद् – he>, नः <m. gen. pl. of 1st person pron. अस्मद् – our>, इंद्रः <m. nom. sg. of इंद्र – Indra>, शिवः <m. nom. sg. of adj. शिव – auspicious>, सखा <m. nom. sg. of strong case सखि – a friend, assistant, companion>, अश्ववत् <n. acc. sg. of adj. अश्ववत् – (each stream) rich in horses>, गोमत् <n. acc. sg. of adj. गोमत् – containing cattle>, यवमत् <n. acc. sg. of adj. यवमत् – containing barley, mixed with barley, abundance of grain>;

उरुधारा <n. acc. pl. of उरुधार (उरु great, large, excellent + धारा n. acc. pl. of धार streams or gushes) – mighty streams, gushes>, इव <ind. – just, indeed, like>, दोहते <pre. middle 3rd person sg. of √दुह् – he yields, gives (acc.)>.

2.5.26 Indra Delights in Soma Drinking

Indra, in his wild delight of Soma, performed the noble deed of slaying Vṛtra (RV viii.32.1). In RV viii.32.1, Indra is described as the receiver of the third press of Soma (ऋजीषी Ṛjīṣī). Cosmic plasmas are discharged and released three times a day; thus, the third press of Soma is the last batch of the day.

RV viii.32.1 author मेधातिथिः काण्वः, to इंद्रः, metre गायत्री छंदः
प्र कृतान्यृजीषिणः कण्वा इंद्रस्य गाथया । मदे सोमस्य वोचत ॥८.३२.१॥
प्र । कृतानि । ऋजीषिणः । कण्वाः । इंद्रस्य । गाथया । मदे । सोमस्य । वोचत ॥८.३२.१॥
O Kaṇvas, praise ye with song the noble deeds of Indra, the receiver of the third press of Soma,
Performed in the Soma's wild delight. (RV viii.32.1)

प्र <ind. prefix to verb वोचत (√वच्)>, कृतानि <n. acc. pl. of कृत – deeds, actions carried out>, ऋजीषिणः <m. gen. sg. of adj. ऋजीषिन् – "receiving the residue of Soma produced by the third pressure of the plant", i.e. receiving the 3rd and the last batch of plasma discharge of the day>, कण्वाः <m. voc. pl. of कण्व – O Kaṇvas (pl.), O Praisers, Roarers>, इंद्रस्य <m. gen. sg. of इंद्र – of Indra>, गाथया <f. inst. sg. of गाथा – with song>;

मदे <m. loc. sg. of मद – in rapture, in delight>, सोमस्य <m. gen. sg. of सोम – of Soma>, प्र-वोचत <ipv. 2nd person pl. of प्र-√वच् – ye (Kaṇvas) praise, commend>.

In RV viii.32.5, the Soma Pond is compared to the stall of the horses and cattle. Cattle represent the celestial lights, the cosmic plasmas; the horses, the forces that generate currents of cosmic plasmas. Indra breaks asunder the Soma Pond as if it were a fortress.

RV viii.32.5 author मेधातिथिः काण्वः, to इंद्रः, metre गायत्री छंदः
स गोरश्वस्य वि व्रजं मंदानः सोम्येभ्यः । पुरं न शूर दर्षसि ॥८.३२.५॥
सः । गोः । अश्वस्य । वि । व्रजं । मंदानः । सोम्येभ्यः । पुरं । न । शूर । दर्षसि ॥८.३२.५॥
Exhilarated by Soma draughts, the stall of horses and cattle,
O Warrior, thou shalt break asunder as if it were a fortress. (RV viii.32.5)

सः <m. nom. sg. of तद् – he>, गोः <m. gen. sg. of गो – of kine, cattle>, अश्वस्य <m. gen. sg. of अश्व – of horse>, वि <ind. – apart, asunder>, व्रजं <m. acc. sg. of व्रज – a stall, cow-shed>, मंदानः <m. nom. sg. of pre. middle pt. मंदान (√मन्द्) – he being exhilarated, intoxicated>, सोम्येभ्यः <mn. abl. pl. of सोम्य – from Soma draughts>;

पुरं <n. acc. sg. of पुर – a fortress>, न <ind. – like, as if>, शूर <m. voc. sg. of शूर – O Warrior, O Hero>, दर्षसि <sub. 2nd person sg. of √दॄ – you shall break asunder, split open>.

In RV iii.36.8, the Soma Pond is compared to Indra's belly; plasma blooms, to the chambers of Indra's belly, which are filled full with the pressed Soma.

RV iii.36.8 author विश्वामित्रः, to इंद्रः, metre त्रिष्टुप् छंदः
ह्रदा इव कुक्षयः सोमधानाः समी विव्याच सवना पुरूणि ।
अन्ना यदिंद्रः प्रथमा व्याश वृत्रं जघन्वाँ अवृणीत सोमं ॥३.३६.८॥
ह्रदाःऽइव । कुक्षयः । सोमऽधानाः । सं । ईमिति । विव्याच । सवना । पुरूणि ।
अन्ना । यत् । इंद्रः । प्रथमा । वि । आश । वृत्रं । जघन्वान् । अवृणीत । सोमं ॥३.३६.८॥
Chambers in his belly are the Soma receptacles. His belly, like lakes, contains pressed Soma libations in abundance. When Indra wanted the foods first, he claimed Soma for himself after slaying Vṛtra. (RV iii.36.8)

ह्रदाः <m. nom. pl. of ह्रद – lakes, pools>, इव <ind. – like, as if>, कुक्षयः <m. nom. pl. कुक्षि – cavities of the abdomen>, सोमधानाः <m. nom. pl. of सोमधान – Soma receptacles>, सं <ind. – in total, altogether>, ईं <ind. – a particle of affirmation>, विव्याच <prf. 3rd person sg. √व्यच् – it (his belly) contains>, सवना <n. acc. pl. of सवन – the pressed out Soma libations>, पुरूणि <n. acc. pl. of adj. पुरु – abundant>;

अन्ना <n. acc. pl. of अन्न – foods (in "a mystical sense", Somas)>, यत् <(यद्) conjunction – when>, इंद्रः <m. nom. sg. of इंद्र – Indra>, प्रथमा <n. acc. pl. of adj. प्रथम – first, earliest>, वि <ind. prefix to verb जघन्वांस् (√हन्)>, आश <prf. 3rd person sg. of √अश् 5P – he gained, obtained>, वृत्रं <m acc sg of वृत्र – Vṛtra>, वि-जघन्वान् <m. nom. sg. of prf. active pt. वि-जघन्वांस् (वि-√हन्) – he having slayed, killed>, अवृणीत <ipf. 3rd person sg. of √वृ – he has chosen, claimed for one's self>, सोमं <m. acc. sg. of सोम – Soma>.

2.5.27 Indra, the Maker of Light and Form

In stanzas RV i.6.3 and 5, one finds quite unusual aspects of Indra. Usually, Tvaṣṭā is called the shaper of all forms (RV viii.91.8, RV x.184.3) and the Stone, the celestial ioniser, is responsible for making celestial lights, cosmic plasmas. In RV i.6.3, Indra is described as a maker of light (केतुः ketuḥ) and shaper of form (पेशः peśaḥ).

RV i.6.3 author मधुच्छंदा वैश्वामित्रः, to Indra इंद्रः, metre गायत्री छंदः
केतुं कृण्वन्नकेतवे पेशो मर्या अपेशसे । समुषद्भिरजायथाः ॥१.६.३॥
केतुं । कृण्वन् । अकेतवे । पेशः । मर्याः । अपेशसे । सं । उषत्ऽभिः । अजायथाः ॥१.६.३॥
Making light where no light was, and shaping form, O men, where form was not,
Together with the Dawns, you were born. (RV i.6.3)

केतुं <m. acc. sg. of केतु – brightness, light, flame>, कृण्वन् <m. nom. sg. of pre. act. pt. कृण्वत् (√कृ) – he making, shaping>, अकेतवे <m. dat. sg. of adj. अकेतु – to where there is no light>, पेशः <n. acc. sg. of पेशस् – shape, form>, मर्याः <m. voc. pl. of मर्य – O men>, अपेशसे <m. dat. sg. of adj. अपेशस् – to where there is no form>;

सं <ind. – together with>, उषद्भिः <f. inst. pl. of उषस् – with the Dawns>, अजायथाः <ipf. middle 2nd person sg. of √जन् 4Ā – you were born>.

In RV i.6.5, Indra, with fires, breaks the ionic bonds and finds cows, that is, Agnis, hiding within the atomic structure. In other words, Indra, with Trita, tritium, breaks the ionic bond of hydrogen and frees Agni and Soma from the bondage. Here, the atomic structure and the nucleus is symbolically described as a stronghold and a cave.

RV i.6.5 author मधुच्छंदा वैश्वामित्रः, to मरुत इंद्रश्च, metre गायत्री छंदः
वीळु चिदारुजत्नुभिर्गुहा चिदिंद्र वह्निभिः । अविंद उस्रिया अनु ॥१.६.५॥
वीळु । चित् । आरुजत्नुऽभिः । गुहा । चित् । इंद्र । वह्निऽभिः । अविंदः । उस्रियाः । अनु ॥१.६.५॥
With fires, by breaking the layered stronghold, O Indra, in the cave,
You found the cows. (RV i.6.5)

वीळु <(=वीडु) n. acc. sg. of वीळु – "anything firmly fixed or strong, stronghold", i.e. the stronghold of the ionic bond, the atomic structure>, चित् <n. acc. sg. of adj. चित् – "forming a layer or stratum", layered>, आरुजत्नुभिः <m. inst. pl. of adj. आरुजत्नु – by breaking>, गुहा <ind. – in a hiding place or a cave, i.e. within the nucleus>, चित् <(चिद्) ind. – even, indeed>, इंद्र <m. voc. sg. – O Indra>, वह्निभिः <m. inst. pl. of वह्नि – with fires, i.e. the fires of the celestial ioniser>;

अनु-अविंदः <ipf. 2nd person sg. of अनु-√विद् – you found, obtained, discovered (with gen.)>, उस्रियाः <f. acc. pl. of उस्रिया – lights, cows>.

Fires (वह्नयः vahnayaḥ, pl. of वह्नि vahni) are the fires of the Stone, that is, Trita (त्रितः), or tritium (^3H), according to the Ṛgveda. Breaking the ionic bonds and discharging the celestial plasmas are the duty of the Stone, the celestial ioniser. In this stanza, it is indicated that Indra, along with Trita, plays a role in cosmic ionisation. It tells us that Indra is a different aspect of Trita.

2.5.28 Saṃvatsarakālacakraṃ, the Soma Pond and the Wheel of Cosmic Ionisation

RV i.164.48 is addressed to Saṃvatsarasaṃsthaṃ Kālacakravarṇanaṃ (संवत्सरसंस्थं कालचक्रवर्णनं),[76] which literally means 'explanation of the wheel of cosmic plasmas that belongs to the Soma Pond'. Saṃvatsarasaṃsthaṃ (संवत्सरसंस्थं), which means 'belonging to the Soma Pond', is a compound word of saṃvat (संवत् f. nom. sg. of संवत् a region, tract) + sara (सरः n. nom. sg. of सरस् a lake, pond) + saṃsthaṃ (संस्थं n. nom. sg. of adj. संस्थ contained in, belonging to). Saṃvatsara (संवत्सरः) is the region abounding with celestial Waters, which is the Soma Pond. In stanzas RV i.164.44, RV iv.33.4, and RV vii.103.1, the 'saṃvatsara' is used to represent the Soma Pond. Śatapathabrāhmaṇaṃ (शतपथब्राह्मणं) also identifies saṃvatsara with the 'world of light' (स्वर्गः लोकः svargaḥ lokaḥ ŚB 8.4.1.24, ŚB 8.6.1.4) or with the 'great Bird' (महासुपर्णः mahāsuparṇaḥ ŚB 12.2.3.7). All these descriptions indicate that Saṃvatsara refers to the Soma Pond.

Kālacakravarṇanaṃ (कालचक्रवर्णनं), which means 'explanation of the wheel of cosmic plasmas', is a compound of kāla (काल a black or dark-blue colour) + cakra (चक्र a wheel) + varṇanaṃ (वर्णनं explanation, delineation). The word kāla (काल), as an adjective, means "black, of a dark colour" and as a noun, "a black or dark-blue colour", the "name of Śiva", and "time" among many other definitions provided. In the Ṛgveda, a black or dark-blue colour indicates Varuṇa, cosmic plasmas in general and the plasma dark discharge in particular; so kālacakraṃ (कालचक्रं) means 'the wheel of cosmic plasmas', which refers to the Stone, the celestial ioniser. Note the Stone is called the Stone Wheel (अश्मचक्रं RV x.101.7) in the Ṛgveda. When you take "time" for the definition of kāla, then kālacakraṃ can be interpreted as "the wheel of time". However, considering that the function of the Heavenly Vault is to turn the wheel of the Stone, the celestial ioniser, for producing and discharging and releasing cosmic plasmas, 'saṃvatsarakālacakraṃ' is translated as 'the Soma Pond and the Wheel of Cosmic Ionisation'.

In stanza RV i.164.48, the anatomical structures of the Soma Pond and the Stone are compared to that of a wheel. The Soma Pond, where the discharged cosmic plasmas gather and bloom and mature, is compared to the nave of the wheel; the firestones of the celestial ioniser laid in a circular formation are compared to the fellies of the wheel.

---

76 F. Max Müller (ed.), *The Hymns of the Rig-Veda in the Samhita and Pada Texts*, 2nd ed., Vol. I, London, Trübner and Co.,1877, p.144 (१४४).

The plasma blooms, maturing in the Soma pond, are compared to the arrows or spears (शंकवः pl.), for they are the weapons Indra uses to slay Vṛtra, the serpent of the Heaven, the plasma sheath (see Figures 6, 7, and 8). "Who has understood it?", the poet asks. Indeed, who has understood it?

> RV i.164.48 author दीर्घतमा औचथ्यः, to संवत्सरसंस्थं कालचक्रवर्णनं, metre त्रिष्टुप् छंदः
> द्वादश प्रधयश्चक्रमेकं त्रीणि नभ्यानि क उ तच्चिकेत ।
> तस्मिन्त्साकं त्रिशता न शंकवोऽर्पिताः षष्टिर्न चलाचलासः ॥१.१६४.४८॥
> द्वादश । प्रऽधयः । चक्रं । एकं । त्रीणि । नभ्यानि । कः । ऊं इति । तत् । चिकेत ।
> तस्मिन् । साकं । त्रिऽशताः । न । शंकवः । अर्पिताः । षष्टिः । न । चलाचलासः ॥१.१६४.४८॥
> Twelve are the fellies, the wheel is single, and three are the naves. Who has understood it?
> Therein are the three hundred and sixty spears cast into, which are ever quivering.
> (RV i.164.48)

द्वादश <m. nom. pl. of द्वादशन् – twelve>, प्रधयः <m. nom. pl. of प्रधि – the fellies of a wheel>, चक्रं <n. nom. sg. of चक्र – the wheel>, एकं <n. nom. sg. of एक – one, single>, त्रीणि <n. nom. pl. of त्रि – three>, नभ्यानि <n. nom. pl. of नभ्य – the centre part of a wheel, the nave>, कः <m. nom. sg. of interrogative pron. क – who?>, उ <ind. written ऊं in padapāṭha – a particle of assent, calling, command; frequently found in interrogative sentences>, तत् <n. acc. sg. of pron. तद् – that>, चिकेत <prf. 3rd person sg. of √चित् 1P – (with कः) who has understood, comprehended>;

तस्मिन् <n. loc. sg. of pron. तद् – within it>, त्रिशताः <m. nom. pl. of त्रिशत – three hundred>, षष्टिः <f. nom. sg. of षष्टि – sixty>, न <ind. – like, as; with another न, it forms strong affirmation>, शंकवः <(शङ्कवः) m. nom. pl. of शंकु (शङ्कु) – arrows, spears>, अर्पिताः <m. nom. pl. of adj. अर्पित – thrown, cast into (loc.)>, साकं <ind. – together, at the same time>, न <see above>, चलाचलासः <m. nom. pl. of adj. चलाचल – ever-moving, tremulous, quivering>.

The number used to describe the firestones is mostly two or ten. When it refers to the upper and lower discs of the celestial ioniser, it is described as 'two Stones' or 'two arms'. When it refers to the number of firestones that are laid in a circular formation, it is described as 'ten maidens' or 'ten fingers' or simply 'stones' in the plural. In this stanza, the number used is twelve rather than ten as in 'twelve fellies'. The numbers of plasma blooms are usually described as a hundred or a thousand (RV ii.41.5, RV i.80.12); here, it is described as 'three hundred and sixty arrows or spears'.

Cosmic plasmas are released three times a day; so, it is said that Indra drank three sacred vessels of Soma (RV i.32.3) and Aśvinā's journey is thrice in the same day (RV i.34.1-8). Fishman and Hartmann (1997) confirm that "about three times a day our sky flashes with a powerful pulse of gamma rays, ..."[77] As cosmic plasmas in the first Soma Pond are fully matured, the Soma Pond bursts to release them. Subsequently, the second and the third ponds are formed and undergo the same process of bursting and releasing the matured plasma blooms, three times a day every day. Thus, in RV i.164.48, it is stated that three

---

[77] Gerald J. Fishman and Dieter H. Hartmann, 'Gamma-Ray Bursts: New Observations Illuminate the Most Powerful Explosions in the Universe', *Scientific American*, July 1997, pp. 46-51.

are the naves. The Soma Pond is compared to 'nave', the hall of the congregation of celestial 'people', the 'people' of cosmic plasmas.

2.5.29. The Thousand-Pillared Abode of Kings Varuṇa and Mitra

In RV ii.41.5, the Soma Pond is described as the eternal, supreme, and thousand-pillared abode of Kings Varuṇa and Mitra.

> RV ii.41.5 author गृत्समदः, to मित्रावरुणौ, metre गायत्री छंदः
> राजानावनभिद्रुहा ध्रुवे सदस्युत्तमे । सहस्रस्थूण आसाते ॥२.४१.५॥
> राजानौ । अनभिऽद्रुहा । ध्रुवे । सदसि । उत्ऽतमे । सहस्रऽस्थूणे । आसाते इति ॥२.४१.५॥
> Kings Mitra and Varuṇa, who are benevolent, dwell in the eternal abode,
> Supreme and thousand-pillared. (RV ii.41.5)

राजानौ <m. nom. du. of राजन् – two kings, i.e. Mitra and Varuṇa>, अनभिद्रुहा <m. inst. sg of adj. अनभिद्रुह् – not malicious>, ध्रुवे <n. loc. sg. of adj. ध्रुव – immovable, lasting, permanent, eternal>, सदसि <n. loc. sg. of सदस् – in abode>, उत्तमे <n. loc. sg. of adj. उत्तम – best, excellent, supreme>;

सहस्रस्थूणे <n. loc. sg. of adj. सहस्रस्थूण – supported by a thousand pillars, thousand-pillared>, आसाते <prf. middle 3rd person du. of √अस् – they (du.) live, dwell>.

2.6 THE SACRIFICIAL POSTS

2.6.1 The Sacrificial Post, the Luminous Pillar of the Heaven

The eleven stanzas of hymn RV iii.8 are addressed to the Sacrificial Post (यूपः yūpaḥ sg. RV iii.8.1-5), to a hundred or a thousand sacrificial posts (यूपाः yūpāḥ pl. RV iii.8.6-10), and to 'the remaining stump of the destroyed sacrificial post (छिन्नयूपस्य मूलभूतः स्थाणुः chinnayūpasya mūlabhūtaḥ sthāṇuḥ RV iii.8.11)'. When the Sacrificial Post is addressed in the singular, it refers to the Soma Pond or the Pillar of the Heaven as a single unit. When they are addressed in the plural, the sacrificial posts refer to the individual plasma blooms, or Soma plants, that grow in the Soma Pond (Figure 7). The Soma Pond, together with the Stone, is the Pillar that props up the Vault of the Heaven and the Milky Way with the unbroken chain of cosmic ionisation and cosmic plasma discharge and release. There are two Pillars in the Heaven, one in the Upper Vault of the Heaven, the other in the Lower Vault of the Heaven (see Figure 6). Each Pillar of the Heaven consists of a hundred or a thousand pillars bound by Vṛtra, the plasma sheath. The Roman fasces, used as an insignia of power and authority of the Roman legions, is a symbolic replica of the bundle of these pillars, or plasma blooms, or Agni's arrows.

Note these plasma blooms, or Agni's arrows, are Indra's weapons used to slay Vṛtra, the serpent of the Heaven. A double-headed or single-headed axe, often attached to the fasces, represents the Stone that produces a bundle of plasma blooms, that is, a bundle of Agni's arrows. The single-headed axe represents the Stone, the celestial ioniser, as a single unit; the double-headed axe, to the two fiery discs of the Stone. In the Ṛgveda, the Stone, the celestial ioniser, is called the Heavenly hatchet (स्वधितिः the Heavenly hatchet RV iii.8.6, परशुः स्वायसः the axe wrought of gold RV x.53.9). The radiant Brahmaṇaspati uses this hatchet wrought of gold to cut the ionic bond asunder (RV x.53.9).

Stanzas RV iii.8.1-5, presented here, are addressed to the Sacrificial Post, or the Pillar of the Heaven (यूपः), which refers to the Soma Pond. In RV iii.8.1, 3, and 6, the Pillar of the Heaven is called the Lord of the Forest (वनस्पतिः), though in other context, the Lord of the Forest can represent the Stone as in RV i.28.6. The Forest refers to the forest of Soma plants, or plasma blooms. They anoint the Lord of the Forest with the heavenly mead, the cosmic plasmas, produced and discharged by the Stone.

In RV iii.8.1, 'this Mother' refers to the Milky Way. The Sacrificial Post, that is, the Soma Pond, is often described as standing upon the lap or bosom of the Mother, the Milky Way.

RV iii.8.1 author विश्वामित्रः, to यूपः, metre त्रिष्टुप् छंदः
अंजंति त्वामध्वरे देवयंतो वनस्पते मधुना दैव्येन ।
यदूर्ध्वस्तिष्ठा द्रविणेह धत्ताद्यद्वा क्षयो मातुरस्या उपस्थे ॥३.८.१॥
अंजंति । त्वां । अध्वरे । देवऽयंतः । वनस्पते । मधुना । दैव्येन ।
यत् । ऊर्ध्वः । तिष्ठाः । द्रविणा । इह । धत्तात् । यत् । वा । क्षयः । मातुः । अस्याः । उपऽस्थे ॥३.८.१॥
They, serving the Devas, with heavenly mead, O Lord of the Forest, anoint you at sacrifice.
Bring thou riches here as thou standest upright while remaining on the lap of this Mother.
(RV iii.8.1)

अंजंति <(अञ्जन्ति) pre. 3rd person pl. of √अंज् (√अञ्ज्) – they anoint, decorate>, त्वां <m. acc. sg. of 2nd person pron. युष्मद् – you>, अध्वरे <m. loc. sg. of अध्वर – at sacrifice (esp. the Soma sacrifice)>, देवयंतः <m. nom. pl. of pre. active. pt. देवयत् of nominal verb √देवय – they serving or adoring the Devas>, वनस्पते <m. voc. sg. of वनस्पति – O Lord of the Forest>, मधुना <n. inst. sg. of मधु – with Soma mead>, दैव्येन <n. inst. sg. of adj. दैव्य – divine, heavenly>;

यत् <(यद्) conjunction – as, when>, ऊर्ध्वः <m. nom. sg. of adj. ऊर्ध्व – erect, upright>, तिष्ठाः <m. nom. sg. of adj. तिष्ठा – standing, stand>, द्रविणा <n. acc. pl. of द्रविण – wealths, riches>, इह <ind. – here, to this place>, धत्तात् <ipv. 2nd person sg. of √धा – you put, place, deposit>, यत् वा <conjunction – and also, as, when>, क्षयः <m. nom. sg. of adj. क्षय – residing, remaining, staying>, मातुः <f. gen. sg. of मातृ – of Mother>, अस्याः <f. gen. sg. of pron. इदं – of this>, उपस्थे <m. loc. sg. of उपस्थ – in the sheltered place, on the lap>.

In RV iii.8.2, the east refers to the direction from where celestial lights, cosmic plasmas, are produced and discharged from the Stone, the celestial ioniser. Once cosmic plasmas are discharged, they grow upwards and mature and bloom in the Soma Pond. Note the mature Soma plants, or plasma blooms, can reach 25,000-30,000 light-years in height.

The prayer is the Speech of cosmic plasmas, and youthful warriors refer to freshly discharged embryonic particles of cosmic plasmas, especially Agnis, the hydrogen ions.

RV iii.8.2 author विश्वामित्रः, to यूपः, metre त्रिष्टुप् छंदः
समिद्धस्य श्रयमाणः पुरस्ताद्ब्रह्म वन्वानो अजरं सुवीरं ।
आरे अस्मदमतिं बाधमान उच्छ्रयस्व महते सौभगाय ॥३.८.२॥
सम्ऽइद्धस्य । श्रयमाणः । पुरस्तात् । ब्रह्म । वन्वानः । अजरं । सुऽवीरं ।
आरे । अस्मत् । अमतिं । बाधमानः । उत् । श्रयस्व । महते । सौभगाय ॥३.८.२॥
From the east, diffusing light of the blazing fire, uttering prayer and procuring ever youthful warriors, Driving destitution far away from us, spread thou upwards for abundant riches.
(RV iii.8.2)

समिद्धस्य <n. gen. sg. of समिद्ध – of the blazing fire>, श्रयमाणः <m. nom. sg. of pre. middle pt. श्रयमाण (√श्रि) – it/he spreading or diffusing (light)>, पुरस्तात् <ind. – from the east>, ब्रह्म <n. acc. sg. of ब्रह्मन् – utterance, prayer, the sacred word>, वन्वानः <m. nom. sg. of pre. middle pt. वन्वान (√वन् 8Ā) – (the verb is doubly used) it/he, (with ब्रह्म) uttering, (with सुवीरं) procuring>, अजरं <m. acc. sg. of adj. अजर – ever young>, सुवीरं <m. acc. sg. of सुवीर – heroes, warriors>;
आरे <ind. – far from, far>, अस्मत् <m. abl. pl. of 1st person pron. अस्मत् – from us>, अमतिं <f. acc. sg. of अमति – poverty, destitution>, वाधमानः <m. nom. sg. of pre. middle pt. वाधमान (√बाध्) – you driving away, removing>, उत् <(उद्) ind. particle – up, upwards>, श्रयस्व <ipv. middle 2nd person sg. of √श्रि – you (the Pillar of the Heaven) spread>, महते <n. dat. sg. of adj. महत् – ample, abundant>, सौभगाय <n. dat. sg. of सौभग – for wealth, riches>.

In RV iii.8.3, the Lord of the Forest is called the rain-bearer. In the Ṛgveda, the streaming charged particles of cosmic plasmas, Agnis and Somas, are called rain. Thus, the Soma Pond, or the Sacrificial Post, within which the rains are collected is called the rain-bearer.

RV iii.8.3 author विश्वामित्रः, to यूपः, metre अनुष्टुप् छंदः
उच्छ्रयस्व वनस्पते वर्ष्मन्पृथिव्या अधि । सुमिती मीयमानो वर्चो धा यज्ञवाहसे ॥३.८.३॥
उत् । श्रयस्व । वनस्पते । वर्ष्मन् । पृथिव्याः । अधि । सुऽमिती । मीयमानः । वर्चः । धाः । यज्ञऽवाहसे ॥३.८.३॥
O Lord of the Forest, the rain-bearer, rise thou upwards high above the Milky Way,
Well established and measured, bringing the radiant light for the offerer of Yajña.
(RV iii.8.3)

उत् <(उद्) ind. particle – upwards>, श्रयस्व <ipv. 2nd person sg. of √श्रि – you turn towards, spread>, वनस्पते <m. voc. sg. of वनस्पति – O Lord of the Forest>, वर्ष्मन् <m. voc. sg. of वर्ष्मन् – O rain-bearer!>, पृतिव्याः <f. abl. sg. of पृथिवी – (with अधि) above Pṛthivī, i.e. above the Milky Way>, अधि <ind. – (with abl.) over, above>;
सुमिती <(=सुमितः?) m. nom. sg. of adj. सुमित – well set up, built>, मीयमानः <m. nom. sg. of pre. passive pt. मीयमान (√मि) – it being measured>, वर्चः <n. acc. sg. of वर्चस् – the illuminating power of fire or the sun, light>, धाः <m. nom. sg. of adj. धा – bringing, bestowing>, यज्ञवाहसे <m. dat. sg. of यज्ञवाहस् – for the offerer of Yajña>.

In RV iii.8.4, the Sacrificial Post is described as the youth who came beautifully robed or well feathered and veiled in a plasma sheath.

RV iii.8.4 author विश्वामित्रः, to यूपः, metre त्रिष्टुप् छंदः
युवा सुवासाः परिवीत आगात्स उ श्रेयान्भवति जायमानः ।
तं धीरासः कवय उन्नयंति स्वाध्यो३ मनसा देवयंतः ॥३.८.४॥

युवा । सुऽवासाः । परिऽवीतः । आ । अगात् । सः । ऊं इति । श्रेयान् । भवति । जायमानः ।
तं । धीरासः । कवयः । उत् । नयंति । सुऽआध्यः । मनसा । देवऽयंतः ॥३.८.४॥
Beautifully robed, veiled, came the youth. Being born, he becomes more splendorous.
Serving the Devas, the skilful and devout singers raise him aloft with eagerness. (RV iii.8.4)

युवा <m. nom. sg. of युवन् – a youth>, सुवासाः <m. nom. sg. of adj. सुवासस् – wearing beautiful garments, well feathered (as an arrow)>, परिवीतः <m. nom. sg. of adj. परिवीत – veiled, enveloped>, आ <ind. prefix to verb अगात् (√गा)>, आ-अगात् <ipf. 3rd person sg. of आ-√गा – he came>, सः <m. nom. sg. of 3rd person pron. तद् – he>, उ <ind. particle – implying assent, calling, command>, श्रेयान् <m. nom. sg. of adj. श्रेयस् – more splendid or beautiful, more excellent>, भवति <pre. 3rd person sg. of √भू – he becomes>, जायमानः <m. nom. sg. of pre. middle pt. जायमान (√जन्) – he being born>;

तं <m. acc. sg. of 3rd person pron. तद् – him>, धीरासः <m. nom. pl. of adj. धीर – wise, skilful>, कवयः <m. nom. pl. of कवि – singers (the firestones of the Stone)>, उन्नयन्ति <pre. 3rd person pl. of √उन्नी (उद्-√नी) – they raise, set up>, स्वाध्यः <m. nom. pl. of adj. स्वाधी – thoughtful, devout>, मनसा <ind. – with all the heart, willingly, with eagerness>, देवयंतः <m. nom. pl. of pre. active. pt. देवयत् of nominal verb √देवय – they adoring or serving the Devas>.

In RV iii.8.5, the deva-longing singer refers to the Stone, the celestial ioniser.

RV iii.8.5 author विश्वामित्रः, to यूपः, metre त्रिष्टुप् छंदः
जातो जायते सुदिनत्वे अह्नां समर्य आ विदथे वर्धमानः ।
पुनंति धीरा अपसो मनीषा देवया विप्र उदियर्ति वाचं ॥३.८.५॥
जातः । जायते । सुदिनऽत्वे । अह्नां । सऽमर्ये । आ । विदथे । वर्धमानः ।
पुनंति । धीराः । अपसः । मनीषा । देवऽयाः । विप्रः । उत् । इयर्ति । वाचं ॥३.८.५॥
Sprung up, he grows during the days' auspicious time, thriving in the sacrificial assembly attended by many. With hymn, they, skilful in sacred work, make him bright. And the singer, longing for the Devas, raises his voice. (RV iii.8.5)

जातः <m. nom. sg. of adj. जात – born, produced>, जायते <pre. middle 3rd person sg. of √जन् 4Ā – he grows (as plants, teeth)>, सुदिनत्वे <n. loc. sg. of सुदिनत्व – in the fair weather, during the auspicious time>, अह्नां <n. gen. pl. of अहन् (अहर्) – of the days>, समर्य <ind.(?) – attended by many, frequented>, आ <ind. prefix to pt. वर्धमानः (√वृध्)>, विदथे <n. loc. sg. of विदथ – in the sacrificial assembly>, आ-वर्धमानः <m. nom. sg. of pre. middle pt. आ-वर्धमान (आ-√वृध्) – he increasing, prospering, thriving>;

पुनंति <pre. 3rd person pl. of √पू – they make clear or pure or bright, purify, illume>, धीराः <m. nom. pl. of adj. धीर – wise, intelligent>, अपसः <m. nom. pl. of adj. अपस् – active, skilful in any art>, मनीषा <f. inst. sg. of मनीषा – with hymn, prayer>, देवयाः <m. nom. sg. of adj. देवया – deva-longing>, विप्रः <m. nom. sg. of विप्र – a singer, poet>, उदियर्ति <pre. 3rd person sg. of √उदृ (उद्-√ऋ) – (with वाचं) he raises voice>, वाचं <f. acc. sg. of वाच् – speech, voice, sound>.

2.6.2 A Hundred or a Thousand Sacrificial Posts

Stanzas RV iii.8.6-10 are addressed to a hundred or a thousand pillars, or sacrificial posts. As noted, the pillars, or sacrificial posts, in the plural denote the Soma plants, the plasma blooms, which are also called bright arrows or spears in other stanzas. Usually, they are described symbolically as a hundred or a thousand pillars; it does not mean there are exactly a hundred or a thousand of them (see Figure 7). In RV iii.8.6, the Stone, the celestial ioniser, is called the Heavenly hatchet.

RV iii.8.6 author विश्वामित्रः, to यूपाः, metre त्रिष्टुप् छंदः
यान्वो नरो देवयंतो निमिम्युर्वनस्पते स्वधितिर्वा ततक्ष ।
ते देवासः स्वरवस्तस्थिवांसः प्रजावदस्मे दिधिषंतु रत्नं ॥३.८.६॥
यान् । वः । नरः । देवऽयंतः । निऽमिम्युः । वनस्पते । स्वऽधितिः । वा । ततक्ष ।
ते । देवासः । स्वरवः । तस्थिऽवांसः । प्रजाऽवत् । अस्मे इति । दिधिषंतु । रत्नं ॥३.८.६॥
Serving the Devas, men have erected sacrificial posts, O Lord of the Forest, which the Heavenly hatchet has fashioned. Let those Heavenly sacrificial posts, which are standing firmly, bring riches upon us with a store of offspring. (RV iii.8.6)

यान् <m. acc. pl. of pron. यद् – whom, which (pl.)>, वः <m. acc. pl. of 2nd person pron. युष्मद् – you (pl.)>, नरः <m. nom. pl. of नृ – men, heroes>, देवयंतः <m. nom. pl. of pre. active. pt. देवयत् of nominal verb √देवय – they serving the Devas>, निमिम्युः <prf. 3rd person pl. of नि-√मि – they have erected, raised (sacrificial posts)>, वनस्पते <m. voc. sg. of वनस्पति – O Lord of the Forest>, स्वधितिः <mf. nom. sg. of स्वधिति – the heavenly axe, hatchet (refers to the Stone, the celestial ioniser)>, वा <ind. – or, as, like, just>, ततक्ष <prf. 3rd person sg. of √तक्ष् – it (hatchet) has formed by cutting or splitting>;

ते <m. nom. pl. of pron. तद् – those>, देवासः <m. nom. pl. of adj. देव – heavenly, divine>, स्वरवः <m. nom. pl. of स्वरु – sacrificial posts>, तस्थिवांसः <m. nom. pl. of prf. active pt. तस्थिवस् (√स्था) – they standing firmly>, प्रजावत् <ind.(?) प्रजावत् – having or granting offspring>, अस्मे <ind. – on us>, दिधिषंतु <ipv. 3rd person pl. of √धा – let them grant, bring>, रत्नं <n. acc. sg. of रत्न – jewel, gem, riches>.

RV iii.8.7 author विश्वामित्रः, to यूपाः, metre अनुष्टुप् छंदः
ये वृक्णासो अधि क्षमि निमितासो यतस्रुचः । ते नो व्यंतु वार्यं देवत्रा क्षेत्रसाधसः ॥३.८.७॥
ये । वृक्णासः । अधि । क्षमि । निऽमितासः । यतऽस्रुचः ।
ते । नः । व्यंतु । वार्यं । देवऽत्रा । क्षेत्रऽसाधसः ॥३.८.७॥
Those who hewed, erected, and raised the sacrificial ladles aloft in the Field,
Let these cultivators of the Field bring wealth to the Devas for us. (RV iii.8.7)

ये <m. nom. pl. of यद् – those who>, वृक्णासः <m. nom. pl. of वृक्ण – ones who cut down, hewed>, अधि <ind. – above, over>, क्षमि <f. loc. sg. of क्षम् – on the ground, in the field (refers to the Soma Pond)>, निमितासः <m. nom. pl. of निमित – ones who measured, erected>, यतस्रुचः <m. nom. pl. of यतस्रुच् – ones who raised or stretched out the sacrificial ladles (sacrificial ladles refer to the sacrificial posts)>;

ते <m. nom. pl. of pron. तद् – they, these>, नः <m. dat. pl. of 1st person pron. अस्मद् – for us>, व्यंतु <ipv. 3rd person pl. of √वी – let them bring, get, procure>, वार्यं <n. acc. sg. of वार्य – treasure, wealth (cosmic plasmas)>, देवत्रा <ind. – among or to the Devas>, क्षेत्रसाधसः <m. nom. pl. of – ones who divide the fields, fix the landmarks, i.e. the cultivators of the fields>.

In RV iii.8.8, Ādityas (आदित्याः ādityāḥ pl.), the sons of Aditi (अदितिः), are a class of the Devas whose chief is Varuṇa. Rudras, the sons of Rudra, refers to the Maruts. Vasus are a class of the Devas whose chief is Indra. The torch of sacrifice refers to the plasma blooms, the eternal flames.

RV iii.8.8 author विश्वामित्रः, to विश्वेदेवा यूपा वा, metre त्रिष्टुप् छंदः
आदित्या रुद्रा वसवः सुनीथा द्यावाक्षामा पृथिवी अंतरिक्षं ।
सजोषसो यज्ञमवंतु देवा ऊर्ध्वं कृण्वंत्वध्वरस्य केतुं ॥३.८.८॥
आदित्याः । रुद्राः । वसवः । सुऽनीथाः । द्यावाक्षामा । पृथिवी । अंतरिक्षं ।
सऽजोषसः । यज्ञं । अवंतु । देवाः । ऊर्ध्वं । कृण्वंतु । अध्वरस्य । केतुं ॥३.८.८॥

The righteous ones are the Ādityas, Rudras, Vasus, the Heaven and the Soma Pond, the Milky Way, and the Innerfield. Let these Devas, together, lead the Yajña and raise the torch of sacrifice aloft. (RV iii.8.8)

आदित्याः <m. nom. pl. of आदित्य – Ādityas, sons of Aditi, a class of the Devas whose chief is Varuṇa>, रुद्राः <m. nom. pl. of रुद्र – sons of Rudra, i.e. the Maruts>, वसवः <m. nom. pl. of वसु – Vasus, a class of Devas whose chief is Indra>, सुनीथाः <m. nom. pl. of adj. सुनीथ – righteous>, द्यावाक्षामा <(=द्यावाक्षमे) f. nom. sg. of द्यावाक्षामा (a dual compound) – the Heaven and the Field, i.e. the Heaven and the Soma Pond>, पृथिवी <f. nom. sg. of पृथिवी – Pṛthivī, "the broad and extended One", the Milky Way>, अंतरिक्षं <n. nom. sg. of अंतरिक्ष – the Innerfield, the Midheaven>;

सजोषसः <m. nom. pl. of adj. सजोषस् – united, together>, यज्ञं <m. acc. sg. of यज्ञ – Yajña, sacrifice>, अवंतु <ipv. 3rd person pl. of √अव् – let them impel, lead>, देवाः <m. nom. pl. of देव – Devas>, ऊर्ध्वं <m. acc. sg. of adj. ऊर्ध्व – elevated, high>, कृण्वंतु <ipv. 3rd person pl. of √कृ – let them make, raise>, अध्वरस्य <m. gen. sg. of अध्वर – of sacrifice (especially the सोम sacrifice)>, केतुं <m. acc. sg. of केतु – flame, torch>.

RV iii.8.9 author विश्वामित्रः, to यूपाः, metre त्रिष्टुप् छंदः
हंसा इव श्रेणिशो यतानाः शुक्रा वसानाः स्वरवो न आगुः ।
उन्नीयमानाः कविभिः पुरस्ताद्देवा देवानामपि यंति पाथः ॥३.८.९॥
हंसाःऽइव । श्रेणिऽशः । यतानाः । शुक्रा । वसानाः । स्वरवः । नः । आ । अगुः ।
उत्ऽनीयमानाः । कविऽभिः । पुरस्तात् । देवाः । देवानां । अपि । यंति । पाथः ॥३.८.९॥
Like swans flying together in flocks, the sacrificial posts came to us assuming radiant form. Raised by the singers from the east, the Devas come to their dwelling place. (RV iii.8.9)

हंसाः <m. nom. pl. of हंस – swans>, इव <ind. – like, in the same manner>, श्रेणिशः <(श्रेणिशस्) ind. – in lines or flocks>, यतानाः <m. nom. pl. of pre. middle pt. यतान (√यत्) – they flying together>, शुक्रा <m. inst. sg of adj. शुक्र – with/in bright or radiant colour or form>, वसानाः <m. nom. pl. of pre. middle pt. वसान (√वस् 2Ā) – they assuming form, wearing garment>, स्वरवः <m. nom. pl. of स्वरु – sacrificial posts>, नः <m. acc. pl. of 1st person pron. अस्मद् – to us>, आगुः <ipf. 3rd person pl. of आ-√गा – they came>;

उन्नीयमानाः <m. nom. pl. of pre. passive pt. उन्नीयमान (√उन्नी (उद्-√नी)) – they being raised>, कविभिः <m. inst. pl. of कवि – by singers>, प्रस्तात् <ind. – from the east>, देवाः <m. nom. pl. of देव – the Devas>, देवानां <m. gen. pl. of देव – of the Devas>, अपि <ind. – very, even>, यंति <pre. 3rd person pl. of √इ – they come to>, पाथः <n. acc. sg. of पाथस् – to the place, i.e. the dwelling place>.

In RV iii.8.10, the sacrificial posts are said to appear as horns of bulls. The battle refers to the cosmic battle between Indra and Vṛtra, the serpent of the Heaven. Indra wins and celestial lights are released.

RV iii.8.10 author विश्वामित्रः, to यूपाः, metre त्रिष्टुप् छंदः
शृंगाणीवेच्छृंगिणां सं ददृश्रे चषालवंतः स्वरवः पृथिव्यां ।
वाघद्भिर्वा विहवे श्रोषमाणा अस्माँ अवंतु पृतनाज्येषु ॥३.८.१०॥
शृंगाणिऽइव । इत् । शृंगिणां । सं । ददृश्रे । चषालऽवंतः । स्वरवः । पृथिव्यां ।
वाघत्ऽभिः । वा । विऽहवे । श्रोषमाणाः । अस्मान् । अवंतु । पृतनाज्येषु ॥३.८.१०॥
With rings adorning them, these sacrificial posts above the Milky Way appear as horns of bulls. In invocation, with the institutors of sacrifice, let these confident ones lead us in the rush to battle. (RV iii.8.10)

शृंगाणि <n. nom. pl. of शृंग (शृङ्ग) – horns>, इव <ind. – like, as>, इत् <(इद्) ind. particle of affirmation – even, just>, शृंगिणां <m. gen. pl. of शृङ्गिण (or शृङ्गिन्) – of bulls (or of wild rams)>, सं <ind. – with, together>, ददृश्रे <prf. middle 3rd

person pl. of √दृश् – (with इव) they appear as>, चषालवंतः <m. nom. pl. of adj. चषालवत् – adorned with rings at the top>, स्वरवः <m. nom. pl. of स्वरु – sacrificial posts>, पृथिव्यां <f. loc. sg. of पृथिवी – on Pṛthivī, on/above the Milky Way>;

वाघद्भिः <m. inst. pl. of वाघत् – with/by the institutors of sacrifice>, वा <ind. – just, even, indeed>, विहवे <m. loc. sg. of विहव – in invocation, at sacrifice>, श्रोषमाणाः <m. nom. pl. of श्रोषमाण – the willing or confident ones (refer to the sacrificial posts)>, अस्मान् <m. acc. pl. of 1st person pron. अस्मद् – us>, अवंतु <ipv. 3rd person pl. of √अव् – let them drive, impel, lead>, पृतनाज्येषु <n. loc. pl. of पृतनाज्य – in the rush to battle>.

2.6.3 The Remaining Stump of the Destroyed Sacrificial Post

The last stanza, RV iii.8.11, is addressed to Chinnayūpasya Mūlabhūta Sthāṇu (छिन्नयूपस्य मूलभूतः स्थाणुः), the 'remaining stump of the cut off sacrificial post'. Chinnayūpasya (छिन्नयूपस्य) is a compound of chinna (छिन्न cut off, cut) + yūpasya (यूपस्य of the sacrificial post), which means 'of the cut off sacrificial post'. Mūlabhūta (मूलभूतः) is a compound of mūla (मूल n. the lowest part) + bhūta (भूतः adj. existing, present, i.e. remaining), which means 'the remaining lowest part'. Sthāṇu (स्थाणुः m.) means 'a stump'. This remaining stump of the cut off sacrificial post is 'the remnant of the sacrifice' and it will become the base of the subsequent growth of Soma plants, or plasma blooms. New shoots of Soma plants will arise from the remaining stumps of the destroyed sacrificial posts.

RV iii.8.11 author विश्वामित्रः, to छिन्नयूपस्य मूलभूतः स्थाणुः, metre त्रिष्टुप् छंदः
वनस्पते शतवल्शो वि रोह सहस्रवल्शा वि वयं रुहेम ।
यं त्वामयं स्वधितिस्तेजमानः प्रणिनाय महते सौभगाय ॥३.८.११॥
वनस्पते । शतऽवल्शः । वि । रोह । सहस्रऽवल्शाः । वि । वयं । रुहेम ।
यं । त्वां । अयं । स्वऽधितिः । तेजमानः । प्रऽनिनाय । महते । सौभगाय ॥३.८.११॥
O Lord of the Forest, shoot forth a hundred branches. With a thousand branches may we grow to greatness. Thou whom this fiery hatchet has brought forth for abundant riches.
(RV iii.8.11)

वनस्पते <m. voc. sg. of वनस्पति – O Lord of the Forest (refers to the Soma Pond)>, शतवल्शः <m. nom. sg. of adj. शतवल्श – having a hundred branches>, वि-रोह <ipv. 2nd person sg. of वि-√रुह् – you grow out, shoot forth, sprout>, सहस्रवल्शाः <m. nom. pl. of adj. सहस्रवल्श – thousand-branched>, वयं <m. nom. pl. of 1st person pron. अस्मद् – we>, वि-रुहेम <opt. 1st person pl. of वि-√रुह् – may we grow out, sprout, bud>;

यं <m. acc. sg. of relative pron. यद् – whom>, त्वां <m. acc. sg. of 2nd person pron. युष्मद् – you (refers to the sacrificial post)>, अयं <m. nom. sg. of pron. इदं – this>, स्वधितिः <m. nom. sg. of स्वधिति – a hatchet>, तेजमानः <m. nom. sg. of pre. middle pt. तेजमान (√तिज्) – it being sharp, fiery>, प्रणिनाय <prf. 3rd person sg. of प्र-√णी (√नी) – it (hatchet) has brought, produced>, महते <n. dat. sg. of adj. महत् – great, abundant>, सौभगाय <n. dat. sg. of सौभग – for wealth, riches>.

As the Lord of the Forest shoots forth a hundred or a thousand branches, the flaming Soma plants, the plasma blooms, will spring up and grow and mature. The Soma Pond will be filled with the flaming Soma plants once more.

2.7 THE STONE, THE CELESTIAL MACHINE OF IONISATION AND COSMIC PLASMA DISCHARGE AND RELEASE

2.7.1 The Stone and the Keystone

The Stone (ग्रावा RV i.84.3, v.36.4, v.40.2, viii.13.32; अद्रिः RV v.43.4, vii.22.1), the Soma Press (सोता sg., सोतारः pl. RV i.28.8), is the celestial machine of ionisation and cosmic plasma discharge and release. This Stone, the celestial ioniser, is the original Philosopher's Stone (रसेन्द्रः rasendraḥ). It is also called the Ring of Fire or Fortress of Fire (वह्निः प्राकारः vahniḥ prākāraḥ). As shown in Figure 8, the Stone consists of two discs of firestones laid in a circular formation. As noted earlier, the 'firestones' refer to the fiery firestones of the celestial ioniser, not the stones used for lining furnaces or ovens. In the Ṛgveda, when the Soma Press, or the Stone, is regarded as a single unit, it is called the Stone, Stone Wheel (अश्मचक्रं RV x.101.7), many-footed Stone (ग्रावा पृथुबुध्नः RV i.28.1), Comb (कंकतः RV i.191.1), Lord of the Forest (वनस्पतिः RV i.28.6), Lord of the Field (क्षेत्रपतिः RV iv.57.1-3), mother milch Cow (धेनुः माता RV iii.55.12), Pṛśni (पृश्निः RV v.60.5, RV vii.56.4), Goddess Iḷā (इळा RV iii.55.13), and so on. In RV i.61.15, the Stone is called the Soma Pressing Etaśa (एतशः सुष्विः), the steed who presses out Soma.

Since the Stone, the celestial ioniser, consists of two circular discs, often it is addressed in dual form as the upper and lower Teeth of Agni (उभा दंष्ट्रा अवरं परं च RV x.87.3), two stones (अद्री du. RV i.109.3), two fire-sticks (अरणी du. RV iii.29.2), mortar and pestle (उलूखलमुसलौ RV i.28.7-8), two lords of the Forest (वनस्पती du. RV i.28.8), two stones of Trita (त्रितस्य पार्षी RV ix.102.2), and so on.

When multiple firestones in each disc of the Ring of Fire are considered (see Figure 8), the celestial ioniser is addressed in the plural as the stones (ग्रावाणः RV vii.104.17, RV x.175.1), Soma pressers (सोतारः RV i.28.8), maidens of Trita (त्रितस्य योषणः f. nom. pl. of योषन् RV ix.32.2, RV ix.38.2), ten fingers (दश क्षिपः RV iii.23.3), the stones with ten water-raising machines (अद्रयः दशयंत्रासः RV x.94.8), Tvaṣṭā's ten maidens (दश त्वष्टुः युवतयः RV i.95.2), lords of the Forest (वनस्पतयः pl. RV v.84.3), and so on.

The images in Figure 8 show the Stone, the celestial ioniser, mapped by Chandra and Hubble scientists. Images 2 and 3 show the two discs of the Stone aligned together as it is explained in the Ṛgveda. These celestial objects are variously named by the scientists who mapped them as shown in the image credits. Images 1 and 5 show the same celestial object, Supernova 1987A. Image 5 shows "Supernova 1987A Debris Disk" after the "explosion of the Supernova". According to the Ṛgveda, however, it is the Stone, the celestial ioniser, the celestial machine of the cosmic ionisation and plasma discharge and release. As we shall see, all of the images of Figure 8 beautifully match the descriptions of the celestial ioniser and its functions as explained in the Ṛgveda.

The Ṛgveda tells us that the fire of the Stone is Trita (त्रित:), tritium (^3H) (RV ix.38.2, RV ix.102.2). Ekata (एकत:), Dvita (द्वित:), and Trita (त्रित:) are the three fires of Agni, proton. Refer to sections 7.1.3 and 7.1.4 for Agni's three forms or fires.

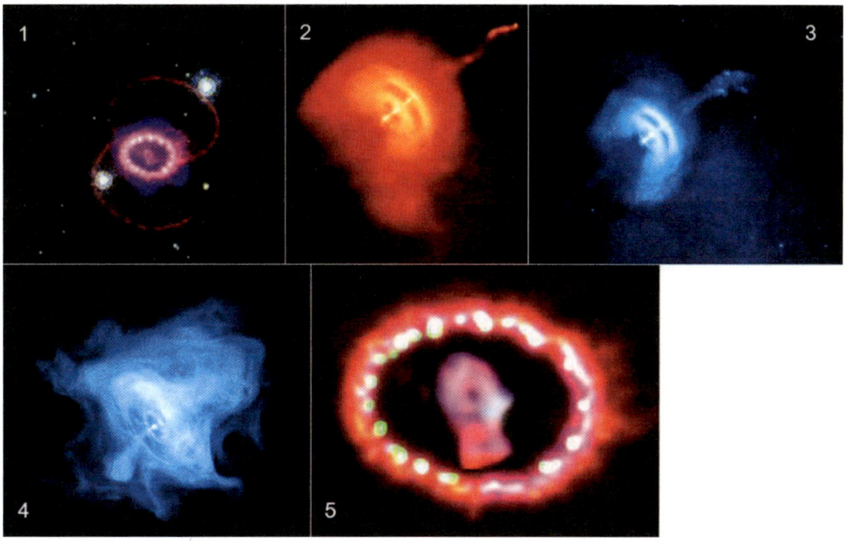

<Figure 8> The Stone, the Celestial Ioniser, Mapped by Chandra and Hubble. Images are NOT on the same scale.
1. *Supernova 1987A*, Composite image of Chandra X-Ray and Hubble Optical (X-Ray: blue-purple; Optical: pink-white), Credit: X-Ray: NASA/CXC/PSU/S. Park & D. Burrows.; Optical: NASA/STScI/ CfA/ P. Challis. Scale: Image is 12 arcsec across. Observation date: 8 hours on Jan 9, 2005. Release date: Feb 22, 2007. **2**. *Vela Pulsar Jet*, Combined 8 Chandra X-Ray Images. Credit: NASA/CXC/PSU/G. Pavlov et al.) Scale: 4 x 3.5 arcmin. **3**. *Vela Pulsar Jet*, Single image of Chandra X-Ray, Credit: NASA/CXC/Univ of Toronto/M.Durant et al. **4**. *Crab Nebula*, Chandra X-Ray, Credit: NASA/CXC/SAO/F. Seward et al. Scale: Image is 5 arcmin across. Observation: 03/14/2001 and 01/27/2004. Release date: Nov 5, 2008. **5**. *Supernova 1987A Debris Disk*, Credit: NASA, ESA, and P. Challis (Harvard-Smithsonian Center for Astrophysics).

In images 1 and 5 of Figure 8, one can identify individual firestones that make up a circular disc. Image 5 is a close-up observation of the same celestial object in image 1. In images 2, 3, and 4, one can see the upward and downward cosmic plasma discharge and release in action. Note the Keystone at the centre of the Ring of Fire in image 5. The Keystone at the centre of the Stone, the celestial ioniser, holds the firestones in a circular formation. This Keystone is the key that unlocks the mystery of the Stone, the celestial machine of ionisation and cosmic plasma discharge and release. One can see that keyholes of old locks were in the shape of this Keystone. In the extant text of the Ṛgveda, this Keystone is called the nave (नभ्यं nabhyaṃ RV x.119.12), the hub of the Stone.

The base of Śiva liṅgaṃ (शिवलिंगं) is a replica of this Keystone. The liṅgaṃ (लिंगं) itself, which stands on the base, represents the Pillar of the Heaven, or Sacrificial Post, that is, the Soma Pond, the womb of cosmic creation. The base of Śiva liṅgaṃ, which is in the shape of this keystone, corresponds to the mons Veneris, the mount of Venus. The long, wavy locks of Venus represent plasma blooms, the flaming Soma plants, which are also called the pillars of the Heaven in the Ṛgveda. The Keystone, located at the centre of the Stone, the celestial ioniser, is the base or the mount of the plasma blooms, the flaming Soma plants, the light-containing plants, for they rise out of the Stone and bloom and mature upon it. The Śiva liṅgaṃ and the mount of Venus are the symbolic representations of the Keystone and the Pillar of the Heaven. As noted, Venus, which is the constellation of cosmic plasmas, is one of the Nakṣatras of the Heavenly Vault. Refer to section 2.5.8 for the Nakṣatras of the Heavenly Vault.

St. Peter's Square in the Vatican City is another replica of this Keystone. Note that the shape of St. Peter's Square is a replica of the Keystone. The obelisk found at the centre represents the Pillar of the Heaven (refer to section 2.6). The flag of the Vatican City bears two crossed keys. The three crowns above the crossed keys represent the Soma Ponds that are destroyed and rebuilt to release celestial lights three times a day every day. Note that the Soma Pond is the crown of King Varuṇa, who is the sovereign of the whole of our galactic system according to the Ṛgveda. Stone henges including Stonehenge in Great Britain, stone circles, and stone wheels found throughout Europe and other parts of the world are the monuments that celebrate and memorialise the Stone, the celestial ioniser.

2.7.2 The Two Stones of Trita

In RV ix.102.2, the Stone, the celestial ioniser, is described as the two stones of Trita, revealing that the upper and lower discs of the Stone are the fire of Trita, tritium (^3H), which is one of three of Agni's fires. The author is Trita himself and the stanza is addressed to Soma Pavamāna, the purified (read ionised) Soma. In the stanza, Soma is called the 'freed One' from the ionic bond.

RV ix.102.2 author त्रितः, to पवमानः सोमः, metre उष्णिक् छंदः
उप त्रितस्य पाष्यो३ रभक्त यद्गुहा पदं । यज्ञस्य सप्त धामभिरध प्रियं ॥९.१०२.२॥
उप । त्रितस्य । पाष्योः । अभक्त । यत् । गुहा । पदं । यज्ञस्य । सप्त । धामऽभिः । अध । प्रियं ॥९.१०२.२॥
O freed One, enter thou the abode, which is hidden above the two stones of Trita,
With Yajña's beloved seven forms of lights. (RV ix.102.2)

उप <ind. – above (with loc.)>, त्रितस्य <m. gen. sg. of त्रित – of Trita (tritium)>, पाष्योः <f. loc. du. of पाषि – (with उप) above the two stones, i.e. above the upper and lower discs of the celestial ioniser>, अभक्त <m. voc. sg. of अभक्त – O freed One (here it refers to Soma freed from the ionic bond)>, यत् <n. nom. sg. of relative pron. यद् – which, that>, गुहा <ind. – in a hiding place, in secret>, पदं <n. acc. sg. of पद – to a station, home, abode (imperative verb enter or come is omitted) – you enter the abode>;

यज्ञस्य <m. gen. sg. of यज्ञ – of Yajña>, सप्त <n. pl. of सप्तन् – seven>, धामभिः <n. inst. pl. of धामन् – with forms or lights>, अध <ind. inceptive particle – now, then, moreover>, प्रियं <ind. – dear, beloved, adored>.

The word 'freed one' (अभक्तः abhaktaḥ) means "not attached to, detached, unconnected with", that is, the one who is freed from the ionic bonds of the atomic structures. When the ionic bond is broken, both Agni and Soma are freed. The abode hidden above the two stones of Trita refers to the Soma Pond. As we can see in Figure 6, the Soma Pond, the crown of Varuṇa, sits right above the two discs of the Stone, the celestial ioniser. When Soma is discharged from the Stone, he enters into the Soma Pond along with Yajña's seven forms of lights.

2.7.3 The Two Stones at the Base of the Soma Pond

Stanza RV i.109.3 states that the two stones of the Stone, the celestial ioniser, are located at the base of the Soma Pond. Stanzas RV i.109.4 and 7 are presented together, for these stanzas explain the sequence of an event. Stanza RV i.109.3 explains that the beams of light, that is, the plasma blooms, will not be cut off until they restore the strength of the forefathers, that is, until they are fully matured. The forefathers are Agnis and Somas who were discharged earlier and gathered along the triply-spun filaments of plasma blooms. The men who prepare for the Soma sacrifice are delighted, for the two press Stones of the celestial ioniser at the base of the Bowl are continuously discharging cosmic plasmas for Indra and Agni. Both Indra and Agni delight in drinking Soma mead.

RV i.109.3 author कुत्स आंगिरसः, to इंद्राग्नी, metre त्रिष्टुप् छंदः
मा च्छेद्म रश्मीँरिति नाधमानाः पितॄणां शक्तीरनुयच्छमानाः ।
इन्द्राग्निभ्यां कं वृषणो मदंति ता ह्यद्री धिषणाया उपस्थे ॥१.१०९.३॥
मा । छेद्म । रश्मीन् । इति । नाधमानाः । पितॄणां । शक्तीः । अनुऽयच्छमानाः ।
इंद्राग्निऽभ्यां । कं । वृषणः । मदंति । ता । हि । अद्री इति । धिषणायाः । उपऽस्थे ॥१.१०९.३॥
Let us not cut off the cords, thus we may seek to restore the powers of our forefathers.
For Indra and Agni, men are delighted, for here at the base of the Bowl are the two press Stones. (RV i.109.3)

मा <ind. – not>, छेद्म <aor. 1st person pl. of √छिद् – we cut off, split>, रश्मीन् <m. acc. pl. of रश्मि – ropes, cords, rays or beams of light, i.e. plasma blooms>, इति <ind. – thus, as, though>, नाधमानाः <m. nom. pl. of pre. middle pt. नाधमान (√नाध्) – we seeking help, asking, begging>, पितॄणां <m. gen. pl. of पितृ – of forefathers, ancestors>, शक्तीः <f. acc. pl. of शक्ति – powers, strengths>, अनुयच्छमानाः <m. nom. pl. of pre. middle pt. अनुयच्छमान (अनु-√दा) – we restore>;

इंद्राग्निभ्यां <m. dat. du. of इन्द्राग्नी (a dual compound) – for Indra-Agni>, कं <ind. – an enclitic with the particle हि>, वृषणः <m. nom. pl. of वृषन् – men, bulls>, मदंति <pre. 3rd person pl. of √मद् – they rejoice, delight or revel in>, ता <(=तौ) m. nom. du. of pron. तद् – these two>, हि <ind. particle – for, because>, अद्री <m. nom. du. of अद्रि – the two press

Stones (two discs of the Stone)>, धिषणायाः <f. gen. sg. of धिषणा – of the Bowl (the Vault of the Heaven or the Soma Pond)>, उपस्थे <m. loc. sg. of उपस्थ – in/at 'the part which is under', i.e. at the base>.

In stanza RV i.109.4, Aśvinā (du.), the twin Jets, are asked to release and bestow sweet mead, the cosmic plasmas, extracted by the celestial Bowl, the Soma Pond. Aśvinā are described as having two hands and beautiful wings. Two hands refer to two Jets that are released at once, one from the Upper Vault of the Heaven and the other from the Lower Vault of the Heaven. The beautiful wings of Aśvinā (du.) refer to plasma blooms that are plucked and released when the twin Jets fly upwards over the summit of the Heaven.

RV i.109.4 author कुत्स आंगिरसः, to इंद्राग्नी, metre त्रिष्टुप् छंदः
युवाभ्यां देवी धिषणा मदायेंद्राग्नी सोममुशती सुनोति ।
तावश्विना भद्रहस्ता सुपाणी आ धावतं मधुना पृंक्तमप्सु ॥१.१०९.४॥
युवाभ्यां । देवी । धिषणा । मदाय । इंद्राग्नी इति । सोमं । उशती । सुनोति ।
तौ । अश्विना । भद्रऽहस्ता । सुपाणी इति सुपाणी । आ । धावतं । मधुना । पृंक्तं । अप्ऽसु ॥१.१०९.४॥
For you both, O Indra and Agni, the celestial Bowl extracts the Soma desiring for your rapture. With two auspicious hands and beautiful wings, O Aśvinā, hasten ye with sweet mead midst the celestial Waters and bestow richly. (RV i.109.4)

युवाभ्यां <m. dat. du. of 2nd person pron. युष्मद् – for you both>, देवी <f. nom. sg. of adj. देव – heavenly, celestial>, धिषणा <f. nom. sg. of धिषणा – the Bowl (here, refers to the Soma Pond)>, मदाय <m. dat. sg. of मद – for rapture, delight>, इंद्राग्नी <m. voc. of the dual compound इंद्राग्नी – O Indra-Agni>, सोमं <m. acc. sg. of सोम – Soma>, उशती <f. nom. sg. of adj. उशत् – wishing, desiring>, सुनोति <pre. 3rd person. sg. of √सु 5P – it (the Bowl) extracts>;

तौ <m. nom. du. of pron. तद् – the two>, अश्विना <(=अश्विनौ) m. voc. du. of अश्विन् – O Aśvinā (du.)>, भद्रहस्ता <(=भद्रहस्तौ) m. voc. du. of adj. भद्रहस्त – with two auspicious hands>, सुपाणी <(=सुपर्णा) m. voc. du. of adj. सुपाणि (=सुपर्ण) – with beautiful wings>, आ <ind. prefix to verb धावतं (√धाव्)>, आ-धावतं <ipv. 2nd person du. of आ-√धाव् 1P – ye (du.) come running, hasten>, मधुना <n. inst. sg. of मधु – with sweet mead, with Soma>, पृंक्तं <(पृङ्क्तं) ipv. 2nd person. du. of √पृच् – ye (du.) bestow richly>, अप्सु <f. loc. pl. of अप् – midst the Waters, i.e. midst cosmic plasmas>.

Now, in RV i.109.7, Indra and Agni, the wielders of thunderbolt, were asked to impel 'us'. Here, 'us' refers to Sūrya's beams of light, that is, plasma blooms.

RV i.109.7 author कुत्स आंगिरसः, to इंद्राग्नी, metre त्रिष्टुप् छंदः
आ भरतं शिक्षतं वज्रबाहू अस्माँ इंद्राग्नी अवतं शचीभिः ।
इमे नु ते रश्मयः सूर्यस्य येभिः सपित्वं इंपतरी न आसन् ॥१.१०९.७॥
आ । भरतं । शिक्षतं । वज्रबाहू । अस्मान् । इंद्राग्नी इति । अवतं । शचीभिः ।
इमे । नु । ते । रश्मयः । सूर्यस्य । येभिः । सऽपित्वं । पितरः । नः । आसन् ॥१.१०९.७॥
O Indra and Agni, the thunder-armed, bring ye riches and bestow, and impel us with the power of your Speeches. These are Sūrya's beams of light wherewith our forefathers gather and dwell. (RV i.109.7)

आ <ind. prefix to verb भरतं (√भृ)>, आ-भरतं <ipv. 2nd person du. of आ-√भृ – ye (du.) bring, carry (wealth, riches)>, शिक्षतं <ipv. 2nd person du. of √शिक्ष् – ye (du.) give, bestow>, वज्रबाहू <m. voc. du. of adj. वज्रबाहु – thunder-armed, wielders of thunderbolt>, अस्मान् <m. acc. pl. of अस्मद् – us>, इंद्राग्नी <m. voc. of a dual compound इंद्राग्नी – O Indra-

Agni>, अवतं <ipv. 2nd person du. of √अव् – ye (du.) drive, impel>, शचीभिः <f. inst. pl. of शची – with Speeches, with the power of Speeches>;

इमे <m. nom. pl. of pron. इदं – these here>, नु <ind. – then, now, indeed, surely>, ते <m. nom. pl. of pron. तद् – these, they>, रश्मयः <m. nom. pl. of रश्मि – cords, rays of light, beams of light>, सूर्यस्य <m. gen. sg. of सूर्य – of Sūrya>, येभिः <m. inst. pl. of यद् – with/by which, wherewith>, सपित्वं <ind(?). – united, gathered>, पितरः <m. nom. pl. of पितृ – deceased forefathers, deceased ancestors>, नः <m. gen. pl. of अस्मद् – our>, आसन् <ipf. 3rd person pl. of √अस् 2P – they abided, dwelt>.

2.7.4 The Stones Are Impelled by Savitā

The four stanzas of hymn RV x.175 are addressed to the stones (ग्रावाणः grāvāṇaḥ), the firestones of the Stone, the celestial ioniser. The author is Ūrdhvagrāvā Ārbudi (ऊर्ध्वग्रावा आर्बुदिः). Ūrdhvagrāvā means "one who raised the stone for pressing the Soma". Ārbudi is a descendent of the serpent-like demon Arbuda (अर्बुदः). In the Ṛgveda, 'pressing the Soma' means pressing out electrons from the atomic structures, which refers to the process of ionisation.

RV x.175.1 states that Savitā (सविता) impels the stones (ग्रावाणः) that break the ionic bonds to produce cosmic plasmas. Savitā represents the impelling, electric force of Tvaṣṭā (त्वष्टा), magnetism in general and magnetic field in particular. Savitā is asked to impel the firestones of the celestial ioniser by the decree of Mitra and Varuṇa. As we shall see, according to the Ṛgveda, Tvaṣṭā is the first born and the origin of cosmic Mind. In the cosmic field, electric phenomena are subsequent to the magnetic field, Tvaṣṭā, and the magnetic forces, the Ṛbhus.

RV x.175.1 author ऊर्ध्वग्रावार्बुदिः, to ग्रावाणः, metre गायत्री छंदः
प्र वो ग्रावाणः सविता देवः सुवतु धर्मणा । धूर्षु युज्यध्वं सुनुत ॥१०.१७५.१॥
प्र । वः । ग्रावाणः । सविता । देवः । सुवतु । धर्मणा । धूःऽसु । युज्यध्वं । सुनुत ॥१०.१७५.१॥
Let Savitā the Deva, O stones, impel you all by the decree of Mitra-Varuṇa.
Fasten ye to the shafts and press the Soma. (RV x.175.1)

प्र <ind. prefix to verb सुवतु (√सू)>, वः <m. acc. pl. of 2nd person pron. युष्मद् – you all>, ग्रावाणः <m. voc. pl. of ग्रावन् – O stones (it refers to the firestones of the celestial ioniser)>, सविता <m. nom. sg. of सवितृ – Savitā, the impeller, vivifier>, देवः <m. nom. sg. of देव – Deva>, प्र-सुवतु <ipv. 3rd person sg. of प्र-√सू – let him (Savitā) incite, impel>, धर्मणा <n. inst. sg of धर्मन् – by the established order of things, decree of Mitra-Varuṇa, i.e. by the law of cosmic plasmas>;

धूर्षु <f. loc. pl. of धुर् – to the shafts, poles>, युज्यध्वं <ipv. 2nd person pl. of √युज् – ye (the stones) yoke, harness, fasten>, सुनुत <ipv. 2nd person. pl. of √सु 5P – ye (the stones) press out (Soma)>.

In RV x.175.2, the ruddy cows, the cosmic plasmas, are called the medicine. The mischief (दुच्छुना) and the wicked (दुर्मतिः) refer to the ionic bonds of the atomic structures. The ionic bonds and plasma sheaths are considered demons and enemies of the Devas for apparent reasons. Mischievous and wicked ionic bonds need to be broken in order to produce cosmic plasmas. Plasma sheaths must be destroyed to release the matured plasma blooms from the Soma Pond. Released celestial lights have sanative properties conducive to the health and well-being of the Heavenly structures and the Milky Way and to the inhabitants of the worlds within the Milky Way. Thus they are called ‘our medicine’.

RV x.175.2 author ऊर्ध्वग्रावार्बुदिः, to ग्रावाणः, metre गायत्री छंदः
ग्रावाणो अप दुच्छुनामप सेधत दुर्मतिं । उस्राः कर्तन भेषजं ॥१०.१७५.२॥
ग्रावाणः । अप । दुच्छुनां । अप । सेधत । दुःऽमतिं । उस्राः । कर्तन । भेषजं ॥१०.१७५.२॥
O stones, drive ye the mischief away, drive away the wicked.
Produce ye the ruddy cows, our medicine. (RV x.175.2)

ग्रावाणः <m. voc. pl. of ग्रावन् – O stones>, अप <ind. prefix to verb सेधत (√सिध्) – (drive ye) away, off (here the verb सेधत is omitted since the verb is used immediately after>, दुच्छुनां <f. acc. sg. of दुच्छुना – the mischief>, अप <ind. prefix to verb सेधत (√सिध्)>, अप-सेधत <ipv. 2nd person pl. of अप-√सिध् 1P – ye (the stones) ward off, drive away, remove>, दुर्मतिं <m. acc. sg. of दुर्मति – the wicked>;

उस्राः <f. acc. pl. of उस्रा – ruddy cows, i.e. celestial lights, cosmic plasmas>, कर्तन <ipv. 2nd person pl. of √कृ 2P – ye (the stones) make, produce>, भेषजं <n. acc. sg. of भेषज – a remedy, medicine, Water>.

RV x.175.3 author ऊर्ध्वग्रावार्बुदिः, to ग्रावाणः, metre गायत्री छंदः
ग्रावाण उपरेष्वा महीयंते सजोषसः । वृष्णे दधतो वृष्ण्यं ॥१०.१७५.३॥
ग्रावाणः । उपरेषु । आ । महीयंते । सऽजोषसः । वृष्णे । दधतः । वृष्ण्यं ॥१०.१७५.३॥
The lower stones rise high with one accord,
Giving the bull his bull-like strength. (RV x.175.3)

ग्रावाणः <m. nom. pl. of ग्रावन् – the stones>, उपरेषु <m. loc. pl. of उपर – on the lower part, i.e. on the lower disc of the Stone>, आ <ind. particle (with a preceding loc.) – in, at, on>, महीयंते <pre. middle 3rd person pl. of nominal verb √महीय – they are joyous or exalted; rise high>, सजोषसः <m. nom. pl. of adj. सजोषस् – united, being in harmony>;

वृष्णे <m. dat. sg. of वृषन् – to the bull or stallion>, दधतः <sub. 3rd person pl. of √धा 3P – they present, impart to (dat.), bestow>, वृष्ण्यं <n. acc. sg. of वृष्ण्य – manliness, virility>.

RV x.175.4 author ऊर्ध्वग्रावार्बुदिः, to ग्रावाणः, metre गायत्री छंदः
ग्रावाणः सविता नु वो देवः सुवतु धर्मणा । यजमानाय सुन्वते ॥१०.१७५.४॥
ग्रावाणः । सविता । नु । वः । देवः । सुवतु । धर्मणा । यजमानाय । सुन्वते ॥१०.१७५.४॥
Now, O stones, let the Deva Savitā impel you all as the law of Mitra-Varuṇa commands,
For him who worships and offers the Soma sacrifice. (RV x.175.4)

ग्रावाणः <m. voc. pl. of ग्रावन् – O stones>, सविता <m. nom. sg. of सवितृ – Savitā, the Impeller>, नु <ind. – so now, then>, वः <m. acc. pl. of 2nd person pron. युष्मद् – you (the stones)>, देवः <m. nom. sg. of देव – the Deva>, सुवतु <ipv. 3rd person sg. of √सू – let him (Savitā) incite, command, impel>, धर्मणा <n. inst. sg. of धर्मन् – by the law, decree of Mitra-Varuṇa>;

यजमानाय <m. dat. sg. of adj. यजमान – sacrificing, worshipping>, सुन्वते <m. dat. sg. of सुन्वत् – the offerer of the Soma sacrifice>.

2.7.5 The Stones Sing and Make Rumbling Sounds

Stanzas RV x.94.1-5 state that the stones, the firestones of the celestial ioniser, sing and make rumbling sounds. Though cosmic plasmas constantly make Speeches and sing, the stones are the most prominent singers (RV x.94.1). They sing loudly as would a hundred or a thousand men (RV x.94.2). The stanzas RV x.94.1-5 are addressed to the stones (ग्रावाणः pl.)

RV x.94.1 author अर्बुदः काद्रवेयः सर्पः, to ग्रावाणः, metre जगती छंदः
प्रैते वदंतु प्र वयं वदाम ग्रावभ्यो वाचं वदता वदद्भ्यः ।
यदद्रयः पर्वताः साकमाशवः श्लोकं घोषं भरथेंद्राय सोमिनः ॥१०.९४.१॥
प्र । एते । वदंतु । प्र । वयं । वदाम । ग्रावऽभ्यः । वाचं । वदत । वदत्ऽभ्यः ।
यत् । अद्रयः । पर्वताः । साकं । आशवः । श्लोकं । घोषं । भरथ । इंद्राय । सोमिनः ॥१०.९४.१॥
Let the stones raise their voices. Let us sing and sing ye all to the stones, the eloquent singers. When offering Soma, O press stones, moving swiftly together, bring the hymn of praise and the roaring thunder to Indra. (RV x.94.1)

प्र <ind. prefix to verb वदंतु (√वद्)>, एते <m. nom. pl. of एतद् – they (the stones)>, प्र-वदंतु <ipv. 3rd person pl. of प्र-√वद् – (with वाचं) let them raise the voice, sing, utter a cry>, प्र <ind. – forward, on, forth>, वयं <m. nom pl. of pron. अस्मद् – we>, वदाम <ipv. 1st person pl. of √वद् – (with वाचं) we shall or let us sing or speak>, ग्रावभ्यः <m. dat. pl. of ग्रावन् – to the stones>, वाचं <f. acc. sg. of वाच् – Speech, sound>, वदत <ipv. 2nd person pl. of √वद् – (with वाचं) ye all sing or speak>, वद्भ्यः <(=वदेभ्यः) m. dat. pl. of वद – to the speakers who speak well or sensibly, to the eloquent singers>;

यत् <conjunction – when>, अद्रयः पर्वताः <m. voc. pl. of अद्रि पर्वत – O press stones; (अद्रि in pl.) the stones for pressing Soma>, साकं <ind. – together, jointly>, आशवः <m. nom. pl. of adj. आशु – fast, going quickly>, श्लोकं <m. acc. sg. of श्लोक – śloka, praise, hymn of praise>, घोषं <m. acc. sg. of घोष – battle-cry, the sound of drum, the roaring of thunder>, भरथ <pre. 2nd person pl. √भृ – ye bear, carry, bring>, इंद्राय <m. dat. sg. of इंद्र – to Indra>, सोमिनः <m. nom. pl. of adj. सोमिन् – offering Soma>.

RV x.94.2 author अर्बुदः काद्रवेयः सर्पः, to ग्रावाणः, metre जगती छंदः
एते वदंति शतवत्सहस्रवदभि क्रंदंति हरितेभिरासभिः ।
विष्ट्वी ग्रावाणः सुकृतः सुकृत्यया होतुश्चित्पूर्वे हविरद्यमाशत ॥१०.९४.२॥
एते । वदंति । शतऽवत् । सहस्रऽवत् । अभि । क्रंदंति । हरितेभिः । आसऽभिः ।
विष्ट्वी । ग्रावाणः । सुऽकृतः । सुऽकृत्यया । होतुः । चित् । पूर्वे । हविःऽअद्यं । आशत ॥१०.९४.२॥
They sing aloud as if accompanied by a hundred, or a thousand men. They roar with their ruddy-coloured jaws. Having performed their duty, the virtuous stones, even before Hotā, got to taste the oblation. (RV x.94.2)

एते <m. nom. pl. of एतद् – they>, वदंति <pre. 3rd person pl. of √वद् – they speak, sing>, शतवत् <n. nom. sg of adj. शतवत् – accompanied by a hundred>, सहस्रवत् <n. nom. sg. adj. of सहस्रवत् – a thousand, thousand-fold>, अभि <ind. prefix to verb क्रंदंति (√क्रंद्)>, अभि-क्रंदंति <pre. 3rd person pl. of अभि-√क्रंद् – they roar, cry out, sound>, हरितेभिः <n. inst. pl. of adj. हरित् – yellowish, reddish>, आसभिः <n. inst. pl. of आसन् – with jaws, mouths>;

विष्ट्वी <ind. pt. of √विष् – having done, performed>, ग्रावाणः <m. nom. pl. of ग्रावन् – the stones>, सुकृतः <m. nom. pl. of adj. सुकृत् – pious, virtuous>, सुकृत्यया <f. inst. sg. of सुकृत्या – with the righteous act, duty>, होतुः <m. abl. sg. of होतृ – before Hotā (the sacrificer)>, चित् <(चिद्) ind. particle – even>, पूर्वे <m. loc. sg. of पूर्व – earlier, being before (with abl.)>, हविरद्यं <n. acc. sg. of हविरद्य – act of eating or tasting the oblation)>, आशत <aor. middle 3rd person pl. of √अश् – they obtained, gained, got>.

Here, in RV x.94.2, it is said that the stones get to taste the oblation before Hotā (होता). Since the stones produce cosmic plasmas, naturally, it is the stones who get to taste the oblation before Hotā, the sacrificer, obtains and releases it.

For celestial Yajñas, four Ṛtvijas (ऋत्विजः ṛtvijaḥ pl. priests) are involved: Hotā, Adhvaryu (अध्वर्युः), Udgātā (उद्गाता), and Brāhmaṇa (ब्राह्मणः m. one who has divine knowledge). Hotā is the Soma sacrificer who releases Soma from the Soma Pond. Adhvaryu is the Stone, the celestial ioniser, who produces celestial plasmas. Udgātā is the singer of hymns. Agni and Soma are Brāhmaṇas (ब्राह्मणाः pl.). Agnis are fiery protons. Somas are the bright drops of light, the electrons. According to the Ṛgveda, Agnis and Somas are the seers and knowers of all. They are the initiators of cosmic Yajñas.

In stanza RV x.94.3, the radiant red Tree refers to the Soma Pond; the branch, to the Soma plants, the plasma blooms; the fire, to the fire of Trita, tritium (^3H); the body, to the body of hydrogen. The bulls refer to the firestones of the celestial ioniser.

RV x.94.3 author अर्बुदः काद्रवेयः सर्पः, to ग्रावाणः, metre जगती छंदः
एते वदंत्यविदन्नना मधु न्यूंखयंते अधि पक्व आमिषि ।
वृक्षस्य शाखामरुणस्य बप्सतस्ते सूभर्वा वृषभाः प्रेमराविषुः ॥१०.९४.३॥
एते । वदंति । अविदन् । अना । मधु । नि । ऊंखयंते । अधि । पक्वे । आमिषि ।
वृक्षस्य । शाखां । अरुणस्य । बप्सतः । ते । सूभर्वाः । वृषभाः । प्र । ईं । अराविषुः ॥१०.९४.३॥
They sing and make a rumbling sound over the body burnt on the fire, and thus obtained sweet mead. These bulls, nourishing the branch of the radiant red Tree, bellowed loudly.
(RV x.94.3)

एते <m. nom. pl. of एतद् – they>, वदंति <pre. 3rd person pl. of √वद् – they speak, sing>, अविदन् <ipf. 3rd person pl. of √विद् 6P – they found, discovered, obtained>, अना <ind. – hereby, thus>, मधु <n. acc. sg. of मधु – sweet mead, i.e. Soma mead, cosmic plasmas>, न्यूंखयंते <(न्यूङ्खयन्ते) pre. middle 3rd person. pl. of nominal verb √न्यूंखय – they make a 'O (ओ)' sound (i.e. humming or rumbling sound)>, अधि <ind. – (with loc.) over, on, at>, पक्वे <n. loc. sg. of adj. पक्व – roasted or prepared on a fire, i.e. prepared on the fire of Trita (tritium)>, आमिषि <n. loc. sg. of आमिस् (cf. आमिष) – over/on flesh, meat, dead body (refer to the body of hydrogen)>;

वृक्षस्य <m. gen. sg. of वृक्ष – of the tree (refers to the Soma Pond)>, शाखां <f. acc. sg. of शाखा – a branch (refers to the plasma bloom)>, अरुणस्य <m. gen. sg. of adj. अरुण – tawny, red>, बप्सतः <m. gen. sg. of pre. active pt. बप्सत् (√भस्) – shining>, ते <m. nom. pl. of pron. तद् – these>, सूभर्वाः <m. nom. pl. of adj. सूभर्व – feeding or nourishing well>, वृषभाः <m. nom. pl. of वृषभ – bulls (firestones of the celestial ioniser)>, प्र <ind. prefix to verb अराविषुः (√रु)>, ईं <ind. – (Vedic) particle of affirmation>, प्र-अराविषुः <aor. 3rd person pl. of प्र-√रु – they roared, bellowed loudly>.

In RV x.94.4, the maidens refer to the firestones of the celestial ioniser that are laid in a circular formation. These firestones are often called Tvaṣṭā's maidens (see section 2.7.7).

RV x.94.4 author अर्बुदः काद्रवेयः सर्पः, to ग्रावाणः, metre जगती छंदः
बृहद्वदंति मदिरेण मंदिनेंद्रं क्रोशंतोऽविदन्नना मधु ।
संरभ्या धीराः स्वसृभिरनर्तिषुराघोषयंतः पृथिवीमुपब्दिभिः ॥१०.९४.४॥
बृहत् । वदंति । मदिरेण । मंदिना । इंद्रं । क्रोशंतः । अविदन् । अना । मधु ।
संऽरभ्य । धीराः । स्वसृऽभिः । अनर्तिषुः । आऽघोषयंतः । पृथिवीं । उपब्दिभिः ॥१०.९४.४॥
The skilful ones sing aloud, calling on Indra with exhilarating drink, for they obtained sweet mead. Laying hold of one another, the skilful ones danced with the maidens, causing the Milky Way to reverberate with sounds. (RV x.94.4)

बृहत् <ind. – great, mighty, loud>, वदंति <pre. 3rd person pl. of √वद् – they speak, sing>, मदिरेण <mn. inst. sg. of adj. मदिर (=मदिन्) – intoxicating, exhilarating>, मंदिना <mn. inst. sg. of मंदिन् (मन्दिन्) – with Soma drink>, इंद्रं <m. acc. sg. of इंद्र – Indra>, क्रोशंतः <m. nom. pl. of pre. active pt. क्रोशंत् (√क्रुश्) – they crying out, calling out>, अविदन् <ipf. 3rd person pl. of √विद् 6P – they found, obtained>, अना <ind. – hereby, thus>, मधु <n. acc. sg. of मधु – sweet mead, i.e. Soma>;

संरभ्य <ind. pt. of संरभ् – they laying hold of one another>, धीराः <m. nom. pl. of धीर – the wise ones, the skilful ones, i.e. the stones>, स्वसृभिः <f. inst. pl. of स्वसृ – with the sisters or fingers, i.e. with the maidens>, अनर्तिषुः <aor. 3rd person pl. of √नृत् – they (the skilful ones, the stones) danced>, आघोषयंतः <m. nom. pl. of pre. cau. active pt. आघोषयत् (आ-√घुष्) – they causing to sound>, पृथिवीं <f. acc. sg. of पृथिवी – Pṛthivī, the Milky Way>, उपब्दिभिः <m. inst. pl. of उपब्दि (=उपब्द) – with sounds>.

Stanza RV x.94.5 explains the downward releases of cosmic plasmas. The 'mystical birds (सुपर्णाः suparṇāḥ pl.)' are identified with Garuḍas (गरुडाः pl.), the firebirds, and refer to plasma blooms in the Soma Pond (see Figure 7). As noted by Monier-Williams, this mystical bird Garuḍa is the "chief of the feathered race" and the "enemy of the serpent race". The feathered race refers to the plasma blooms; they are called the enemy of the serpent race, for they are used as the weapons by Indra to slay Vṛtra, the serpent of the Heaven, to release cosmic plasmas from the Soma Pond. The serpent race refers to the plasma sheaths. The Greek Phoenix, Slavic Firebird, Egyptian Bennu, and so on are analogues of Garuḍa. In most of the existing Ṛgveda translations, this mystical bird is translated as a vulture, a bird of prey, or an eagle.

According to the Ṛgveda, the word niṣkṛtaṃ (निष्कृतं) refers to both the Place of the Atoned and to the Place of Atonement. The Place of the Atoned is where cosmic plasmas gather. The Place of Atonement is where the cosmic ionisation occurs. In RV x.94.5, niṣkṛtaṃ refers to the Place of the Atoned, Mt. Meru, the Galactic Bar. The celestial lights, released downward, fall upon Mt. Meru, the Galactic Bar, at the centre of the Milky Way. Mt. Meru is called the Place of the Atoned, for cosmic plasmas released downward, both from the Upper Vault and from the Lower Vault of the Heaven, fall and gather on the Galactic Bar (see Figure 13). Note that Mt. Meru, the Galactic Bar, the Place of the Atoned, is located right below the lower disc of the Stone, the celestial ioniser. So in RV x.94.5, it is said that the Place of the Atoned, the Galactic Bar, is 'below the nether Stone'.

In the Ṛgveda, three places are called the Place of the Atoned: the two Soma Ponds and the Galactic Bar. In the Soma Ponds, the discharged cosmic plasmas gather and bloom. On Mt. Meru, the Galactic Bar, cosmic plasmas that are released downward gather there before they are divided into four and flow down along the Arms of the Milky Way. On the other hand, the Stone, the celestial ioniser, is the Place of Atonement where atonement for the sin of the ionic bond occurs. Note that cosmic plasmas are the celestial 'people' atoned for their sin (आगः āgaḥ RV vii.93.7, RV ii.28.5; एनः enaḥ RV i.24.14) of the ionic bonds.

RV x.94.5 author अर्बुदः काद्रवेयः सर्पः, to ग्रावाणः, metre त्रिष्टुप् छंदः
सुपर्णा वाचमक्रतोप द्यव्याखरे कृष्णा इषिरा अनर्तिषुः ।
न्य१ँड्गिन यंत्युपरस्य निष्कृतं पुरू रेतो दधिरे सूर्यश्वितः ॥१०.९४.५॥
सुऽपर्णाः । वाचं । अक्रत । उप । द्यवि । आऽखरे । कृष्णाः । इषिराः । अनर्तिषुः ।
न्यक् । नि । यंति । उपरस्य । निःऽकृतं । पुरु । रेतः । दधिरे । सूर्यऽश्वितः ॥१०.९४.५॥
The birds made their Speech in the Heaven. In the Soma Pond, the spirited dark ones danced. Then downwards they fall and impart an abundant stream and the all-surpassing Sūrya to the Place of the Atoned below the nether Stone. (RV x.94.5)

सुपर्णाः <m. nom. pl. of सुपर्ण – mythical birds (Garuḍas), refer to plasma blooms>, वाचं <f. acc. sg. of वाच् – Speech, sound>, अक्रत <ipf. 3rd person pl. of √कृ – they made, performed>, उप <ind. – in, above, up to (with loc.)>, द्यवि <m. loc sg. of दिव् – in the Heaven>, आखरे <m. loc. sg. of आखर – in the lair, i.e. in the Soma Pond or Soma Receptacle>, कृष्णाः <m. nom. pl. of कृष्ण – Kṛṣṇas, the dark ones (=Varuṇas, cosmic plasmas)>, इषिराः <m. nom. pl. of adj. इषिर – fresh, flourishing, vigorous>, अनर्तिषुः <aor. 3rd person pl. of √नृत् – they danced>;

न्यक् <ind. – downwards, down>, नि <ind. prefix to gen. noun उपरस्य – down, below>, यंति <pre. 3rd person pl. of √इ – they go, flow, fall>, उपरस्य <m. gen. sg. of उपर – of the lower Stone, i.e. the lower disc of the celestial ioniser>, निष्कृतं <n. acc. sg. of निष्कृत – the Place of the Atoned (here it refers to Mt. Meru, the Galactic Bar)>, पुरु <n. acc. sg. of adj. पुरु – much, abundant>, रेतः <n. acc. sg. of रेतस् – a flow, stream, current of seminal fluid, i.e. a draught of Soma or a stream of cosmic plasmas>, दधिरे <prf. 3rd person pl. of √धा – they bestowed on, imparted to>, सूर्यश्वितः <n. acc. sg. of सूर्यविश्वतस् (=सूर्यविश्वतुर्) – the all-surpassing Sūrya>.

In stanza RV x.94.5, the Kṛṣṇas (कृष्णाः Kṛṣṇāḥ pl.), the spirited dark Ones, who dance in the Soma Pond refer to the discharged cosmic plasmas. In the Ṛgveda, the discharged cosmic plasmas are represented by Varuṇa, the plasma dark discharge in particular and cosmic plasmas in general. The colour black or dark royal blue is considered the colour of authority, for it is the colour of King Varuṇa, the monarch of the Heaven and the Milky Way. Thus Kṛṣṇa appears in black or dark-blue skin. Kṛṣṇa of Purāṇic literature carries a flute, has a feather of a peacock on his headband, and is accompanied by a deer or a cow. The flute is a symbolic representation of the sound, the Speech of cosmic plasmas; the feathers of a peacock, plasma blooms. The cow or deer represents the Soma Pond, the womb of cosmic creation, with celestial lights blooming within it. Note the medicine Buddha also appears in royal blue skin.

Devī Kālī (देवी काली), with ten or four arms and a garland of human skulls around her neck, is a symbolic representation of the Stone, the celestial ioniser, the Immolator, who dismembers and burns the bodies of hydrogen to produce cosmic plasmas. Kālī is the Mother of death and creation, for she dismembers the bodies of hydrogen to produce cosmic plasmas, the lights and fires of cosmic creation. Mother Kālī also appears in black or royal blue skin. Śiva is the masculine counterpart of Kālī, thus he represents the Stone, the celestial ioniser, that immolates hydrogen. He destroys the atomic structures to yield cosmic plasmas. The Ḍamaru, the sacred drum of Śiva, is a symbolic representation of the twofold Heavenly Vault, the marvellous Creator's machine. Śiva's matted hair represents cosmic plasmas blooming in the Soma Pond (see Figures 7 and 14). Holding the Ḍamaru in his right hand and the flames of fire in his left hand, Śiva dances to produce and discharge celestial lights. The trident of Śiva represents the triply-spun filament of a plasma bloom, a Soma plant, which is the weapon Indra uses to slay Vṛtra, the serpent of the Heaven, the plasma sheath, to release celestial lights.

The name of the dancing Śiva is Naṭarāja (नटराजः), the Lord of dance. His dance is the dance of the firestones of the Stone, the celestial ioniser. His flying locks represent discharged plasmas blooming in the Soma Pond. The infant-like demon figure trampled under his right foot represents the body of hydrogen, which is crushed and destroyed by his dance and discharged as cosmic plasmas. Note, in the Ṛgveda, the ionic bond of hydrogen is described as a wicked demon that needs to be destroyed in order to produce celestial lights, cosmic plasmas; and Agni is often called an infant. The mount, upon which he stands, represents the Stone, the celestial ioniser. The arched sheath, with three-tongued flames planted on it, that surrounds Śiva represents the Soma Pond, the abode of celestial lights, where discharged cosmic plasmas gather, bloom, and mature. Each three-tongued flame represents the three fires or bodies of Agni: Ekata (protium), Dvita (deuterium), and Trita (tritium). The five-tongued flame in his left hand represents the fire of the celestial ioniser, the fire that crushes and destroys the atomic structures to produce celestial lights. It appears that, among the many Naṭarāja statues and sculptures that exist, the tenth century Chola dynasty bronze sculpture, “Shiva as the Lord of Dance”, included in the Southeast Asian Art collection at the Los Angeles County Museum of Art, is the most authentic representation of Śiva (see Figure 14). This gracefully poised majestic sculpture of Śiva corresponds beautifully to what is described in the Ṛgveda about the celestial ionisation and cosmic plasma discharge.

In the extant text of the Ṛgveda, we find no stanzas addressed to Śiva. However, we find one of Śiva's names, Aghora (अ not + घोरः venerable, sublime), in a few stanzas. In RV x.85.44, Sūryā (सूर्या f.), the feminine counterpart of Sūrya (सूर्यः), is called 'the eye of Śiva the terrible (अघोरचक्षुः aghoracakṣuḥ)'. Aghoracakṣu is a compound word of 'aghora (अघोर Śiva the terrible) + cakṣuḥ (चक्षुः the eye)', thus, the eye of Śiva the terrible. On the other hand, in RV x.37.1, the masculine Sūrya (सूर्यः m.) is called the eye of Mitra and Varuṇa (मित्रस्य वरुणस्य चक्षः).

2.7.6 The Two Arms and Ten Fingers of the Stone

In RV v.43.4, the Stone, the celestial ioniser, is called Adri (अद्रिः adriḥ, a stone for pressing Soma). Adri, a masculine noun, is described in the feminine forms of nouns, pronouns, and adjectives. The upper and lower discs of the Stone, the celestial ioniser, are compared to the two arms of the Stone. The firestones of the Stone, arranged in a circular formation, are compared to the ten fingers (see Figure 8). The filament of the Soma plant refers to the triply-spun filament at the centre of each Soma plant. Beautiful branches refer to blooming plasmas along the surface of the filament.

RV v.43.4 author अत्रिः, to विश्वे देवाः, metre त्रिष्टुप् छंदः
दश क्षिपो युंजते बाहू अद्रिं सोमस्य या शमितारा सुहस्ता ।
मध्वो रसं सुगभस्तिर्गिरिष्ठां चनिश्चदद्दुदुहे शुक्रमंशुः ॥५.४३.४॥
दश । क्षिपः । युंजते । बाहू इति । अद्रिं । सोमस्य । या । शमितारा । सुऽहस्ता ।
मध्वः । रसं । सुऽगभस्तिः । गिरिऽस्थां । चनिश्चदत् ।दुदुहे । शुक्रं । अंशुः ॥५.४३.४॥
Two arms and ten fingers set the Stone, Soma's dexterous immolator, ready.
The Stone yields the bright juice of Soma dwelling on the mountains. And the filament of the Soma plant with beautiful branches shines brilliantly. (RV v.43.4)

दश <f. nom. pl. of दशन् – ten>, क्षिपः <f. nom. pl. of क्षिप् – fingers (a symbolic representation of the firestones of the celestial ioniser)>, युंजते <(युञ्जते) pre. middle 3rd person pl. of √युज् – they fasten, harness; make ready, prepare, set to work>, बाहू <f. nom. du. of बाहु – two arms, i.e. the two stones, the upper and lower discs of the celestial ioniser>, अद्रिं <f. acc. sg. of अद्रि – a stone for pressing Soma, i.e. the Stone, the celestial ioniser>, सोमस्य <m. gen. sg. of सोम – of Soma>, या <f. nom. sg. of यद् – who, which>, शमितारा <f. nom. sg. of शमितार (cf. शमितृ) – the immolator, preparer>, सुहस्ता <f. nom. sg. of adj. सुहस्त – skillful, dexterous with the hands>;

मध्वः <m. gen. sg of मधु – of sweet mead, of Soma>, रसं <m. acc. sg. of रस – the sap or juice of plants>, सुगभस्तिः <m. nom. sg. of adj. सुगभस्ति – having beautiful branches>, गिरिष्ठां <m. acc. sg. of adj. गिरिष्ठा – dwelling on the mountains>, चनिश्चदत् <intensive pt. of श्चन्द् – shining brilliantly>, दुदुहे <prf. middle 3rd person sg. of √दुह् – the Stone gives, yields>, शुक्रं <m. acc. sg. of adj. शुक्र – clear, bright>, अंशुः <m. nom. sg. of अंशु – "a filament (especially of the Soma plant)">.

2.7.7 Tvaṣṭā's Ten Maidens

In stanza RV i.95.2, the firestones of the Stone are called the ten maidens of Tvaṣṭā. This indicates that Tvaṣṭā, the magnetic field, plays a role in keeping the firestones of Trita, tritium (^3H), in a circular formation. Tvaṣṭā's ten maidens produce an embryonic particle of cosmic plasmas, in this case Agni, and lead him to the people. Note again, in the Ṛgveda, men or people or heroes usually refer to the Devas and Agnis in particular.

RV i.95.2 author कुत्स आंगिरसः, to अग्निरग्निरौषसो वा, metre त्रिष्टुप् छंदः
दशेमं त्वष्टुर्जनयंत गर्भमतंद्रासो युवतयो विभृत्रं ।
तिग्मानीकं स्वयशसं जनेषु विरोचमानं परि षीं नयंति ॥१.९५.२॥
दश । इमं । त्वष्टुः । जनयंत । गर्भं । अतंद्रासः । युवतयः । विऽभृत्रं ।
तिग्मऽअनीकं । स्वऽयशसं । जनेषु । विऽरोचमानं । परि । सीं । नयंति ॥१.९५.२॥
Tvaṣṭā's ten maidens, alert and vigilant, produce this embryonic particle borne in various directions. They guide this fiery, glorious, fulgent embryonic particle to the people of all races. (RV i.95.2)

दश <f. nom. pl. of दशन् – ten>, इमं <m. acc. sg. of pron. इदं – this>, त्वष्टुः <m. gen. sg. of त्वष्टृ – of Tvaṣṭā>, जनयंत <inj. middle 3rd person pl. of √जन् 10Ā – they beget, generate>, गर्भं <m. acc. sg. of गर्भ – an embryonic particle, in this case Agni>, अतंद्रासः <f. nom. pl. of adj. अतंद्र – alert, wakeful, watchful>, युवतयः <f. nom. pl. of युवति – maidens>, विभृत्रं <m. acc. sg. of adj. विभृत्र – being borne about or in various directions>;

तिग्मानीकं <m. acc. sg. of adj. तिग्मानीक (=तिग्मभृष्टि) – sharp-pointed, fiery>, स्वयशसं <m. acc. sg. of adj. स्वयशस् – innately glorious or illustrious>, जनेषु <m. loc. pl. of जन – among/to the living, men, races, people>, विरोचमानं <m. acc. sg. of adj. विरोचमान – most shining, bright, splendid, fulgent>, परि <ind. preposition – about, around>, षीं <(सीं) ind. – enclitic particle>, नयंति <pre. 3rd person pl. of √नी – they lead, guide, direct towards or to (loc.)>.

2.7.8 The Stone Wheel

In RV x.101.7, the Soma Press, the Stone, the celestial ioniser, is called the Stone Wheel (अश्मचक्रं). The stanza explains that the Soma Pond has long troughs as Soma vessels. In the stanza, Soma vessels refer to plasma blooms. The Soma Pond is armoured with the plasma sheath and furnished with the Stone Wheel, the celestial ioniser (see Figures 6, 7, and 8). The Author is Budha (बुधः), a descendant of Soma. The stanza is addressed to the Viśve Devas and the Ṛtvijas (विश्वे देवा ऋत्विजो वा). The Ṛtvijas (ऋत्विजः pl.) are the priests, the sacrificers.

RV x.101.7 author बुधः सौम्यः, to विश्वे देवा ऋत्विजो वा, metre त्रिष्टुप् छंदः
प्रीणीताश्वान्हितं जयाथ स्वस्तिवाहं रथमित्कृणुध्वं ।
द्रोणाहावमवतमश्मचक्रमंसत्रकोशं सिंचता नृपाणं ॥१०.१०१.७॥
प्रीणीत । अश्वान् । हितं । जयाथ । स्वस्तिऽवाहं । रथं । इत् । कृणुध्वं ।
द्रोणऽआहावं । अवतं । अश्मऽचक्रं । अंसत्रकोशं । सिंचत । नृऽपानं ॥१०.१०१.७॥
Make ye the horses satisfied and ready; ye shall acquire a chariot laden with riches.
Ye pour out the drink of men to the reservoir that has long troughs as Soma vessels and the sheath for armour, and is furnished with the Stone Wheel. (RV x.101.7)

प्रीणीत <ipv. 2nd person pl. of √प्री – ye (Ṛtvijas, sacrificers) please, cheer, comfort>, अश्वान् <m. acc. pl. of अश्व – horses>, हितं <m. acc. sg. of adj. हित – prepared, made ready>, जयाथ <sub. 2nd person pl. of √जि 1P – ye shall acquire, win>, स्वस्तिवाहं <m. acc. sg. of adj. स्वस्तिवाह – carrying wealth, fortune>, रथं <m. acc. sg. of रथ – chariot>, इत् <(इद्) ind. particle of affirmation – only, indeed>, कृणुध्वं <ipv. 2nd person pl. of √कृ – ye make>;

द्रोणाहावं <m. acc. sg of adj. द्रोणाहाव – having long troughs as Soma vessels (refer to plasma blooms)>, अवतं <m. acc. sg. of अवत – to cistern or reservoir, i.e. to the Soma Pond>, अश्मचक्रं <m. acc. sg. of adj. अश्मचक्र – furnished with stone wheel (i.e. the Stone)>, अंसत्रकोशं <m. acc. sg. of adj. अंसत्रकोश – having sheath for armour>, सिंचत <सिञ्चत ipv. 2nd person pl. of √सिच् – ye pour out, discharge>, नृपाणं <n. acc. sg of नृपाण – drink of men, i.e. drink of Agnis>.

Here, the reservoir refers to the Soma pond; long troughs, to plasma blooms. The sheath refers to Vṛtra, the plasma sheath. At the growing and maturing stage of plasma blooms, the plasma sheath, Vṛtra, serves as an armour to protect the discharged cosmic plasmas within the Soma Pond until they are fully matured and ready to be released.

2.7.9 The Two Stones, Mortar and Pestle

The nine stanzas of RV i.28 are addressed to Indra (इंद्रः RV i.28.1-4), Mortar (उलूखलं RV i.28.5-6), and Mortar and Pestle (उलूखलमुसलौ RV i.28.7-8). The last stanza (RV i.28.9) is addressed to the several different Devas: Prajāpati (प्रजापतिः), Ruddy Moon (हरिश्चंद्रः), Soma Press and Shield (अधिषवणचर्मः), and Soma (सोमः). Even though the stanzas are addressed to the various Devas, the entire hymn describes the Stone, the celestial ioniser, and specific aspects of the disks and the firestones of the Stone.

Prajāpati is the lord of all created beings. The Ruddy Moon refers to the Soma Pond; the Soma press, to the Stone, the celestial ioniser; the shield, to the plasma sheath. The poet is Śunaḥśepa (शुनःशेपः), which literally means 'the tail (शेपः) of the auspicious One (शुनः)'. Here, the auspicious One refers to the Stone, the celestial ioniser, not Vāyu as indicated by Monier-Williams. The tail refers to the Soma Pond with plasma blooms growing and maturing in it.

In stanza RV i.28.1, the Stone is described as 'many-footed', for both upper and lower discs of the Stone consist of many firestones of Trita, tritium (^3H), as shown in images 1 and 5 of Figure 8.

RV i.28.1 author शुनःशेप आजीगर्तिः, to इंद्रः, metre अनुष्टुप् छंदः
यत्र ग्रावा पृथुबुध्न ऊर्ध्वो भवति सोतवे । उलूखलसुतानामवेद्विंद्र जल्गुलः ॥१.२८.१॥
यत्र । ग्रावा । पृथुऽबुध्नः । ऊर्ध्वः । भवति । सोतवे । उलूखलऽसुतानां । अव । इत् । ऊं इति । इंद्र । जल्गुलः ॥१.२८.१॥
There where the many-footed Stone is raised to press out Soma,
Drink eagerly, O Indra, the Soma pressed in the mortar. (RV i.28.1)

यत्र <ind. – where, wherein, in which place>, ग्रावा <m. nom. sg. of ग्रावन् – the Stone, the celestial ioniser>, पृथुबुध्नः <m. nom. sg. of adj. पृथुबुध्न – many-footed (पृथु adj. manifold + बुध्न mn. lowest part of anything as root, foot)>, ऊर्ध्वः <m. nom. sg. of adj. ऊर्ध्व – raised, erected>, भवति <pre. 3rd person sg. of √भू – it (the Stone) is>, सोतवे <m. dat. sg. of सोतु – (सोतवे dat. as infinitive) to press out Soma>;

उलूखलसुतानां <m. gen. pl. of adj. उलूखलसुत – pressed out in a mortar (as the Soma)>, अव <ind. prefix to verb जल्गुलः (√गॄ)>, इत् <(इद्) ind. particle of affirmation – just, even (emphasising preceding words)>, उ <ind. particle implying assent, command>, इंद्र <m. voc. sg. of इंद्र – O Indra>, अव-जल्गुलः <sub. 2nd person sg. of अव-√गॄ 6P – you eagerly swallow down, drink (with gen.)>.

In stanza RV i.28.2, the two discs of the Stone are compared to the two mounts (जघना jaghanā du.), upon which the Soma Pond stands (see Figure 6). For jaghana (जघन sg.), definitions of "the hinder part", "the hinder part of an altar", and so on are provided by Monier-Williams. As can be seen in Figure 6, the two discs of the Stone are the hinder part or the mount of the Soma Pond; and the Soma Pond is the altar of Soma sacrifices. Thus, the definitions provided by Monier-Williams for jaghana (जघन sg.) are descriptive and accurate. Here, 'jaghanā (जघना)', the dual form of jaghana, is interpreted as the 'two mounts' of the Soma Pond.

RV i.28.2 author शुनःशेप आजीगर्तिः, to इंद्रः, metre अनुष्टुप् छंदः
यत्र द्वाविव जघनाधिषवण्या कृता । उलूखलसुतानामवेद्विंद्र जल्गुलः ॥१.२८.२॥
यत्र । द्वौऽइव । जघना । अधिऽसवन्या । कृता । उलूखलऽसुतानां । अव । इत् । ऊं इति । इंद्र । जल्गुलः ॥१.२८.२॥

There where, the two mounts, the upper and lower discs of the Stone, press out Soma,
Drink eagerly, O Indra, the Soma pressed in the mortar. (RV i.28.2)

यत्र <ind. – where, wherein, in which place>, द्वौ <m. nom. du. of pron. द्वि – the two>, इव <ind. – like, in the same manner as>, जघना <(=जघनौ) m. nom. du. of जघन – two hinder parts of an altar (refer to the two discs of the Stone; considering the context of the stanza it is interpreted as two mounts)>, अधिषवण्या <m. nom. du. of अधिषवण्य – "the two parts of the hand-press for extracting and straining Soma juice", i.e. the upper and lower discs of the Soma press, the Stone, the celestial ioniser>, कृता <m. nom. du. of adj. कृत् – performing, doing (pressing out Soma)>;

उलूखलसुतानां अव इत् उ इंद्र जल्गुलः <see RV i.28.1 – Drink eagerly, O Indra, the Soma pressed in the mortar.>.

RV i.28.3 author शुनःशेप आजीगर्तिः, to इंद्रः, metre अनुष्टुप् छंदः
यत्र नार्यपच्यवमुपच्यवं च शिक्षते । उलूखलसुतानामवेद्विंद्र जल्गुलः ॥१.२८.३॥
यत्र । नारी । अपऽच्यवं । उपऽच्यवं । च । शिक्षते । उलूखलऽसुतानां । अव । इत् । ऊं इति । इंद्र । जल्गुलः
॥१.२८.३॥
There where the maiden undertakes the shaking of the firestones to and fro,
Drink eagerly, O Indra, the Soma pressed in the mortar. (RV i.28.3)

यत्र <ind. – where, wherein, in which place>, नारी <f. nom. sg. of नारी - a woman, i.e. the maiden of Tvaṣṭā>, अपच्यवं <m. acc. sg. of अपच्यव – the act of moving or shaking away (the firestones of the celestial ioniser)>, उपच्यवं <m. acc. sg. of उपच्यव – the act of shaking or moving towards>, च <ind. – and>, शिक्षते <pre. middle 3rd person sg. of √शिक्ष् – she (the maiden) undertakes>;

उलूखलसुतानां अव इत् उ इंद्र जल्गुलः <see RV i.28.1 – Drink eagerly, O Indra, the Soma pressed in the mortar.>.

RV i.28.4 author शुनःशेप आजीगर्तिः, to इंद्रः, metre अनुष्टुप् छंदः
यत्र मंथां विबध्नते रश्मीन्यमितवा इव । उलूखलसुतानामवेद्विंद्र जल्गुलः ॥१.२८.४॥
यत्र । मंथां । विऽबध्नते । रश्मीन् । यमितवैऽइव । उलूखलऽसुतानां । अव । इत् । ऊं इति । इंद्र । जल्गुलः
॥१.२८.४॥
Where they fasten the beams of light together as if to hold up the churning staff,
Drink eagerly, O Indra, the Soma pressed in the mortar. (RV i.28.4)

यत्र <ind. – where, wherein, in which place>, मंथां <f. acc. sg. of मंथ (मन्थ) – churning-staff>, विबध्नते <pre. middle 3rd person pl. of वि-√बन्ध् – they bind, tie, fix, fasten>, रश्मीन् <m. acc. pl. of रश्मि – ropes, cords, rays of light, beams>, यमितवै <infinitive of √यं – to sustain, hold up, raise>, इव <ind. – like, as it were, as if>;

उलूखलसुतानां अव इत् उ इंद्र जल्गुलः <see RV i.28.1 – Drink eagerly, O Indra, the Soma pressed in the mortar.>.

RV i.28.5 author शुनःशेप आजीगर्तिः, to उलूखलं, metre अनुष्टुप् छंदः
यच्चिद्धि त्वं गृहेगृह उलूखलक युज्यसे । इह द्युमत्तमं वद जयतामिव दुंदुभिः ॥१.२८.५॥
यत् । चित् । हि । त्वं । गृहेऽगृहे । उलूखलक । युज्यसे । इह । द्युमत्ऽतमं । वद । जयतांऽइव । दुंदुभिः
॥१.२८.५॥
O Mortar, when thou art indeed yoked to all assistants,
Utter thou the loudest sound as the drum of conquerors. (RV i.28.5)

यच्चिद् <(यत् + चित्) if indeed, when indeed>, हि <ind. particle – surely>, त्वं <m. nom. sg. of 2nd person pron. युष्मद् – you>, गृहेगृहे <m. loc. sg. of गृहगृह – in/among every assistant, all assistants>, उलूखलक <n. voc. sg. of उलूखलक – O Mortar>, युज्यसे <pre. passive 2nd person sg. of √युज् – you are yoked to, joined (loc.)>;

इह <ind. – here, now>, द्युमत्तमं <n. acc. sg. of adj. द्युमत्तम (द्युमत् + तम) – the loudest (sound)>, वद <ipv. 2nd person sg. of √वद् – you utter>, जयतां <m. gen. pl. of जयत् – of conquerors>, इव <ind. – like, as>, दुंदुभिः <mf. nom. sg. of दुंदुभि – a large drum>.

RV i.28.6 author शुनःशेप आजीगर्तिः, to उलूखलं, metre अनुष्टुप् छंदः
उत स्मं ते वनस्पते वातो वि वात्यग्रमित् । अथो इंद्राय पातवे सुनु सोममुलूखल ॥१.२८.६॥
उत । स्म । ते । वनस्पते । वातः । वि । वाति । अग्रं । इत् । अथो इति । इंद्राय । पातवे । सुनु । सोमं । उलूखल ॥१.२८.६॥
O Lord of the Forest, Vāta just blew forth all of your pressed Soma.
Thus, press thou the Soma, O Mortar, for Indra to drink. (RV i.28.6)

उत <ind. – and, also, even>, स्म <ind. particle joined with pre. verb or pre. pt. to make them a past tense>, ते <m. gen. sg. of 2nd person pron. युष्मद् – of you, yours>, वनस्पते <m. voc. sg. of वनस्पति – O Lord of the Forest (refers to the Stone, the celestial ioniser)>, वातः <m. nom. sg. of वात – Vāta>, वि-वाति <pre. 3rd person sg. of वि-√वा – (with स्म) he blew in every direction>, अग्रं <n. acc. sg. of अग्र – "a measure of food given as alms; (also) rest, remainder", i.e. the remainder or all of the food (pressed Soma)>, इत् <(इद्) ind. particle – even, just>;

अथो <ind. – now, likewise, therefore>, इंद्राय <m. dat. sg. of इंद्र – for Indra>, पातवे <infinitive of √पा – to drink>, सुनु <ipv. 2nd person sg. of √सु – you press out>, सोमं <m. acc. sg. of सोम – Soma juice>, उलूखल <n. voc. sg. of उलूखल – O Mortar (also refers to the Stone)>.

In stanza RV i.28.7, the two sacrificers refer to the upper and the lower discs of the Stone. The sacrificial food refers to the Soma discharged by the Stone, the celestial ioniser.

RV i.28.7 author शुनःशेप आजीगर्तिः, to उलूखलमुसलौ, metre गायत्री छंदः
आयजी वाजसातमा ता ह्यु१च्चा विजर्भृतः । हरी इवांधांसि बप्सता ॥१.२८.७॥
आयजी इत्याऽयजी । वाजऽसातमा । ता । हि । उच्चा । विऽजर्भृतः । हरी इवेति हरीऽइव । अंधांसि । बप्सता ॥१.२८.७॥
Two sacrificers, procuring the best sacrificial food, carry it up into the Heaven,
Like two bay horses champing herbs. (RV i.28.7)

आयजी <m. nom. du. of आयजि – two worshippers, sacrificers (refer to the mortar and pestle, i.e. the upper and lower discs of the Stone, the celestial ioniser)>, वाजसातमा <m. nom. du. of वाजसातम (superlative of वाजसा=वाजसनि=वाज sacrificial food + सनि gaining procuring) – procuring the best sacrificial food>, ता <m. nom. du. of 3rd person pron. तद् – two, both>, हि <ind. particle – indeed, surely>, उच्चा <ind. – above, upwards (to the Heaven)>, वि-जर्भृतः <int. 3rd person du. of वि-√भृ – they (du.) raise, lift up, carry>;

हरी <m. nom. du. of हरि – two bay or ruddy horses>, इव <ind. – like, as>, अंधांसि <n. acc. pl. of अंधस् – herbs>, बप्सता <m. nom. du. of pre. active pt. बप्सत् (√भस्) – the two bay horses chewing, masticating>.

In RV i.28.8, the two discs of the Stone, the Soma Press, are called the two lords of the Forest.

RV i.28.8 author शुनःशेप आजीगर्तिः, to उलूखलमुसलौ, metre गायत्री छंदः
ता नो अद्य वनस्पती ऋष्वावृष्वेभिः सोतृभिः । इंद्राय मधुमत्सुतं ॥१.२८.८॥
ता । नः । अद्य । वनस्पती इति । ऋष्वौ । ऋष्वेभिः । सोतृऽभिः । इंद्राय । मधुऽमत् । सुतं ॥१.२८.८॥
Ye two lords of the Forest, both swift, with the swiftly moving presser stones, press today
The Soma rich in honey for Indra. (RV i.28.8)

ता <m. nom. du. of 3rd person pron. तद् – two, both>, नः <m. dat. pl. of 1st person pron. अस्मद् – for us>, अद्य <ind. – today>, वनस्पती <m. nom. du. of वनस्पति – two lords of the Forest (refer to the upper and lower discs of the Stone, the celestial ioniser>, ऋष्वौ <m. nom. du. of adj. ऋष्व – both swift>, ऋष्वेभिः <m. inst. pl. of adj. ऋष्व – swiftly moving>, सोतृभिः <m. inst. pl. of सोतृ – with Soma pressers, i.e. the firestones of the celestial ioniser>;

इंद्राय <m. dat. sg. of इंद्र – for Indra>, मधुमत् <n. acc. sg. of adj. मधुमत् – sweet, rich in honey>, सुतं <m. acc. sg. of सुत – the pressed out, extracted (Soma juice)>.

In the stanza, RV i.28.9, the strainer refers to the filamentary structure of plasma blooms, or Soma plants; and the Cow, to the Soma Pond.

RV i.28.9 author शुनःशेप आजीगर्तिः, to प्रजापतिर्हरिश्चंद्रोऽधिषवणचर्म सोमो वा, metre गायत्री छंदः
उच्छि॒ष्टं च॒म्वो᳚र्भ॒र सोमं॑ प॒वित्र॒ आ सृ॑ज । नि धे॑हि॒ गोरधि॑ त्व॒चि ॥१.२८.९॥
उत् । शि॒ष्टं । च॒म्वोः᳚ । भ॒र । सोमं॑ । प॒वित्रे॑ । आ । सृ॒ज॒ । नि । धे॒हि॒ । गोः । अधि॑ । त्व॒चि ॥१.२८.९॥
Fetch thou any remnant and pour the Soma on the strainer in the two Vessels,
Place a sheath over the Cow. (RV i.28.9)

उत् <(उद्) ind. prefix to verb सृज (√सृज्)>, शिष्टं <n. acc. sg. of शिष्ट – anything that remains or is left, remnant>, चम्वोः <f. loc. du. of चमू – within the two Vessels (here, the two Vessels refer to the Vault of the Heaven and the Soma Pond within>, आ-भर <ipv. 2nd person sg. of आ-√भृ – you bring, carry, fetch>, सोमं <m. acc. sg. of सोम – Soma>, पवित्रे <n. loc. sg. of पवित्र – on the strainer>, उत्-सृज <ipv. 2nd person sg. of उत्-√सृज् – you pour out, send forth>;

नि-धेहि <ipv. 2nd person sg. of नि-√धा – you put, place>, गोः <m. abl. sg. of गो – (with अधि) over the cow>, अधि <ind. – (with abl.) over>, त्वचि <f. loc. sg. of त्वच् – skin, hide, i.e. plasma sheath>.

2.7.10 Like a Real Comb

In stanza RV i.191.1, the Stone, the celestial ioniser, is compared to a comb (see Figure 9). This may sound far-fetched on the face of it. However, when we consider that, in the Ṛgveda, the creation of the twofold Heavenly Vault and cosmic plasma discharge and the spinning of plasma blooms are compared to weaving, that the 'teeth' of a Noble Comber are laid in a circular formation just as the firestones of the celestial ioniser are, and that the plasma blooms growing in the Soma Pond resemble the combed wool tops protruding

<Figure 9> A Noble Comb at Bradford Industrial Museum.[78]
Image Credit: Linda Spashett Storye_book. Licensed under the Creative Commons Attribution 3.0 Unported License.

[78] Wikimedia, Bradford Industrial Museum, *Spinning Machine at Bradford Industrial Museum, Bradford, West Yorkshire, England. Understood to be a Noble comb*, https://commons.wikimedia.org/wiki/Category:Bradford_Industrial_Museum#/media/File:Bradford_Industrial_Museum_073.jpg Original image without change (Accesed 2 November 2018).

from the circular comber, then it is easier to see why the Stone, the celestial ioniser, is compared to a 'real comb (सतीनकंकतः)' (compare Figure 9 to Figures 7, 8).

The author of hymn RV i.191 is Viṣaśaṃkāvān Agastya (विषशंकावान् अगस्त्यः). Agastya is the 'caster of mountain'. Viṣaśaṃkāvān (√विष् to consume, eat + अशङ्क adj. fearless + अवान् m. nom. sg. of pre. active pt. अवत् (√अव्) he driving, impelling) may mean 'driving bull' or 'fearless eater and impeller' depending upon how you read the compound word 'viṣaśaṃkāvān'. Note the mountain refers to the Soma Pond. So, the 'caster of mountain' is the Stone, the celestial ioniser. The hymn is addressed to Water-Grass-Sūrya. Water refers to cosmic plasmas; Grass, to plasma blooms, or Soma plants; and Sūrya, to the ignited Agni. Of the sixteen stanzas of RV i.191, stanzas RV i.191.1-5 presented here explain the celestial ionisation process.

In stanzas RV i.191.1 and 2, the Invisibles refer to hydrogen atoms. In stanza RV i.191.3, the Invisibles refer to ionised, charged particles, Agnis and Somas. Both hydrogen atoms and ionised particles are invisible to the naked eye; thus, they are described as such.

RV i.191.1 author विषशंकावानगस्त्यः, to अमृणसूर्याः, metre अनुष्टुप् छंदः
कंकतो न कंकतोऽथो सतीनकंकतः । द्वाविति प्लुषी इति न्य१ ॒ दृष्टा अलिप्सत ॥१.१९१.१॥
कंकतः । न । कंकतः । अथो इति । सतीनऽकंकतः ।
द्वौ । इति । प्लुषी इति । इति । नि । अदृष्टाः । अलिप्सत ॥१.१९१.१॥
A comb and a comb, certainly, like a real comb,
The two burn, thus the Invisibles are now on fire. (RV i.191.1)

कंकतः <(कङ्कतः) m. nom. sg. of कंकत – a comb>, न <ind. – like>, कंकतः <m. nom. sg. of कंकत – a comb>, अथो <(=अथ) ind. – now, then, certainly>, सतीनकंकतः <m. nom. sg. of सतीनकंकत (सतीन real + कंकत comb) – a real comb>;

द्वौ <m. nom. du. of द्वि – the two combs (refer to the upper and lower discs of the Stone, the celestial ioniser)>, इति <ind. – in this manner, thus>, प्लुषी <pre. 3rd person du. of √प्लुष् 4P – (the two) burn, scorch, singe>, इति <ind. – thus>, नि <ind. prefix to verb अलिप्सत (√लिप्)>, अदृष्टाः <m. nom. pl. of अदृष्ट – the Invisibles (here, the Invisibles refer to hydrogen atoms)>, नि-अलिप्सत <aor. middle 3rd person pl. of नि-√लिप् – (they) have been kindled, inflamed, burnt>.

In RV i.191.2, the ionisation process, the process of breaking up the ionic bond of hydrogen by the fire of Trita is explained. Hydrogen atoms, the Invisibles, which can be described as a hermaphrodite or an androgyne, having the two opposite charges of proton (male) and electron (female) in its 'body', is pounded, crushed, and destroyed. As a result, the ionic bond is broken and the charges are separated. Soma (electron) is pressed out, Agni (proton) is left alone, and the ionisation process is completed. When the produced cosmic plasmas are discharged into the Soma Pond, they bloom and mature. In stanza RV i.191.3, these blooming plasmas, or Soma plants, are called grasses: Śara, Kuśara, Darbha, Sairya, Muñja, and Vīraṇa. They are all tall reedy grasses.

In stanza RV i.191.2, the comb, the Stone, the celestial ioniser, takes a feminine form. Note that all the adjectives and participles are in the feminine declension forms.

RV i.191.2 author विषशंकावानगस्त्यः, to अमृणसूर्याः, metre अनुष्टुप् छंदः
अदृष्टान्हंत्यायत्यथो हंति परायती । अथो अवघ्नती हंत्यथो पिनष्टि पिंषती ॥१.१९१.२॥
अदृष्टान् । हंति । आऽयती । अथो इति । हंति । पराऽयती । अथो इति । अवऽघ्नती । हंति । अथो इति । पिनष्टि । पिंषती ॥१.१९१.२॥
Coming towards, it strikes the Invisibles, now moving way, it strikes them.
Then threshing them, it pounds. Now it grinds and crushes them. (RV i.191.2)

अदृष्टान् <m. acc. pl. of अदृष्ट – the Invisibles, i.e. hydrogen atoms>, हंति <pre. 3rd person sg. of √हन् – it (the comb) strikes>, आयती <f. nom. sg. of adj. आयत् – it (the comb) coming near>, अथो <(=अथ) ind. – now, then>, हंति <pre. 3rd person sg. of √हन् – it (the comb) strikes>, परायती <f. nom. sg. of pre. active pt. of परायत् (परा-√इ) – it (the comb) moving away>;

अयो <(=अथ) ind. – then, now>, अवघ्नती <f. nom. sg. of pre. active pt. of अवघ्नत् (अव-√हन्) – it (the comb) threshing>, हंति <pre. 3rd person sg. of √हन् – it strikes, pounds>, अथो <(=अथ) ind. auspicious inceptive particle – now, then>, पिनष्टि <pre. 3rd person sg. of √पिष् – it grinds, destroys>, पिंषती <f. nom. sg. of pre. active pt. of √पिष् – it (the comb) crushes>.

RV i.191.3 author विषशंकावानगस्त्यः, to अमृणसूर्याः, metre अनुष्टुप् छंदः
शरासः कुशरासो दुर्भासः सैर्या उत । मौंजा अदृष्टा बैरिणाः सर्वे साकं न्यलिप्सत ॥१.१९१.३॥
शरासः । कुशरासः । दुर्भासः । सैर्याः । उत । मौंजाः । अदृष्टाः । वैरिणाः । सर्वे । साकं । नि । अलिप्सत ॥१.१९१.३॥
Śara, Kuśara, Darbha, Sairya grasses,
Muñja and Vīraṇa grasses, and the Invisibles, altogether, lit up in flames.
(RV i.191.3)

शरासः <m. nom. pl. of शर – Śara grass>, कुशरासः <m. nom. pl. of कुशर – Kuśara grass>, दर्भासः <m. nom. pl. of दर्भ – Darbha grass>, सैर्याः <m. nom. pl. of सैर्य – Sairya grass>, उत <ind. – and, also>;

मौंजाः <(=मौञ्जाः) m. nom. pl. of मौंज (मौञ्ज) – Muñja grass>, अदृष्टाः <m. nom. pl. of अदृष्ट – the Invisibles (here they are charged particles of cosmic plasmas)>, वैरिणाः <m. nom. pl. of वैरिण (=वीरण) – Vīraṇa grass>, सर्वे <m. nom. pl. of pron. सर्व – all, every>, साकं <ind. – together, at the same time>, नि <ind. prefix to verb अलिप्सत (√लिप्)>, नि-अलिप्सत <aor. 3rd person pl. of नि-√लिप् – they are inflamed, kindled, burnt, lit up in flames>.

In RV i.191.4, the cows and wild beasts refer to the celestial lights, or charged particles of cosmic plasmas, especially Agnis; the cowshed and home, to the Soma Pond.

RV i.191.4 author विषशंकावानगस्त्यः, to अमृणसूर्याः, metre अनुष्टुप् छंदः
नि गावो गोष्ठे असदन्नि मृगासो अविक्षत । नि केतवो जनानां न्य१ दृष्टा अलिप्सत ॥१.१९४.४॥
नि । गावः । गोऽस्थे । असदन् । नि । मृगासः । अविक्षत । नि । केतवः । जनानां । नि । अदृष्टाः । अलिप्सत ॥१.१९१.४॥
The cows settled in the cowshed, the wild beasts settled in their lairs.
Casting the lights of men, the Invisibles lit up in flames. (RV i.191.4)

नि <ind. prefix to verb असदन् (√सद्)>, गावः <m. nom. pl. of गो – cows, i.e. the celestial lights, esp. Agnis>, गोष्ठे <m. loc. sg. of गोष्ठ – in the cow-house, cowshed>, नि-असदन् <ipf. 3rd person pl. of नि-√सद् (नि-√षद्) – they sat down, lay

down>, नि <ind. prefix to verb अविक्षत (√विश्)>, मृगासः <m. nom. pl. मृग – forest animals, wild beasts>, नि-अविक्षत <aor. 3rd person pl. of नि-√विश् – they entered into, settled down>;

नि <ind. (here, it expresses क्षेप or दान) – throwing, casting, tossing>, केतवः <m. nom. pl. of केतु – rays of light, flames, torches>, जनानां <m. gen. pl. of जन – of men, people (the Devas)>, नि <ind. prefix to verb अलिप्सत (√लिप्)>, अदृष्टाः <m. nom. pl. of अदृष्ट – the Invisibles (charged particles of cosmic plasmas)>, नि-अलिप्सत <aor. 3rd person pl. of नि-√लिप् – they lit up in flames>.

RV i.191.5 author विषशंकावानगस्त्यः, to अमृणसूर्याः, metre अनुष्टुप् छंदः
एत उ त्ये प्रत्यदृश्रन्प्रदोषं तस्करा इव । अदृष्टा विश्वदृष्टाः प्रतिबुद्धा अभूतन ॥१.१९१.५॥
एते । ऊं इति । त्ये । प्रति । अदृश्रन् । प्रऽदोषं । तस्कराःऽइव । अदृष्टाः । विश्वऽदृष्टाः ।
प्रतिऽबुद्धाः । अभूतन ॥१.१९१.५॥
And these, these flames appear as torches lit up in the dark.
The Invisibles, the seers of all, have become illuminated. (RV i.191.5)

एते <m. nom. pl. of pron. एतद् – these (केतवः flames)>, उ <ind. particle – and>, त्ये <m. nom. pl. of त्यद् – these>, प्रति <ind. prefix to verb अदृश्रन् (√दृश्)>, प्रति-अदृश्रन् <pre. middle. 3rd person pl. of प्रति-√दृश् – they appear as>, प्रदोषं <ind. – in the evening, in the dark>, तस्कराः <m. nom. pl. of तस्कर – (in pl.) "name of particular Ketus" (ketus केतवः pl. of केतु), flames, torches>, इव <ind. – like, as>;

अदृष्टाः <m. nom. pl. of अदृष्ट – the Invisibles, unmanifested, i.e. freed, ionised Agnis and Somas>, विश्वदृष्टाः <m. nom. pl. of विश्वदृष्ट (विश्व all + दृष्ट perception, observation) – perceiving, observing all, i.e. the seers of all (not "seen by all")>, प्रतिबुद्धाः <m. nom. pl. of adj. प्रतिबुद्ध – awakened, enlightened, illuminated>, अभूतन <aor. 3rd person pl. of √भू – they have been, become>.

Refer to Figure 7 and see these flames, which appear as torches lit up in the dark. In RV i.191.5, the Invisibles, the charged cosmic plasma particles, that is, Agnis and Somas, are called the seers of all. They, Agnis (protons) and Somas (electrons), see all and know all.

2.7.11 Kṣetrapati, the Lord of the Field

Hymn RV iv.57 compares the cosmic plasma discharge process to the ploughing and making of furrows. Soma plants, or plasma blooms, are compared to the furrows drawn in the field. The author is Ṛṣi Gautama Vāmadeva (गौतमः वामदेवः), a descendant of Gotama (गोतमः), who belongs to the Aṅgirās (अङ्गिरसः pl.), the messengers of celestial lights or of the Heaven. Vāmadeva refers to Rudra (रुद्रः), the guardian of the celestial lights and of the Soma Pond. The adjective word vāma means "beautiful, splendid, noble".

In stanzas RV iv.57.1-3, the Stone, the Soma press, the celestial ioniser, is called Kṣetrapati (क्षेत्रपतिः or क्षेत्रस्य पतिः), Lord of the Field. The field (क्षेत्रं) refers to the Soma Pond, the field ploughed by the Stone, the celestial ioniser.

RV iv.57.1 author वामदेवः, to क्षेत्रपतिः, metre अनुष्टुप् छंदः
क्षेत्रस्य पतिना वयं हितेनेव जयामसि । गामश्वं पोषयित्वा स नो मृळातीदृशे ॥४.५७.१॥

क्षेत्रस्य । पतिना । वयं । हितेनऽइव । जयामसि । गां । अश्वं । पोषयित्नु । आ । सः । नः । मृळाति । ईदृशे ॥४.५७.१॥

Through the Lord of the Field, as through the benefactor, we obtain
That which nourishes our cattle and steeds. In such, may he be gracious to us. (RV iv.57.1)

क्षेत्रस्य <n. gen. sg. of क्षेत्र – of the Field>, पतिना <m. inst. sg. of पति – through the Lord, master>, वयं <m. nom. pl. of 1st person pron. अस्मद् – we>, हितेन <m. inst. sg. of हित – through benefactor>, इव <ind. – like, as, as if>, जयामसि <pre. 1st person pl. of √जि – we obtain>;

गां <m. acc. sg. of गो – the cow, the celestial light>, अश्वं <m. acc. sg. of अश्व – the horse, the force behind current of cosmic plasmas>, पोषयित्नु <n. acc. sg. of पोषयित्नु – that which cause to grow or thrive, nourishes>, आ <ind. particle – (after adj.) 'as, like' or simply strengthens the sense of the proceeding word>, सः <m. nom. sg. of pron. तद् – he>, नः <m. dat. pl. of 1st person pron. अस्मद् – to us>, मृळाति <sub. 3rd person sg. of √मृळ् (=√मृड्) – may he be gracious (dat.)>, ईदृशे <f. dat. sg. of adj. ईदृश् – in such, such occasion>.

RV iv.57.2 author गौतमो वामदेव ऋषिः, to क्षेत्रपतिः, metre त्रिष्टुप् छंदः
क्षेत्रस्य पते मधुमंतमूर्मिं धेनुरिव पयो अस्मासु धुक्ष्व । मधुश्चुतं घृतमिव सुपूतमृतस्य नः पतयो मृळयंतु ॥४.५७.२॥
क्षेत्रस्य । पते । मधुऽमंतं । ऊर्मिं । धेनुःऽइव । पयः । अस्मासु । धुक्ष्व ।
मधुऽश्चुतं । घृतंऽइव । सुऽपूतं । ऋतस्य । नः । पतयः । मृळयंतु ॥४.५७.२॥

Like a milch cow, O Lord of the Field, yield thou the wave that bears sweetness for us,
The wave overflowing with sweets like well-clarified ghee. Let the lords of sacrifice be gracious to us. (RV iv.57.2)

क्षेत्रस्य <n. gen. sg. of क्षेत्र – of the Field>, पते <m. voc. sg. of पति – O Lord>, मधुमंतं <m. acc. sg. of adj. मधुमत् – possessing sweetness, sweet>, ऊर्मिं <m. acc. sg. of ऊर्मि – a wave, billow>, धेनुः <f. nom. sg. of धेनु – a milch cow>, इव <ind. – like>, पयः <n. acc. sg of पयस् – fluid, juice, milk>, अस्मासु <m. loc. pl. of 1st person pron. अस्मद् – on us>, धुक्ष्व <ipv. 2nd person sg. of √दुह् – you yield, produce>;

मधुश्चुतं <m. acc. sg. of adj. मधुश्चुत् – (the wave) distilling sweetness, overflowing with sweets>, घृतं <n. acc. sg. of घृत – clarified butter, ghee>, इव <ind. – like>, सुपूतं <n. acc. sg. of adj. सुपुत – well clarified>, ऋतस्य <m. gen. sg. of ऋत – of sacrifice>, नः <m. dat. pl. of 1st person pron. अस्मद् – to us>, पतयः <m. nom. pl. of पति – lords, masters>, मृळयंतु <ipv. 3rd person pl. of √मृळ् (=√मृड्) – let them be gracious>.

Though all the Devas, who are involved in the celestial Yajña, can be the lords of sacrifice, in RV iv.57.2, the lords of sacrifice refer to the firestones of the Stone. Stanza RV iv.57.2 is a supplication to the Lord of the Field, the Stone, the celestial ioniser, for abundantly yielding the wave that bears sweetness.

RV iv.57.3 author वामदेवः, to क्षेत्रपतिः, metre त्रिष्टुप् छंदः
मधुमतीरोषधीर्द्याव आपो मधुमन्नो भवत्वंतरिक्षं ।
क्षेत्रस्य पतिर्मधुमान्नो अस्त्वरिष्यंतो अन्वेनं चरेम ॥४.५७.३॥
मधुऽमतीः । ओषधीः । द्यावः । आपः । मधुऽमत् । नः । भवतु । अंतरिक्षं ।
क्षेत्रस्य । पतिः । मधुऽमान । नः । अस्तु । अरिष्यंतः । अनु । एनं । चरेम ॥४.५७.३॥

The Heavens and the Waters, let the Midheaven be full of light-containing plants rich in sweet mead. Let the Lord of the Field be full of sweet mead for us, and may we seek after the sweet mead without fail. (RV iv.57.3)

मधुमतीः <f. acc. pl. of adj. मधुमत् – containing sweetness, sweet>, ओषधीः <f. acc. pl. of ओषधि – light containing plants>, द्यावः <m. nom. pl. of दिव् – the Heavens>, आपः <f. nom. pl. of अप् – celestial Waters, i.e. cosmic plasmas>,

मधुमत् <n. nom. sg. of adj. मधुमत् – containing sweetness, sweet>, नः <m. dat. pl. of 1st person pron. अस्मद् – for us>, भवतु <ipv. 3rd person sg. of √भू – let it (the Innerfield or the Midheaven) be>, अंतरिक्षं <n. nom. sg. of अन्तरिक्ष – the Innerfield, the Midheaven>;

क्षेत्रस्य <n. gen. sg. of क्षेत्र – of the Field>, पतिः <m. nom. sg. of पति – Lord>, मधुमान् <m. nom. sg. of adj. मधुमत् – containing sweetness, sweet>, नः <m. dat. pl. of 1st person pron. अस्मद् – for us>, अस्तु <ipv. 3rd person sg. of √अस् – let (the Lord of the field) be>, अरिष्यंतः <m. nom. pl. of adj. अरिष्यत् (cf. √रिष्) – uninjured, without being lost or fail>, अनु <ind. prefix to verb चरेम (√चर्)>, एनं <m. acc. sg. of pron. एतद् – it (sweet mead)>, अनु-चरेम <opt. 1st person pl. of अनु-√चर् – may we go, move along; pursue, seek after>.

Stanza RV iv.57.4 is addressed to Śuna (शुनः the auspicious One), the Stone, the celestial ioniser.

RV iv.57.4 author वामदेवः, to शुनः, metre अनुष्टुप् छंदः
शुनं वाहाः शुनं नरः शुनं कृषतु लांगलं । शुनं वरत्रा बध्यंतां शुनमष्ट्रामुदिंगय ॥४.५७.४॥
शुनं । वाहाः । शुनं । नरः । शुनं । कृषतु । लांगलं । शुनं । वरत्राः । बध्यंतां । शुनं ।
अष्ट्रां । उत् । इंगय ॥४.५७.४॥
Happily, O men, happily Bulls! Let the plough make furrows happily.
Happily, let them fasten the straps, make thou the goad swing happily. (RV iv.57.4)

शुनं <ind. – happily, auspiciously, for growth or prosperity>, वाहाः <m. voc. pl. of वाह – O bulls!>, शुनं <ind. – happily>, नरः <m. voc. pl. of नृ – O men, heroes>, शुनं <ind. – happily>, कृषतु <ipv. 3rd person sg. of √कृष् – let it (the plough) draw, make furrows>, लांगलं <n. nom. sg. of लांगल (लाङ्गल) – the plough>;

शुनं <ind. – happily>, वरत्राः <f. acc. pl. of वरत्रा – straps>, बध्यंतां <ipv. 3rd person pl. of बन्ध् – let them bind, tie, fasten>, शुनं <ind. – happily>, अष्ट्रां <f. acc. sg. of अष्ट्रा – a prick, goad>, उदिंगय <(उदिङ्गय) ipv. cau. 2nd person sg. of √उदिंग् (उदिङ्ग्) – you cause to impart a tremulous motion, to vibrate (sound), to swing>.

Stanzas RV iv.57.5 and 8 are addressed to Śunā and Sīra (शुनासीरौ), 'ploughshare (शुना) and plough (सीरः)'. The plough refers to the Stone, the celestial ioniser. The ploughshares, the cutting blades of the plough, represent the firestones of the Stone, that cut asunder the atomic structure of hydrogen to yield cosmic plasmas, which are described as milk.

RV iv.57.5 author वामदेवः, to शुनासीरौ, metre पुर उष्णिक् छंदः
शुनासीराविमां वाचं जुषेथां यद्दिवि चक्रथुः पयः । तेनेमामुप सिंचतं ॥४.५७.५॥
शुनासीरौ । इमां । वाचं । जुषेथां । यत् । दिवि । चक्रथुः । पयः । तेन । इमां । उप । सिंचतं ॥४.५७.५॥
Śunā and Sīra, delight ye in this Speech and with the milk which ye have made in the Heaven,
With that milk bedew ye this field. (RV iv.57.5)

शुनासीरौ <m. nom. du. compound of शुना (f. a ploughshare, the main cutting blade of a plough + सीर m. a plough) – Śuna and Sīra, ploughshare and plough>, इमां <f. acc. sg. of इदं – this>, वाचं <f. acc. sg. of वाच् – Speech, sound>, जुषेथां <ipv. 2nd person du. of √जुष् – you both delight in>, यत् <n. acc. sg. of यद् – that, which>, दिवि <m. loc. sg. of दिव् – in the Heaven>, चक्रथुः <prf. 2nd person du. of कृ – ye (du.) have made>, पयः <n. acc. sg. of पयस् – juice, (esp.) milk, water>;

तेन <n. inst. sg. of तद् – with that (milk, water)>, इमां <f. acc. sg. of इदं – this here, this field>, उप-सिंचतं <ipv. 2nd person du. of उप-√सिच् – ye (du.) besprinkle or moisten with (inst.)>.

RV iv.57.8 author वामदेवः, to शुनासीरौ, metre त्रिष्टुप् छंदः
शुनं नः फाला वि कृषंतु भूमिं शुनं कीनाशा अभि यंतु वाहैः ।
शुनं पर्जन्यो मधुना पयोभिः शुनासीरा शुनमस्मासु धत्तं ॥४.५७.८॥
शुनं । नः । फालाः । वि । कृषंतु । भूमिं । शुनं । कीनाशाः । अभि । यंतु । वाहैः ।
शुनं । पर्जन्यः । मधुना । पयःऽभिः । शुनासीरा । शुनं । अस्मासु । धत्तं ॥४.५७.८॥
Happily let the ploughshares draw furrows in the field for us, let the cultivators run along happily with the oxen. Happily Parjanya. With sweet mead and milk, Śunā and Sīra, happily, bestow ye upon us. (RV iv.57.8)

शुनं <ind. – happily>, नः <m. dat. pl. of 1st person pron. अस्मद् – for us>, फालाः <m. nom. pl. of फाल – ploughshares>, वि-कृषंतु <ipv. 3rd person pl. of वि-√कृष् – let them draw furrows>, भूमिं <f. acc. sg. of भूमि – the field, ground>, शुनं <ind. – happily>, कीनाशाः <m. nom. pl. of कीनाश – cultivators of the soil>, अभि <ind. prefix to verb यंतु (√इ)>, अभि-यंतु <ipv. 3rd person pl. of अभि-√इ – let them go along or after>, वाहैः <mn. inst. pl. of वाह – with oxen>;

शुनं <ind. – happily, auspiciously>, पर्जन्यः <m. nom. sg. of पर्जन्य – Parjanya, the lord of rain, rain personified>, मधुना <n. inst. sg. of मधु – sweet, mead>, पयोभिः <n. inst. pl. of पयस् – with juice, (esp.) milk, waters>, शुनासीरा <m. nom. of dual compound शुनासीर (शुना f. a ploughshare + सीर m. a plough) – Śuna and Sīra>, शुनं <ind. – happily>, अस्मासु <m. loc. pl. of 1st person pron. अस्मद् – on us>, धत्तं <ipv. 2nd person du. of √धा – ye (du.) bestow>.

2.7.12 The Adhvaryus Make Sweet Soma Mead

The Adhvaryus make the sweet Soma mead and Vāyu is asked to be the first to drink it as Hotā, one of the officiating priests (ऋत्विजः pl.). In RV v.43.3, the Stone, the celestial ioniser, is called Adhvaryu. In the stanza, Adhvaryu is addressed in the plural as Adhvaryus and they refer to the firestones of the Stone, the celestial ioniser. Monier-Williams notes, under the head word 'ṛtvij (ऋत्विज्)', that Adhvaryu has three assistants: Pratiprasthātā (प्रतिप्रस्थाता), Neṣṭā (नेष्टा), and Unnetā (उन्नेता). Pratiprasthātā drives the firestones of the celestial ioniser, Neṣṭā leads the hydrogen atoms to the Stone to immolate them, and Unnetā pours the produced Soma juice into the receptacles, that is, the plasma blooms of the Soma Pond. Together, they are called the Adhvaryus. These three assistant priests represent three major functions of the firestones of the Stone, the celestial ioniser.

RV v.43.3 author अत्रिः, to विश्वे देवाः, metre त्रिष्टुप् छंदः
अध्वर्यवश्चकृवांसो मधूनि प्र वायवे भरत चारु शुक्रं ।
होतेव नः प्रथमः पाह्यस्य देव मध्वो ररिमा ते मदाय ॥५.४३.३॥
अध्वर्यवः । चकृऽवांसः । मधूनि । प्र । वायवे । भरत । चारु । शुक्रं ।
होताऽइव । नः । प्रथमः । पाहि । अस्य । देव । मध्वः । ररिम । ते । मदाय ॥५.४३.३॥
The Adhvaryus make the sweet Soma mead. Offer ye the pleasing bright juice to Vāyu.
As our foremost Hotā, O Deva, drink thou this sweet mead, which we present for your rapture. (RV v.43.3)

अध्वर्यवः <m. nom. pl. of अध्वर्यु – the Adhvaryus, Adhvaryu priests>, चकृवांसः <m. nom. pl. of prf. active pt. चकृवः (√कृ) – they make>, मधूनि <n. acc. pl. of मधु – sweet mead, drinks (Soma)>, प्र <ind. prefix to verb भरत (√भृ)>, वायवे <m. dat. sg. of वायु – to Vāyu>, प्र-भरत <ipv. 2nd person. pl. of प्र-√भृ – ye bring forth, place before, offer>, चारु <n. acc. sg. of चारु – agreeable, pleasing, beautiful>, शुक्रं <n. acc. sg. of शुक्र – clear, bright liquid (as Soma), juice>;

होता <m. nom. sg. of होतृ – Hotā, the sacrificer, one of the four officiating priests who are involved in the ionisation, discharge, and release of the cosmic plasmas>, इव <ind. – like, as>, नः <m. gen. pl. of pron. अस्मद् – our>, प्रथमः <m. nom. sg. of adj. प्रथम – first, foremost>, पाहि <ipv. 2nd person sg. of √पा 1P – you drink (gen.)>, अस्य <n. gen. sg. of इदं – of this>, देव <m. voc. sg. of देव – O Deva (refers to Vāyu)>, मध्वः <m. gen. sg. of मधु – of sweet mead>, ररिम <prf. 1st person pl. of √रा – we (Viśve Devas) grant, give, bestow>, ते <m. dat. sg. of 2nd person pron. युष्मद् – for your>, मदाय <m. dat. sg. of मद – for rapture>.

2.8 PṚTHIVĪ, THE MILKY WAY

2.8.1 The Milky Way Has Four Major Spiral Arms

Pṛthivī is one of the most misunderstood celestial objects of the Ṛgveda. Translated as 'earth', it has been known to be the planet earth by Vedic pundits and scholars. When read in the context of the Ṛgveda, Pṛthivī, which literally means "the broad and extended One", is revealed to be the Milky Way. According to the Ṛgveda, the Milky Way has four major Arms (RV i.164.42). The four major Arms of the Milky Way are called the four pointed-out regions (प्रदिशः चतस्रः RV i.164.42). The Arms of the Milky Way are also called mountains ranges (पर्वताः pl. RV v.84.1), downward sloping mountain range (प्रवत्वान् m. RV v.84.1), six divine expanses (देवीः षट् उर्वीः RV x.128.5), and so on.

In the Sūryasiddhānta and the Siddhāntaśiromaṇi, and in the Purāṇic literature, the spiral Arms of the Milky Way are called Dvīpas (द्वीपाः pl. peninsulas) or Varṣas (वर्षाः pl. divisions). Dvīpas are the peninsulas that project out from Mt. Meru (मेरुः), the Galactic Bar, at the centre of the Milky Way, and wrap around Mt. Meru. The Dvīpas are "separated from each other by the concentric circumambient oceans".[79]

Presented here are stanzas RV i.164.41 and 42. Stanza RV i.164.41 describes the growth of Soma plants, the plasma blooms, in the Soma Pond. A growing plasma bloom is compared to a foot of the mother milch Cow, the Stone, the celestial ioniser. One footed means that one Soma plant, or plasma bloom, has grown. One-footed, two-footed, four-footed, and so on, and then at its peak, the mother milch Cow became the thousand-syllabled. By using the phrase 'thousand-syllabled' at its peak instead of 'thousand-footed', it is indicated that the Speeches of cosmic plasmas play an important role in the growing and maturing of plasma blooms, or Soma plants. It does not mean exactly a thousand plasma blooms have grown; it just means that the Soma Pond is filled full with matured Soma plants.

---

[79] See Monier Monier-Williams, *A Sanskrit-English Dictionary*, 1899, op. cit., p. 507, under द्वीप dvīpa; P. Gangooly (ed.), *Translation of the Sūrya-siddhānta,* 1935, op. cit., p. 287.

RV i.164.41 author दीर्घतमा औचथ्यः, to विश्वे देवाः, metre जगती छंदः
गौरीर्मिमाय सलिलानि तक्षत्येकपदी द्विपदी सा चतुष्पदी ।
अष्टापदी नवपदी बभूवुषी सहस्राक्षरा परमे व्योमन् ॥१.१६४.४१॥
गौरीः । मिमाय । सलिलानि । तक्षति । एकऽपदी । द्विऽपदी । सा । चतुःऽपदी ।
अष्टाऽपदी । नवऽपदी । बभूवुषी । सहस्रऽअक्षरा । परमे । विऽओमन् ॥१.१६४.४१॥
She hurls the celestial lights and fashions the waves of celestial Water. One-footed, two-footed, four-footed, She has become eight-footed, nine-footed, then thousand-syllabled at its peak, O Vyomā. (RV i.164.41)

गौरीः <f. acc. pl. of गौरी – bisons, water buffaloes (refer to celestial lights, cosmic plasmas)>, मिमाय <prf. 3rd person sg. of √मि – she (mother milch Cow, the Stone) casts, throws>, सलिलानि <n. acc. pl. सलिलं – surges, waves (of celestial Water, cosmic plasmas)>, तक्षति <pre. 3rd person sg. of √तक्ष् – she fashions by cutting, chiselling>, एकपदी <f. nom. sg. of adj. एकपद – having only one foot, one-footed>, द्विपदी <f. nom. sg. of adj. द्विपद – two-footed>, सा <f. nom. sg. of 3rd person pron. √तद् – she (mother milch Cow)>, चतुष्पदी <f. nom. sg. of adj. चतुष्पद – four-footed>;

अष्टापदी <f. nom. sg. of adj. अष्टापद – eight-footed>, नवपदी <f. nom. sg. of adj. नवपद – nine-footed>, बभूवुषी <f. nom. sg. of prf. active pt. of बभूवुषी (√भू) – she has become>, सहस्राक्षरा <f. nom. sg. of adj. सहस्राक्षर – having a thousand syllables or Speeches>, परमे <n. loc. sg. of adj. परम – at the highest point, extreme limit, the highest degree>, व्योमन् <m. voc. sg. of व्योमन् – O Vyomā, O Vault of the Heaven>.

Stanza RV i.164.42 states that there are four major Arms (प्रदिशः चतस्रः) of the Milky Way. And by the flows of the imperishable Water, that is, cosmic plasmas, the Milky Way is sustained. The first line of the stanza is addressed to Vāk (वाक् Speech) and the second line, to the celestial Waters (आपः āpaḥ pl.), the cosmic plasmas.

RV i.164.42 author दीर्घतमा औचथ्यः, to वाक् (पूर्वार्धस्य) and आपः (उत्तरार्धस्य), metre प्रस्तारपंक्तिः छंदः
तस्याः समुद्रा अधि वि क्षरंति तेन जीवंति प्रदिशश्चतस्रः ।
ततः क्षरत्यक्षरं तद्विश्वमुप जीवति ॥१.१६४.४२॥
तस्याः । समुद्राः । अधि । वि । क्षरंति । तेन । जीवंति । प्रऽदिशः । चतस्रः ।
ततः । क्षरति । अक्षरं । तत् । विश्वं । उप । जीवति ॥१.१६४.४२॥
From her, the celestial oceans flow out, thereby the four Arms of the Milky Way remain alive. Thence, flows the imperishable Water, by which the whole World is sustained. (RV i.164.42)

तस्याः <f. abl. sg. of तद् – from her (the mother milch Cow, i.e. from the Stone)>, समुद्राः <m. nom. pl. of समुद्र – celestial oceans (here, the oceans in the plural refer to the individual plasma blooms)>, अधि <ind. preposition – from>, वि-क्षरंति <pre. 3rd person pl. of वि-√क्षर् – they flow away>, तेन <mn. inst. sg. of pron. तद् – by that, thereby>, जीवंति <(जीवन्ति) pre. 3rd person pl. of √जीव् – they live, remain alive>, प्रदिशः <f. nom. pl. of प्रदिश् – the pointed-out regions, i.e. the Arms of the Milky Way>, चतस्रः <f. nom. pl. of चतुर् – four>;

ततः <(ततस्) ind. – from that place, thence>, क्षरति <pre. 3rd person sg. of √क्षर् – it (the imperishable Water) flows>, अक्षरं <n. nom. sg. of अक्षर – imperishable Water, i.e. cosmic plasmas>, तत् <n. nom. sg. of pron. तद् – this>, विश्वं <n. nom. sg. of विश्व – the whole World, i.e. the Milky Way>, उप-जीवति <pre. 3rd person sg. of उप-√जीव् – it lives, is supported by>.

Stanzas 37 and 38 of Bhuvanakośa (भुवनकोशः) of Siddhāntaśiromaṇi (सिद्धान्तशिरोमणिः) support the view of the Ṛgveda that there are four major spiral Arms. Stanza 37 states that the sacred Gaṅgā (गंगा), which has emerged from Viṣṇu's foot (विष्णुपदी), falls upon

Mt. Meru, the Galactic Bar, and is divided into four branches. Stanza 38 explains that the four divided branches of Gaṅgā flow along the four Arms of the Milky Way. According to the Siddhāntaśiromaṇi, the four Arms of the Milky Way are Bhadrāśvavarṣa (भद्राश्ववर्षः), Bhāratavarṣa (भारतवर्षः), Ketumālāvarṣa (केतुमालावर्षः), and Uttarakuruvarṣa (उत्तरकुरुवर्षः Upperkuruvarṣa). Bhadrāśvavarṣa is likely the innermost Arm and Uttarakuruvarṣa the outermost Arm of the Milky Way. See section 6.2.6 for the translated stanzas of 37 and 38 of Bhuvanakośa, Siddhāntaśiromaṇi. In the extant text of Bhūgolādhyāya (भूगोलाध्यायः) of Sūryasiddhānta (सूर्यसिद्धान्तः), we do not find the detailed information about the spiral Arms of the Milky Way.

2.8.2 Different Regions of the Milky Way

The Ṛgveda also states that the Milky Way has five (RV ix 86.29), six (RV x.128.5), seven (RV i.22.16), or eight (RV i.35.8) regions. Figure 10 is the "Annotated Roadmap to the Milky Way (artist's conception)" by R. Hurt at Caltech Spitzer Science Center.[80] Of the available maps of the Milky Way, this map corresponds fairly well to the Milky Way described in the Ṛgveda and in the Siddhāntaśiromaṇi.

<Figure 10> "Annotated Roadmap to the Milky Way (artist's conception)". Image credit: NASA/JPL-Caltech/R. Hurt (SSC-Caltech)

From this NASA Milky Way map, one can see that, depending on how one counts, the divisions of the Milky Way can be described as four, five, six, seven, or eight. When we

80 NASA/Jet Propulsion Laboratory, *A Road Map to the Milky Way (Annotated)*, [website] http://www.spitzer.caltech.edu/images/1925-ssc2008-10b-A-Roadmap-to-the-Milky-Way-Annotated (accessed 26 June 2019).

count only the major Arms of the Milky Way, there are four major regions. When we count the Outer Arm as the separate region from Norma Arm, then we have five regions, and so on.

2.8.3 Five Regions

Stanza RV ix.86.29 states that the five spiral Arms (पंच प्रदिशः) of the Milky Way are governed by the laws of Pavamāna Soma (पवमानः सोमः). The stanza is addressed to Pavamāna Soma. Pavamāna Soma is the bright, flowing-clear Soma; here, in the stanza, it is called Samudra, the Ocean of celestial Waters. The poet's name is Pṛśnaya Ajā (पृश्नयः अजाः), 'the herds of dappled cows (पृश्नयः f. nom. pl. of पृश्नि cows, अजाः m. nom. pl. of अज herds, leaders)', 'the spotted rams (पृश्नयः m. nom. pl. of adj. पृश्नि spotted, अजाः m. nom. pl. of अज he-goats)', or 'the spotted she-goats (पृश्नयः f. nom. pl. of adj. पृश्नि spotted, अजाः f. nom. pl. of अजा she-goats)' depending on how you interpret it. Whichever one you choose, it refers to the Soma plants, or plasma blooms, growing in the Soma Pond. According to the stanza, the celestial lights (ज्योतींषि pl.) and Sūrya are of Pavamāna Soma.

RV ix.86.29 author पृश्नयोऽजाः, to पवमानः सोमः, metre जगती छंदः
त्वं समुद्रो असि विश्ववित्कवे तवेमाः पंच प्रदिशो विधर्मणि ।
त्वं द्यां च पृथिवीं चाति जभ्रिषे तव ज्योतींषि पवमान सूर्यः ॥९.८६.२९॥
त्वं । समुद्रः । असि । विश्वऽवित् । कवे । तव । इमाः । पंच । प्रऽदिशः । विऽधर्मणि ।
त्वं । द्यां । च । पृथिवीं । च । अति । जभ्रिषे । तव । ज्योतींषि । पवमान । सूर्यः ॥९.८६.२९॥
Thou art the Ocean of celestial Waters, the knower of all, O Seer. Under thy laws are these five Arms of the Milky Way. Thou reachest beyond the Heaven and beyond the Milky Way. Thine are the lights, O Pavamāna, thine is Sūrya. (RV ix.86.29)

त्वं <m. nom. sg. of 2nd person pron. युष्मद् – you>, समुद्रः <m. nom. sg. of समुद्र – Samudra, the Ocean of celestial Waters>, असि <pre. 2nd person sg. of √अस् 2P – you are>, विश्ववित् <m. nom. sg. of विश्वविद् – the omniscient one, the knower of all>, कवे <m. voc. sg. of कवि – O seer, singer>, तव <m. gen. sg. of 2nd person pron. युष्मद् – your, yours>, इमाः <f. nom. pl. of pron. इदं – these>, पंच <mfn. nom. pl of पञ्चन् – five>, प्रदिशः <f. nom. pl. of प्रदिश् – the pointed-out regions, quarters of the "sky", i.e. Arms of the Milky Way>, विधर्मणि <n. loc. sg. of विधर्मन् – under the laws, rules>;

त्वं <m. nom. sg. of 2nd person pron. युष्मद् – you>, द्यां <m. acc. sg. of दिव् – the Heaven>, च <ind. – and, both, also>, पृथिवीं <f. acc. sg. of पृथिवी – Pṛthivī, the Milky Way>, च <ind. – and, both, also>, अति <ind. prefix to verb जभ्रिषे (√भृ)>, अति-जभ्रिषे <prf. 2nd person sg. of अति-√भृ 2P – you reach beyond>, तव <m. gen. sg. of 2nd person pron. युष्मद् – yours, thine>, ज्योतींषि <n. nom. pl. of ज्योतिस् – celestial lights, i.e. cosmic plasmas>, पवमान <m. voc. sg. of पवमान – O Pavamāna>, सूर्यः <m. nom. sg. of सूर्य – Sūrya>.

2.8.4 Six Divine Expanses

In RV x.128.5, the regions of the Milky Way are described as six divine Expanses (देवीः षट् उर्वीः). The poet's name is Vihavya Āṅgirasa (विहव्यः आङ्गिरसः vihavyaḥ āṅgirasaḥ), 'the invoked descendant of Aṅgirā (अङ्गिराः)'. The stanza is addressed to the Viśve Devas. The Viśve Devas represent the different aspects of cosmic plasmas that are discharged into the Soma Pond and grow, bloom, and mature within it until they are released.

RV x.128.5 author विहव्यः, to विश्वे देवाः, metre त्रिष्टुप् छंदः
देवीः षळुर्वीरुरु नः कृणोत विश्वे देवास इह वीरयध्वं ।
मा हास्महि प्रजया मा तनूभिर्मा रधाम द्विषते सोम राजन् ॥१०.१२८.५॥
देवीः । षट् । उर्वीः । उरु । नः । कृणोत । विश्वे । देवासः । इह । वीरयध्वं ।
मा । हास्महि । प्रऽजया । मा । तनूभिः । मा । रधाम । द्विषते । सोम । राजन् ॥१०.१२८.५॥
O Viśve Devas, for us, build ye the six divine expanses of the Milky Way and divide them far apart. Let us not leave behind our offspring. O King Soma, let us not succumb to our foe. (RV x.128.5)

देवीः <f. acc. pl. of adj. देव – heavenly, divine>, षट् <f. acc. pl. of षष् – six>, उर्वीः <f. acc. pl. of उर्वी – (with षष्) six spaces or rivers of Pṛthivī (the Milky Way)>, उरु <ind. – far, widely, far off>, नः <m. dat. pl. of अस्मद् – for us>, कृणोत <ipv. 2nd person pl. of √कृ 5P – ye place, put, lay, prepare, build>, विश्वे देवासः <m. voc. pl. of विश्वे देव – O ye Viśve Devas>, इह <ind. – here, now, at this time>, वीरयध्वं <ipv. 2nd person pl. of √वीर् (वि-√ईर्) – ye split, divide asunder>;

मा <ind. a particle of prohibition – not>, हास्महि <sub. 1st person pl. of √हा – (with मा) let us not abandon, leave behind (inst.)>, प्रजया <f. inst. sg. of प्रजा – offspring, children, aftergrowth, i.e. newly born cosmic plasmas>, मा <ind. particle – not>, तनूभिः <f. inst. pl. of तनू – with/by selves, themselves (used as a reflexive pron.)>, मा <ind. particle – not>, रधाम <sub. 1st person pl. √रध् – (with मा) let us not succumb to (dat.)>, द्विषते <mn. dat. sg. of द्विषत् – to the foe, enemy>, सोम <m. voc. sg. of सोम – O Soma>, राजन् <m. voc. sg. of राजन् – O King>.

2.8.5 Seven Domains

According to stanza RV i.22.16, there are seven domains of the Milky Way. The poet's name is Medhātithi Kāṇva (मेधातिथिः काण्वः), 'the drink of Agni, descendant of Kaṇva (कण्वः)'. Kaṇva belongs to the family of the Aṅgirās (अङ्गिरसः pl.), the angels, the messengers of celestial lights. Medhātithi (मेधः drink, juice + अतिथिः a name of Agni) is the 'drink of Agni', which is Soma. The stanza is addressed to Viṣṇu and other Devas.

RV i.22.16 author मेधातिथिः काण्वः, to विष्णुर्देवा वा, metre गयत्री छंदः
अतो देवा अवंतु नो यतो विष्णुर्विचक्रमे । पृथिव्याः सप्त धामभिः ॥१.२२.१६॥
अतः । देवाः । अवंतु । नः । यतः । विष्णुः । विऽचक्रमे । पृथिव्याः । सप्त । धामऽभिः ॥१.२२.१६॥
From here, whence Viṣṇu strides, let the Devas lead us
Through the seven domains of the Milky Way. (RV i.22.16)

अतः <(अतस्) ind. – from here, hence, henceforth>, देवाः <m. nom. pl. of देव – the Devas>, अवंतु <ipv. 3rd person pl. of √अव् – let (the Devas) drive, impel, lead>, नः <m. acc. pl. of 1st person pron. अस्मद् – us>, यतः <(यतस्) ind. – from where, wherefrom, whence>, विष्णुः <m. nom. sg. of विष्णु – Viṣṇu>, विचक्रमे <prf. middle 3rd. person sg. of वि-√क्रम् – he strides through, traverses>;

पृथिव्याः <f. gen. sg. of पृथिवी – of Pṛthivī, of the Milky Way>, सप्त <n. pl. of सप्तन् – seven>, धामभिः <n. inst. pl. of धामन् – through dwelling places, domains>.

The 'place whence Viṣṇu strides' is the Soma Pond. Note the cosmic plasmas are produced and discharged into the Soma Pond and released from the Soma Pond to sustain and to give life to the worlds of the Milky Way.

Stanzas 21-25 of Bhuvanakośa of Siddhāntaśiromaṇi also list the seven Dvīpas, the 'peninsulas or islands' of the Milky Way.[81] The seven Dvīpas include Jambu (जम्बुः), Gomedaka (गोमेदकः or गोमेदकं), Śālmalī (शाल्मली) or Śālmala (शाल्मलः), Kuśa (कुशः) or Kauśa (कौशः), Krauñca (क्रौञ्चः), Śāka (शाकः), and Puṣkara (पुष्करः or पुष्करं). However, it is difficult to map these Dvīpas to the names of the Milky Way Arms on the NASA map. Only a few of these names can be mapped on NASA's map of the Milky Way based on the descriptions provided. In stanza 27 of Bhuvanakośa, it is stated that 'this valley here (इदं)' is called 'Bhāratavarṣa (भारतवर्षं)', indicating that Bhāratavarṣa is the branch of the Milky Way where our solar system is located. Stanza 31 of Bhuvanakośa states that 'in the middle of Ilāvṛtavarṣa (इलावृतवर्षं) stands Mt. Meru'. So, it is clear that two bow-shaped Arms Near 3kpc and Far 3kpc that surround the Galactic Bar at the centre of the Milky Way are called Ilāvṛtavarṣa.

2.8.6 Eight Regions

According to RV i.35.8, Savitā illumines the Milky Way's eight regions, three bow-shaped paths, and seven rivers. The stanza indicates that there are eight regions, three bow-shaped paths, and seven rivers in the Milky Way. Savitā is the electric, impelling aspect of Tvaṣṭā. As can be seen on NASA's map, in the Milky Way, currently, two bow-shaped paths (Near 3kpc and Far 3kpc Arms) are identified and mapped. The 'aquatic' jewels (रत्ना वार्याणि) refer to the celestial Waters, the cosmic plasmas.

RV i.35.8 author हिरण्यस्तूप आंगिरसः, to सविता, metre त्रिष्टुप् छंदः
अष्टौ व्यख्यत्ककुभः पृथिव्यास्त्री धन्व योजना सप्त सिंधून् ।
हिरण्याक्षः सविता देव आगाद्दधद्रत्ना दाशुषे वार्याणि ॥१.३५.८॥
अष्टौ । वि । अख्यत् । ककुभः । पृथिव्याः । त्री । धन्व । योजना । सप्त । सिंधून् ।
हिरण्यऽअक्षः । सविता । देवः । आ । अगात् । दधत् । रत्ना । दाशुषे । वार्याणि ॥१.३५.८॥
The eight regions of the Milky Way, he illumined, the three bow-paths, and the seven rivers.
The Deva Savitā, the golden-eyed, came, bringing aquatic jewels to the worshipper.
(RV i.35.8)

अष्टौ <f. acc. pl. of अष्टन् – eight>, वि <ind. prefix to verb अख्यत् (√ख्या)>, वि-अख्यत् <ipf. or aor. 3rd person sg. of वि-√ख्या – he (Savitā) illumined>, ककुभः <f. acc. pl. of ककुभ् – spaces, regions>, पृथिव्याः <f. gen. sg. of पृथिवी – of Pṛthivī, of the Milky Way>, त्री <(=त्रीणि) n. nom. acc. pl. of त्रि – three>, धन्व <n. acc. sg. of धन्वन् – a bow>, योजना <योजनानि n. acc. pl. of योजन – (with धन्व) bow-shaped paths, bow-paths>, सप्त <m. acc. pl. of सप्तन् – seven>, सिंधून् <m. acc. pl. of सिंधू – sindhus (pl.), celestial rivers>;

हिरण्याक्षः <m. nom. sg. of adj. हिरण्याक्ष – golden-eyed>, सविता <m. nom. sg. of सवितृ – Savitā>, देवः <m. nom. sg. of देव – the Deva>, आ <ind. prefix to verb अगात् (√गा)>, आ-अगात् <aor. 3rd person sg. of आ-√गा – he (Savitā) came, arrived>, दधत् <sub. 3rd person sg. of √धा – he brings>, रत्ना <n. acc. pl. of रत्न – gemstones, jewels>, दाशुषे <m. dat. sg. of दाशु – for/to the worshipper, sacrificer>, वार्याणि <n. acc. pl. of adj. वार्य – watery, aquatic>.

---

81 Bāpū Deva Sāstri (ed., trans.), *Translation of the Sūryasiddhānta by Bāpū Deva Sāstri, and of the Siddhānta Śiromani by the late Lancelot Wilkinson*, Calcutta, Baptist Mission Press, 1861, pp.116-119.

2.8.7 A Hymn to Pṛthivī

We find only one short hymn with three stanzas, RV v.84.1-3, addressed to Pṛthivī, the Milky Way, in the extant text of the Ṛgveda. The author is Bhauma Atri (भौमः अत्रिः), Bhauma means 'related to Bhūmi'. Atri is the devourer, the Stone. Bhūmi translates as "earth, soil, ground" or "territory". In the Heavenly Spheres, the Milky Way is the ground or territory, on which life is created and nurtured. The Stone, the celestial ioniser, feeds the Milky Way with cosmic plasmas.

In stanza RV v.84.1, the mountain ranges that slope downwards refer to the Arms of the Milky Way.

RV v.84.1 author भौमः अत्रिः, to पृथिवी, metre अनुष्टुप् छंदः
बळित्था पर्वतानां खिद्रं बिभर्षि पृथिवि । प्र या भूमिं प्रवत्वति मह्ना जिनोषि महिनि ॥५.८४.१॥
बट् । इत्था । पर्वतानां । खिद्रं । बिभर्षि । पृथिवि ।
प्र । या । भूमिं । प्रवत्वति । मह्ना । जिनोषि । महिनि ॥५.८४.१॥
Thou, O Milky Way, indeed bearest the weight of the mountain ranges.
O Mighty One, thou who with might procurest the territory along the downward sloping mountain range. (RV v.84.1)

बट् <ind. – in truth, certainly>, इत्था <ind. – thus, truly, really>, पर्वतानां <m. gen. pl. of पर्वत – of the mountain ranges, mountains, i.e. the Arms of the Milky Way>, खिद्रं <n. acc. sg. of खिद्र – (from √खिद् to press down) burden, weight>, बिभर्षि <pre. 2nd person sg. of √भृ – you bear, carry>, पृथिवि <f. voc. sg. of पृथिवी – O Pṛthivī, O Milky Way>;

प्र <ind. prefix to verb जिनोषि (√जि)>, या <f. nom. sg. of यद् – who>, भूमिं <f. acc. sg. of भूमि – soil, ground, territory>, प्रवत्वति <mn. loc. sg. of प्रवत्वत् – along the downwards sloping mountain range>, महना <n. inst. sg. of महन् – with might, power>, प्र-जिनोषि <pre. 2nd person sg. of प्र-√जि – you win, conquer, acquire>, महिनि <f. voc. sg. of महिन् (f. महिनी) – O Mighty One>.

In stanza RV v.84.2, both Wanderer and Arjunī refer to the Milky Way; the sacrificers, to the Devas who participate in the Soma Yajña to release cosmic plasmas to sustain the Milky Way.

RV v.84.2 author भौमः अत्रिः, to पृथिवी, metre अनुष्टुप् छंदः
स्तोमासस्त्वा विचारिणि प्रति ष्टोभंत्यक्तुभिः । प्र या वाजं न हेषंतं पेरुमस्यस्यर्जुनि ॥५.८४.२॥
स्तोमासः । त्वा । विऽचारिणि । प्रति । स्तोभंति । अक्तुऽभिः ।
प्र । या । वाजं । न । हेषंतं । अस्यसि । अर्जुनि ॥५.८४.२॥
Upon thee, O Wanderer, the sacrificers utter the joyful hymn with celestial lights,
As thou, O Arjunī, hurlest forth the swift neighing steed. (RV v.84.2)

स्तोमासः <m. nom. pl. of स्तोम – sacrificers>, त्वा <f. acc. sg. of 2nd person pron. युष्मद् – you, thee>, विचारिणि <f. voc. sg. of विचारिन् – O Wanderer (refers to Pṛthivī, the Milky Way)>, प्रति <ind. – (as a preposition with proceeding acc.) upon, on>, स्तोभंति <pre. 3rd person. pl. of √स्तुभ् – they hum, utter a joyful sound, make a succession of exclamations>, अक्तुभिः <m. inst. pl. of अक्तु – with rays, lights>;

प्र <ind. prefix to verb अस्यसि (√अस्)>, या <f. nom. sg. of pron. यद् – who, the person>, वाजं <m. acc. sg. of वाज – a swift horse>, न <ind. – like, as>, हेषंतं <m. acc. sg. of pre. active pt. हेषत् (√हेष्) – it/he neighing>, प्र-अस्यसि <pre. 2nd

sperson sg. of प्र-√अस् 4P – you throw, hurl forth, cast>, अर्जुनि <f. voc. sg. of अर्जुनी – O Arjunī (literally 'white' or 'of the milk', i.e. Pṛthivī, the Milky Way)>.

RV v.84.3 author भौमः अत्रिः, to पृथिवी, metre अनुष्टुप् छंदः
दृळ्हा चिद्या वनस्पतीन्क्ष्मया दर्धर्ष्योजसा । यत्ते अभ्रस्य विद्युतो दिवो वर्षंति वृष्टयः ॥५.८४.३॥
दृळ्हा । चित् । या । वनस्पतीन् । क्ष्मया । दर्धर्षि । ओजसा ।
यत् । ते । अभ्रस्य । विऽद्युतः । दिवः । वर्षंति । वृष्टयः ॥५.८४.३॥
She, who indeed is stupendous, with her might, holds the lords of the Forest on the ground,
When those rain-floods pour down from the flashing thunderbolt of the Heaven's water-bearer.
(RV v.84.3)

दृळ्हा <f. nom. sg. of adj. दृढ (दृळ्ह) – firm, strong, solid, massive>, चित् <(चिद्) ind. – even, indeed, also>, या <f. nom. sg. of यद् – she who (Pṛthivī, the Milky Way)>, वनस्पतीन् <m. acc. pl. of वनस्पति – the lords of the Forest (refer to the firestones of the Stone, the celestial ioniser)>, क्ष्मया <ind. – "on the earth", i.e. on the ground>, दर्धर्षि <int. 3rd person sg. of √धृ – she holds, bears>, ओजसा <n. inst. sg. of ओजस् – with strength, power>;

यत् <ind. conjunction – when>, ते <m. nom. pl. of pron. तद् – they, those (rains)>, अभ्रस्य <m. gen. sg. of अभ्र – of the "water-bearer", here it refers to the Soma Pond>, विद्युतः <f. abl. sg. of विद्युत् – from the lightning, flashing thunderbolt>, दिवः <m. gen. sg. of दिव् – of the Heaven>, वर्षंति <pre. 3rd person pl. of √वृष् – they pour down>, वृष्टयः <mf. nom. pl. of वृष्टि – rains, rain-floods, i.e. the floods of cosmic plasmas>.

In the stanza RV v.84.3, the lords of the Forest (वनस्पतयः m. nom. pl. of वनस्पति) refer to the firestones of the Stone, which are laid in a circular formation (see Figure 8). The Heaven's water-bearer is the Soma Pond; the flashing thunderbolt refers to the plasma blooms filled full with the rain-floods. Here, the rain-floods refer to the cosmic plasmas that are released through the base of the Soma Pond flooding the Milky Way.

2.9 AŚVATTHA, THE INVERTED TREE, THE VEDIC TREE OF LIFE

In RV i.135.8 and RV x.97.5, the Heavenly structure, which includes the Vault of the Heaven, the pair of the Stone and the Soma Pond, and the Milky Way, is called Aśvattha (अश्वत्थः), *Ficus religiosa*, the Holy Fig Tree. Aśvattha is the inverted tree with its root standing above and its branches and leaves hanging below (RV i.24.7). The Vault of the Heaven and the pair of the Stone and the Soma Pond are the root system that provides the celestial lights, the cosmic plasmas, and sustain the branches and leaves that spread downwards, that is, the Milky Way, the regions of life. This tree is the Tree of the World, the sacred Vedic Tree of Life (see Figure 11).

In the extant text of the Ṛgveda, RV i.24.7 is the only stanza that explains what the Aśvattha tree is. The author is Śunaḥśepa (शुनःशेपः), 'the tail (शेपः) of the auspicious One (शुनः)'. The tail is the Soma Pond with plasma blooms, or Soma plants, growing in it. The auspicious One is the Stone. Thus the author's name is 'the tail of the Stone', the tail that grows on the Stone, the celestial ioniser. The stanza is addressed to Varuṇa.

RV i.24.7 author शुनःशेपः आजीगर्तिः (कृत्रिमो वैश्वामित्रो देवरातः), to वरुणः, metre त्रिष्टुप् छंदः
अबुध्ने राजा वरुणो वनस्योर्ध्वं स्तूपं ददते पूतदक्षः ।
नीचीनाः स्थुरुपरि बुध्न एषामस्मे अंतर्निहिताः केतवः स्युः ॥१.२४.७॥
अबुध्ने । राजा । वरुणः । वनस्य । ऊर्ध्वं । स्तूपं । ददते । पूतऽदक्षः ।
नीचीनाः । स्थुः । उपरि । बुध्नः । एषां । अस्मे इति । अंतः । निऽहिताः । केतवः । स्युरिति स्युः ॥१.२४.७॥
Varuṇa, the King of hallowed might, sustains the raised canopy of the tree in the bottomless region. The rays of light, whose root stands high above, duly flow downward. May they abide kept among us. (RV i.24.7)

अबुध्ने <mn. loc. sg. of adj. अबुध्न – in the bottomless region>, राजा <m. nom. sg. of राजन् – King>, वरुणः <m. nom. sg. of वरुण – Varuṇa>, वनस्य <n. gen. sg. of वन – of the tree>, ऊर्ध्वं <m. acc. sg. of adj. ऊर्ध्व – erected, raised>, स्तूपं <m. acc. sg. of स्तूप – a tuft of hair, the upper part of the head, top, i.e. the canopy>, ददते <pre. middle 3rd person sg. of √दद् 1Ā (=√दा) – he keeps, holds, preserves>, पूतदक्षः <m. nom. sg. of adj. पूतदक्ष – hallowed might, purified force>;

नीचीनाः <m. nom. pl. of adj. नीचीन – hanging or flowing down>, स्थुः <(=सुष्ठु) ind. – fitly, duly>, उपरि <ind. – above>, बुध्नः <m. nom. sg. of बुध्न – the root>, एषां <m. gen. pl. of इदं – of these, whose>, अस्मे <mf. loc. pl. of 1st person pron. अस्मद् – among us>, अंतः <(अन्तर्) ind. – in the middle, among>, निहिताः <m. nom. pl. of adj. निहित – laid, deposited, kept in, given>, केतवः <m. nom. pl. of केतु – rays of light>, स्युः <opt. 3rd person pl. of √अस् 2P – may they abide, stay>.

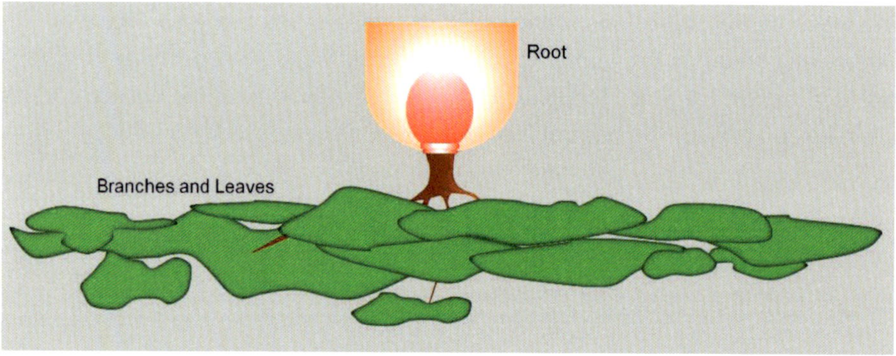

<Figure 11> Holy Fig Tree (अश्वत्थः aśvatthaḥ), *Ficus religiosa*, the Inverted Tree, the Vedic Tree of Life. The root is the Heavenly Vault, with the Stone and the Soma Pond, which produces and pours out celestial Waters, cosmic plasmas. The Branches and Leaves represent the Milky Way, the Regions of Life. Schematic diagram, NOT to scale.

We find another stanza that explains the Aśvattha tree in the Kaṭha Upaniṣad (कठोपनिषत् kaṭhopaniṣat ii.3.1). The galactic structure itself, according to the Ṛgveda and the Kaṭha Upaniṣad, is eternal and imperishable. Within it rest all the worlds (लोकाः lokāḥ pl.), the innumerable solar systems like ours.

Kaṭhopaniṣad ii.3.1 (कठोपनिषत् २.३.१)
ऊर्ध्वमूलोऽवाक्शाख एषोऽश्वत्थः सनातनः । तदेव शुक्रं तद्ब्रह्म तदेवामृतमुच्यते ।
तस्मिँल्लोकाः श्रिताः सर्वे तदु नात्येति कश्चन । एतद्वै तत् ॥कठोपनिषत् २.३.१॥

With the root high above, the branches turned downwards, this Aśvattha is eternal.
It is said that this alone is Brahma, imperishable and resplendent. In this Aśvattha tree rest all the worlds, and nothing can surpass it. This indeed is that. (Kaṭhopaniṣad ii.3.1)

ऊर्ध्वमूलः <m. nom. sg. of adj. ऊर्ध्वमूल – having the root above>, अवाक्शाखः <m. nom. sg. of adj. अवाक्शाख – having branches turned downwards>, एषः <m. nom. sg. of pron. एतद् – this>, अश्वत्थः <m. nom. sg. of अश्वत्थ – Aśvattha, the holy fig tree, Ficus religiosa>, सनातनः <m. nom. sg. of adj. सनातन – eternal, perpetual, primaeval, ancient>;

तत् <n. nom. sg. of pron. तद् – that, this, (Aśvattha, the holy fig tree)>, एव <ind. – indeed, truly, only>, शुक्रं <n. nom. sg. of adj. शुक्र – bright, resplendent>, तत् <n. nom. sg. of pron. तद् – that, this>, ब्रह्म <n. nom. sg. of ब्रह्मन् – brahma, sacred hymn, prayer (as explained below, the masculine form Brahmā should have been used instead of the neuter form brahma)>, तत् <n. nom. sg. of pron. तद् – that, this>, एव <ind. – truly, only, alone>, अमृतं <n. nom. sg. of adj. अमृत – immortal, imperishable>, उच्यते <pre. passive 3rd person sg. of √वच् – it is said>;

तस्मिन् <m. loc. sg. of pron. तद् – in it (in Aśvattha)>, लोकाः <m. nom. pl. of लोक – divisions, regions, the worlds>, श्रिताः <m. nom. pl. of adj. श्रित – situated in, contained>, सर्वे <m. nom. pl. of pron. सर्व – all>, तत् <n. acc. sg. of pron. तद् – that>, उ <ind. – used as a particle for strengthening the particle न>, न <ind. particle of negation – not, no>, अत्येति <pre. 3rd person sg. of √अती – it overtakes, overshadows, surpasses>, कश्चन <interrogative pron. कस् (कः) + particle चन – anyone, (with न) no one, nothing>;

एतत् <n. nom. sg. of एतद् – this>, वै <ind. particle of emphasis and affirmation – indeed, truly, verily>, तत् <n. acc. sg. of pron. तद् – that (Brahma ब्रह्म, from which everthing orginates)>.

Note, in the Kaṭha Upaniṣad ii.3.1, it is stated that Aśvattha, the inverted tree, alone is 'brahma (ब्रह्म n. nom. sg. of ब्रह्मन्)'. Aśvattha, the inverted tree, symbolically represents the Vault of the Heaven and its substructures, the marvellous Creator's machine, and the Milky Way. Thus, in the context of this stanza, 'Brahmā (ब्रह्मा m. nom. sg. of ब्रह्मन्)', the masculine form of brahman (ब्रह्मन्), should have been chosen instead of 'brahma (ब्रह्म n. nom. sg. of ब्रह्मन्)', which is the neuter form of brahman (ब्रह्मन्). According to Monier-Williams, the neuter form 'brahma' translates as "pious utterance", "prayer", or "the sacred word". The masculine form 'Brahmā', the Creator, translates as "the Absolute", "the Eternal", or "one divine essence or source from which all created things emanate ... and to which they return". In the Ṛgveda, the neuter form 'brahma' is always used to mean 'the sacred hymn', or 'prayer' (RV x.61.7, RV i.52.7, RV iii.8.2, RV vii.103.8). On the other hand, the masculine form 'Brahmā' is used for 'the Vault of the Heaven' or 'the Creator'. For example, in RV i.163.35, the Speeches (वचाः pl.) and the Vault of the Heaven (व्योम n. nom. sg. of व्योमन्) are called Brahmā.

In addition, as noted, in stanza 90 of Bhūgolādhyāya of Sūryasiddhānta, the Vault of the Heaven is called 'Brahmā's Egg, the Hemispherical Bowl (ब्रह्माण्डसंपुटः)'. In stanza 67 of Bhuvanakośa of Siddhāntaśiromaṇi, the Vault of the Heaven is also called 'Brahmā's Egg, the Hemispherical Cauldron Bowl (ब्रह्माण्डकटाहः संपुटः)'. Thus, the inverted tree Aśvattha, which represents the Vault of the Heaven and the Milky Way, should be called 'Brahmā', not 'brahma'. Aśvattha, the imperishable and resplendent Tree, is Brahmā, the maker and discharger of celestial lights and the weaver of the nebulous Milky Way.

III. The Divine Architects

3.1 Tvaṣṭā, Magnetism in General and the Magnetic Field in Particular

Deva Tvaṣṭā is magnetism in general and magnetic field in particular. The Ṛbhus (ऋभवः ṛbhavaḥ pl.) are the magnetic forces. Tvaṣṭā and the Ṛbhus are called the divine architects, artisans, or carpenters. Tvaṣṭā is the first-born (अग्रियः agriyaḥ RV i.13.10), the shaper of all forms (रूपतक्षा RV viii.102.8), and the divine architect (विश्वकर्मा viśvakarmā RV x.81, RV x.82, RV x.170.4). He is the possessor of the World (सक्षणिः sakṣaṇiḥ RV ii.31.4). Tvaṣṭā, with his knowledge of magic arts, also brings the sacred goblets (पात्रा n. nom. pl.) that hold the drinks of the Devas (RV x.53.9).

3.1.1 Tvaṣṭā, the First Born

Tvaṣṭā is the first born, emerging from the ocean of celestial hydrogen. He, along with the Ṛbhus, subsequently builds the Vault of the Heaven and its substructures.

RV i.13.10 author मेधातिथिः काण्वः, to त्वष्टा, metre गायत्री छंदः
इह त्वष्टारमग्रियं विश्वरूपमुप ह्वये । अस्माकमस्तु केवलः ॥१.१३.१०॥
इह । त्वष्टारं । अग्रियं । विश्वऽरूपं । उप । ह्वये । अस्माकं । अस्तु । केवलः ॥१.१३.१०॥
Now, I invoke Tvaṣṭā, the first-born, the wearer of all forms.
Let him be ours and ours alone. (RV i.13.10)

इह <ind. – here, now, at this time>, त्वष्टारं <m. acc. sg. of त्वष्टृ – Tvaṣṭā>, अग्रियं <m. acc. sg. of adj. अग्रिय – oldest, the first-born>, विश्वरूपं <m. acc. sg. of adj. विश्वरूप – wearing all forms>, उप <ind. prefix to verb ह्वये (√ह्वे)>, उप-ह्वये <pre. middle 1st person sg. of उप-√ह्वे – I call, invoke>;

अस्माकं <m. gen. pl. of 1st person pron. अस्मद् – ours>, अस्तु <ipv. 3rd person sg. of √अस् – let him be>, केवलः <m. nom. sg. of adj. केवल – exclusively one's own>.

3.1.2 The Shaper of All Forms

Tvaṣṭā shapes all forms (RV x.110.9, RV viii.102.8). He fashions the Heavenly structures and all beings into their forms (RV x.110.9). The sacrifice at which Hotā offers oblation to honour Tvaṣṭā is the Agni Yajña, the fire sacrifice, that breaks up the ionic bonds of hydrogen and discharges celestial lights and fires, cosmic plasmas, into the Soma Pond.

Stanza RV x.110.9 is addressed to Tvaṣṭā. The author is Ṛṣi Jamadagni Vāsuta vā Rāma (जमदग्निस्वासुतो वा रामः). Jamadagni is 'Agni the blazing fire'. Rāma is a "name of Varuṇa" and, as an adjective, it means "dark, dark-coloured".

RV x.110.9 author जमदग्निस्वासुतो वा रामः, to त्वष्टा, metre त्रिष्टुप् छंदः
य इमे द्यावापृथिवी जनित्री रूपैरपिंशद्भुवनानि विश्वा ।
तमद्य होतरिषितो यजीयान्देवं त्वष्टारमिह यक्षि विद्वान् ॥१०.११०.९॥
यः । इमे इति । द्यावापृथिवी इति । जनित्री इति । रूपैः । अपिंशत् । भुवनानि । विश्वा ।
तं । अद्य । होतः । इषितः । यजीयान् । देवं । त्वष्टारं । इह । यक्षि । विद्वान् ॥१०.११०.९॥
He who fashioned the twofold Heaven and the Milky Way, our progenitors, and all beings with their forms, Today, to that Deva Tvaṣṭā, let the wise Hotā skilled in sacrifice offer oblations with speed. (RV x.110.9)

यः <m. nom. sg. of pron. यद् – he who>, इमे <f. acc. du. of pron. इदं – these two>, द्यावापृथिवी <f. acc. of dual compound द्यावापृथिवी – the twofold Heaven and the Milky Way (Pṛthivī)>, जनित्री <m. acc. du. of जनित्रि (जनितृ?) – progenitors, parents>, रूपैः <n. inst. pl. of रूप – with forms>, अपिंशत् <ipf. 3rd person sg. of √पिश् (=√पिंश्) – he carved, moulded, fashioned>, भुवनानि <n. acc. pl. of भुवन – all beings>, विश्वा <n. acc. pl. of adj. विश्व – all, every>;

तं <m. acc. sg. of pron. तद् – him, that>, अद्य <ind. – today, now>, होतः <m. nom. sg. of adj. होतृ – Hotā, one who sacrifices, the sacrificer>, इषितः <m. nom. sg. of adj. इषित – animated, quick, speedy>, यजीयान् <m. nom. sg. of adj. यजीयस् – sacrificing excellently, skilled in sacrifice>, देवं <m. acc. sg. of देव – the Deva>, त्वष्टारं <m. acc. sg. of त्वष्टृ – Tvaṣṭā>, इह <ind. – here, in this place, now, at this time>, यक्षि <ipv. 3rd person sg. of √यज् – let him (Hotā) worship or offer (with sacrifice or oblations)>, विद्वान् <m. nom. sg. of adj. विद्वत् (=विद्वस्) – wise, mindful>.

In RV viii.102.8, Tvaṣṭā is asked to come to assist 'us' to shape the forms according to the design of Agni, the magnificent One. The stanza tells us that the magnetic field shapes all forms according to the design of the magnificent hydrogen ions, the protons. The stanza is addressed to Agni.

RV viii.102.8 author प्रयोगो भार्गव अग्निर्वा पावको बार्हस्पत्यः वा अथवाग्नी गृहपतियविष्ठौ सहसः सुतौ तयोर्वान्यतरः, to अग्निः, metre गायत्री छंदः
अयं यथा न आभुवत्त्वष्टा रूपेव तक्ष्या । अस्य क्रत्वा यशस्वतः ॥८.१०२.८॥
अयं । यथा । नः । आऽभुवत् । त्वष्टा । रूपाऽइव । तक्ष्या । अस्य । क्रत्वा । यशस्वतः ॥८.१०२.८॥
May this Tvaṣṭā come to assist us thus the forms of ours to be shaped,
By the design of this magnificent One. (RV viii.102.8)

अयं <m. nom. sg. of pron. इदं – this, this here>, यथा <ind. – in which way, as, as it is, so that, in order that>, नः <m. gen. pl. of 1st person pron. अस्मद् – our>, आभुवत् <sub. 3rd person sg. of आ-√भू – may (he) dwell, assist>, त्वष्टा <m.

nom. sg. of त्वष्टृ – Tvaṣṭā>, रूपा <n. acc. pl. of रूप – forms>, इव <ind. – so, indeed, very>, तक्ष्या <n. acc. pl. of adj. तक्ष्य – to be formed, shaped>;

अस्य <m. gen. sg. of pron. इदं – of this>, क्रत्वा <m. inst. sg. of क्रतु – by plan, design, intention>, यशस्वतः <m. gen. sg. of adj. यशस्वत् – of glorious, splendid, magnificent One (refers to Agni)>.

3.1.3 The Possessor of the World

In stanza RV ii.31.4, Tvaṣṭā is described as the possessor of the World. The implication is that the magnetic field fashions all forms of beings and holds the entire manifested World together. Tvaṣṭā, acting in harmony with Speeches, impels the chariot for cosmic Yajña to succeed. This cosmic Yajña is the Soma sacrifice of releasing cosmic plasmas from the Soma Pond. The second line of the stanza lists the names of the auspicious ones who assist Tvaṣṭā. They are Iḷā, the milch Cow, the Stone; Bhaga, the dispenser of the pressed Soma; Bṛhaddivā, the full-grown heavenly Marut deified as goddess; Rodasī, the pair of the Stone and the Soma Pond; Puraṃdhi, the city of the Devas, the Soma Pond, deified as goddess of abundance; Pūṣā, who guides the cosmic plasmas on the journey; and Aśvinau, the twin Jets, who carry the released cosmic plasmas.

RV ii.31.4 author गृत्समदः, to विश्वे देवाः, metre जगती छंदः
उत स्य देवो भुवनस्य सक्षणिस्त्वष्टा ग्नाभिः सजोषा जूजुवद्रथं ।
इळा भगो बृहद्दिवोत रोदसी पूषा पुरंधिरश्विनावधा पती ॥२.३१.४॥
उत । स्यः । देवः । भुवनस्य । सक्षणिः । त्वष्टा । ग्नाभिः । सऽजोषाः । जूजुवत् । रथं ।
इळा । भगः । बृहत्ऽदिवा । उत । रोदसी इति । पूषा । पुरंऽधिः । अश्विनौ । अध । पती इति ॥२.३१.४॥
And the Deva Tvaṣṭā, the possessor of the World, acting in harmony with Speeches, impelled the chariot. Iḷā, Bhaga, Bṛhaddivā, Rodasī, Puraṃdhi, Pūṣā, and Aśvinau the twin lords.
(RV ii.31.4)

उत <ind. – and, also>, स्यः <m. nom. sg. of pron. त्यद् – that (used as an article)>, देवः <m. nom. sg. of देव – the Deva>, भुवनस्य <n. gen. sg. of भुवन – of the World, i.e. the whole galactic system>, सक्षणिः <m. nom. sg. of सक्षणि – a possessor>, त्वष्टा <m. nom. sg. of त्वष्टृ – Tvaṣṭā, magnetism in general and magnetic field in particular>, ग्नाभिः <f. inst. pl. of ग्ना (=वाच्) – with Speeches>, सजोषाः <m. nom. sg. of adj. सजोषस् – being or acting in harmony with (inst.)>, जूजुवत् <(=जूजवत्) aor. 3rd person sg. of √जू – he impelled quickly, urged on>, रथं <m. acc. sg. of रथ – the chariot>;

इळा <f. nom. sg. of इळा – Devī Iḷā, the mother milch Cow, the Stone>, भगः <m. nom. sg. of भग – Bhaga, the Dispenser>, बृहद्दिवा <f. nom. sg. of बृहद्दिवा – Bṛhaddivā, the full-grown heavenly Marut deified as goddess>, उत <ind. – and, also>, रोदसी <n. nom. du of रोदस् – Rodasī, the pair of the Stone and the Soma Pond>, पूषा <m. nom. sg. of पूषन् – Pūṣā, the conductor on journey (of celestial lights, cosmic plasmas)>, पुरंधिः <f. nom. sg. of पुरंधि (from पुर् f. a fortress, wall, castle, city, the body) – Puraṃdhi, the fortress of the Soma Pond deified>, अश्विनौ <m. nom. du. of अश्विन् – Aśvinau, the twin Jets>, अध <ind. particle – now>, पती <m. nom. du. of पति – the twin lords>.

3.1.4 Tvaṣṭā Holds the Goblets That Serve the Drink of the Devas

Stanza RV x.53.9, though addressed to Agni, explains Tvaṣṭā's work. The authors are the Devas. Tvaṣṭā, proficient in the art of magic, holds the goblets that serve the drinks of the Devas, and sharpens the hatchet, with which Brahmaṇaspati, the lord of prayer, uses to cut the atomic structure asunder to produce cosmic plasmas. Brahmaṇaspati is the name of the Stone deified, the counterpart of Devī Iḷā (इळा). In this stanza, the goblets, in the plural, refer to the individual plasma blooms, or Soma plants, that contain celestial ambrosia, the Soma juice.

RV x.53.9 author देवाः, to अग्निः सौवीकः, metre जगती छंदः
त्वष्टा माया वेदपसामपस्तमो बिभ्रत् पात्रा देवपानानि शंतमा ।
शिशीते नूनं परशुं स्वायसं येन वृश्चादेतशो ब्रह्मणस्पतिः ॥१०.५३.९॥
त्वष्टा । मायाः । वेत् । अपसां । अपःऽतमः । बिभ्रत् । पात्रा । देवऽपानानि । शंऽतमा ।
शिशीते । नूनं । परशुं । सुऽआयसं । येन । वृश्चात् । एतशः । ब्रह्मणः । पतिः ॥१०.५३.९॥
Tvaṣṭā, most skilful in sacred works, well versed in magic arts, holding the goblets that serve salutary drinks of the Devas, Sharpens the hatchet wrought of gold, with which the radiant Brahmaṇaspati shall cut asunder. (RV x.53.9)

त्वष्टा <m. nom. sg. of त्वष्टृ – Tvaṣṭā, magnetism in general and magnetic field in particular>, मायाः <f. acc. pl. of माया – māyā, magic arts, sorcery, supernatural powers>, वेत् <ipf. 3rd person sg. of √विद् – he knew, understood, acquainted with>, अपसां <n. gen. pl. of अपस् – of sacred or sacrificial acts>, अपस्तमः <m. nom. sg. of adj. अपस्तम – most skilful in any art>, बिभ्रत् <m. nom. sg. of adj. बिभ्रत् – bearing, carrying, holding>, पात्रा <n. acc. pl. of पात्र – goblets, cups>, देवपानानि <n. acc. pl. of देवपान – drinks of the Devas>, शंतमा <n. acc. pl. of adj. शंतम – most beneficent, salutary (beneficial, health-giving)>;

शिशीते <pre. middle 3rd person sg. of √शो – he sharpens>, नूनं <ind. – now, at present>, परशुं <m. acc. sg. of परशु – a hatchet, axe>, स्वायसं <m. acc. sg. of स्वायस (स्व्-आयस). cf. आयस adj. from अयस् iron, metal, gold – made of good metal or gold>, येन <m. inst. sg. of यद् – with which>, वृश्चात् <sub. 3rd person sg. of √व्रश्च् – he will/shall cleave or cut asunder (the ionic bonds)>, एतशः <m. nom. sg. of adj. एतश – shining, brilliant>, ब्रह्मणस्पतिः <m. nom. sg. of ब्रह्मणस्पति (=बृहस्पति) – Brahmaṇaspati, the Lord of prayers and sacrifices. He is the Stone, the celestial ioniser, deified>.

3.1.5 Tvaṣṭā, the Divine Architect, the All-Seeing One with Two Winged Arms

Hymn RV x.81 is one of the two hymns addressed to Viśvakarmā in the extant text of the Ṛgveda. The author is Viśvakarmā Bhauvana (विश्वकर्मा भौवनः), 'the divine architect who belongs to the World'. Bhauvana (भौवनः), the patronymic of Viśvakarmā, translates as 'belonging to the World (भुवनं bhuvanaṃ)'. In the hymn RV x.81, Tvaṣṭā is called Viśvakarmā, the divine architect. He brought the World and all beings into existence (RV x.81.1-7, RV x.170.4). Tvaṣṭā, the divine architect, is Ṛṣi, the sacrificer (Hotā), the Father. Seeking the archetypal essence, that is, the essence of all beings or forms, he obtains the embryonic particles, the fundamental building blocks of all beings (RV x.81.1). According to the Ṛgveda, Agnis and Somas are the fundamental building blocks. Tvaṣṭā is the mighty all-seeing One who created the Milky Way and illumined the Heaven (RV x.81.2). He has his eyes, face, arms, and feet everywhere. Creating the twofold Heaven and the Milky Way, he welds them together with his two winged arms (RV x.81.3).

RV x.81.1 author विश्वकर्मा भौवनः, to विश्वकर्मा, metre त्रिष्टुप् छंदः
य इमा विश्वा भुवनानि जुह्वदृषिर्होता न्यसीदत्पिता नः ।
स आशिषा द्रविणमिच्छमानः प्रथमच्छदवराँ आ विवेश ॥१०.८१.१॥
यः । इमा । विश्वा । भुवनानि । जुह्वत् । ऋषिः । होता । नि । असीदत् । पिता । नः ।
सः । आशिषा । द्रविणं । इच्छमानः । प्रथमच्छत् । अवरान् । आ । विवेश ॥१०.८१.१॥
He, Ṛṣi, Hotā, our Father, who sat at sacrifice offering oblation to the worlds and all beings,
Seeking the archetypal essence, with prayer, obtained the fundamental embryonic particles.
(RV x.81.1)

यः <m. nom. sg. of यद् – he who>, इमा <n. acc. pl. of pron. इदं – these>, विश्वा <n. acc. pl. of adj. विश्व – all, every>, भुवनानि <n. acc. pl. of भुवन – to the worlds and all beings>, जुह्वत् <m. nom. sg. of adj. जुह्वत् (pre. pt. of √हु used as adj.) – offering an oblation>, ऋषिः <m. nom. sg. of ऋषि – Ṛṣi, a singer of sacred hymns>, होता <m. nom. sg. of होतृ – Hotā, the sacrificer, the priest>, नि <ind. – down, back, into>, असीदत् <ipf. 3rd person sg. of √सद् – he sat down, sat at>, पिता <m. nom. sg. of पितृ – a father>, नः <m. gen. pl. of 1st person pron. अस्मद् – our>;

सः <m. nom. sg. of 3rd person pron. तद् – he>, आशिषा <f. inst. sg. of आशिस् – with prayer, i.e. with hymn>, द्रविणं <n. acc. sg. of द्रविण – essence, substantiality, i.e. the essence of all beings or forms>, इच्छमानः <m. nom. sg. of pre. middle pt. इच्छमान (√इष् 6Ā) – he striving, seeking>, प्रथमच्छत् <n. acc. sg. of adj. प्रथमच्छद् – primary, typical, initial, archetypal>, अवरान् <m. acc. pl. of अवर (at the end of a compound) the least, the lowest degree – (द्रविणं प्रथमच्छदवरान्) the archetypal essence of the lowest degree, i.e. the fundamental embryonic particles, which are the fundamental building blocks of all beings (refer to Agnis and Somas)>, आ <ind. prefix to verb विवेश (√विश्)>, आ-विवेश <prf. 3rd person sg. of आ-√विश् – he obtained, took possession of>.

RV x.81.2 author विश्वकर्मा भौवनः, to विश्वकर्मा, metre त्रिष्टुप् छंदः
किं स्विदासीदधिष्ठानमारंभणं कतमत्स्वित्कथासीत् ।
यतो भूमिं जनयन्विश्वकर्मा वि द्यामौर्णोन्महिना विश्वचक्षाः ॥१०.८१.२॥
किं । स्वित् । आसीत् । अधिऽस्थानं । आऽरंभणं । कतमत् । स्वित् । कथा । आसीत् ।
यतः । भूमिं । जनयन् । विश्वऽकर्मा । वि । द्यां । और्णोत् । महिना । विश्वऽचक्षाः ॥१०.८१.२॥
Whence was this power derived? Whence was it seized? How?
From what Viśvakarmā, the divine architect, the mighty all-seeing one, manifested the Heaven and begot the Milky Way? (RV x.81.2)

किं <ind. particle of interrogation (originally n. nom. sg. of क) – what, how, whence, wherefore?>, स्वित् <स्विद् ind. particle used with interrogative pron. – "do you think?" "perhaps", "indeed">, आसीत् <ipf. 3rd person sg. of √अस् – it was>, अधिष्ठानं <n. nom. sg. of अधिष्ठान – authority, power>, आरंभणं <n. nom. sg. of आरंभण (आरम्भण) – the act of taking hold of, the place of seizing>, कतमत् <n. nom. sg. of कतम – who, which, what?>, स्वित् <(स्विद्) ind. particle used with interrogative pron.>, कथा <ind. – how?>, आसीत् <ipf. 3rd person sg. of √अस् – it was>;

यतः <(यतस्) ind. – from what, whence>, भूमिं <f. acc. sg. of भूमि – soil, ground, territory (here, refers to the Milky Way)>, जनयन् <m. nom. sg. of pre. active pt. of जनयत् (√जन् 10P) – he begets, creates>, विश्वकर्मा <mf. nom. sg. of विश्वकर्मन् – Viśvakarmā, all-creator, the divine architect>, वि <ind. prefix to verb और्णोत् (√वृ)>, द्यां <m. acc. sg. of दिव् – the Heaven>, वि-और्णोत् <ipf. 3rd person sg. of वि-√वृ – he revealed, manifested, illumined>, महिना <ind. – mighty>, विश्वचक्षाः <m. nom. sg. of विश्वचक्षस् (=विश्वचक्ष) – the all-seeing one>.

RV x.81.3 author विश्वकर्मा भौवनः, to विश्वकर्मा, metre त्रिष्टुप् छंदः
विश्वतश्चक्षुरुत विश्वतोमुखो विश्वतोबाहुरुत विश्वतस्पात् ।
सं बाहुभ्यां धमति सं पतत्रैर्द्यावाभूमी जनयन्देव एकः ॥१०.८१.३॥
विश्वतःऽचक्षुः । उत । विश्वतःऽमुखः । विश्वतःऽबाहुः । उत । विश्वतःऽपात् ।
सं । बाहुऽभ्यां । धमति । सं । पतत्रैः । द्यावाभूमी इति । जनयन् । देवः । एकः ॥१०.८१.३॥

He, who has eyes on all sides, whose face is on all sides, has arms and feet on all sides,
The Deva, only he alone, creates the twofold Heaven and the Milky Way, welds them together with two arms and wings. (RV x.81.3)

विश्वतश्चक्षुः <m. nom. sg. of विश्वतश्चक्षुस् – one who has eyes on all sides>, उत <ind. – and, also>, विश्वतोमुखः <m. nom. sg. of विश्वतोमुख – one whose face is on all sides>, विश्वतोबाहुः <m. nom. sg. of विश्वतोबाहु – one who has arms on all sides>, उत <ind. – and, also>, विश्वतस्पात् <(=विश्वतस्पत्) m. nom. sg. of विश्वतस्पद् – one who has feet on all sides>;

सं <ind. prefix to verb धमति √धं (=√ध्मा)>, बाहुभ्यां <m. inst. du. of बाहु – with two arms>, सं-धमति <pre. 3rd person sg. of सं-√धं (=√ध्मा) – he welds or fuses together>, सं <ind. prefix to pt. जनयन्>, पतत्रैः <m. inst. pl. of पतत्र – with wings>, द्यावाभूमी <f. acc. of dual compound द्यावाभूमि (=द्यावापृथिवी) –the twofold Heaven and the Milky Way>, सं-जनयन् <m. nom. sg. of pre. active pt. of सं-जनयत् (सं-√जन् 10P) – he brings forth, creates>, देवः <m. nom. sg. of देव – the Deva>, एकः <m. nom. sg. of adj. एक – one, alone, that one only>.

RV x.81.4 author विश्वकर्मा भौवनः, to विश्वकर्मा, metre त्रिष्टुप् छंदः
किं स्विद्वनं क उ स वृक्ष आस यतो द्यावापृथिवी निष्टतक्षुः ।
मनीषिणो मनसा पृच्छतेदु तद्यदध्यतिष्ठद्भुवनानि धारयन् ॥१०.८१.४॥
किं । स्वित् । वनं । कः । ऊं इति । सः । वृक्षः । आस । यतः । द्यावापृथिवी इति । निःऽततक्षुः ।
मनीषिणः । मनसा । पृच्छत । इत् । ऊं इति । तत् । यत् । अधिऽअतिष्ठत् । भुवनानि । धारयन् ॥१०.८१.४॥
Whence this Forest? What was this tree from which they fashioned the Heaven and the Milky Way? O wise men, with your whole heart, inquire ye about upon which they stood and about that which sustains the worlds. (RV x.81.4)

किं <ind. particle of interrogation (originally n. nom. sg. of क) – what, whence, wherefore?>, स्वित् <(स्विद्) ind. enclitic particle (with interrogative pron.) – do you think?, perhaps, indeed>, वनं <n. nom. sg. of वन – the forest (refers to the Soma Pond)>, कः <m. nom. sg. of किम् – what?>, उ <ind. particle – and, also>, सः <m. nom. sg. of 3rd person pron. तद् – this>, वृक्षः <m. nom. sg. of वृक्ष – tree (refers to the individual Soma plants)>, आस <prf. 3rd person sg. of √अस् 2P – it was>, यतः <(यतस्) ind. – from which>, द्यावापृथिवी <f. acc. of dual compound द्यावापृथिवि - Dyāvāpṛthivī, the twofold Heaven and the Milky Way>, निष्टतक्षुः <prf. 3rd person pl. of निस्-√तक्ष् – they carved out, fashioned>;

मनीषिणः <m. voc. pl. of मनीषिन् – O wise men, O sages>, मनसा <ind. – in the mind, with all the heart>, पृच्छत <ipv. 2nd person pl. of √प्रच्छ् – ye (pl.) inquire, ask>, इत् <(इद्) ind. particle – even, indeed>, उ <ind. particle – and, also>, तत् <m. nom. sg. of 3rd person. pron. तद् – that>, यत् <m. nom. sg. of pron. यद् – which>, अध्यतिष्ठन् <ipf. 3rd person pl. of √अधिष्ठा (अधि-√स्था) – they stood upon, depend upon>, भुवनानि <n. acc. pl. of भुवन – the worlds>, धारयन् <m. nom. sg. of pre. active cau. pt. धारयत् (√धृ) – it/that holds, bears, supports, maintains>.

RV x.81.5 author विश्वकर्मा भौवनः, to विश्वकर्मा, metre त्रिष्टुप् छंदः
या ते धामानि परमाणि यावमा या मध्यमा विश्वकर्मन्नुतेमा ।
शिक्षा सखिभ्यो हविषि स्वधावः स्वयं यजस्व तन्वं वृधानः ॥१०.८१.५॥
या । ते । धामानि । परमाणि । या । अवमा । या । मध्यमा । विश्वऽकर्मन् । उत । इमा ।
शिक्ष । सखिऽभ्यः । हविषि । स्वधाऽवः । स्वयं । यजस्व । तन्वं । वृधानः ॥१०.८१.५॥
Whether your seats are in the Upper Heaven, in the Lower Heaven, or here in the Midheaven, O Viśvakarmā, Help thy companions, O self-powered one, honour thine own self with oblations and be thou exalted. (RV x.81.5)

या <n. nom. pl. of pron. यद् – which, what, whatever>, ते <m. gen. sg. of 2nd person pron. युष्मद् – your>, धामानि <n. nom. pl. of धामन् – dwelling-places, abodes, domains (esp. seats of the Devas)>, परमाणि <n. nom. pl. of adj. परम – uppermost places, i.e. places in the Upper Heaven>, या <n. nom. pl. of pron. यद् – which, what, whatever>, अवमा <n. nom. pl. of adj. अवम – undermost places, i.e. places in the Lower Heaven>, या <n. nom. pl. of pron. यद् – which, what,

whatever>, मध्यमा <n. nom. pl. of adj. मध्यम – middlemost places, i.e. places in the Midheaven>, विश्वकर्मन् <m. voc. sg. of विश्वकर्मन् – O Viśvakarmā, O Divine Architect>, उत <ind. – and, also>, इमा <n. nom. pl. of pron. इदं – these here>;

शिक्ष <ipv. 2nd person sg. of √शिक्ष् – you help, aid (dat.)>, सखिभ्यः <m. dat. pl. of सखि – friends, companions>, हविषि <n. loc. sg. of हविस् – in oblation, offering>, स्वधावः <m. voc. sg. of स्वधाव (cf. स्वधा self-power) – O self-powered one>, स्वयं <ind. – one's self (applicable to all persons e.g. myself, thyself, themselves)>, यजस्व <ipv. 2nd person sg. of √यज् – you worship, honour (esp. with sacrifice or oblations)>, तन्वं <f. acc. sg. of तनू – own self (used as a reflexive pron.)>, वृधानः <m. nom. sg. of prf. middle pt. वृधान (√वृध्) – being/become elevated, exalted (esp. with praise or sacrifice)>.

RV x.81.6 author विश्वकर्मा भौवनः, to विश्वकर्मा, metre त्रिष्टुप् छंदः

विश्वकर्मन्हविषा वावृधानः स्वयं यजस्व पृथिवीमुत द्यां ।
मुह्यन्त्वन्ये अभितो जनास इहास्माकं मघवा सूरिरस्तु ॥१०.८१.६॥
विश्वऽकर्मन् । हविषा । ववृधानः । स्वयं । यजस्व । पृथिवीं । उत । द्यां ।
मुह्यंतु । अन्ये । अभितः । जनासः । इह । अस्माकं । मघऽवा । सूरिः । अस्तु ॥१०.८१.६॥

O Viśvakarmā, now thyself exalted, worship thou the Heaven and the Milky Way with oblation. Let other men around us be stupefied, but here in the Midheaven, let our Soma sacrificer be generous in offering oblations. (RV x.81.6)

विश्वकर्मन् <m. voc. sg. of विश्वकर्मन् – O Viśvakarmā, O Divine Architect>, हविषा <n. inst. sg. of हविस् – with oblation, sacrifice>, ववृधानः <m. nom. sg. of prf. middle pt. वृधान (√वृध्) – he being exhilarated, exalted (esp. the gods, with praise or sacrifice)>, स्वयं <ind. – self, one's self>, यजस्व <ipv. 2nd person sg. of √यज् – you worship, honour>, पृथिवीं <f. acc. sg. of पृथिवी – Pṛthivī, the Milky Way>, उत <ind. – and, also>, द्यां <m. acc. sg. of दिव् – the Heaven>;

मुह्यंतु <ipv. 3rd person pl. of √मुह् – let them be stupefied (astonished, shocked), be bewildered>, अन्ये <m. nom. pl. of pron. अन्य – others>, अभितः <(अभितस्) ind. – in the proximity, around us>, जनासः <m. nom. pl. of जन – men, people>, इह <ind. – in this place, here (refers to the Midheaven)>, अस्माकं <m. gen. pl. of pron. अस्मद् – our>, मघवा <m. nom. sg. of adj. मघवन् – munificent, generous (in distributing oblations)>, सूरिः <m. nom. sg. सूरि – the institutor of sacrifice, presser of Soma, Soma sacrificer>, अस्तु <ipv. 3rd person sg. of √अस् – let him be>.

RV x.81.7 author विश्वकर्मा भौवनः, to विश्वकर्मा, metre त्रिष्टुप् छंदः

वाचस्पतिं विश्वकर्माणमूतये मनोजुवं वाजे अद्या हुवेम ।
स नो विश्वानि हवनानि जोषद्विश्वशंभूरवसे साधुकर्मा ॥१०.८१.७॥
वाचः । पतिं । विश्वऽकर्माणं । ऊतये । मनःऽजुवं । वाजे । अद्य । हुवेम ।
सः । नः । विश्वानि । हवनानि । जोषत् । विश्वऽशंभूः । अवसे । साधुऽकर्मा ॥१०.८१.७॥

Today, let us invoke the Speeches and the lord Viśvakarmā, who is swift as thought, for help in battle. May he, the beneficent Viśvaśaṃbhū, be pleased with all our invocations for assistance. (RV x.81.7)

वाचः <f. acc. pl. of वाच् – Speeches>, पतिं <m. acc. sg. of पति – the master, ruler, lord>, विश्वकर्माणं <m. acc. sg. of विश्वकर्मन् – Viśvakarmā>, ऊतये <f. dat. sg. of ऊति – for help, protection>, मनोजुवं <m. acc. sg. of मनोजू – swift as thought>, वाजे <m. loc. sg. of वाज – in war, battle (against Vṛtra, the serpent of the Heaven)>, अद्य <ind. – now, today>, हुवेम <opt. 1st person pl. of √हू (weak form of √ह्वे) – may we invoke>;

सः <m. nom. sg. of 3rd person pron. तद् – he>, नः <m. gen. of 1st person pron. अस्मद् – our>, विश्वानि <n. acc. pl of adj. विश्व – all, every>, हवनानि <n. acc. pl. of हवन – calling, invocations>, जोषत् <sub. 3rd person sg. of √जुष् – may he be pleased or be fond of (acc.)>, विश्वशंभूः <m. nom. sg. of विश्वशंभू –Viśvaśaṃbhū, he who is the source of all prosperity, a name of Viśvakarmā (refers to Tvaṣṭā)>, अवसे <n. dat. sg. of अवस् – for favour, assistance>, साधुकर्मा <m. nom. sg. of adj. साधुकर्मन् – righteous, beneficent>.

3.2 THE ṚBHUS, THE MAGNETIC FORCES

The Ṛbhus (ऋभवः ṛbhavaḥ, pl.), the magnetic forces, form a triad; they are Ṛbhu (ऋभुः) or Ṛbhukṣā (ऋभुक्षाः), Vāja (वाजः), and Vibhvā (विभ्वा m. nom. sg. of विभ्वन्). Ṛbhu is an artisan, builder. The word ṛbhu (ऋभु) is derived from the verb √rabh (√रभ्), which means "to take hold of, grasp, clasp, embrace". So, Ṛbhu (ऋभु) is the embracing, attracting force.

Vāja is a swift steed. The word vāja (वाज) means "swift horse", "steed", "spirit", "energy", and is derived from the verb √vaj (√वज्), which means "to go". Thus, Vāja is the running, impelling or repelling force.

Vibhvā is the all-pervading, omnipresent force. The word vibhvan (विभ्वन्) means "far-reaching, penetrating, pervading", which is originated from the verb vi-√bhū (वि-√भू). The verb vi-√bhū means "to expand", "to pervade, fill", "to exist". As an adjective, vibhū (विभू) means "being everywhere", "far-extending", "all-pervading", "eternal", "mighty". Thus, the Ṛgveda tells us that magnetism involves these three forces. These forces work on all and everything and interact synergistically with māyā, the magic force of Varuṇa, to produce a resulting effect. For māyā, see section 4.9. According to the Ṛgveda, these forces of the Ṛbhus are the exerting magnetic forces born of the innate forces contained within the atomic structure of hydrogen (RV x.129.5). These exerting magnetic forces were the builders of the Heavenly structures in the beginning of cosmic creation.

Compare these unified forces of the Ṛbhus to the dissociated forces of the standard model that are carried by different force carrier particles and act on different groups of particles with the exception of the gravity force. The forces of the standard model include gravity force, electromagnetic force, strong force, and weak force. The gravity force acts on all. The carrier particle of gravity, which is named 'graviton', is yet to be observed. The electromagnetic force is carried by photon and acts on quarks and charged leptons and on W^+ and W^- particles. The strong force is carried by gluon and acts on quarks and gluons. The weak force is carried by W^+, W^-, and Z^0 particles and acts on quarks and leptons.[82]

The eleven stanzas of hymn RV iv.33 presented below best explain the Ṛbhus and their key attributes. The author of the hymn is Vāmadeva (वामदेवः), which literally means 'the Deva of Splendour'. His patronymic Gautama (गौतमः) tells us that he is a descendent of Gotama (गोतमः). Gotama, the superlative of 'go (गो)', translates as 'the best of the herds,

82 Particle Data Group of Lawrence Berkeley National Laboratory, 'The Standard Model: the Theory of Fundamental Particles and Forces', *The Particle Adventure*, 2014. [website] https://www.particleadventure.org/standard-model.html (accessed on January 31, 2020).

the celestial lights' and belongs to the family of Aṅgirā (अङ्गिराः sg.), the angel, the messenger of celestial lights. Aṅgirās (अङ्गिरसः aṅgirasaḥ pl.) are the "personifications of luminous objects", the personifications of blooming cosmic plasmas, the celestial lights and fires.

3.2.1 The Ṛbhus Revive Their Old Parents and Impel the Yajña

Vāmadeva sends his Speech to the Ṛbhus and implores the brilliant milch Cow, the Stone, the celestial ioniser, to pour out the pressed Soma. The flowing Waters refer to the streaming current of cosmic plasmas (RV iv.33.1). According to RV iv.33.1, the flowing celestial Waters, that is, the current of streaming cosmic plasmas, travel around the Vault of the Heaven, which is more than 25,000 light-years in height, in an instant.

RV iv.33.1 author वामदेवः, to ऋवभः, metre त्रिष्टुप् छंदः
प्र ऋभुभ्यो दूतमिव वाचमिष्य उपस्तिरे श्वैतरीं धेनुमीळे ।
ये वातजूतास्तरणिभिरेवैः परि द्यां सद्यो अपसो बभूवुः ॥४.३३.१॥
प्र । ऋभुऽभ्यः । दूतंऽइव । वाचं । इष्ये । उपऽस्तिरे । श्वैतरीं । धेनुं । ईळे ।
ये । वातजूताः । तरणिऽभिः । एवैः । परि । द्यां । सद्यः । अपसः । बभूवुः ॥४.३३.१॥
I sent forth my Speech as an envoy to the Ṛbhus, and I implore the brilliant milch Cow to pour out. Driven by Vāta, the flowing Waters, with swift steeds, go around the Heaven in an instant. (RV iv.33.1)

प्र <ind. prefix to verb इष्ये (√इष्)>, ऋभुभ्यः <m. dat. pl. of ऋभु – to the Ṛbhus>, दूतं <m. acc. sg. of दूत – a messenger, envoy>, इव <ind. – like, as>, वाचं <f. acc. sg. of वाच् – Speech, sound>, प्र-इष्ये <ipf. 1st person sg. of प्र-√इष् (√प्रेष्) – I impelled, sent forth>, उपस्तिरे <f. dat. sg. of उपस्तिर् – (dat. as infinitive) to spread over (Waters, cosmic plasmas)>, श्वैतरीं <f. acc. sg. of श्वैतरी – a white, brilliant cow abounding in milk>, धेनुं <f. acc. sg. adj. धेनु – milch>, ईळे <(ईडे) pre. middle 1st person sg. of √ईळ् (√ईड्) – I request, implore>;

ये <m. nom. pl. of pron. यद् – they who>, वातजूताः <m. nom. pl. of adj. वातजूत – driven by Vāta>, तरणिभिः <m. inst. pl. of adj. तरणि – quick, energetic>, एवैः <m. inst. pl. of एव – with horses, steeds>, परि <ind. prefix to verb बभूवुः (√भू)>, द्यां <m. acc. sg. of दिव् – the Heaven>, सद्यः <(सद्यस्) ind. – at once, in an instant>, अपसः <mf. nom. pl. of अपस् – the running Waters, i.e. the current of Waters (cosmic plasmas)>, परि-बभूवुः <prf. 3rd person pl. of परि-√भू – they go or fly around>.

RV iv.33.2 author वामदेवः, to ऋवभः, metre त्रिष्टुप् छंदः
यदारमक्रन्नृभवः पितृभ्यां परिविष्टी वेषणा दंसनाभिः ।
आदिद्देवानामुप सख्यमायन्धीरासः पुष्टिमवहन्मनायै ॥४.३३.२॥
यदा । अरं । अक्रन् । ऋभवः । पितृऽभ्यां । परिऽविष्टी । वेषणा । दंसनाभिः ।
आत् । इत् । देवानां । उप । सख्यं । आयन् । धीरासः । पुष्टिं । अवहन् । मनायै ॥४.३३.२॥
When the Ṛbhus provided suitable services to their parents with attendance and marvellous skills, The skilful ones brought prosperity and attained the companionship of the Devas for their devotion. (RV iv.33.2)

यदा <ind. – when>, अरं <ind. – suitable, proper>, अक्रन् <ipf. 3rd person pl. of √कृ – they provided>, ऋभवः <m. nom. pl. of ऋभु – Ṛbhus (pl.)>, पितृभ्यां <m. dat. du. of पितृ – to parents>, परिविष्टी <f. acc. du. of परिविष्टि – services (to both

parents)>, वेषणा <n. inst. sg. of वेषण – with attendance>, दंसनाभिः <f. inst. pl. of दंसना – with marvellous powers or skills>;

आत् <ind. – afterwards, then>, इत् <(इद्) ind. a particle of affirmation – indeed>, देवानां <m. gen. pl. of देव – of the Devas>, उप <ind. – by the side of, together with>, सख्यं <n. acc. sg. of सख्य – friendship, companionship>, आयन् <ipf. 3rd person pl. of √ए – they reached, attained>, धीरासः <m. nom. pl. of धीर – the wise, skilful ones (the Ṛbhus)>, पुष्टिं <f. acc. sg. of पुष्टि – well-nourished condition, wealth, prosperity>, अवहन् <ipf. 3rd person pl. of √वह् – they brought>, मनायै <f. dat. sg. of मना – for devotion>.

The Ṛbhus make their parents who were lying like two decaying posts young again and impel the celestial Yajña (RV iv.33.3).

RV iv.33.3 author वामदेवः, to ऋभवः, metre त्रिष्टुप् छंदः
पुनर्ये चक्रुः पितरा युवाना सना यूपेव जरणा शयाना ।
ते वाजो विभ्वाँ ऋभुरिन्द्रवंतो मधुप्सरसो नोऽवंतु यज्ञं ॥४.३३.३॥
पुनः । ये । चक्रुः । पितरा । युवाना । सना । यूपाऽइव । जरणा । शयाना ।
ते । वाजः । विऽभ्वा । ऋभुः । इंद्रऽवंतः । मधुऽप्सरसः । नः । अवंतु । यज्ञं ॥४.३३.३॥
They who made their old parents, who were lying like two decaying posts, young again,
Let these Soma lovers, Vāja, Vibhvā, and Ṛbhu, accompanied by Indra, lead the Yajña for us.
(RV iv.33.3)

पुनः <(पुनर्) ind. – again>, ये <m. nom. pl. of pron. यद् – they who>, चक्रुः <prf. 3rd person pl. of √कृ – they made>, पितरा <m. acc. du of पितृ – parents>, युवाना <m. acc. du of adj. युवन् – young, youthful>, सना <ind. – old>, यूपाइव <यूपा m. acc. du. of यूप – two posts>, इव <ind. – like>, जरणा <m. acc. du. of adj. जरण – decayed, decaying>, शयाना <m. acc. du. of adj. शयान – lying down, resting>;

ते <m. nom. pl. of 3rd person pron. तद् – they, these>, वाजः <m. nom. sg. of वाज – Vāja (one of the three Ṛbhus)>, विभ्वा <m. nom. sg. of विभ्वन् – Vibhvā (a Ṛbhu)>, ऋभुः <m. nom. sg. of ऋभु – Ṛbhu (one of the three Ṛbhus)>, इंद्रवंतः <m. nom. pl. of adj. इंद्रवत् – accompanied by Indra>, मधुप्सरसः <m. nom. pl. of adj. मधुप्सरस् – who are fond of Soma mead, Soma lovers>, नः <m. dat. pl. of 1st person pron. अस्मद् – for us>, अवंतु <ipv. 3rd person pl. of √अव् – let them drive, impel, lead>, यज्ञं <m. acc. sg. of यज्ञ – Yajña, sacrifice>.

3.2.2 The Ṛbhus Guard the Cow and Make the Four Celestial Bowls

For a year, the Ṛbhus, the magnetic forces, guard the Cow, sustain her light, and fashion the fetters, the plasma sheaths, of the Soma Pond and of the individual plasma blooms (RV iv.33.4). Through these endeavours, the Ṛbhus attain immortality. In the Ṛgveda, the celestial plasmas and associated plasma phenomena are called the Devas, the immortals. Note that, in the Ṛgveda, a cycle of celestial ionisation and cosmic plasma discharge and release is described as a year, a season, or a rainy season.

RV iv.33.4 author वामदेवः, to ऋभवः, metre त्रिष्टुप् छंदः
यत्संवत्समृभवो गामरक्षन्यत्संवत्समृभवो मा अपिंशन् ।
यत्संवत्समभरन्भासो अस्यास्ताभिः शमीभिरमृतत्वमाशुः ॥४.३३.४॥
यत् । संऽवत्सं । ऋभवः । गां । अरक्षन् । यत् । संऽवत्सं । ऋभवः । माः । अपिंशन् ।

यत् । संऽवत्सं । अभरन् । भासः । अस्याः । ताभिः । शमीभिः । अमृतऽत्वं । आशुः ॥४.३३.४॥
For a year, when the Ṛbhus guarded the Cow, they fashioned the fetters.
Throughout the year, they sustained her light, and through these endeavours, the Ṛbhus have attained immortality. (RV iv.33.4)

यत् <(यद्) conjunction – when, as>, संवत्सं <ind. – for a year (here, a cycle of plasma discharge and release is compared to a year)>, ऋभवः <m. nom. pl. of ऋभु – Ṛbhus (pl.)>, गां <m. acc. sg. of गो – the cow (refers to the Soma Pond)>, अरक्षन् <ipf. 3rd person pl. of √रक्ष् – they guarded, protected>, यत् <(यद्) n. nom. sg. – that, which>, संवत्सं <ind. – throughout a year>, ऋभवः <m. nom. pl. of ऋभु – Ṛbhus (pl.)>, माः <f. acc. pl. of मा – bindings, fetters, i.e. the plasma sheath of the Soma Pond and the plasma sheaths of individual plasma blooms>, अपिंशन् <ipf. 3rd person pl. of √पिश् – they formed, fashioned>;

यत् <(यद्) adverb conjunction – so that, that>, संवत्सं <ind. – throughout a year>, अभरन् <ipf. 3rd person pl. of √भृ – they supported, fostered, maintained>, भासः <n. acc. sg. of भासस् – brightness, light>, अस्याः <f. gen. sg. of इदं – of her>, ताभिः <f. inst. pl. of pron. तद् – through these>, शमीभिः <f. inst. pl. of शमी – with/through efforts or labours>, अमृतत्वं <n. acc. sg. of अमृतत्व (=अमृतता) – immortality>, आशुः <prf. 3rd person pl. of √अश् – they have gained, obtained>.

In RV iv.33.5, the Ṛbhus discuss how many celestial Bowls they should make. With the acknowledgement of Tvaṣṭā, they settle on making four. The Ṛbhus, the magnetic forces, set out to make the four Bowls. Once the four Bowls are made by the Ṛbhus, Tvaṣṭā, the magnetic field, cares for them (RV iv.33.6). Four Bowls refer to the two Heavenly Vaults and the two Soma Ponds (see Figure 6).

RV iv.33.5 author वामदेवः, to ऋवभः, metre त्रिष्टुप् छंदः
ज्येष्ठ आह चमसा द्वा करेति कनीयान्त्रीन्कृणवामेत्याह ।
कनिष्ठ आह चतुरस्करेति त्वष्ट ऋभवस्तत्पनयद्वचो वः ॥४.३३.५॥
ज्येष्ठः । आह । चमसा । द्वा । कर । इति । कनीयान् । त्रीन् । कृणवाम । इति । आह ।
कनिष्ठः । आह । चतुरः । कर । इति । त्वष्टा । ऋभवः । तत् । पनयत् । वचः । वः ॥४.३३.५॥
Make thou two celestial bowls, thus said the eldest Ṛbhu. Let us make three, thus the younger one spoke. Make four bowls, thus spoke the youngest. Tvaṣṭā acknowledged the latter counsel of yours, O Ṛbhus. (RV iv.33.5)

ज्येष्ठः <m. nom. sg. of ज्येष्ठ – the eldest (the chief of the three Ṛbhus)>, आह <prf. 3rd person sg. √अह् – he spoke>, चमसा <m. acc. du. of चमस – two celestial bowls>, द्वा <m. nom. du. of द्वि (original stem द्व) – two>, कर <ipv. 2nd person sg. of √कृ – you make>, इति <ind. – thus>, कनीयान् <m. nom. sg. कनीयस् – the younger one (of the three Ṛbhus)>, त्रीन् <m. acc. pl. of त्रि – three>, कृणवाम <ipv.1st person pl. of √कृ – let us make>, इति <ind. – thus>, आह <prf. 3rd person sg. √अह् – he spoke>;

कनिष्ठः <m. nom. sg. of कनिष्ठ – the youngest (of the three Ṛbhus)>, आह <prf. 3rd person sg. √अह् – he spoke>, चतुरः <m. acc. pl. of चतुर् – four>, कर <ipv. 2nd person sg. of √कृ – you make>, इति <ind. – thus>, त्वष्टा <m. nom. sg. of त्वष्टृ – Tvaṣṭā, magnetism in general and magnetic field in particular>, ऋभवः <m. voc. pl. of ऋभु – O Ṛbhus (pl.)>, तत् <n. acc. sg. of 3rd person pron. तद् – that, this, i.e. the latter>, पनयत् <ipf. cau. 3rd person sg. of √पन् – he acknowledged>, वचः <n. acc. sg. of वचस् – speech, word, advice>, वः <m. gen. pl. of 2nd person pron. युष्मद् – your>.

RV iv.33.6 author वामदेवः, to ऋवभः, metre त्रिष्टुप् छंदः
सत्यमूचुर्नर एवा हि चक्रुरनु स्वधामृभवो जग्मुरेतां ।
विभ्राजमानांश्चमसाँ अहेवावेनत्त्वष्टा चतुरो ददृश्वान् ॥४.३३.६॥

सत्यं । ऊचुः । नरः । एव । हि । चक्रुः । अनु । स्वधां । ऋभवः । जग्मुः । एतां ।
विऽभ्राजमानान् । चमसान् । अहाऽइव । अवेनत् । त्वष्टा । चतुरः । ददृश्वान् ॥४.३३.६॥
The Ṛbhus indeed have spoken, and of their own accord, they set out and made this a reality. Tvaṣṭā, beholding the bowls that are resplendent as the day, tended the four celestial bowls. (RV iv.33.6)

सत्यं <n. acc. sg. of सत्य – truth, reality>, ऊचुः <prf. 3rd person pl. of √वच् – they have spoken>, नरः <m. nom. pl. of नृ – men, heroes, i.e. the Ṛbhus>, एव <ind. – indeed>, हि <ind. – surely>, चक्रुः <prf. 3rd person pl. of √कृ – (with सत्यं) they made it a reality, fulfilled>, अनु <ind. – with, according to>, स्वधां <f. acc. sg. of स्वधा – (with अनु) willingly, of their own accord>, ऋभवः <m. nom. pl. of ऋभु – the Ṛbhus (pl.)>, जग्मुः <prf. 3rd person pl. of √गं – they set out>, एतां <f. acc. sg. of एतद् – this>;

विभ्राजमानान् <m. acc. pl. of pre. middle pt. of विभ्राजमान (वि-√भ्राज्) – they being bright, radiant>, चमसान् <m. acc. pl. of चमस – vessels, bowls>, अहा <n. nom. pl. of अह – the days>, इव <ind. – like>, अवेनत् <ipf. 3rd person sg. of √वेन् – he tended, cared for>, त्वष्टा <m. nom. sg. of त्वष्टृ – Tvaṣṭā>, चतुरः <m. acc. pl. of चतुर् – four (bowls)>, ददृश्वान् <m. nom. sg. of prf. active pt. ददृश्वः (√दृश्) – he seeing, beholding>.

3.2.3 The Skilful Architects Repose at Agni's Rite of Soma Reception

While the Ṛbhus repose at Agni's rite of receiving Soma for twelve days, they prepare the beautiful fields, lead the celestial rivers to the dry soil, and direct the celestial Waters to the river channels (RV iv.33.7). According to RV iv.33.7, the Ṛbhus, the magnetic forces, are responsible for laying out beautiful fields and leading the discharged cosmic plasmas into the river channels. The entire time, they stand by the light-containing plants, the plasma blooms, until they are released. The stanza makes it clear that these maturing plasma blooms are magnetic field-aligned filamentary structures.

It is puzzling why the Ṛbhus repose at Agni's rite of receiving Soma for twelve days. It cannot be taken in the literal sense, for Agni's Soma reception occurs three times a day. So, it is reasonable to assume that the 'twelve days' symbolically represent the time it takes for the plasma blooms in the Soma Pond to fully grow and mature, just as the earth's new moon takes about fourteen days to become a full moon.

RV iv.33.7 author वामदेवः, to ऋवभः, metre त्रिष्टुप् छंदः
द्वादश द्यून्यदगोह्यस्यातिथ्ये रणन्नृभवः ससंतः ।
सुक्षेत्राकृणवन्ननयंत सिंधून्धन्वातिष्ठन्नोषधीर्निम्नमापः ॥४.३३.७॥
द्वादश । द्यून् । यत् । अगोह्यस्य । आतिथ्ये । रणन् । ऋभवः । ससंतः ।
सुऽक्षेत्रा । अकृणवन् । अनयंत । सिंधून् । धन्व । आ । अतिष्ठन् । ओषधीः । निम्नं । आपः ॥४.३३.७॥
For twelve days, while the Ṛbhus joyed reposing at Agni's rite of receiving Soma,
They laid out beautiful fields, led the rivers to the dry soil and the celestial Water to the lower ground, and stood by the light-containing plants. (RV iv.33.7)

द्वादश <m. acc. pl. of द्वादशन् – twelve>, द्यून् <m. acc. pl. of द्यु – days>, यत् <conjunction – when, as>, अगोह्यस्य <m. gen. sg. of अगोह्य – of the one not concealed or covered (refers to ionised Agni of cosmic plasmas who is not bound to the atomic structure)>, आतिथ्ये <m. loc. sg. of आतिथ्य – at the rite (the reception of the Soma when it is brought to

the place of sacrifice)>, रणन् <ipf. 3rd person pl. of √रण् – they joyed, delighted in>, ऋभवः <m. nom. pl. of ऋभु – the Ṛbhus (pl.)>, ससंतः <m. nom. pl. of pre. active pt. ससत् (√सस्) – they reposing>;

सुक्षेत्रा <n. acc. pl. of सुक्षेत्र – beautiful fields, dwelling-places (refer to individual plasma blooms)>, अकृण्वन् <ipf. 3rd person. pl. √कृ – they made, established>, अनयंत <ipf. 3rd person pl. √नी – they led>, सिंधून् <m. acc. pl. सिंधु – sindhus, celestial rivers>, धन्व <mn. acc. sg. of धन्वन् – to dry soil, the desert (i.e. the drained field after the release of the matured cosmic plasmas)>, आ <ind. prefix to verb अतिष्ठन् (√स्था)>, आ-अतिष्ठन् <ipf. 3rd person pl. of आ-√स्था – they stood by>, ओषधीः <f. acc. pl. of ओषधि – "light-containing" plants, flaming or burning plants, i.e. plasma blooms>, निम्नं <n. acc. sg. of निम्न – to the low ground, cavity, i.e. to the channel of a plasma bloom>, आपः <n. acc. sg of आपस् – celestial Water, cosmic plasmas>.

RV iv.33.8 author वामदेवः, to ऋवभः, metre त्रिष्टुप् छंदः
रथं ये चक्रुः सुवृतं नरेष्ठां ये धेनुं विश्वजुवं विश्वरूपां ।
त आ तक्षंत्वृभवो रयिं नः स्ववसः स्वपसः सुहस्ताः ॥४.३३.८॥
रथं । ये । चक्रुः । सुऽवृतं । नरेऽस्थां । ये । धेनुं । विश्वऽजुवं । विश्वऽरूपां ।
ते । आ । तक्षंतु । ऋभवः । रयिं । नः । सुऽअवसः । सुऽअपसः । सुऽहस्ताः ॥४.३३.८॥
They who fashioned the beautifully-running chariot, which stands by men and the swift and dappled cow, Let these Ṛbhus fashion our Soma plants. The Ṛbhus, the good protectors, skilful architects, dexterous-handed. (RV iv.33.8)

रथं <m. acc. sg. of रथ – chariot>, ये <m. nom. pl. of pron. यद् – they who>, चक्रुः <prf. 3rd person pl. √कृ – they made>, सुवृतं <m. acc. sg. of adj. सुवृत् – running well (as a chariot)>, नरेष्ठां <m. acc. sg. of adj. नरेष्ठा – standing by heroes or men>, ये <m. nom. pl. of pron. यद् – who>, धेनुं <f. acc. sg. of धेनु – the cow (refers to the plasma bloom)>, विश्वजुवं <f. acc. sg. of adj. विश्वजू – all-quick, speedy>, विश्वरूपां <f. acc. sg. of adj. विश्वरूप – variegated, dappled>;

ते<m. nom. pl. of 3rd person pron. तद् – they>, आ <ind. prefix to verb तक्षंतु √तक्ष्>, आ-तक्षंतु <ipv. 3rd person pl. of आ-√तक्ष् – let them chisel, fashion>, ऋभवः <m. nom. pl. of ऋभु – the Ṛbhus (pl.)>, रयिं <m. acc. sg. of रयि – treasure, wealth (refers to the Soma plants)>, नः <m. gen. pl. of 1st person pron. अस्मद् – our>, स्ववसः <m. nom. pl. of स्ववस् – good protectors>, स्वपसः <m. nom. pl. of स्वपस् – skilful artificers, architects>, सुहस्ताः <m. nom. pl. of सुहस्त – beautiful-handed, dexterous-handed>.

According to stanza RV iv.33.9, the Devas are pleased with the sacred work the Ṛbhus have completed with their thoughtful planning. Thus, Vāja, the swift steed, becomes the property of all the Devas. Ṛbhukṣā or Ṛbhu, the skilled architect, becomes the property of Indra. And the all-pervading, all-seeing Vibhvā becomes the property of Varuṇa.

RV iv.33.9 author वामदेवः, to ऋवभः, metre त्रिष्टुप् छंदः
अपो ह्येषामजुषंत देवा अभि क्रत्वा मनसा दीध्यानाः ।
वाजो देवानामभवत्सुकर्मेंद्रस्य ऋभुक्षा वरुणस्य विभ्वा ॥४.३३.९॥
अपः । हि । एषां । अजुषंत । देवाः । अभि । क्रत्वा । मनसा । दीध्यानाः ।
वाजः । देवानां । अभवत् । सुऽकर्मा । इंद्रस्य । ऋभुक्षाः । वरुणस्य । विऽभ्वा ॥४.३३.९॥
Reflecting upon the Ṛbhus' sacred work, completed with design and intelligence, the Devas were delighted. Vāja became of the Devas, the expert architect Ṛbhukṣā of Indra, Vibhvā of Varuṇa. (RV iv.33.9)

अपः <n. acc. sg. of अपस् – sacred work, sacrificial work>, हि <ind. – on account of, indeed, surely>, एषां <m. gen. pl. of pron. इदं – their (of the Ṛbhus)>, अजुषंत <ipf. 3rd person pl. of √जुष् – they were pleased, delighted>, देवाः <m. nom. pl. of देव – the Devas>, अभि <ind. prefix to pt. दीध्यान (√धी)>, क्रत्वा <f. inst. sg. of क्रतु – with plan, design, intention>,

मनसा <n. inst. sg. of मनस् – with intelligence>, अभि-दीध्यानाः <m. nom. pl. of pre. middle pt. अभि-दीध्यान (अभि-√धी) – they (the Devas) considering, reflecting upon>;

वाजः <m. nom. sg. of वाज – Vāja, the name of one of the three Ṛbhus>, देवानां <m. gen. pl. of देव – of the Devas>, अभवत् <ipf. 3rd person sg. of √भू – became, was>, सुकर्मा <m. nom. sg. of सुकर्मन् – an expert artificer or architect>, इंद्रस्य <m. gen. sg. of इंद्र – of Indra>, ऋभुक्षाः <m. nom. sg. of ऋभुक्षिन् – Ṛbhukṣā (=Ṛbhu), the first of the Ṛbhus>, वरुणस्य <m. gen. sg. of वरुण – of Varuṇa>, विभ्वा <m. nom. sg. of विभ्वन् – Vibhvā, the name of one of the Ṛbhus>.

3.2.4 The Ṛbhus Make Two Bay Steeds for Indra

Exhilarated by the oblation and praise, the Ṛbhus made two bay steeds for Indra (RV iv.33.10). At the third Soma sacrifice, which is the last Soma sacrifice of releasing the cosmic plasmas for the day, the author, Vāmadeva (वामदेवः), who is the Soma Pond deified, supplicates the Ṛbhus to bestow prosperity with treasures and the essence of matter, that is, Agnis, upon 'us' (RV iv.33.10-11).

RV iv.33.10 author वामदेवः, to ऋबभः, metre त्रिष्टुप् छंदः
ये हरी मेधयोक्था मदंत इंद्राय चक्रुः सुयुजा ये अश्वा ।
ते रायस्पोषं द्रविणान्यस्मे धत्त ऋभवः क्षेमयंतो न मित्रं ॥४.३३.१०॥
ये । हरी इति । मेधया । उक्था । मदंतः । इंद्राय । चक्रुः । सुऽयुजा । ये । अश्वा ।
ते । रायः । पोषं । द्रविणानि । अस्मे इति । धत्त । ऋभवः । क्षेमऽयंतः । न । मित्रं ॥४.३३.१०॥
They, exhilarated by the oblation and praise, who fashioned the two well-yoked steeds for Indra, Provide safety as an ally. O Ṛbhus, bestow ye the riches, prosperity, and the essence of matter upon us. (RV iv.33.10)

ये <m. nom. pl. of pron. यद् – they who (the Ṛbhus)>, हरी <m. acc. du. of adj. हरि – tawny, bay>, मेधया <m. inst sg. of मेधा (मेध) – with oblation, offering>, उक्था <n. inst. sg. of उक्थ – with eulogy, praise>, मदंतः <m. nom. pl. of pre. active pt. मदत् (√मद् 1P) – they being exhilarated>, इंद्राय <m. dat. sg. of इंद्र – for Indra>, चक्रुः <prf. 3rd person pl. of √कृ – they made>, सुयुजा <m. acc. du. of adj. सुयुज् – well-yoked>, ये <m. nom. pl. of pron. यद् – repetition of the previous ये>, अश्वा <m. acc. du. of अश्व – two steeds>;

ते <m. nom. pl. of 3rd person pron. तद् – they, these (note ये - ते format)>, रायः <m. acc. pl. of रै – wealth, riches>, पोषं <m. acc. sg. of पोष – prosperity, abundance>, द्रविणानि <n. acc. pl. of द्रविण – substance, essence, i.e. the essence of matter (Agnis)>, अस्मे <mf. loc. pl. of 1st person pron. अस्मद् – upon us>, धत्त <ipv. 2nd person pl. of √दा – ye bestow, grant>, ऋभवः <m. voc. pl. of ऋभु – O Ṛbhus (pl.)>, क्षेमयंतः <m. voc. pl. of adj. क्षेमयत् – providing safety, peace, rest, security, secure state (cf. क्षेम)>, न <ind. – like, as>, मित्रं <n. nom. sg. of मित्र – a friend, an ally>.

RV iv.33.11 author वामदेवः, to ऋबभः, metre त्रिष्टुप् छंदः
इदाह्नः पीतिमुत वो मदं धुर्न ऋते श्रांतस्य सख्याय देवाः ।
ते नूनमस्मे ऋभवो वसूनि तृतीये अस्मिन्त्सवने दधात ॥४.३३.११॥
इदा । अह्नः । पीतिं । उत । वः । मदं । धुः । न । ऋते । श्रांतस्य । सख्याय । देवाः ।
ते । नूनं । अस्मे इति । ऋभवः । वसूनि । तृतीये । अस्मिन् । सवने । दधात ॥४.३३.११॥
At the sacrifice of this reposeful day, the Devas brought your exhilarating drink for the community. Now, O Ṛbhus, bestow ye the treasures upon us at this third Soma sacrifice. (RV iv.33.11)

इदा <ind. – now, at this moment>, अह्नः <n. gen. sg. of अहन् – (with इदा) of this present day>, पीतिं <f. acc. sg. of पीति – drink, a draught>, उत <ind. – also, and>, वः <m. gen. pl. of 2nd person pron. युष्मद् – your (pl.)>, मदं <m. acc. sg.

of मद – rapturous, exhilarating drink, i.e. Soma>, धुः <inj. 3rd person pl. of √धा – they brought forth>, न <ind. – like, as>, ऋते <n. loc. sg. of ऋत – at the sacrifice>, श्रांतस्य <n. gen. sg. of adj. श्रांत – calmed, tranquil>, सख्याय <n. dat. sg. of सख्य – for friendship, fellowship, community>, देवाः <m. nom. pl. of देव – the Devas>;

ते <n. loc. sg. of 3rd pron. base त – at this very>, नूनं <ind. – now, therefore>, अस्मे <mf. loc. pl. of 1st person pron. अस्मद् – upon us>, ऋभवः <m. voc. pl. of ऋभु – O Ṛbhus (pl.)>, वसूनि <n. acc. pl. of वसु – riches, treasures>, तृतीये <n. loc. sg. of adj. तृतीय – at the third>, अस्मिन् <n. loc. sg. of इदं – at this>, सवने <n. loc. sg. of सवन – at the Soma festival, at the Soma sacrifice>, दधात <ipv. 2nd person pl. of √धा – ye bestow>.

IV. Celestial Ionisation and Other Plasma Phenomena

4.1 CELESTIAL IONISATION

Once hydrogen atoms are pulled into the Stone, the celestial ioniser, they throw their bodies off, and become the immortals, the cosmic plasmas. Hymn RV x.34 describes the celestial ionisation from two different perspectives. Of the fourteen stanzas of the hymn, stanzas RV x.34.1, 7, 9, 12, and 13 are addressed to Akṣakṛṣipraśaṃsā (अक्षकृषिप्रशंसा), 'Praise (प्रशंसा) of the Ploughing Wheel (अक्षकृषि:)'. The Ploughing Wheel refers to the Stone that shatters the ionic bonds. Thus, Akṣakṛṣipraśaṃsā means 'the praise of the Stone', the celestial ioniser, for producing cosmic plasmas. Stanzas RV x.34.2-6, 8, 10, 11, and 14 are addressed to Akṣakitavanindā (अक्षकितवनिंदा), 'Gamester's Reproach (कितवनिंदा) of the Ploughing Wheel (अक्ष:)'. The gamester refers to Agni who is rebellious and reluctant to be burnt in the celestial ioniser and enter into the Soma Pond even after his body is burnt and destroyed.

The author of the hymn is Kavaṣa Ailūṣa Akṣa vā Maujavān (कवष: ऐलूष: अक्ष: वा मौजवान्), 'Kavaṣa Akṣa (कवष: अक्ष:), the descendant of Ilūṣa (ऐलूष:), produced on Mt. Mūjavān (मौजवान्)'. Of the fourteen stanzas of RV x.34, stanzas 1, 7, 9 and 2-6 are presented.

4.1.1 Praise of the Ploughing Wheel

Stanza RV x.34.1 explains how delightful the tall and swaying plasma bloom, the flaming Vibhīdaka tree, is to the poet. The flaming Vibhīdaka tree here refer to individual plasma blooms. The tawny Deer (हरिण: hariṇaḥ) refers to the Soma Pond. Thus, in RV x.34.1, the plasma blooms are described as whirling on the tawny Deer. One may compare the flaming Vibhīdaka tree, the tree of blooming cosmic plasmas, to the tree of knowledge the Lord God caused to grow, which is pleasing to the sight and good for food (Genesis 2:9).

RV x.34.1 author कवष ऐलूष अक्षो वा मौजवान्, to अक्षकृषिप्रशंसा, metre त्रिष्टुप् छन्दः
प्रावेपा मा बृहतो मादयंति प्रवातेजा इरिणे वर्वृतानाः ।
सोमस्येव मौजवतस्य भक्षो विभीदको जागृविर्मह्यमच्छान् ॥१०.३४.१॥
प्रावेपाः । मा । बृहतः । मादयंति । प्रवातेऽजाः । इरिणे । वर्वृतानाः ।
सोमस्यऽइव । मौजऽवतस्य । भक्षः । विऽभीदकः । जागृविः । मह्यं । अच्छान् ॥१०.३४.१॥
Swaying, tall, born of a draught of Vāta, whirling on the tawny Deer, they delight me.
The Vibhīdaka tree, feeding on the Soma of Mt. Mūjavān, to me, indeed appears flaming.
(RV x.34.1)

प्रावेपाः <m. nom. pl. of adj. प्रावेप (=वेपथु) – quivering, trembling, swaying>, मा <m. acc. sg. of 1st person pron. अस्मद् – me>, बृहतः <m. nom. pl. of adj. बृहत् – lofty, high, tall>, मादयंति <pre. cau. 3rd person pl. of √मद् – they delight, gladden, exhilarate>, प्रवातेजाः <m. nom. pl. of adj. प्रवातेज – born or produced from a draught of Vāta (ionised plasma particles, i.e. Agnis and Somas)>, हरिणे <m. loc. sg. of हरिण – on the tawny deer, i.e. in the Soma Pond (in the Ṛgveda, plasma blooms are compared to horns of bulls)>, वर्वृतानाः <m. nom. pl. of pre. middle pt. वर्वृतान (√वृत्) – they turning, revolving, whirling>;

सोमस्य <m. gen. sg. of सोम – of Soma>, इव <ind. – like, as if, just, indeed>, मौजवतस्य <m. gen. sg. of adj. मौजवत – coming from Mt. Mūjavān (मूजवान् - m. nom. sg. of मूजवत्)>, भक्षः <m. nom. sg. of adj. भक्ष – having something for food, living upon>, विभीदकः <m. nom. sg. of विभीदक (=विभीतक) – the Vibhīdaka tree (the plasma bloom is compared to a burning Vibhīdaka tree)>, जागृविः <m. nom. sg. of adj. जागृवि – going on burning>, मह्यं <m. dat. sg. of 1st person pron. अस्मद् – to me>, अच्छान् <aor. 3rd person sg. of √छद् cl.10 – it (the Vibhīdaka tree) seemed, appeared>.

Stanzas RV x.34.7 and 9 explain how, the wheels, the firestones of the Stone, the celestial ioniser, destroy the bodies of hydrogen to produce cosmic plasmas. Stanza RV x.34.7 states that the wheels with deceitful, piercing, tormenting, burning hooks destroy the atomic structures to acquire youths, that is, Agnis as gifts.

RV x.34.7 author कवष ऐलूष अक्षो वा मौजवान्, to अक्षकृषिप्रशंसा, metre जगती छंदः
अक्षास इदंकुशिनो नितोदिनो निकृत्वानस्तपनास्तापयिष्णवः।
कुमारदेष्णा जयतः पुनर्हणो मध्वा संपृक्ताः कितवस्य बर्हणा ॥१०.३४.७॥
अक्षासः । इत् । अंकुशिनः । निऽतोदिनः । निऽकृत्वानः । तपनाः । तापयिष्णवः ।
कुमारऽदेष्णाः । जयतः । पुनःऽहनः । मध्वा । संऽपृक्ताः । कितवस्य । बर्हणा ॥१०.३४.७॥
The wheels with deceitful, piercing, tormenting, burning hooks,
Repeatedly slaying, winning Agnis as gifts, are certainly filled with the sweet mead of the gamester. (RV x.34.7)

अक्षासः <m. nom. pl. of अक्ष – the wheels (refers to the firestones of the celestial ioniser)>, इत् <इद् ind. – even, just, only, indeed>, अंकुशिनः <(अङ्कुशिनः) m. nom. pl. of adj. अंकुशिन् – having aṅkuśas, hooks>, नितोदिनः <m. nom. pl. of adj. नितोदिन् – piercing, penetrating>, निकृत्वानः <m. nom. pl. of adj. निकृत्वन् – deceitful>, तपनाः <m. nom. pl. of adj. तपन – causing pain, tormenting>, तापयिष्णवः <m. nom. pl. of adj. तापयिष्णु – heating, burning>;

कुमारदेष्णाः <m. nom. pl. of adj. कुमारदेष्ण (कुमार a child, boy, a name of Agni + देष्ण giving, a gift) – children or youths as gifts (here, Agnis, the embryonic particles of cosmic plasmas, are called children or youths>, जयतः <m. nom. pl. of

adj. जयत् – winning, acquiring, conquering>, पुनर्हणः <(=पुनर्हनः) m. nom. pl. of adj. पुनर्हन् – repeatedly striking, killing, slaying>, मध्वा <n. inst. sg. of मधु – with sweet mead, nectar, Soma>, संपृक्ताः <m. nom. pl. of adj. संपृक्त (सम्पृक्त) – (with inst.) came into contact with, filled with>, कितवस्य <m. gen. sg. of कितव – of the gamester (refers to the rebellious Agni)>, बर्हणा <ind. – really, certainly>.

In stanza RV x.34.9, the firestones of the celestial ioniser are compared to 'heavenly charcoals'. Though the charcoals are 'cold', they can shatter the ionic bonds and burn down the atomic structures completely. The Heavenly charcoals are 'cold' fire. In the stanza, the firestones of the celestial ioniser are described as 'cast on the tawny Deer', that is, cast on the Soma Pond. One may compare the heavenly charcoals in the Ṛgveda to the fire of the burning coals, into which the wicked are cast (Psalm 140:10).

RV x.34.9 author कवष ऐलूष अक्षो वा मौजवान्, to अक्षकृषिप्रशंसा, metre त्रिष्टुप् छंदः
नीचा वर्तंत उपरि स्फुरंत्यहस्तासो हस्तवंतं सहंते ।
दिव्या अंगारा इरिणे न्युप्ताः शीताः संतो हृदयं निर्दहंति ॥१०.३४.९॥
नीचा । वर्तंति । उपरि । स्फुरंति । अहस्तासः । हस्तऽवंतं । सहंते ।
दिव्याः । अंगाराः । हरिणे । निऽउप्ताः । शीताः । संतः । हृदयं । निः । दहंति ॥१०.३४.९॥
Downwards they roll, then rebound upwards. Handless yet dexterous as though having hands, they vanquish the enemies. The heavenly charcoals cast on the tawny Deer, though cold themselves, they burn down the body completely. (RV x.34.9)

नीचा <ind. – down, downwards>, वर्तंत <sub. middle 3rd person pl. of √वृत् – they (the wheels) roll or turn>, उपरि <ind. – above, upon, upwards>, स्फुरंति <pre. 3rd person pl. of √स्फर् (=√स्फुर्) – they (the wheels) spring, rebound>, अहस्तोसः <(=अहस्ताः) m. nom. pl. of adj. अहस्त – handless>, हस्तवंतं <m. nom. pl. of adj. हस्तवत् – dexterous as though having hands>, सहंते <pre. middle 3rd person pl. of √सह् – they (the wheels) vanquish, conquer, defeat enemies (the enemies are the ionic bonds or atomic structures)>;

दिव्याः <m. nom. pl. of adj. दिव्य – celestial, heavenly>, अंगाराः <(अङ्गाराः), m. nom. pl. of अंगार – charcoals>, हरिणे <m. loc. sg. of हरिण – on the tawny deer, i.e. on the Soma Pond>, न्युप्ताः <m. nom. pl. of adj. न्युप्त – thrown down, cast>, शीताः <m. nom. pl. of adj. शीत – cold, cool>, संतः <m. nom. pl. of pre. active pt. सत् (√अस्) – they being>, हृदयं <n. acc. sg. of हृदय – the heart or interior of the body (refers to the atomic structure)>, निः <ind. prefix to verb दहंति (√दह्)>, निर्-दहंति <pre. 3rd person pl. of निर्-√दह् – they (the heavenly charcoals) burn down, destroy completely>.

As in stanzas RV i.191.1-5 (section 2.7.10) and here in RV x.34.7 and 9, the process of cosmic ionisation is described as violent. We find a similar statement that describes the cosmic ionisation process in one of the ancient alchemical texts. The "priest of the sanctuary" declares that he is "ion" and "survived intolerable violence", including "burning" and "dismembering". He was torn "asunder according to the rigour of harmony".[83] This alchemical text shows that, during the ancient and early medieval periods, the knowledge related to cosmic creation were still shared among the old ruling oligarchs and mystics,

83 Stanton J. Linden (ed.), 'Zosimos of Panopolis (fl. c. 300AD): Of Virtue, Lessons 1-3', *The Alchemy Reader: From Hermes Trismegistus to Isaac Newton*, 6th printing, New York, Cambridge University Press, 2014, p. 51.

though the true meaning was long lost and mystified. When we consider that St. Peter's Square in the Vatican City, which was built during the mid-seventeenth century, replicated the Keystone of the Stone, one can see that the symbolism associated with cosmic creation was practiced among ruling oligarchs in secrecy during the late medieval and early renaissance periods. Apparently, the practice still continues to this day. Practicing symbolism is not inherently wrong, yet the utterly corrupted and perverted practice and display of symbolism associated with cosmic creation by oligarchs throughout history is a common indication of the decaying souls of the oligarchs involved.

4.1.2 The Gamester's Reproach of the Ploughing Wheel That Shatters the Ionic Bond

In stanzas RV x.34.2-6, Agni and Soma, bound in the atomic structure of hydrogen, are described as husband and wife (RV x.34.2-4). This is the story of Agni who expels his wife, Soma, on account of the Stone (RV x.34.2). He is hated by her and her family; he finds himself wretched (RV x.34.3). Agni did not want to go to the Place of Atonement, the celestial ioniser, to be burnt with his friends, and thus he drifts away from them. However, while listening to the Speech of the ruddy cows, who were cast into the Soma Pond after going through the ionisation process, Agni changes his mind and eventually goes to the Place of Atonement, as would a woman filled with love for her paramour (RV x.34.5). And Agni is rewarded with the Soma draughts he enjoys (RV x.34.6).

In the Ṛgveda, the ionic bond of hydrogen is described as 'sin' (see sections 4.3.3, 4.6.1, 4.6.2). It is the original sin that must be expiated by the fire sacrifice to produce celestial lights, that is, cosmic plasmas, for the creation and the sustenance of the World and life.

RV x.34.2 author कवष ऐलूष अक्षो वा मौजवान्, to अक्षकितवनिंदा, metre त्रिष्टुप् छंदः
न मा मिमेथ न जिहीळ एषा शिवा सखिभ्य उत मह्यमासीत् ।
अक्षस्याहमेकपरस्य हेतोरनुव्रतामप जायामरोधं ॥१०.३४.२॥
न । मा । मिमेथ । न । जिहीळे । एषा । शिवा । सखिऽभ्यः । उत । मह्यं । आसीत् ।
अक्षस्य । अहं । एकऽपरस्य । हेतोः । अनुऽव्रतां । अप । जायां । अरोधं ॥१०.३४.२॥
She never vexed me nor was angry with me, but to me and my companions was ever gracious.
On account of the wheel of singular importance, I drove out my devoted wife. (RV x.34.2)

न <ind. – not, no, nor>, मा <m. acc. sg. of 1st person pron. अस्मद् – me, i.e. Agni>, मिमेथ <prf. 3rd person sg. of √मिथ् – she engaged in altercation>, न <ind. – not, no, nor>, जिहीळे <pre. middle 3rd person sg. of √हीळ् (√हीड्) – she was angry>, एषा <f. nom. sg. of pron. एतद् – she (wife, Soma)>, शिवा <f. nom. sg. of adj. शिव – auspicious, gracious>, सखिभ्यः <m. dat. pl. of सखि – to friends, assistants, companions>, उत <ind. – and, also, even, or>, मह्यं <m. dat. sg. of 1st person pron. अस्मद् – to me>, आसीत् <ipf. 3rd person sg. of √अस् – she was>;

अक्षस्य <m. gen. sg. of अक्ष – of the wheel, i.e. of the celestial ioniser>, अहं <m. nom. sg. of 1st person pron. अस्मद् – I, Agni>, एकपरस्य <m. gen. sg. of adj. एकपर – of singular importance, more important than any other>, हेतोः <m. abl. sg. of हेतु – by reason of, on account of (with gen.)>, अनुव्रतां <f. acc. sg. of adj. अनुव्रत – devoted, faithful, attached>,

अप <ind. prefix to verb अरोधं (√रुध्)>, जायां <f. acc. sg. of जाया – wife, Soma (electron)>, अप-अरोधं <ipf. 1st person sg. of अप-√रुध् – I expelled, drove out>.

RV x.34.3 author कवष ऐलूष अक्षो वा मौजवान्, to अक्षकितवनिंदा, metre त्रिष्टुप् छंदः
द्वेष्टि श्वश्रूरप जाया रुणद्धि न नाथितो विंदते मर्डितारं ।
अश्वस्येव जरतो वस्न्यस्य नाहं विंदामि कितवस्य भोगं ॥१०.३४.३॥
द्वेष्टि । श्वश्रूः । अप । जाया । रुणद्धि । न । नाथितः । विंदते । मर्डितारं ।
अश्वस्यऽइव । जरतः । वस्न्यस्य । न । अहं । विंदामि । कितवस्य । भोगं ॥१०.३४.३॥
My mother-in-law hates me, my wife expels me, this wretched man finds none who shows sympathy. As a valuable horse grown old and becomes of no use, I find no pleasure in being a wretched gamester. (RV x.34.3)

द्वेष्टि <pre. 3rd person sg. of √द्विष् – she hates, is hostile>, श्वश्रूः <f. nom. sg. of श्वश्रू – a mother-in-law>, अप <ind. prefix to verb रुणद्धि (√रुध्)>, जाया <f. nom. sg. of जाया – a wife, i.e. Soma>, अप-रुणद्धि <pre. 3rd person sg. of अप-√रुध् – she expels, drives out>, न <ind. – not, no, nor>, नाथितः <m. nom. sg. of adj. नाथित – the oppressed, one who needs help, wretched (refers to Agni himself)>, विंदते <pre. middle 3rd person sg. of √विद् 6Ā – he (Agni) finds, discovers>, मर्डितारं <m. acc. sg. of मर्डितृ – one who shows compassion>;

अश्वस्य <m. gen. sg. of अश्व – of a horse>, इव <ind. – like, as if>, जरतः <m. gen. sg. of adj. जरत् – old, no longer in use>, वस्न्यस्य <m. gen. sg. of adj. वस्न्य – precious, valuable>, न <ind. – not, no, nor>, अहं <m. nom. sg. of 1st person pron. अस्मद् – I, Agni>, विंदामि <pre. 1st person sg. of √विद् 6P – I find, discover>, कितवस्य <m. gen. sg. of कितव – of a gamester, cheat>, भोगं <m. acc. sg. of भोग – advantage, pleasure, delight>.

RV x.34.4 author कवष ऐलूष अक्षो वा मौजवान्, to अक्षकितवनिंदा, metre त्रिष्टुप् छंदः
अन्ये जायां परि मृशंत्यस्य यस्यागृधद्वेदने वाज्य१ क्षः ।
पिता माता भ्रातर एनमाहुर्न जानीमो नयता बद्धमेतं ॥१०.३४.४॥
अन्ये । जायां । परि । मृशंति । अस्य । यस्य । अगृधत् । वेदने । वाजी । अक्षः ।
पिता । माता । भ्रातरः । एनं । आहुः । न । जानीमः । नयत । बद्धं । एतं ॥१०.३४.४॥
Others seize his wife whose property that impetuous wheel endeavoured to acquire.
His father, mother, brothers speak of him, We know him not, Carry ye him away fettered. (RV x.34.4)

अन्ये <m. nom. pl. of अन्य – others (other Agnis)>, जायां <f. acc. sg. of जाया – wife (Soma)>, परि <ind. prefix to verb मृशंति (√मृश्)>, परि-मृशंति <pre. 3rd person pl. of परि-√मृश् – they grasp, seize>, अस्य <m. gen. sg. of pron. इदं – his>, यस्य <m. gen. sg. of relative pron. यद् – whose>, अगृधत् <aor. 3rd person sg. of √गृध् – it (the wheel) endeavoured to gain, covet, eager for (loc.)>, वेदने <n. loc. sg. of वेदन – knowledge, property, goods>, वाजी <m. nom. sg. of adj. वाजिन् – swift, impetuous, heroic>, अक्षः <m. nom. sg. of अक्ष – the wheel, i.e. the celestial ioniser>;

पिता <m. nom. sg. of पितृ – father>, माता <f. nom. sg. of मातृ – mother>, भ्रातरः <m. nom. pl. of भ्रातृ – brothers>, एनं <m. acc. sg. of एतद् – him>, आहुः <(=ब्रुवन्ति) pre. 3rd person pl. of √ब्रू – they speak about person or thing>, न <ind. – no, not, nor>, जानीमः <pre. 1st person pl. of √ज्ञा – we know, recognise>, नयत <ipv. 2nd person pl. of √नी – ye take or carry away>, बद्धं <m. acc. sg. of adj. बद्ध – chained, fettered>, एतं <(=एनं) m. acc. sg. of pron. एतद् – him>.

RV x.34.5 author कवष ऐलूष अक्षो वा मौजवान्, to अक्षकितवनिंदा, metre त्रिष्टुप् छंदः
यदादीध्ये न दविषाण्येभिः परायद्भ्योऽव हीये सखिभ्यः ।
न्युप्ताश्च बभ्रवो वाचमक्रतँ एमीदेषां निष्कृतं जारिणीव ॥१०.३४.५॥
यत् । आऽदीध्ये । न । दविषाणि । एभिः । परायत्ऽभ्यः । अव । हीये । सखिभ्यः ।
निऽउप्ताः । च । बभ्रवः । वाचं । अक्रत । एमि । इत् । एषां । निःऽकृतं । जारिणीऽइव ॥१०.३४.५॥

When I chose not to be burnt with them, I am abandoned by my companions who are willing. O cast ruddy cows, ye made Speech. And I go to the Place of Atonement like a woman enamoured of her paramour. (RV x.34.5)

यत् <conjunction – that, when>, आदीध्ये <pre. middle1st person sg. of आ-√धी – I think, wish, desire>, न <ind. – not, no, nor, neither>, दविषाणि <sub. 1st person sg. of √दु (=√दू) cl.5 – I am to be burnt>, एभिः <m. inst. pl. of pron. इदं – with these, with them (friends, companions)>, परायद्भ्यः <m. abl. pl. of adj. परायद् (परा + आयत्) – entering, arriving, adhering, abiding>, अव <ind. prefix to verb हीये (√हा)>, अव-हीये <pre. passive 1st person sg. of अव-√हा – I am abandoned, left behind (abl.)>, सखिभ्यः <m. abl. pl. of सखि – from companions, friends>;

न्युप्ताः <mf. voc. pl. of adj. न्युप्त – thrown down, cast, offered, i.e. discharged or poured out from the celestial ioniser>, च <ind. – and, also>, बभ्रवः <f. voc. pl. of बभ्रु – O ruddy cows (celestial lights)>, वाचं <f. acc. sg. of वाच् – Speech, sound>, अक्रत <ipf. 2nd person pl. of √कृ – ye made>, एमि <pre. 1st person sg. of √इ – I go, arrive at>, इत् <(इद्) ind. – just, only, indeed>, एषां <m. gen. pl. of pron. इदं – of these (of the ruddy cows)>, निष्कृतं <n. acc. sg. of निष्कृत – to the Place of Atonement to be expiated for the sin of the ionic bond (the place of the atonement refers to the Stone, the celestial ioniser)>, जारिणी <f. nom. sg. of जारिणी – a woman who has a paramour>, इव <ind. – like, as>.

RV x.34.6 author कवष ऐलूष अक्षो वा मौजवान्, to अक्षकितवनिंदा, metre त्रिष्टुप् छंदः
सभामेति कितवः पृच्छमानो जेष्यामीति तन्वा३ शूशुजानः ।
अक्षासो अस्य वि तिरंति कामं प्रतिदीव्ने दधत आ कृतानि ॥१०.३४.६॥
सभां । एति । कितवः । पृच्छमानः । जेष्यामीति । इति । तन्वा । शूशुजानः ।
अक्षासः । अस्य । वि । तिरंति । कामं । प्रतिऽदीव्ने । दधतः । आ । कृतानि ॥१०.३४.६॥

The gamester goes to the hall of atonement, with his thin body puffed up, asks thus, Shall I succeed? The wheels carry through his wish and bring the prize to this adversarial gamester. (RV x.34.6)

सभां <f. acc. sg. of सभा – to the congregation, a large hall, a court of a king or of justice (refers to the Stone, the celestial ioniser)>, एति <pre. 3rd person sg. of √इ – he goes to>, कितवः <m. nom. sg. of कितव – a gamester, cheater>, पृच्छमानः <m. nom. sg. of pre. middle pt. पृच्छमान (√प्रछ्) – he asking>, जेष्यामि <fut. 1st person sg. of √जि – shall I win or succeed?>, इति <ind. – thus>, तन्वा <m. inst. sg. of तनु (=तनू) – with (his) thin, minute body (note that his body is hydrogen)>, शूशुजानः <m. nom. sg. of prf. middle pt. शूशुजान (√शुज्) – it/he being puffed up, swollen>;

अक्षासः <m. nom. pl. of अक्ष – the wheels, cars, i.e. the firestones of the celestial ioniser>, अस्य <m. gen. sg. of pron. इदं – his>, वि <ind. prefix to verb तिरंति (√तॄ)>, वि-तिरंति <pre. 3rd person pl. of वि-√तॄ – they (the wheels) grant, carry through>, कामं <m. acc. sg. of काम – wish, desire>, प्रतिदीव्ने <m. dat. sg. of प्रतिदीवन् – to an adversary at play (refers to the gamester Agni who has been reluctant to go through the ionisation process and is calculating about a prospect)>, दधतः <m. nom. pl. of adj. दधत् (√धा) – bringing (loc. or dat.); interpreted as an adjective, not as accusative plural of pre. active pt. दधत् (√धा)>, आ <ind. particle – moreover, further, and>, कृतानि <n. acc. pl. of कृत –the prize or booty gained (refers to Soma draughts Agni enjoys, the booty gained after the ionisation process)>.

4.1.3 Cosmic Plasma Discharge

The seven stanzas of RV x.135 describe cosmic plasma discharge phenomena. The discharge of the freshly produced cosmic plasmas into the Some Pond is described as 'the tubular stalks of Soma plants being blown accompanied by songs of praise (RV x.135.7)'. In the hymn, the cosmic plasma discharge is also described as the rollout of the chariot and as the birth of Kumāra, the child, that is, the birth of Agni. The discharged cosmic plasmas are also called the funeral gift (अनुदेयी anudeyī RV x.135.5-6), the gift followed by the burning of the bodies of hydrogen; thus, it is the funeral gift.

The author of the hymn is Kumāra Yāmāyana (कुमारः यामायनः), 'Kumāra's path of journey'. Kumāra means "a child, boy, youth", and refers to newly born Agni, the proton. All seven stanzas of the hymn are addressed to Yama (यमः) who presides over the 'ancient sires' and "rules the spirits of the dead". Note that the spirits of the dead refer to the spirits of discharged cosmic plasmas. Yama is the lord of the house (विश्पतिः RV x.135.1). The House here refers to the Soma Pond, Yama's seat, which is also called the house of the Devas (देवमानं RV x.135.7).

In RV x.135.1, the Tree clothed with beautiful leaves refers to the Soma Pond filled with plasma blooms. The plasma blooms are compared to the beautiful leaves of the Tree, the Soma Pond.

RV x.135.1 author कुमारो यामायनः, to यमः, metre अनुष्टुप् छंदः
यस्मिन्वृक्षे सुपलाशे देवैः संपिबते यमः । अत्रा नो विश्पतिः पिता पुराणाँ अनु वेनति ॥१०.१३५.१॥
यस्मिन् । वृक्षे । सुऽपलाशे । देवैः । संऽपिबते । यमः ।
अत्र । नः । विश्पतिः । पिता । पुराणान् । अनु । वेनति ॥१०.१३५.१॥
In the Tree clothed with beautiful leaves, where Yama drinks together with the Devas,
Here in this place, the father, the lord of the house, cares for our ancient sires.
(RV x.135.1)

यस्मिन् <m. loc. sg. of यद् – in which, where>, वृक्षे <m. loc. sg. of वृक्ष – in the tree>, सुपलाशे <m. loc. sg. of adj. सुपलाश – having beautiful leaves>, देवैः <m. inst. pl. of देव – with the Devas>, संपिबते <pre. middle 3rd person sg. of सम्-√पा – he drinks together>, यमः <m. nom. sg. of यम – Yama, a twin of the pair of Yama and Yamī. Yama represents Agni, Yamī represents Soma>;

अत्र <ind. – here, in this place, i.e. the Soma Pond, the abode of the Devas>, नः <m. gen. pl. of 1st person pron. अस्मद् – our>, विश्पतिः <m. nom. sg. of विश्पति – the lord of the house, Yama>, पिता <m. nom. sg. of पितृ – father (Yama)>, पुराणान् <m. acc. pl. of पुराण – the old, ancient ones, i.e. previously discharged Agnis>, अनु <ind. prefix to verb वेनति (√वेन्)>, अनु-वेनति <pre. 3rd person sg. of अनु-√वेन् – he cares for, looks after, tends>.

RV x.135.2 author कुमारो यामायनः, to यमः, metre अनुष्टुप् छंदः
पुराणाँ अनुवेनंतं चरंतं पापयामुया । असूयन्नभ्यचाकशं तस्मा अस्पृहयं पुनः ॥१०.१३५.२॥
पुराणान् । अनुऽवेनंतं । चरंतं । पापया । अमुया ।
असूयन् । अभि । अचाकशं । तस्मै । अस्पृहयं । पुनरिति ॥१०.१३५.२॥
Him running around and looking after our ancient sires with the protector,
I looked on, murmuring at him, I longed for having him back again. (RV x.135.2)

पुराणान् <m. acc. pl. of पुराण – the old or ancient ones, i.e. previously discharged Agnis>, अनुवेनंतं <m. acc. sg. of pre. active pt. अनुवेनत् (अनु after, along + √वेन् to care for) – him caring for, looking after>, चरंतं <m. acc. sg. of pre. active pt. चरत् (√चर्) – him running around>, पापया <f. inst. sg. of पाप (from √पा 2P watch, keep, protect) – with the watcher, protector>, अमुया <f. inst. sg. of pron. अदस् – a certain, that>;

असूयन् <m. nom. sg. of pre. active pt. असूयत् (nominal verb √असूय) – murmuring at (with dat.)>, अभि <ind. prefix to verb अचाकशं (√काश्)>, अभि-अचाकशं <ipf. 1st person sg. of अभि-√काश् – I looked on, perceived>, तस्मै <m. dat. sg. of तद् – for him>, अस्पृहयं <ipf. 1st person sg. of √स्पृह् – I desired, longed for (dat.)>, पुनः <(पुनर्) ind. – back, again, repeatedly, once more>.

In RV x.135.3, the new shoot of a Soma plant, or plasma bloom, is compared to a new wheelless chariot, and it is called a one-poled chariot, which turns in all directions.

RV x.135.3 author कुमारो यामायनः, to यमः, metre अनुष्टुप् छंदः
यं कुमार नवं रथमचक्रं मनसाकृणोः । एकेषं विश्वतः प्रांचमपश्यन्नधि तिष्ठसि ॥१०.१३५.३॥
यं । कुमार । नवं । रथं । अचक्रं । मनसा । अकृणोः ।
एकऽईषं । विश्वतः । प्रांचं । अपश्यन् । अधि । तिष्ठसि ॥१०.१३५.३॥
The new wheelless chariot, O child, which thou willingly tookest up of thine own accord,
They beheld the one-poled chariot turning to all directions, upon which thou standest firmly.
(RV x.135.3)

यं <m. acc. sg. of यद् – that, which>, कुमार <m. voc. sg. of कुमार – O child, O son>, नवं <m. acc. sg. of adj. नव – new, fresh, recent>, रथं <m. acc. sg. of रथ – a car, chariot>, अचक्रं <m. acc. sg. of adj. अचक्र – having no wheels, moving by itself>, मनसा <ind. – with all the heart, willingly>, अकृणोः <ipf. 2nd person sg. of √कृ – you made, undertook>;

एकेषं <m. acc. sg. of adj. एकेष – furnished with only one pole (chariot)>, विश्वतः <(विश्वतस्) ind. – everywhere>, प्रांचं <m. acc. sg. of adj. प्रांचम् (प्राञ्च्) – (with विश्वतः) turned to all directions>, अपश्यन् <ipf. 3rd person pl. of √पश् or √दृश् – they (the ancient sires) beheld, looked at>, अधि <ind. adverb or preposition – (with acc.) over, upon>, तिष्ठसि <pre. 2nd person sg. of √स्था – you stand firmly, station yourself>.

RV x.135.4 author कुमारो यामायनः, to यमः, metre अनुष्टुप् छंदः
यं कुमार प्रावर्तयो रथं विप्रेभ्यस्परि । तं सामानु प्रावर्तत समितो नाव्याहितं ॥१०.१३५.४॥
यं । कुमार । प्र । अवर्तयः । रथं । विप्रेभ्यः । परि ।
तं । साम । अनु । प्र । अवर्तत । सं । इतः । नावि । आऽहितं ॥१०.१३५.४॥
The chariot which thou, O child, madest to roll out from the singers,
That chariot thou madest to roll out and the Sāma hymn added on it travel together in a boat.
(RV x.135.4)

यं <m. acc. sg. of यद् – that, which>, कुमार <m. voc. sg. of कुमार – O child, O son (i.e. newborn Agni)>, प्र <ind. prefix to verb अवर्तयः (√वृत्)>, प्र-अवर्तयः <ipf. cau. 2nd person sg. of प्र-√वृत् – you made (chariot) to move or roll onwards>, रथं <m. acc. sg. of रथ – a car, chariot>, विप्रेभ्यः <m. abl. pl. of विप्र – from the singers, i.e. the firestones of the celestial ioniser (note that the firestones are the most prominent singers>, परि <ind. – (with abl.) from, away from, out of>;

तं <m. acc. sg. of 3rd person pron. तद् – that (chariot)>, साम <n. acc. sg. of सामन् – Sāma hymn>, अनु <ind. preposition – (with acc.) after, along with>, प्र <ind. prefix to verb अवर्तयः (√वृत्)>, प्र-अवर्तयः <ipf. cau. 2nd person sg. of प्र-√वृत् – you made to roll forwards>, सं <ind. prefix to verb इतः (√इ)>, सं-इतः <pre. 3rd person du. of सं-√इ – they (du.) travel or flow together>, नावि <f. loc. sg. of नौ – in a boat (the boat refers to each plasma bloom, which is a sheath of plasma)>, आहितं <n. acc. sg. of adj. आहित – placed on, deposited, put on, added>.

RV x.135.5 author कुमारो यामायनः, to यमः, metre अनुष्टुप् छंदः
कः कुमारमजनयद्रथं को निरवर्तयत् । कः स्वित्तदद्य नो ब्रूयादनुदेयी यथाभवत् ॥१०.१३५.५॥
कः । कुमारं । अजनयत् । रथं । कः । निः । अवर्तयत् ।
कः । स्वित् । तत् । अद्य । नः । ब्रूयात् । अनुऽदेयी । यथा । अभवत् ॥१०.१३५.५॥

Who made the child to be born? Who made the chariot to roll out?
Who will this day indeed proclaim for us how the funeral gift came into being? (RV x.135.5)

कः <m. nom. sg. of interrogative pron. क (or किं) – who?>, कुमारं <m. acc. sg. of कुमार – the child, youth>, अजनयत् <ipf. cau. 3rd person sg. of √जन् – caused to be born>, रथं <m. acc. sg. of रथ – the chariot>, कः <m. nom. sg. of interrogative pron. क (or किं) – who?>, निरवर्तयत् <ipf. cau. 3rd person sg. of निर्-√वृत् – caused to roll out>;
कः <m. nom. sg. of interrogative pron. क (or किं) – who?>, स्वित् <(स्विद्) ind. particle of enquiry or doubt – who do you think, pray, indeed>, तत् <n. nom. sg. of pron. तद् – it, that, this>, अद्य <ind. – today, now>, नः <m. dat. of 1st person pron. अस्मद् – to/for us>, ब्रूयात् <opt. 3rd person sg. of √ब्रू – (who) will declare, pronounce>, अनुदेयी <f. nom. sg. of अनुदेयी – after-gift, i.e. the gift obtained after the burning of the bodies of hydrogen, the funeral gift>, यथा <ind. – as, like, in which manner>, अभवत् <ipf. 3rd person sg. of √भू – it (the funeral gift) came into being, arose>.

Stanza RV x.135.6 explains that the tip of the flame of cosmic plasmas arises from the east where the celestial ioniser is located at the base of the Soma Pond and grows westwards toward the top of the Heavenly Vault where it is released.

RV x.135.6 author कुमारो यामायनः, to यमः, metre अनुष्टुप् छंदः
यथाभंवदनुदेयी ततो अग्रमजायत । पुरस्ताद्बुध्न आततः पश्चान्निरयणं कृतं ॥१०.१३५.६॥
यथा । अभंवात् । अनुऽदेयी । ततः । अग्रं । अजायत ।
पुरस्तात् । बुध्नः । आऽततः । पश्चात् । निःऽअयनं । कृतं ॥१०.१३५.६॥
As the funeral gift arose, thereupon the foremost tip of the flame appeared.
From the east, the root of the flame was extended westwards, and a foward-path was made ready. (RV x.135.6)

यथा <ind. – as, like, in which way>, अभवत् <ipf. 3rd person sg. of √भू – it (the funeral gift) came into being, arose>, अनुदेयी <f. nom. sg. of अनुदेयी – the funeral gift, celestial lights (cosmic plasmas) that arose followed by the burning of the bodies of hydrogen>, ततः <(ततस्) ind. – from that place, thence, thereupon>, अग्रं <n. nom. sg. of अग्र – point, tip, the foremost point (of the flame)>, अजायत <ipf. middle 3rd person sg. of √जन् cl.4 – the tip of flame was born, was produced>;
पुरस्तात् <ind. – from the east (from where the celestial lights arise)>, बुध्नः <m. nom. sg. of बुध्न – the root, ground, base (of the flame)>, आततः <m. nom. sg. of past passive pt. of आ-तत (आ-√तन्) – was extended, stretched>, पश्चात् <ind. – westwards (towards the top of the Heavenly Vault)>, निरयणं <n. nom. sg. of निरयण – a forward-path, i.e. a path to grow forward>, कृतं <n. nom. sg. of past passive pt. कृत (√कृ cl.5 or 8) – it was made ready>.

According to RV x.135.7, the individual stalks of Soma plants are in tubular forms. These tubular stalks of Soma plants are blown, accompanied by songs of praise. The tubular stalks are often called tongues of Agni's flames, thunderbolts, arrows, spears, and so on.

RV x.135.7 author कुमारो यामायनः, to यमः, metre अनुष्टुप् छंदः
इदं यमस्य सादनं देवमानं यदुच्यते । इयमस्य धम्यते नाळीरयं गीर्भिः परिष्कृतः ॥१०.१३५.७॥
इदं । यमस्य । सदनं । देवऽमानं । यत् । उच्यते ।
इयं । अस्य । धम्यते । नाळीः । अयं । गीःऽभिः । परिऽकृतः ॥१०.१३५.७॥
This place is the seat of Yama, which is called the house of the Devas.
Here, his tubular stalk of Soma plant is blown accompanied by songs of praise.
(RV x.135.7)

इदं <ind. – here, this place>, यमस्य <m. gen. sg. of यम – of Yama>, सदनं <n. nom. sg. of सदन – seat, dwelling>, देवमानं <n. nom. sg. of देवमान – the house of the Devas>, यत् <n. nom. sg. of relative pron.यद् – which>, उच्यते <pre. passive 3rd person sg. of √वच् – it is called, named>;

इयं <f. nom. sg. of इदं – here>, अस्य <m. gen. sg. of इदं – of this, his>, धम्यते <pre. passive 3rd person sg. of √धं – it (the stalk of Soma plant) is blown>, नाळीः <(नाडीः) m. nom. sg. of नाळी (नाडी) – the tubular stalk (of Soma plant)>, अयं <m. nom. sg. of इदं – this, i.e. the blowing of the Soma plant>, गीर्भिः <f. inst. pl. of गिर् – by songs of praise>, परिष्कृतः <m. nom. sg. of adj. परिष्कृत – adorned, embellished, accompanied by (inst.)>.

4.2 CELESTIAL WATERS, COSMIC PLASMAS

Āpaḥ (आपः pl. of अप्), the celestial Waters, refers to cosmic plasmas. Cosmic plasmas are described in various names in the Ṛgveda: celestial Waters, ambrosia (अमृतं amṛtaṃ), bright drops of Soma (इंदुः induḥ), Soma mead (सोम्यं मधु somyaṃ madhu), milk, drinks of the Devas, celestial lights, and so on. Each of these descriptions represents a certain aspect of the discharged cosmic plasmas. The Adhvaryus press and pour out the wave of celestial Waters (RV x.30.2, 3, and 15) and Indra cuts the channel for them (RV vii.47.4). And the channels, in which the celestial Waters flow, are called the celestial rivers in the Ṛgveda.

4.2.1 The Adhvaryus Press Out the Celestial Waters

The Adhvaryus (अध्वर्यवः adhvaryavaḥ pl.) represent the firestones of the Stone, the celestial ioniser, personified as priests. They press out the wave of Waters (ऊर्मिः ūrmiḥ), that is, the wave of cosmic plasmas (RV x.30.2, 3). In RV x.30.2, the individual plasma blooms, or Soma plants, are called a mythical red bird (अरुणः सुपर्णः aruṇaḥ suparṇaḥ). The word suparṇa (सुपर्णः) is often translated as a beautiful leaf, eagle, or vulture. Suparṇa is the mythical bird Garuḍa (गरुडः), the firebird, and it is the equivalent of Phoenix in classical mythology. The discharged celestial Waters are carried by this mythical bird for Indra.

The Adhvaryus are called the beautiful-handed (RV x.30.2). The fiery firestones of the Stone are described as the 'beautiful hands' of the Adhvaryus (Figure 8). These firestones, described also as fingers or maidens in the Ṛgveda, are laid in a circular formation, making up each of the upper and the lower discs of the Stone. The event of releasing the matured plasma blooms from the Soma Pond is called Soma sacrifice (अध्वरः).

RV x.30.2 author कवष ऐलूषः, to आप अपंनपाद्वा, metre त्रिष्टुप् छंदः
अध्वर्यवो हविष्मंतो हि भूताच्छाप इतोशतीरुशंतः ।
अव याश्चष्टे अरुणः सुपर्णस्तमास्यध्वमूर्मिमद्या सुहस्ताः ॥१०.३०.२॥
अध्वर्यवः । हविष्मंतः । हि । भूत । अच्छ । अपः । इत । उशतीः । उशंतः ।
अव । याः । चष्टे । अरुणः । सुऽपर्णः । तं । आ । अस्यध्वं । ऊर्मिं । अद्य । सुऽहस्ताः ॥१०.३०.२॥

O longing Adhvaryus, the offerers of oblation, be ye ready and obtain the longing Waters,
The Waters which the ruddy Bird keeps its eye on. Pour out the wave of Waters this day,
O ye beautiful-handed. (RV x.30.2)

अध्वर्यवः <m. voc. pl. of अध्वर्यु – O Adhvaryus (Adhvaryus, the officiating priests, refer to the firestones of the Stone)>, हविष्मंतः <m. voc. pl. of हविष्मत् – O the offerers of oblation>, हि <ind. – just, do>, भूत <inj. 2nd person pl. of √भू – be ye ready>, अच्छ <ind. prefix to verb इत (√इ)>, अपः <f. acc. pl. of अप् – Waters, i.e. cosmic plasmas>, अच्छ-इत <ipv. 2nd person pl. of अच्छ-√इ (=अच्छा-√इ) – ye reach, obtain, attain>, उशतीः <f. acc. pl. of adj. उशत् – wishing, desiring>, उशंतः <m. voc. pl. of adj. उशत् – wishing, desiring>;

अव <ind. prefix to अस्यध्वं (√अस्)>, याः <f. acc. pl. of यद् – them (Waters) which>, आ-चष्टे <pre. middle 3rd person sg. of आचक्ष् (आ-√चक्ष्) – he looks at>, अरुणः <m. nom. sg. of adj. अरुण – red, ruddy>, सुपर्णः <m. nom. sg. of सुपर्ण – eagle, a mythical bird, Garuḍa>, तं <m. acc. sg. of pron. तद् – that (wave)>, आ <ind. prefix to verb चष्टे (√चक्ष्)>, अव-अस्यध्वं <ipv. middle 2nd person pl. of अव-√अस् 4P – ye throw, cast away, pour out>, ऊर्मिं <m. acc. sg. of ऊर्मि – wave, billow (of Waters)>, अद्य <ind. – today, now>, सुहस्ताः <m. voc. pl. of सुहस्त – O ye beautiful-handed (refers to the Adhvaryus)>.

RV x.30.3 author कवष ऐलूषः, to आप अपंनपाद्वा, metre त्रिष्टुप् छंदः
अध्वर्यवोऽप इता समुद्रमपां नपातं हविषा यजध्वं ।
स वो ददद्रूर्मिमद्या सुपूतं तस्मै सोमं मधुमंतं सुनोत ॥१०.३०.३॥
अध्वर्यवः । अपः । इत । समुद्रं । अपां । नपातं । हविषा । यजध्वं ।
सः । वः । ददत् । ऊर्मिं । अद्य । सुऽपूतं । तस्मै । सोमं । मधुऽमंतं । सुनोत ॥१०.३०.३॥
O Adhvaryus, blow ye the Waters out to the celestial Ocean and honour the son of Waters with your oblation. The billowy wave of yours, which he will offer today, press ye for him the well-purified Soma rich in sweetness. (RV x.30.3)

अध्वर्यवः <m. voc. pl. of अध्वर्यु – Adhvaryus, the firestones of the Stone>, अपः <f. acc. pl. of अप् – Waters>, इत <ipv. 2nd person pl. of √इ – ye blow, spread>, समुद्रं <m. acc. sg. of समुद्र – to Samudra, the Ocean of celestial Waters (it refers to the Soma Pond)>, अपां <f. gen. pl. of अप् – of Waters, i.e. of cosmic plasmas>, नपातं <m. acc. sg. of नपात् – the son (of Waters), i.e. Agni)>, हविषा <n. inst. sg. of हविस् – with oblation, with offering>, यजध्वं <ipv. middle 2nd person pl. of √यज् – ye worship, honour>;

सः <m. nom. sg. of 3rd person pron. तद् – he>, वः <m. gen. pl. of 2nd person pron. युष्मद् – of yours, your (pl.)>, ददत् <sub. 3rd person sg. of √दा – he will give, bestow, present, offer>, ऊर्मिं <m. acc. sg. of ऊर्मि – a wave, billow, i.e. billowy plasma blooms>, अद्य <ind. – today, now>, सुपूतं <m. acc. sg. of adj. सुपूत – well-purified>, तस्मै <m. dat. sg. of 3rd person pron. तद् – for him>, सोमं <m. acc. sg. of सोम – Soma>, मधुमंतं <m. acc. sg. of adj. मधुमत् – containing sweetness>, सुनोत <ipv. 2nd person pl. √सु – ye press out>.

RV x.30.15 author कवष ऐलूषः, to आप अपंनपाद्वा, metre त्रिष्टुप् छंदः
आग्मन्नाप उशतीर्बर्हिरेदं न्यध्वरे असदन्देवयंतीः ।
अध्वर्यवः सुनुतेंद्राय सोममभूदु वः सुशका देवयज्या ॥१०.३०.१५॥
आ । अग्मन् । आपः । उशतीः । बर्हिः । आ । इदं । नि । अध्वरे । असदन् । देवऽयंतीः ।
अध्वर्यवः । सुनुत । इंद्राय । सोमं । अभूत् । ऊं इति । वः । सुऽशका । देवऽयज्या ॥१०.३०.१५॥
Waters reached the sacrificial grass and those who wish to serve the Devas were seated at the Soma sacrifice. O Adhvaryus, press ye Soma for Indra so will your worship of the Devas be effortless. (RV x.30.15)

आ <ind. prefix to verb अग्मन् (√गं)>, आ-अग्मन् <(=आ-अगमन्) ipf. 3rd person pl. of आ-√गं – they came to, arrived at, reached>, आपः <f. nom. pl. of अप् – the celestial Waters, cosmic plasmas>, उशतीः <f. acc. pl. of उशत् – ones who wish, desire>, बर्हिः <n. acc. sg. of बर्हिस् – the sacrificial grass>, आ <ind. prefix to verb असदन् (√सद्)>, इदं <ind. – now, even,

here>, नि <ind. – down, back>, अध्वरे <m. loc. sg. of अध्वर – at the Soma sacrifice>, आ-असदन् <ipf. 3rd person pl. of आ-√सद् – they seated (someone), they were seated>, देवयंतीः <f. acc. pl. of pre. active pt. देवयंती (nominal verb √देवय) – them serving the Devas>;

अध्वर्यवः <m. voc. pl. of अध्वर्यु – O Adhvaryus (refer to the firestones of the Stone)>, सुनुत <ipv. 2nd person pl. of √सु – ye press>, इंद्राय <m. dat. sg. of इंद्र – for Indra>, सोमं <m. acc. sg. of सोम – Soma>, अभूत् <aor. 3rd person sg. of √भू – it became>, उ <ind. particle – and, also, further>, वः <m. gen. pl. of 2nd person pron. युष्मद् – your>, सुशका <f. nom. sg. of adj. सुशक – easy to be done>, देवयज्या <f. nom. sg. of देवयज्या – worship of the Devas, a sacrifice>.

4.2.2 Celestial Waters Rise Up and Are Released Day After Day

Stanzas RV i.164.51 and 52 inform us that cosmic plasmas rise up day after day and are released again and again. Rains of celestial plasmas reanimate all the worlds in the Milky Way, and the fires of Agni reinvigorate the Heavenly structure.

RV i.164.51 author दीर्घतमा औचथ्यः, to सूर्यः पर्जन्याग्नी वा, metre अनुष्टुप् छंदः
समानमेतदुदकमुच्चैत्यव चाहभिः । भूमिं पर्जन्या जिन्वंति दिवं जिन्वंत्यग्नयः ॥१.१६४.५१॥
समानं । एतत् । उदकं । उत् । च । एति । अव । च । अहऽभिः ।
भूमिं । पर्जन्याः । जिन्वंति । दिवं । जिन्वंति । अग्नयः ॥१.१६४.५१॥
This same Water rises up and departs day after day.
Rains reanimate the Milky Way, Agni's fires give renewed vigour to the Heaven.
(RV i.164.51)

समानं <n. nom. sg. of समान – same, identical>, एतत् <n. nom. sg. of pron. एतद् – this>, उदकं <n. nom. sg. of उदक – Water, i.e. cosmic plasma>, उत् <(उद्) ind. prefix to verb एति (√इ)>, च <ind. – and, also>, उद्-एति <pre. 3rd person sg. of उद्-√इ – it (Water) rises up, comes up>, अव <ind. – off, away, (with एति (√इ), departs or move away)>, च <ind. – and, also>, अहभिः <(=अहैः) n. inst. pl. of अह – with passing days, day after day>;

भूमिं <f. acc. sg. of भूमि – ground, territory, the manifested world, i.e. the Soma Pond and the Milky Way>, पर्जन्याः <m. nom. pl. of पर्जन्य – rains, i.e. cosmic plasmas>, जिन्वंति <pre. 3rd person pl. of √जिन्व् – they (rains) refresh, animate>, दिवं <m. acc. sg. of दिव् – the Heaven>, जिन्वंति <pre. 3rd person pl. of √जिन्व् – they (Agni's fires) impel, incite, animate>, अग्नयः <m. nom. pl. of अग्नि – Agnis, i.e. fires of Agni, sacrificial fires>.

RV i.164.52 author दीर्घतमा औचथ्यः, to सरस्वान् सूर्यो वा, metre त्रिष्टुप्
दिव्यं सुपर्णं वायसं बृहंतमपां गर्भं दर्शतमोषधीनां ।
अभीपतो वृष्टिभिस्तर्पयंतं सरस्वंतमवसे जोहवीमि ॥१.१६४.५२॥
दिव्यं । सुऽपर्णं । वायसं । बृहंतं । अपां । गर्भं । दर्शतं । ओषधीनां ।
अभीपतः । वृष्टिऽभिः । तर्पयंतं । सरस्वंतं । अवसे । जोहवीमि ॥१.१६४.५२॥
The bird celestial, lofty with beautiful wings, the magnificient womb of the Waters' flaming plants, Him who delights us with rains at the right time, I invoke Sarasvān for the release of Waters. (RV i.164.52)

दिव्यं <m. acc. sg. of दिव्य – divine, celestial>, सुपर्णं <m. acc. sg. of adj. सुपर्ण – with beautiful wings (with plasma blooms)>, वायसं <m. acc. sg. of वायस – the bird (refers to the Soma Pond)>, बृहंतं <m. acc. sg. of adj. बृहंत (=बृहत्) – lofty, high, tall>, अपां <f. gen. pl. of अप् – of Waters, i.e. of cosmic plasmas>, गर्भं <m. acc. sg. of गर्भ – womb, any interior chamber (refers to the Soma Pond)>, दर्शतं <m. acc. sg. of adj. दर्शत – visible, striking the eye, beautiful>,

ओषधीनां <f. gen. pl. of ओषधि – (cf. ओष burning, light-containing) of burning, flaming, light-containing plants (i.e. of plasma blooms)>;

अभीपतः <(अभीपतस्) ind. – at the right time>, वृष्टिभिः <f. inst. pl of वृष्टि – with rains, i.e. with cosmic plasmas>, तर्पयंतं <m. acc. sg. of pre. active cau. pt. of तर्पयत् (√तृप्) – (him who) delights, pleases (us)>, सरस्वंतं <m. acc. sg. of सरस्वत् – Sarasvān (सरस्वान्), the masculine form of Sarasvatī (the Soma Pond)>, अवसे <n. dat. sg. of अवस् – for wish, desire of the Waters (cosmic plasmas), i.e. wish for the release of the Waters>, जोहवीमि <int. 1st person sg. of √हू – I call upon, invoke>.

4.2.3 Sūrya Spins the Celestial Waters with His Beams of Light

Hymn RV vii.47, composed of four stanzas, is addressed to the celestial Waters (आपः āpaḥ), the cosmic plasmas. The wave (ऊर्मिः ūrmiḥ) refers to the current of discharged plasmas. In this hymn, celestial Waters, the discharged cosmic plasmas, are called the Devīs, the goddesses. The last stanza of the hymn, RV vii.47.4, states that Sūrya weaves the discharged Waters, the cosmic plasmas, into threads of plasma blooms and Indra cuts a channel for the Waters to flow. Note that these plasma blooms, or light-containing plants, are stupendously long, tens of thousands of light-years long.

RV vii.47.1 author वसिष्ठः, to आपः, metre त्रिष्टुप् छंदः
आपो यं वः प्रथमं देवयंत इंद्रपानमूर्मिमकृणवतेळः ।
तं वो वयं शुचिमरिप्रमद्य घृतप्रुषं मधुमंतं वनेम ॥७.४७.१॥
आपः । यं । वः । प्रथमं । देवऽयंतः । इंद्रऽपानं । ऊर्मिं । अकृणवत । इळः ।
तं । वः । वयं । शुचिं । अरिप्रं । अद्य । घृतऽप्रुषं । मधुऽमंतं । वनेम ॥७.४७.१॥
Your first wave, O Waters, the refreshing draughts, worthy to be Indra's drink, which the deva-serving men produced. May we obtain this day that wave of yours, bright, clear, rich in sweets, and dripping ghee. (RV vii.47.1)

आपः <f. voc. pl. of अप् – O celestial Waters>, यं <m. acc. sg. of pron. यद् – which>, वः <m. gen. pl. of 2nd person pron. युष्मद् – your (pl.)>, प्रथमं <m. acc. sg. of प्रथम – first>, देवयंतः <m. nom. pl. of pre. active pt. देवयत् nominal verb √देवय – they (the firestones of the celestial ioniser) serving the Devas>, इंद्रपानं <m. acc. sg. of इंद्रपान – worthy to be Indra's drink>, ऊर्मिं <m. acc. sg. of ऊर्मि – the wave (of Waters)>, अकृण्वत <ipf. 3rd person pl. of √कृ – they made>, इळः <(इडः) m. acc. pl. of इळ् (इड्) – refreshing draughts, libations>;

तं <m. acc. sg. of 3rd person pron. तद् – that (wave)>, वः <m. gen. pl. of 2nd person pron. युष्मद् – your (pl.)>, वयं <m. nom. pl. of 1st person pron. अस्मद् – we>, शुचिं <m. acc. sg. of adj. शुचि – radiant, bright>, अरिप्रं <m. acc. sg. of adj. अरिप्र – spotless, clear>, अद्य <ind. – today, now>, घृतप्रुषं <m. acc. sg. adj. घृतप्रुष् – dripping ghee>, मधुमंतं <m. acc. sg. of adj. मधुमत् – containing sweetness, rich in sweets>, वनेम <opt. 1st person pl. of √वन् – may we gain, obtain>.

RV vii.47.2 author वसिष्ठः, to आपः, metre त्रिष्टुप् छंदः
तमूर्मिमापो मधुमत्तमं वोऽपां नपादवत्वाशुहेमा ।
यस्मिन्निंद्रो वसुभिर्मादयाते तमश्याम देवयंतो वो अद्य ॥७.४७.२॥
तं । ऊर्मिं । आपः । मधुमत्ऽतमं । वः । अपां । नपात् । अवतु । आशुऽहेमा ।
यस्मिन् । इंद्रः । वसुऽभिः । मादयाते । तं । अश्याम । देवऽयंतः । वः । अद्य ॥७.४७.२॥
O Waters, let the son of Waters, urging his horses, impel that wave of yours most rich in sweets,
That wave, which makes Indra and the Vasus elated, may we pious obtain today. (RV vii.47.2)

तं <m. acc. sg. of 3rd person pron. तद् – that>, ऊर्मिं <m. acc. sg. of ऊर्मि – the wave>, आपः <f. voc. pl. of अप् – O Waters>, मधुमत्तमं <m. acc. sg. of मधुमत्तम – most rich in sweets>, वः <m. gen. pl. of 2nd person pron. युष्मद् – your (pl.)>, अपांनपात् <m. nom. sg. of अपांनपात् – the son of Waters, i.e. Agni>, अवतु <ipv. 3rd person sg. of √अव् – let him drive, impel>, आशुहेमा <m. nom. sg. of आशुहेमन् – “inciting his horses”, the name of Agni, especially when regarded as Apāṃnapāt (अपांनपात्)>;

यस्मिन् <m. loc. sg. of pron. यद् – in which>, इंद्रः <m. nom. sg. of इंद्र – Indra>, वसुभिः <m. inst. pl. of वसु – with the Vasus>, मादयाते <sub. cau. middle 3rd person sg. of √मद् – it (the wave) will make (Indra) elated>, तं <m. acc. sg. of 3rd person pron. तद् – that (wave)>, अश्याम <ipv. 1st person pl. of √अश् – let us gain, obtain>, देवयंतः <m. nom. pl. of pre. active pt. देवयत् nominal verb √देवय – the god-serving, pious>, वः <m. gen. pl. of 2nd person pron. युष्मद् – yours (pl.)>, अद्य <ind. – today, now>.

RV vii.47.3 author वसिष्ठः, to आपः, metre त्रिष्टुप् छंदः
शतपवित्राः स्वधया मदंतिर्देवीर्देवानामपि यंति पाथः ।
ता इंद्रस्य न मिनंति व्रतानि सिंधुभ्यो हव्यं घृतवज्जुहोत ॥७.४७.३॥
शतऽपवित्राः । स्वधया । मदंतीः । देवीः । देवानां । अपि । यंति । पाथः ।
ताः । इंद्रस्य । न । मिनंति । व्रतानि । सिंधुऽभ्यः । हव्यं । घृतऽवत् । जुहोत ॥७.४७.३॥
A hundred purifying filaments, with self-power, join to pour out exulting Waters to paths of the Devas. They never violate the ordinances of Indra. Offer ye the oblation abounding in ghee to the rivers. (RV vii.47.3)

शतपवित्राः <mf. nom. pl. of शतपवित्र (शत a hundred + पवित्र filters, strainers) – a hundred strainers (refer to the filaments of plasma blooms or Soma plants)>, स्वधया <f. inst. sg. of स्वधा – by self-power, with inherent power>, मदंतीः <f. acc. pl. of pre. active pt. मदंती (√मद्) – them rejoicing, exulting>, देवीः <f. acc. pl. of देवी – the Devīs (pl.) (here, the Devīs refer to celestial Waters)>, देवानां <m. gen. pl. of देव – of the Devas>, अपि <ind. prefix to verb यंति (√इ) >, अपि-यंति <pre. 3rd person pl. of √अपी (अपि-√इ) – they (a hundred strainers or purifying filaments) “join to pour out”>, पाथः <(=पथः) m. acc. pl. of पथिन् – to paths, roads>;

ताः <f. nom. pl. of 3rd person pron. तद् – they (the Devīs or celestial Waters)>, इंद्रस्य <m. gen. sg. of इंद्र – of Indra>, न <ind. – not, no, never>, मिनंति <pre. 3rd person pl. of √मी – they (a hundred filaments) transgress, violate>, व्रतानि <n. acc. pl. of व्रत – commands, laws, ordinances>, सिंधुभ्यः <m. dat. pl. of सिंधु – to the sindhus, to the rivers>, हव्यं <n. acc. sg. of हव्य – sacrificial gifts, oblation>, घृतवत् <n. acc. sg. of adj. घृतवत् – abounding in ghee>, जुहोत <(=जुहुत) ipv. 2nd person pl. of √हु – ye (celestial Waters) present, offer>.

RV vii.47.4 author वसिष्ठः, to आपः, metre त्रिष्टुप् छंदः
याः सूर्यो रश्मिभिराततान याभ्य इंद्रो अरदद्गातुमूर्मिं ।
ते सिंधवो वरिवो धातना नो यूयं पात स्वस्तिभिः सदा नः ॥७.४७.४॥
याः । सूर्यः । रश्मिऽभिः । आऽततान । याभ्यः । इंद्रः । अरदत् । गातुं । ऊर्मिं ।
ते । सिंधवः । वरिवः । धातन । नः । यूयं । पात । स्वस्तिऽभिः । सदा । नः ॥७.४७.४॥
Whom Sūrya with his beams of light has spun, for whom Indra cuts a path to flow and leads the wave of Waters into it. O ye Rivers, clear the path for us and attend to us evermore with prosperity. (RV vii.47.4)

याः <f. acc. pl. of यद् – whom (Waters, cosmic plasmas)>, सूर्यः <m. nom. sg. of सूर्य – Sūrya>, रश्मिभिः <m. inst. pl. of रश्मि – with beams of light>, आततान <prf. 3rd person sg. of आ-√तन् – he spins, weaves>, याभ्यः <f. dat. pl. of pron. यद् – for whom (Waters)>, इंद्रः <m. nom. sg. of इंद्र – Indra>, अरदत् <ipf. 3rd person sg. of √रद् – he (Indra) digs, cuts (a road or path) or leads (a river) into a channel>, गातुं <infinitive of √गा 2P – to go, to flow, to walk (on a path acc.)>, ऊर्मिं <m. acc. sg. of ऊर्मि – the wave (of Waters)>;

ते <m. voc. pl. of 3rd person pron. तद् – O ye>, सिंधवः <m. voc. pl. of सिंधु – O Sindhus, O Rivers>, वरिवः <n. acc. sg. of वरिवस् – room, space, (with √धा) clear the path to>, धातन <ipv. 2nd person pl. of √धा 4Ā – ye clear the path to (with acc. and dat.)>, नः <m. dat. pl. of 1st person pron. अस्मद् – for us>, यूयं <m. nom. pl. of 2nd person pron. युष्मद् – ye>, पात <ipv. 2nd person pl. of √पा 2P – ye watch, protect, attend to>, स्वस्तिभिः <nf. inst. pl. of स्वस्ति – with prosperity, with blessings>, सदा <ind. – always, ever>, नः <m. acc. pl. of 1st person pron. अस्मद् – to us>.

4.2.4 Varuṇa, Mitra, Aryamā: Cosmic Plasma Discharge Modes

Varuṇa (वरुणः), Mitra (मित्रः), and Aryamā (अर्यमा) represent cosmic plasma discharge modes. Though Varuṇa represents cosmic plasmas in general, when the different plasma discharges are considered, Varuṇa represents the plasma dark discharge in particular; Mitra, the plasma light discharge; and Aryamā, the plasma fire discharge. Stanzas RV v.3.1 and 2 provide important clues that unveil their identities. They are addressed to Agni, the cosmic hydrogen ion.

According to these stanzas, Agni is Varuṇa, the plasma dark discharge, when born as the son of Waters, that is, as the son of cosmic plasmas. When ignited, he becomes Mitra, the plasma light discharge (RV v.3.1). When Agni becomes Aryamā of Trita's maidens, he bears a mystical mark (RV v.3.2). Agni, as the fire of Aryamā, forms the firestones of the Stone, the celestial ioniser. According to hymn RV v.3, Agni is Mitra, Varuṇa, and Aryamā; they are the different manifestations of Agni.

RV v.3.1 author वसुश्रुत आत्रेयः, to अग्निः, metre त्रिष्टुप् छंदः
त्वमग्ने वरुणो जायसे यत्त्वं मित्रो भवसि यत्समिद्धः ।
त्वे विश्वे सहसस्पुत्र देवास्त्वमिंद्रो दाशुषे मर्त्याय ॥५.३.१॥
त्वं । अग्ने । वरुणः । जायसे । यत् । त्वं । मित्रः । भवसि । यत् । संऽइद्धः ।
त्वे इति । विश्वे । सहसः । पुत्र । देवाः । त्वं । इंद्रः । दाशुषे । मर्त्याय ॥५.३.१॥
When born, O Agni, you are Varuṇa. When you are ignited, you become Mitra.
You are Indra to the deceased who brings oblation. In thee, O son of Water, are all the Devas.
(RV v.3.1)

त्वं <m. nom. sg. of 2nd person pron. युष्मद् – you>, अग्ने <m. voc. sg. of अग्नि – O Agni>, वरुणः <m. nom. sg. of वरुण – Varuṇa>, जायसे <pre. middle 2nd person sg. of √जन् 4Ā – you are born>, यत् <conjunction – when>, त्वं <m. nom. sg. of 2nd person pron. युष्मद् – you>, मित्रः <m. nom. sg. of मित्र – Mitra>, भवसि <pre. 2nd person sg. of √भू – you are, you become>, यत् <conjunction – when>, समिद्धः <m. nom. sg. of adj. समिद्ध – on fire, lighted, kindled, ignited>;

त्वे <loc. of pronominal form of त्व (युष्मद्) – in thee, in you>, विश्वे <m. nom. pl. of adj. विश्व – all, every>, सहसः <n. gen. sg. of सहस् – of power, force, water>, पुत्र <m. voc. sg. of पुत्र – O son>, देवाः <m. nom. pl. of देव – the Devas>, त्वं <m. nom. sg. of 2nd person pron. युष्मद् – you>, इंद्रः <m. nom. sg. of इंद्र – Indra>, दाशुषे < m. dat. sg. of prf. active pt. ददाश्वस् (√दाश्) – to the one offering (oblation)>, मर्त्याय <m. dat. sg. of मर्त्य – to the deceased man (the ionised)>.

RV v.3.2 author वसुश्रुत आत्रेयः, to अग्निः, metre त्रिष्टुप् छंदः
त्वमर्यमा भवसि यत्कनीनां नाम स्वधावन्गुह्यं बभर्षि ।
अंजंति मित्रं सुधितं न गोभिर्यद्दंपती समनसा कृणोषि ॥५.३.२॥

त्वं । अ॒र्य॒मा । भ॒व॒सि॒ । यत् । क॒नीनां । नाम॑ । स्व॒धा॒ऽव॒न् । गुह्यं॑ । बि॒भ॒र्षि॒ ।
अं॒जंति॑ । मि॒त्रं । सु॒ऽधि॑तं । न । गोभिः॑ । यत् । दंप॑ती इति॒ दंऽप॑ती । सऽम॑नसा । कृ॒णोषि॑ ॥५.३.२॥
O self-powered One, when you become Aryamā of Trita's maidens, you bear a mystical mark. When you make Trita and his maiden one-minded, they anoint well-disposed Mitra with streams of light. (RV v.3.2)

त्वं <m. nom. sg. of 2nd person pron. युष्मद् – you>, अर्यमा <m. nom. sg. of अर्यमन् – Aryamā>, भवसि <pre. 2nd person sg. of √भू – you are, you become>, यत् <conjunction – when>, कनीनां <f. gen. pl. of कनी – of maidens (Trita's)>, नाम <n. acc. sg. of नामन् – a characteristic mark or sign>, स्वधावन् <m. voc. sg. of स्वधावन् – O self-powered One>, गुह्यं <n. acc. sg. of adj. गुह्य – secret, mysterious, mystical>, बिभर्षि <pre. 2nd person sg. of √भृ – you bear, carry, possess>;

अंजंति <pre. 3rd person pl. of √अंज् (√अञ्ज्) – they anoint, honour, celebrate>, मित्रं <m. acc. sg. of मित्र – Mitra>, सुधितं <m. acc. sg. of adj. सुधित – well-disposed, kind, benevolent>, न <ind. – like, as, as it were>, गोभिः <m. inst. pl. of गो – with cows, with rays of light, with streams of milk>, यत् <conjunction – when>, दंपती <m. acc. du. of दंपति (दम्पति) – the lord and his maiden, i.e. Trita and his maiden>, समनसा <m. acc. du. of adj. समनस् – being one mind>, कृणोषि <pre. 2nd person sg. of √कृ – you (Agni) make>.

In stanza RV v.3.2, Agni is called self-powered One (स्वधावान् svadhāvān). According to the Ṛgveda, in the beginning, the magnetic field and forces, emanated from the ocean of celestial hydrogen, set the cosmic creation in motion. Subsequently, the two Vaults of the Heaven were built by the magnetic field and forces. Then the celestial ionisation was initiated, followed by the cosmic plasma discharge and release and the weaving of the Milky Way (पृथिवी pṛthivī), which provides 'terrestrial fields' for the creation and sustenance of life. The celestial ionisation and cosmic plasma discharge and release and the sustenance of the galactic system are self-powered; no external power input is needed.

4.2.5 Varuṇa, the Supreme Monarch

Varuṇa knows everything. Varuṇa, the supreme Monarch, holds up the Heaven and traverses the wide expanse of the Milky Way. He presides over all the worlds in the Milky Way (RV viii.42.1). Varuṇa, the herdsman of the world immortal, sustains our threefold abode: the Upper Heaven, the Lower Heaven, and the Midheaven, the Innerfield, where the Milky Way is formed and sustained (RV viii.42.2).

Stanzas RV viii.42.1 and 2 are addressed to Varuṇa. In stanza RV viii.42.1, Varuṇa is called Asura. The author is Nābhāka Kāṇva (नाभाकः काण्वः), 'the descendant of Nabhāka (नभाकः) and Kaṇva (कण्वः)'. Nabhāka (नभाकः) is the cloud of Soma; Kaṇva is the praiser who belongs to the family of Aṅgirā (अङ्गिराः aṅgirāḥ m. nom. sg. of अङ्गिरस्), the messenger of celestial lights.

RV viii.42.1 author नाभाकः काण्वः, to वरुणः, metre त्रिष्टुप् छंदः
अस्त॑भ्ना॒द्द्याम॑सु॑रो वि॒श्ववे॑दा॒ अमि॑मीत वरि॒माणं॑ पृथि॒व्याः ।
आसी॑दु॒द्विश्वा॒ भुव॑नानि स॒म्राड्विश्वेत्तानि॒ वरु॑णस्य व्र॒तानि॑ ॥८.४२.१॥

अस्तभ्नात् । द्यां । असुरः । विश्वऽवेदाः । अमिमीत । वरिमाणं । पृथिव्याः ।
आ । असीदत् । विश्वा । भुवनानि । संऽराट् । विश्वा । इत् । तानि । वरुणस्य । व्रतानि ॥८.४२.१॥

All-knowing Asura propped the Heaven and traversed the wide expanse of the Milky Way. He, the supreme Monarch, presided over the worlds and all beings. All these are Varuṇa's spheres of action. (RV viii.42.1)

अस्तभ्नात् <ipf. 3rd person sg. of √स्तभ् – he fixed firmly, propped>, द्यां <m. acc. sg. of दिव् – the Heaven>, असुरः <m. nom. sg of असुर – Asura, a supreme spirit, i.e. Varuṇa>, विश्ववेदाः <m. nom. sg. of adj. विश्ववेदस् (=विश्वविद्) – knowing everything, omniscient>, अमिमीत <ipf. 3rd person sg. of √मा – he measured across, he traversed>, वरिमाणं <m. acc. sg. of वरिमन् – wide expanse>, पृथिव्याः <f. gen. sg. of पृथिवि – of Pṛthivī, of the Milky Way>;

आ <ind. prefix to verb असीदत् (√सद्)>, आ-असीदत् <ipf. 3rd person sg. of आ-√सद् – he presided over>, विश्वा <n. acc. pl. of adj. विश्व – all, every>, भुवनानि <n. acc. pl. of भुवन – all the worlds and beings>, संराट् <m. nom. sg. of संराज् – a universal or supreme ruler, a name of Varuṇa>, विश्वा <n. nom. pl. of adj. विश्व – all, every>, इत् <(इद्) ind. – a particle of affirmation, indeed>, तानि <n. nom. pl. of तद् – these>, वरुणस्य <m. gen. sg. of वरुण – of Varuṇa>, व्रतानि <n. nom. pl. of व्रत – dominions, spheres of action, laws, actions>.

RV viii.42.2 author नाभाकः काण्वः, to वरुणः, metre त्रिष्टुप् छंदः
एवा वंदस्व वरुणं बृहंतं नमस्या धीरममृतस्य गोपां ।
स नः शर्म त्रिवरूथं वि यंसत्पातं नो द्यावापृथिवी उपस्थे ॥८.४२.२॥
एव । वंदस्व । वरुणं । बृहंतं । नमस्य । धीरं । अमृतस्य । गोपां ।
सः । नः । शर्म । त्रिऽवरूथं । वि । यंसत् । पातं । नः । द्यावापृथिवी इति । उपऽस्थे ॥८.४२.२॥

Thus, laud thou the mighty Varuṇa and worship him, the wise herdsman of the World immortal. He shall sustain our threefold abode prosperous. Protect us, O the Heaven and the Milky Way, in your lap. (RV viii.42.2)

एव <ind. – so, truly>, वंदस्व <ipv. 2nd person sg. of √वंद् (√वन्द्) – you praise, laud>, वरुणं <m. acc. sg. of वरुण – Varuṇa>, बृहंतं <m. acc. sg. of adj. बृहंत (=बृहत्) – large, great, mighty>, नमस्य <ipv. 2nd person sg. of √नमस्य (nominal verb) – you worship, revere>, धीरं <m. acc. sg. of adj. धीर – intelligent, wise>, अमृतस्य <n. gen. sg. of अमृत – of world of immortal, heaven, eternity>, गोपां <m. acc. sg. of गोपा – a herdsman, guardian>;

सः <m. nom. sg. of 3rd person pron. तद् – he, Varuṇa>, नः <m. gen. pl. of 1st person pron. अस्मद् – our>, शर्म <n. acc. sg. of adj. शर्मन् – happy, prosperous>, त्रिवरूथं <n. acc. sg of त्रिवरूथ (त्रि three + वरूथ shelter, dwelling, abode) – three-fold abode (the Upper Heaven, the Lower Heaven, the Midheaven)>, वि <ind. prefix to verb यंसत् (√यं)>, वि-यंसत् <sub. 3rd person sg. of वि-√यं – he shall hold, support, sustain>, पातं <ipv. 2nd person du. of √पा 2P – ye (du.) watch, preserve>, नः <m. acc. pl. of 1st person pron. अस्मद् – us>, द्यावापृथिवी <f. voc. du. – O twofold Heaven and Pṛthivī (the Milky Way)>, उपस्थे <m. loc. sg. of उपस्थ – in the sheltered place, in the lap>.

4.2.6 Within Varuṇa the Three Heavens Are Placed

Stanza RV vii.87.5 states that within Varuṇa the three Heavens and three Fields are placed. The three Heavens (द्यावः pl.) are the Upper, Mid, and Lower Heavens. The three Fields (भूमिः bhūmiḥ pl.) are the two Soma Ponds and the Milky Way. The three Heavens are placed first and the three Fields are laid subsequently forming an order of six.

Note one Soma Pond is located in the Upper Vault, the other in the Lower Vault of the Heaven (see Figure 6). The Milky Way is in the Innerfield, the Midheaven. The stanza is addressed to Varuṇa. The 'Roamer who abounds in gold' (RV vii.87.5) is Agni.

RV vii.87.5 author वसिष्ठः, to वरुणः, metre त्रिष्टुप् छंदः
तिस्रो द्यावो निहिता अंतरस्मिन्तिस्रो भूमीरुपराः षड्विधानाः ।
गृत्सो राजा वरुणश्चक्र एतं दिवि प्रेंखं हिरण्ययं शुभे कं ॥७.८७.५॥
तिस्रः । द्यावः । निऽहिताः । अंतः । अस्मिन् । तिस्रः । भूमीः । उपराः । षट्ऽविधानाः ।
गृत्सः । राजा । वरुणः । चक्रे । एतं । दिवि । प्रऽईंखं । हिरण्ययं । शुभे । कं ॥७.८७.५॥
Within him the three Heavens are placed, and the three Fields are laid subsequently forming an order of six. The wise King Varuṇa has made this roamer abounding in gold to shine in the Heaven. (RV vii.87.5)

तिस्रः <f. nom. pl. of त्रि – three>, द्यावः <f. nom. pl. of दिव् – Heavens (the Upper, Mid, and Lower)>, निहिताः <f. nom. pl. of adj. निहित – laid, placed>, अंतः <(अन्तर्) ind. – within>, अस्मिन् <m. loc. sg. of pron. इदं – in him (Varuṇa)>, तिस्रः <f. nom. pl. of त्रि – three>, भूमीः <f. nom. pl. of भूमि – fields or grounds (two Soma Ponds, one in the Upper and the other in the Lower Vaults of the Heaven, and the Milky Way)>, उपराः <f. nom. pl. of adj. उपर – posterior, later>, षड्विधानाः <f. nom. pl. of adj. षड्विधान – forming an order or series of six>;

गृत्सः <m. nom. sg. of adj. गृत्स – judicious, wise>, राजा <m. nom. sg. of राजन् – Rājā, King>, वरुणः <m. nom. sg. of वरुण – Varuṇa>, चक्रे <prf. 3rd person sg. of √कृ – he has made>, एतं <m. acc. sg. of pron. एतद् – this>, दिवि <m. loc. sg. of दिव् – in the Heaven>, प्रेंखं <(प्रेङ्खं) m. acc. sg. of प्रंख (प्रेङ्ख) – a wanderer, roamer (refers to Agni)>, हिरण्ययं <m. acc. sg. of adj. हिरण्यय – golden, abounding in gold>, शुभे <f. dat. sg. of शुभ् – (dat. as infinitive) to shine, flash>, कं <ind. particle – (a particle attached to the dative case शुभे to give a stronger meaning) yes, well>.

4.2.7 Mitra and Varuṇa Sustain the Cosmic Sacrifice Through the Cosmic Sacrifice

Cosmic Yajñas, or cosmic sacrifices, are the celestial events involving celestial ionisation and cosmic plasma discharge and release. Stanza RV i.23.5 explains that the cosmic sacrifice is sustained and strengthened by Mitra and Varuṇa through the cosmic sacrifice itself. Once the process of cosmic creation is set in motion, it is sustained through the unbroken chain of celestial ionisation and cosmic plasma discharge and release.

The author of the stanza is Medhātithiḥ Kāṇvaḥ (मेधातिथिः काण्वः). Medhātithiḥ (मेध "nourishing drink" + अतिथिः the "name of Agni") means the 'nourishing drink of Agni'. It is addressed to Mitra and Varuṇa.

RV i.23.5 author मेधातिथिः काण्वः, to मित्रावरुणौ, metre गायत्री छंदः
ऋतेन यावृतावृधावृतस्य ज्योतिषस्पती । ता मित्रावरुणा हुवे ॥१.२३.५॥
ऋतेन । यौ । ऋतऽवृधौ । ऋतस्य । ज्योतिषः । पती इति । ता । मित्रावरुणा । हुवे ॥१.२३.५॥
The two, lords of light and sacrifice, who through the cosmic sacrifice sustain and strengthen the cosmic sacrifice, I invoke the two, Mitra and Varuṇa. (RV i.23.5)

ऋतेन <n. inst. sg. of ऋत – through the cosmic sacrifice (Yajña)>, यौ <m. nom. du of यद् – the two who>, ऋतावृधौ <m. nom. du of adj. ऋतावृध् – strengthening the sacrifice, causing the sacrifice to prosper>, ऋतस्य <n. gen. sg. of ऋत – of sacrifice>, ज्योतिषः <mn. gen. sg. of ज्योतिस् – of celestial light>, पती <m. nom. du. of पति – two masters or lords>;

ता <m. acc. du. of तद् – the two>, मित्रावरुणा <m. acc. of dual compound मित्रावरुण – Mitra and Varuṇa>, हुवे <pre. 1st person sg. of √हू (weak form of √ह्वे) – I call, invoke>.

4.2.8 Mitra Beholds the Tillers of the Field with an Unwinking Eye

Mitra is almost always addressed with Varuṇa and hardly mentioned alone. In the extant text of the Ṛgveda, only one hymn (RV iii.59) is addressed to Mitra. Three stanzas of the hymn (RV iii.59.1, 8, and 9) provide some insights into key attributes of Mitra, the plasma light discharge. Stanza RV iii.59.1 is in triṣṭup metre (त्रिष्टुप् छंदः), and stanzas 8 and 9 are in gāyatrī metre (गायत्री छंदः).

Mitra, making Speech, brings the 'people' of cosmic plasmas together, holds the Heaven and the Milky Way, and watches the tillers of the field with an unwinking eye (RV iii.59.1). The field refers to the Soma Pond; the tillers of the field, to the firestones of the celestial ioniser who pour out cosmic plasmas, making furrows in the field.

RV iii.59.1 author विश्वामित्रः, to मित्रः, metre त्रिष्टुप् छंदः
मित्रो जनान्यातयति ब्रुवाणो मित्रो दाधार पृथिवीमुत द्यां ।
मित्रः कृष्टीरनिमिषाभि चष्टे मित्राय हव्यं घृतवज्जुहोत ॥३.५९.१॥
मित्रः । जनान् । यातयति । ब्रुवाणः । मित्रः । दाधार । पृथिवीं । उत । द्यां ।
मित्रः । कृष्टीः । अनिऽमिषा । अभि । चष्टे । मित्राय । हव्यं । घृतऽवत् । जुहोत ॥३.५९.१॥
Mitra, uttering the sound, brings people together and holds the Heaven and the Milky Way.
Mitra beholds the tillers of the field with an unwinking eye. To this Mitra, offer ye an oblation mixed with ghee. (RV iii.59.1)

मित्रः <m. nom. sg. of मित्र – Mitra>, जनान् <m. acc. pl. of जन – men, race, people, i.e. the people of cosmic plasmas or different races of the Devas>, यातयति <pre. cau. 3rd person sg. of √यत् – he joins, unites>, ब्रुवाणः <m. nom. sg. of adj. ब्रुवाण – speaking, saying>, मित्रः <m. nom. sg. of मित्र – Mitra>, दाधार <prf. 3rd person sg. of √धृ – he holds, carries, maintains>, पृथिवीं <f. acc. sg. of पृथिवी – Pṛthivī, the Milky Way>, उत <ind. – and, also>, द्यां <m. acc. sg. of दिव् – the Heaven>;

मित्रः <m. nom. sg. of मित्र – Mitra>, कृष्टीः <f. acc. pl. of कृष्टि – tillers, ploughers (of the field)>, अनिमिषा <m. inst. sg. of adj. अनिमिष् – without winking, with an unwinking eye>, अभि <ind. prefix to verb चष्टे (√चक्ष्)>, अभि-चष्टे <pre. middle 3rd person sg. of अभि-√चक्ष् – he looks at, views>, मित्राय <m. dat. sg. of मित्र – to Mitra>, हव्यं <n. acc. sg. of हव्य – oblation>, घृतवत् <n. acc. sg of adj. घृतवत् – mixed with ghee (clarified butter)>, जुहोत <(=जुहुत) ipv. 2nd person pl. of √हु – ye offer>.

All five races of men, the Devas, that is, five different classes of celestial plasma 'people', obey Mitra, for Mitra sustains all the Devas (RV iii.59.8). Mitra prepares sought-after draughts for the man who gathers and spreads the sacrificial grass. By the draughts Mitra prepares, all the good works are accomplished.

RV iii.59.8 author विश्वामित्रः, to मित्रः, metre गायत्री छंदः
मित्राय पंच येमिरे जना अभिष्टिशवसे । स देवान्विश्वान्बिभर्ति ॥३.५९.८॥
मित्राय । पंच । येमिरे । जनाः । अभिष्टिऽशवसे । सः । देवान् । विश्वान् । बिभर्ति ॥३.५९.८॥
All five races of the Devas are faithful to Mitra who renders powerful assistance
And cherishes all the Devas. (RV iii.59.8)

मित्राय <m. dat. sg. of मित्र – to Mitra>, पंच <m. nom. pl of पंचन् (पञ्चन्) – five>, येमिरे <prf. middle 3rd person pl. of √यं – they give themselves up to, obey, faithful to (dat.)>, जनाः <m. nom. pl. of जन – races of Devas or men>, अभिष्टिशवसे <m. dat. sg. of adj. अभिष्टिशवस् – rendering powerful assistance>;

सः <m. nom. sg. of pron. तद् – he>, देवान् <m. acc. pl. of देव – Devas>, विश्वान् <m. acc. pl. of adj. विश्व – all, every>, बिभर्ति <pre. 3rd person sg. of √भृ – he supports, maintains, cherishes>.

RV iii.59.9 author विश्वामित्रः, to मित्रः, metre गायत्री छंदः
मित्रो देवेष्वायुषु जनाय वृक्तबर्हिषे । इषं इष्टव्रता अकः ॥३.५९.९॥
मित्रः । देवेषु । आयुषु । जनाय । वृक्तऽबर्हिषे । इषः । इष्टऽव्रताः । अकरित्यकः ॥३.५९.९॥
Mitra, of lineages of the Devas, for the man who gathers and spreads the sacrificial grass,
Prepared refreshing draughts by which good works are accomplished. (RV iii.59.9)

मित्रः <m. nom. sg. of मित्र – Mitra>, देवेषु <m. loc. pl. of देव – among the Devas>, आयुषु <m. loc. pl. of adj. आयु – among families, lineages>, जनाय <m. dat. sg. of जन – for the man>, वृक्तबर्हिषे <m. dat. sg. of adj. वृक्तबर्हिस् – for the one who has gathered and spread the sacrificial grass>;

इषः <f. acc. pl. of इष् – refreshing draughts, libations>, इष्टव्रताः <f. acc. pl. of adj. इष्टव्रत – "that by which good (इष्ट) works (व्रत) succeed">, अकः <ipf. 3rd person sg. of √कृ – he made, prepared>.

4.3 ADITI, THE DOUBLE HEADED

Aditi (अदितिः) is the mother of the Ādityas. In the extant text of the Ṛgveda, we find no hymn addressed to Aditi. However, several stanzas of the Ṛgveda provide keys for identifying who Aditi is. Aditi is the Milch Cow (RV i.153.3). Aditi unbinds the sin of the ionic bond (RV vii.93.7). Aditi is the indestructible light (अवध्रं ज्योतिः RV vii.82.10). These stanzas clarify that Aditi is another name for the Stone, the celestial ioniser. Aditi is the mother of all the Devas, the shining ones, and the associated cosmic plasma phenomena.

Aditi is called Dhenu (धेनुः a milch cow RV i.153.3), Mātā Dhenu (माता धेनुः mother milch Cow RV iii.55.12), Dakṣiṇā (दक्षिणा a prolific milch cow ŚB iii.2.4.6), Mahī Mātā (मही माता mighty mother RV viii.25.3), Rājaputrā (राजपुत्रा mother of kings RV ii.27.7), and Kṣatriyā (क्षत्रिया f. an empress, monarch B.Yajur i.2.4). Aditi is the 'double headed' (उभयतः शीर्ष्णी B.Yajur i.2.4, ŚB iii.2.4.16), for she has two heads, the upper and the lower discs of the Stone. She is the mighty Mother, the Mother of creation, the Mother of all created beings. She is the monarch of the Bhuvanatrayaṃ (भुवनत्रयं), the threefold World, that is, the whole of the galactic system with the Upper Heaven, the Lower Heaven, and the Innerfield where the Milky Way is formed and sustained.

Aditi produces inexhaustible cosmic plasmas and discharges them. The definitions of Aditi (अदितिः) as a feminine noun provided by Monier-Williams include "name of one of the most ancient of the Indian goddesses", "freedom", "boundlessness", "inexhaustible abundance", "creative power". As a masculine noun, Aditi is the "devourer" and "death".

Aditi is the Terrible who devours, slays, and dismembers the bodies of hydrogen. Aditi frees Somas and Agnis from the bondage of atomic structures to become the embryonic particles of cosmic plasmas. Once freed, Agnis and Somas will have boundless potential to create other elements, through recombination, during the cosmic plasma discharge and release processes. As we can see, Aditi, the Stone, the celestial ioniser, is associated with all the characteristic meanings provided by Monier-Williams.

4.3.1 Aditi, the Mother of Kings Mitra and Varuṇa

Aditi is the mother of Kings Mitra and Varuṇa (RV ii.27.7). Here, Aryamā is implored to let Aditi deliver 'us', passing beyond the foes, that is, passing beyond the ionic bonds of hydrogen. The lofty shelter refers to the Soma Pond or the individual plasma blooms. The stanza indicates that Aryamā, the plasma fire discharge, is associated with the fire of the Stone, the Trita. The male offspring refers to Agnis. Stanza RV ii.27.7 is addressed to the Ādityas, the sons of Aditi.

RV ii.27.7 author कूर्मो गार्त्समदो गृत्समदो वा, to आदित्याः, metre त्रिष्टुप् छंदः
पिपर्तु नो अदिती राजपुत्राति द्वेषांस्यर्यमा सुगेभिः ।
बृहन्मित्रस्य वरुणस्य शर्मोप स्याम पुरुवीरा अरिष्टाः ॥२.२७.७॥
पिपर्तु । नः । अदितिः । राजऽपुत्रा । अति । द्वेषांसि । अर्यमा । सुऽगेभिः ।
बृहत् । मित्रस्य । वरुणस्य । शर्म । उप । स्याम । पुरुऽवीराः । अरिष्टाः ॥२.२७.७॥
Aryamā, let Aditi, mother of Kings, deliver us beyond the foes through safe paths.
May we, possessed of many male offspring, without fail, go to the lofty shelter of Mitra and Varuṇa. (RV ii.27.7)

पिपर्तु <ipv. 3rd person sg. of √पृ – let (Aditi) bring over, deliver>, नः <m. acc. pl. of 1st person pron. अस्मद् – us>, अदितिः <f. nom. sg of अदिति – Aditi>, राजपुत्रा <f. nom. sg. of राजपुत्रा – having kings for sons, mother of kings>, अति <ind. preposition – beyond, over>, द्वेषांसि <n. acc. pl. of द्वेषस् – enemies, foes (refers to the ionic bonds)>, अर्यमा <m. nom. sg. of अर्यमन् – Aryamā>, सुगेभिः <(=सुगैः) n. inst. pl. of सुग – through/by good paths, easy or successful courses>;

बृहत् <n. acc. sg. of बृहत् – lofty, tall>, मित्रस्य <m. gen. sg. of मित्र – of Mitra>, वरुणस्य <m. gen. sg. of वरुण – of Varuṇa>, शर्म <n. acc. sg. of शर्मन् – to the shelter, house>, उप <ind. prefix to verb स्याम (√अस्)>, उप-स्याम <opt. 1st person pl. of उप-√अस् – may we approach, go to>, पुरुवीराः <m. nom pl. of adj. पुरुवीर – possessed of many men or male offspring>, अरिष्टाः <m. nom pl. of adj. अरिष्ट – uninjured, without fail>.

Aditi, the mighty Mother, performing sacrificial works of cosmic ionisation, brought forth Mitra and Varuṇa for supreme rule (RV viii.25.3). Stanza RV viii.25.3 is addressed to Mitra and Varuṇa.

RV viii.25.3 author विश्वमना वैयश्वः, to मित्रावरुणौ, metre उष्णिक् छंदः
ता माता विश्ववेदसासुर्याय प्रमहसा । मही जजानादितिर्ऋतावरी ॥८.२५.३॥
ता । माता । विश्वऽवेदसा । असुर्याय । प्रऽमहसा । मही । जजान । अदितिः । ऋतऽवरी ॥८.२५.३॥
These two, the two all-knowers of great might, for supreme rule,
Aditi, mighty Mother, performing sacred work, brought forth. (RV viii.25.3)

ता <m. acc. du. of pron. तद् – these Two (Mitra and Varuṇa)>, माता <f. nom. sg. of मातृ – Mother>, विश्ववेदसा <(=विश्ववेदसौ) m. acc. du. of विश्ववेदस् – two sages, all-knowers>, असुर्याय <m. dat. sg. of असुर्य – for supreme dominion, rule>, प्रमहसा <m. acc. du. of adj. प्रमहस् – of great might or splendour>;

मही <f. nom. sg. of महि – great, mighty>, जजान <prf. 3rd person sg. of √जन् – she (Aditi) brought forth>, अदितिः <f. nom. sg. of अदिति – Aditi>, ऋतावरी <f. nom. sg. of adj. ऋतावन् – performing sacred work (refers to the work of cosmic ionisation)>.

4.3.2 Aditi, the Milch Cow

Aditi is the milch Cow who pours out milk, the oblations, offered to the Devas for the Soma sacrifice (RV i.153.3). Stanza RV i.153.3 is addressed to Mitra and Varuṇa. Hotā is the chief priest of the Soma sacrifice, the event of releasing the matured cosmic plasmas from the Soma Pond. The author is Dīrghatamā Aucathya (दीर्घतमाः औचथ्यः), the 'lofty one who is worthy of praise, the son of Ucathya (उचथ्यः)'. Ucathya is the name of an Aṅgirā (अङ्गिराः m. nom. sg. of अङ्गिरस्), the messenger of the Heaven.

RV i.153.3 author दीर्घतमा औचथ्यः, to मित्रावरुणौ, metre त्रिष्टुप् छंदः
पीपायं धेनुरदितिर्ऋताय जनाय मित्रावरुणा हविर्दे ।
हिनोति यद्वां विदथे सपर्यन्त्स रातहव्यो मानुषो न होता ॥१.१५३.३॥
पीपायं । धेनुः । अदितिः । ऋतायं । जनाय । मित्रावरुणा । हविःऽदे ।
हिनोति । यत् । वां । विदथे । सपर्यन् । सः । रातऽहव्यः । मानुषः । न । होता ॥१.१५३.३॥
Aditi the milch Cow overflows for the sacrifice, O Mitra-Varuṇa, for the man who offers oblation. She discharges you both when he, who is favourable to men, serving as Hotā, receives the oblation in the assembly. (RV i.153.3)

पीपाय <prf. 3rd person sg. of √पि (=√पी) – she overflows, swells>, धेनुः <f. nom. sg. of धेनु – a milch Cow>, अदितिः <f. nom. sg. of अदितिः – Aditi>, ऋताय <n. dat. sg. of ऋत – for the sacrifice (refers to Soma sacrifice)>, जनाय <m. dat. sg. of जन – for man, people>, मित्रावरुणा <m. voc. du. compound – Mitra and Varuṇa>, हविर्दे <m. dat. sg. of adj. हविर्दा – giving or offering oblation>;

हिनोति <pre. 3rd person sg. of √हि – she (Aditi) discharges, hurls, casts>, यत् <conjunction – when>, वां <m. acc. du. of युष्मद् – you two (Mitra and Varuṇa)>, विदथे <n. loc. sg. of विदथ – in the assembly>, सपर्यन् <m. nom. sg. of pre. active pt. of सपर्यत् (nominal verb √सपर्य) – he serving>, सः <m. nom. sg. of 3rd person pron. तद् – he>, रातहव्यः <m. nom. sg. of रातहव्य – one who receives the oblation>, मानुषः <m. nom. sg. of adj. मानुष – favourable or propitious to men>, न <ind. – like, as>, होता <m. nom. sg. of होतृ – Hotā, the sacrificer, offerer of an oblation>.

4.3.3 Aditi Unbinds the Sin of the Ionic Bond

In stanza RV vii.93.7, Aryamā and Aditi are implored to have the ten maidens of Trita unbind the sin of the ionic bond. Stanza RV vii.93.7 is addressed to Indra and Agni.

In the Ṛgveda, the ionic bond of hydrogen is referred to as the 'sin' that must be expiated by the fire sacrifice (read ionisation) for the cosmic creation and sustenance of the galactic system. Mother Aditi, the Stone, the celestial ioniser, unbinds the ionic bonds to produce and discharge cosmic plasmas into the Soma Pond. And Indra slays the serpent of the

Heaven, the plasma sheath, to release celestial lights, cosmic plasmas, from the Soma Pond.

RV vii.93.7 author वसिष्ठः, to इंद्राग्नी, metre त्रिष्टुप् छंदः
सो अग्न एना नमसा समिद्धोऽच्छा मित्रं वरुणमिन्द्रं वोचेः ।
यत्सीमागश्चकृमा तत्सु मृळ तदर्यमादितिः शिश्रथंतु ॥७.९३.७॥
सः । अग्ने । एना । नमसा । संऽइद्धः । अच्छ । मित्रं । वरुणं । इंद्रं । वोचेः ।
यत् । सीं । आगः । चकृम । तत् । सु । मृळ । तत् । अर्यमा । अदितिः । शिश्रथंतु ॥७.९३.७॥
O Agni, thus kindled by the adoration, mayest thou invite Mitra, Varuṇa, and Indra.
Pardon thou whatever sin we have committed. Aryamā and Aditi, have them quickly unbind it. (RV vii.93.7)

सः <m. nom. sg. of pron. तद् – he, thou, that very person>, अग्ने <m. voc. sg. of अग्नि – O Agni>, एना <ind. – here, in this manner, thus, then>, नमसा <n. inst. sg. of नमस् – by salutation, worship, adoration (with sacred sound)>, समिद्धः <m. nom. sg. of adj. समिद्ध – lighted, ignited, kindled>, अच्छ <(=अच्छा) ind. prefix to verb वोचेः (√वच्)>, मित्रं <m. acc. sg. of मित्र – Mitra>, वरुणं <m. acc. sg. of वरुण – Varuṇa>, इंद्रं <m. acc. sg. of इंद्र – Indra>, अच्छा-वोचेः <opt. 2nd person sg. of अच्छा-√वच् – may you invite>;

यत् <n. acc. sg of यद् – what, that, whichever>, सीं <ind. – him, her, it, them (employed for all genders, numbers, persons) and often a particle of emphasising>, आगः <n. acc. sg of आगस् – transgression, offence, sin>, चकृम <prf. 1st person pl. of √कृ – we have made, committed>, तत् <n. acc. sg. of तद् – that, it (the sin)>, सु <ind. – good, quickly>, मृळ <(=मृड) ipv. 2nd person sg. of √मृळ् (√मृड्) – you pardon, spare>, तत् <n. acc. sg. of तद् – that, it, (the sin)>, अर्यमा <m. nom. sg. of अर्यमन् – Aryamā>, अदितिः <f. nom. sg. of अदिति – Aditi>, शिश्रथंतु <ipv. cau. 3rd person pl. of √श्रथ् – let them (ten maidens or the firestones of the Stone) loosen, untie, unbind>.

4.3.4 Aditi, Daughter of Dakṣa

Stanzas RV x.72.1-5 explain the genesis of the Devas and Aditi (अदितिः) and Dakṣa (दक्षः) and the relationship between them.

RV x.72.1 author बृहस्ततिर्बृहस्पतिर्वा लौक्य अदितिर्वा दाक्षायणी, to देवाः, metre अनुष्टुप् छंदः
देवानां नु वयं जाना प्र वोचाम विपन्यया । उक्थेषु शस्यमानेषु यः पश्यादुत्तरे युगे ॥१०.७२.१॥
देवानां । नु । वयं । जाना । प्र । वोचाम । विपन्यया ।
उक्थेषु । शस्यमानेषु । यः । पश्यात् । उत्ऽतरे । युगे ॥१०.७२.१॥
Now, we will proclaim with delight the origins of the Devas,
That one may behold them in these hymns when sung in a future age. (RV x.72.1)

देवानां <m. gen. pl. of देव – of the Devas>, नु <ind. – now, at once>, वयं <m. nom. pl. of 1st person pron. अस्मद् – we>, जाना <n. acc. pl. of जान – origins, birth places>, प्र <ind. prefix to verb वोचाम (√वच्)>, प्र-वोचाम <sub. 1st person pl. of प्र-√वच् – we will speak, declare, recite, describe>, विपन्यया <ind. – joyfully, wonderfully>;

उक्थेषु <n. loc. pl. of उक्थ – in praises, hymns>, शस्यमानेषु <n. loc. pl. of adj. शस्यमान – be recited, be sung, be praised>, यः <m. nom. sg. of pron. यद् – he who, a man, one, any one>, पश्यात् <sub. 3rd person sg. of √पश् – he/she shall see, behold, look at, observe>, उत्तरे <n. loc. sg. of adj. उत्तर – later, future>, युगे <n. loc. sg. of युग – in an age>.

Brahmaṇaspati, the lord of prayer, the lord of Brāhmaṇa (ब्राह्मणः brāhmaṇaḥ), blasted and smelt hydrogen, the 'non-real (असत् asat)', to produce cosmic plasmas, the 'real (सत् sat)' (RV x.72.2). Brahmaṇaspati is the Stone deified, the masculine counterpart of Aditi. In the

Ṛgveda, the Devas are the cosmic plasmas and the associated phenomena deified, and they belong to the immortals (RV x.72.5). Agnis and Somas are called Brāhmaṇas (ब्राह्मणाः pl.) who have divine knowledge.

RV x.72.2 author बृहस्ततिर्बृहस्पतिर्वा लौक्य अदितिर्वा दाक्षायणी, to देवाः, metre अनुष्टुप् छंदः
ब्रह्मणस्पतिरेता सं कर्मार इवाधमत् । देवानां पूर्व्ये युगेऽसतः सदजायत ॥१०.७२.२॥
ब्रह्मणः । पतिः । एता । सं । कर्मारःऽइव । अधमत् ।
देवानां । पूर्व्ये । युगे । असतः । सत् । अजायत ॥१०.७२.२॥
Brahmaṇaspati, like a blacksmith, blasted and smelted them.
From Non-Real in a previous cycle of the Devas, Real was born. (RV x.72.2)

ब्रह्मणस्पतिः <m. nom. sg. of ब्रह्मणस्पति (=बृहस्पति) – Brahmaṇaspati, lord of prayer or of Brāhmaṇa (ब्राह्मणः); Agni and Soma are called Brāhmaṇa>, एता < n. acc. pl. of एतद् – them, i.e. hydrogen atoms>, सं <ind. prefix to verb अधमत् (√धं)>, कर्मारः <m. nom. sg. of कर्मार – an artificer, blacksmith>, इव <ind. – like, as if>, अधमत् <ipf. 3rd person sg. सं-√धं – he smelted by blasting>;

देवानां <m. gen. pl. of देव – of the Devas>, पूर्व्ये <m. loc. sg. of adj. पूर्व्य – former, previous>, युगे <n. loc. sg. of युग – in an age, period, cycle>, असतः <n. abl. sg. of असत् – from non-existence, non-real, i.e. the unmanifested, undifferentiated hydrogen>, सत् <n. nom. sg. of सत् – that which is real, existence, reality, i.e. cosmic plasmas>, अजायत <ipf. middle 3rd person sg. of √जन् – he/it was born, produced, came into existence>.

RV x.72.3 author बृहस्ततिर्बृहस्पतिर्वा लौक्य अदितिर्वा दाक्षायणी, to देवाः, metre अनुष्टुप् छंदः
देवानां युगे प्रथमेऽसतः सदजायत । तदाशा अन्वजायंत तदुत्तानपदस्परि ॥१०.७२.३॥
देवानां । युगे । प्रथमे । असतः । सत् । अजायत ।
तत् । आशाः । अनु । अजायंत । तत् । उत्तानऽपदः । परि ॥१०.७२.३॥
From Non-Real in an earlier cycle of the Devas, Real was born.
Thus, thereupon, foods were produced, then Soma plants after Soma plants. (RV x.72.3)

देवानां <m. gen. pl. of देव – of the Devas>, युगे <n. loc. sg. of युग – in an age, period, cycle>, प्रथमे <n. loc. sg. of adj. प्रथम – earliest, prior, former>, असतः <n. abl. sg. of असत् – from non-existence, non-real, i.e. hydrogen>, सत् <n. nom. sg. of सत् – that which is real, existence, reality, i.e. cosmic plasmas>, अजायत <ipf. middle 3rd person sg. of √जन् – it was born, produced, came into existence>;

तत् <ind. – thus, in this manner>, आशाः <m. nom. pl. of आश – foods, i.e. Somas (note that Soma is food of the Devas)>, अनु <ind. – after, along, thereupon, then>, अजायंत <ipf. middle 3rd person pl. of √जन् – they were born, produced>, तत् <ind. clause-connecting particle – now, then>, उत्तानपदः <f. nom. pl. of उत्तानपद् – upward-germinating plants (refer to Soma plants or plasma blooms)>, परि <ind. – successively, one after another>.

Stanza RV x.72.4 clarifies that Aditi, the celestial ioniser, is born of Dakṣa. And once born, she then becomes the mother of Dakṣa.

RV x.72.4 author बृहस्ततिर्बृहस्पतिर्वा लौक्य अदितिर्वा दाक्षायणी, to देवाः, metre अनुष्टुप् छंदः
भूर्जज्ञ उत्तानपदो भुव आशा अजायंत । अदितेर्दक्षो अजायत दक्षाद्वदितिः परि ॥१०.७२.४॥
भूः । जज्ञे । उत्तानऽपदः । भुवः । आशाः । अजायंत ।
अदितेः । दक्षः । अजायत । दक्षात् । ऊं इति । अदितिः । परि ॥१०.७२.४॥
Sacrificial fire was born. Then Agni and his food and upward-germinating Soma plants were produced. From Aditi, Dakṣa was born, then from Dakṣa, Aditi in turn. (RV x.72.4)

भूः <f. nom. sg. of भू – becoming, substance, sacrificial fire>, जज्ञे <prf. middle 3rd person sg. of √जन् – it was produced, born>, उत्तानपदः <f. nom. pl. of उत्तानपद् – upward-germinating plants, i.e. the Soma plants or plasma blooms>, भुवः <m. nom. sg. of भुव – the name of Agni>, आशाः <m. nom. pl. of आश – foods, i.e. Somas (note that Soma is food of the Devas)>, अजायंत <ipf. middle 3rd person pl. of √जन् – they were born, produced>;

अदितेः <f. abl. sg. of अदिति – from Aditi>, दक्षः <m. nom. sg. of दक्ष – Dakṣa>, अजायत <ipf. middle 3rd person sg. of √जन् – he/it was born, produced, came into existence>, दक्षात् <m. abl. sg. of दक्ष – from Dakṣa>, ऊं <ind. particle of assent, interrogation>, अदितिः <f. nom. sg. of अदिति – Aditi>, परि <ind. – successively>.

RV x.72.5 author बृहस्ततिर्बृहस्पतिर्वा लौक्य अदितिर्वा दाक्षायणी, to देवाः, metre अनुष्टुप् छंदः
अदितिर्ह्यजनिष्ट दक्ष या दुहिता तव । तां देवा अन्वजायंत भद्रा अमृतबंधवः ॥१०.७२.५॥
अदितिः । हि । अजनिष्ट । दक्ष । या । दुहिता । तव ।
तां । देवाः । अनु । अजायंत । भद्रः । अमृतऽबंधवः ॥१०.७२.५॥
Indeed, O Dakṣa, Aditi, who is your daughter, was brought forth.
After Aditi were the blessed Devas born, who belong to the immortals. (RV x.72.5)

अदितिः <f. nom. sg. of अदिति – Aditi>, हि <ind. – on account of, indeed, surely>, अजनिष्ट <aor. Middle 3rd person sg. of √जन् – she (Aditi) was born or produced, came into existence>, दक्ष <m. voc. sg. of दक्ष – O Dakṣa>, या <f. nom. sg. of pron. यद् – she who>, दुहिता <f. nom. sg. of दुहितृ – daughter>, तव <m. gen. sg. of 2nd person pron. of युष्मद् – your>;

तां <f. acc. sg. of 3rd person pron. तद् – her (Aditi)>, देवाः <m. nom. pl. of देव – the Devas>, अनु <ind. – after, next>, अजायंत <ipf. middle 3rd person pl. of √जन् – they were born, produced>, भद्राः <m. nom. pl. of adj. भद्र – blessed>, अमृतबंधवः <m. nom. pl. of अमृतबंधु (अमृत m. an immortal + बंधु m. relation, belonging to) – belonging to the immortals>.

4.3.5 Aditi and Diti

Though none of the hymns is addressed to Aditi in the Ṛgveda, the name of Aditi is invoked numerous times mostly along with her sons, the Ādityas (आदित्याः, pl.). In a few stanzas, Aditi is mentioned with Diti (RV v.62.8, RV iv.2.11).

When we take the name of Diti (दितिः) to be an antithesis to that of Aditi (अदितिः), then Diti should be the 'binder' or to be 'bound'. Diti refers to the Soma Pond, the abode of the blessed Devas, into which cosmic plasmas are discharged (RV iv.2.11). Thus, Aditi and Diti refer to the pair of the Stone and the Soma Pond. We shall see below the context in which Aditi and Diti are invoked together in stanzas RV v.62.8 and RV iv.2.11.

RV v.62.8 author श्रुतविदात्रेयः, to मित्रावरुणौ, metre त्रिष्टुप्
हिरण्यरूपमुषसो व्युष्टावयःस्थूणमुदिता सूर्यस्य ।
आ रोहथो वरुण मित्र गर्तमतश्चक्षाथे अदितिं दितिं च ॥५.६२.८॥
हिरण्यऽरूपं । उषसः । विऽउष्टौ । अयःऽस्थूणं । उत्ऽइता । सूर्यस्य ।
आ । रोहथः । वरुण । मित्र । गर्तं । अतः । चक्षाथे इति । अदितिं । दितिं । च ॥५.६२.८॥
At the first gleam of morning light, a gold-hued pillar of Sūrya has risen.
O Mitra and Varuṇa, mount ye the seat of your chariot. From there, behold Aditi and Diti.
(RV v.62.8)

हिरण्यरूपं <n. nom. sg. of adj. हिरण्यरूप – golden, made of gold, gold-hued>, उषसः <f. gen. sg. of उषस् – of morning light, of dawn>, व्युष्टौ <f. loc. sg. of व्युष्टि – at the first gleam, at the breaking of dawn>, अयःस्थूणं <n. nom. sg. of अयःस्थूण (अयस् n. gold + स्थूण n. a post, pillar) – a golden pillar (refers to the individual plasma blooms, which is called a pillar of

the Heaven or a sacrificial post)>, उदिता <f. nom. sg. of adj. उदित – risen, ascended>, सूर्यस्य <m. gen. sg. of सूर्य – of Sūrya>;

आ <ind. prefix of verb रोहथः (√रुह्)>, आ-रोहथः <pre. 2nd person du. of आ-√रुह् – ye (du.) ascend, mount>, वरुण <m. voc. sg. of वरुण – O Varuṇa>, मित्र <m. voc. sg. of मित्र – O Mitra>, गर्तं <m. acc. sg. of गर्त – a high seat, the seat of a war-chariot>, अतः <(अतस्) ind. – from that time, from there (from the seat of the chariot)>, चक्षाथे <pre. 2nd person du. of √चक्ष् – ye (du.) behold>, अदितिं <f. acc. sg. of अदिति – Aditi, the Milch Cow, the celestial ioniser>, दितिं <f. acc. sg. of दिति – Diti, the womb of cosmic creation, the Soma Pond>>, च <ind. – and>.

RV iv.2.11 author वामदेवः, to अग्निः, metre त्रिष्टुप्
चित्तिमचित्तिं चिनवद्वि विद्वान्पृष्ठेव वीता वृजिना च मर्तान् ।
राये च नः स्वपत्याय देव दितिं च रास्वादितिमुरुष्य ॥४.२.११॥
चित्तं । अचित्तं । चिनवत् । वि । विद्वान् । पृष्ठाऽइव । वीता । वृजिना । च । मर्तान् ।
राये । च । नः । सुऽअपत्याय । देव । दितिं । च । रास्व । अदितिं । उरुष्य ॥४.२.११॥
The wise will discern sense and folly of men, like straight and crooked backs of horses.
Bestow thou, O Deva, riches and noble offspring upon Diti, and secure Aditi for us. (RV iv.2.11)

चित्तं <mn. acc. sg of चित्त – intellect, sense, reason>, अचित्तं <n. acc. sg. of अचित्त – destitute of intellect or sense>, वि-चिनवत् <sub. 3rd person sg. of वि-√चि – he will discern, distinguish>, वि <ind. prefix to verb चिनवत् √चि>, विद्वान् <m. nom. sg. of विद्वत् (=विद्वस्) – the wise man, seer>, पृष्ठा <n. acc. pl. of पृष्ठ – backs (of the horses or animals)>, इव <ind. – like, as>, वीता <n. acc. pl. of adj. वीत – straight>, वृजिना <n. acc. pl. of adj. वृजिन – bent, crooked>, च <ind. – and>, मर्तान् <m. acc. pl. of मर्त – men>;

राये <m. dat. sg. of रै – for wealth, riches>, च <ind. – and>, नः <m. dat. pl. of 1st person pron. अस्मद् – for us>, सुअपत्याय <n. dat. sg. of सुअपत्य – for noble offspring>, देव <m. voc. sg. of देव – O Deva (here, the Deva is Agni as Trita)>, दितिं <f. acc. sg. of दिति – Diti>, च <ind. – and>, रास्व <ipv. middle 2nd person sg. of √रा – you give, bestow, impart (dat.)>, अदितिं <f. acc. sg. of अदिति – Aditi>, उरुष्य <ipv. 2nd person sg. nominal verb √उरुष्य – you secure, protect>.

4.3.6 The Ādityas, the Sons of Aditi

The Ādityas (आदित्याः pl.) are the sons of Aditi. Stanza RV ix.114.3, though their names are not enumerated, states that there are seven Ādityas. In RV ii.27.1, six Ādityas are enumerated: Varuṇa, Mitra, Aryamā, Bhaga, Dakṣa, and Aṃśa. Savitā is known to be the seventh Āditya. The chief of the Ādityas is Varuṇa. The Ādityas are cosmic plasmas and their associated forces that are produced and discharged into Diti, the Soma Pond, by Aditi, the Stone, the celestial ioniser.

Varuṇa is cosmic plasmas in general and plasma dark discharge in particular; Mitra, plasma light discharge; and Aryamā, plasma fire discharge. It appears that Dakṣa (दक्षः) is the cosmic force associated with Soma. Aṃśa (अंशः) is the cosmic force associated with Agni. Bhaga (भगः) is another name for Rudra and the dispenser of cosmic plasmas. Note, Rudra is the protector of the hearth, the Soma Pond, within which the flames of cosmic plasmas are ablaze. Stanza RV ii.27.1 is addressed to the Ādityas.

RV ii.27.1 author कूर्मो गार्त्समदो गृत्समदो वा, to आदित्याः, metre त्रिष्टुप्
इमा गिर आदित्येभ्यो घृतस्नूः सनाद्राजभ्यो जुह्वा जुहोमि ।
शृणोतु मित्रो अर्यमा भगो नस्तुविजातो वरुणो दक्षो अंशः ॥२.२७.१॥
इमाः । गिरः । आदित्येभ्यः । घृतऽस्नूः । सनात् । राजभ्यः । जुह्वा । जुहोमि ।
शृणोतु । मित्रः । अर्यमा । भगः । नः । तुविऽजातः । वरुणः । दक्षः । अंशः ॥२.२७.१॥
These songs that drip brilliant ghee, with the sacrificial ladle, I ever offer to the Kings Ādityas.
May Mitra, Aryamā, Bhaga, hear us, the mighty Varuṇa, Dakṣa, and Aṃśa. (RV ii.27.1)

इमाः <f. acc. pl. of pron. इदं – these>, गिरः <f. acc. pl. of गिर् – songs>, आदित्येभ्यः <m. dat. pl. of आदित्य – to Ādityas>, घृतस्नूः <f. acc. pl. of adj. घृतस्नु – dripping ghee>, सनात् <ind. – always, for ever>, राजभ्यः <m. dat. pl. of राजन् – to the kings>, जुह्वा <f. inst. sg. of जुहू – with a tongue of flame, sacrificial ladle (refers to the individual plasma blooms)>, जुहोमि <pre.1st person sg. of √हु – I offer>;

शृणोतु <ipv. 3rd person sg. of √श्रु – let them hear, may they hear>, मित्रः <m. nom. sg. of मित्र – Mitra, the plasma light discharge>, अर्यमा <m. nom. sg. of अर्यमन् – Aryamā, the plasma fire discharge>, भगः <m. nom. sg. of भग – Bhaga, the 'dispenser', gracious lord>, नः <m. acc. pl. of 1st person pron. अस्मद् – us>, तुविजातः <m. nom. sg. of adj. तुविजात – born mighty>, वरुणः <m. nom. sg. of वरुण – Varuṇa>, दक्षः <m. nom. sg. of दक्ष – Dakṣa>, अंशः <m. nom. sg. of अंश – Aṃśa>.

4.4 SAVITĀ

4.4.1 Savitā, Tvaṣṭā, Viśvarūpa

In stanza RV x.10.5, Savitā, the vivifier, and Viśvarūpa, the creator of all forms, are used as epithets of Tvaṣṭā, indicating that Savitā and Viśvarūpa represent different aspects of Tvaṣṭā, the divine architect. The fourteen stanzas of hymn RV x.10 are a dialogue between Yama (यमः Agni, proton) and his twin-sister Yamī (यमी Soma, electron). The author of stanzas RV x.10.1, 3, 5-7, 11, and 13 is Yamī Vaivasvatī (यमी वैवस्वती), 'Yamī, a female child of Vivasvān (विवस्वान् m. nom. sg. of विवस्वत्)'. Vivasvān is "the Brilliant One", the Stone, the celestial ioniser. The rest of the stanzas are authored by Yama Vaivasvata (यमः वैवस्वतः), 'Yama, a male child of Vivasvān'. To read this hymn, we need to understand that, within the atomic structure, Yamas, the protons, are intimately joined with one another within the nucleus of an atom and Yamīs orbit the nucleus as the outer shell of the atomic structure. The stanzas of this hymn explain this relationship between protons and electrons within the atomic structure.

Of the fourteen stanzas of hymn RV x.10, presented here is stanza RV x.10.5 that clarifies the identity of Savitā. The author of the stanza is Yamī Vaivasvatī and it is addressed to Yama. In this stanza, Yama and Yamī are called 'the lord and the lady of the house'. Here, the house refers to the Soma Pond. However, in other context, it can refer to the atomic structure.

RV x.10.5 author यमी वैवस्वती, to यमो वैवस्वतः, metre त्रिष्टुप् छंदः
गर्भे नु नौ जनिता दंपती कर्देवस्त्वष्टा सविता विश्वरूपः ।
नकिरस्य प्र मिनंति व्रतानि वेद नावस्य पृथिवी उत द्यौः ॥१०.१०.५॥
गर्भे । नु । नौ । जनिता । दंपती इति दंऽपती । कः । देवः । त्वष्टा । सविता । विश्वऽरूपः ।
नकिः । अस्य । प्र । मिनंति । व्रतानि । वेद । नौ । अस्य । पृथिवी । उत । द्यौः ॥१०.१०.५॥
In the womb, the Deva Tvaṣṭā, the vivifier and shaper of forms, made us the lord and the lady of the house. No one violates his laws and the Heaven, even the Milky Way, acknowledges us two as his. (RV x.10.5)

गर्भे <m. loc. sg. of गर्भ – in the womb, i.e. in the Soma Pond>, नु <ind. – now, just, then, certainly>, नौ <m. acc. du. of 1st person pron. अस्मद् – us (du.), Yama and Yamī>, जनिता <prf. 3rd person sg. of √जन् – he generated, produced>, दंपती <m. acc. du. of दंपति – two lords, two masters, husband and wife of the house, i.e. Yama (Agni) and Yamī (Soma)>, कः <m. nom. sg. of interrogative pron. क – used as a pronoun referring to Deva Tvaṣṭā>, देवः <m. nom. sg. of देव – Deva>, त्वष्टा <m. nom. sg. of त्वष्टृ – Tvaṣṭā, magnetism in general and magnetic field in particular, the divine architect, carpenter>, सविता <m. nom. sg. of सवितृ – Savitā, the vivifier; the impelling aspect of Tvaṣṭā>, विश्वरूपः <m. nom. sg. of विश्वरूप – Viśvarūpa, a wearer of all forms, a shaper of all forms>;

नकिः <(नकिस्) ind. – no one, nobody>, अस्य <m. gen. sg. of इदं – of this, his>, प्र <ind. prefix to verb मिनंति (√मी)>, प्र-मिनंति <pre. 3rd person pl. of प्र-√मी – they (refers to 'no one') violate, transgress>, व्रतानि <n. acc. pl. of व्रत – laws, commands>, वेद <pre. or prf. 3rd person sg. of √विद् – he (the Heaven) knows, understands>, नौ <m. acc. du. of 1st person pron. अस्मद् – us two>, अस्य <m. gen. sg. of इदं – his>, पृथिवी <f. nom. sg. of पृथिवी – Pṛthivī, the Milky Way>, उत <ind. – even, or>, द्यौः <m. nom. sg. of दिव् – the Heaven>.

4.4.2 By Savitā's Propulsion Celestial Waters Flow

Stanza RV iii.33.6 explains that, when Indra opens a path by slaying Vṛtra who encloses cosmic plasmas within the Soma Pond, Savitā propels the cosmic plasmas towards the six expanses of the Milky Way. The six expanses of the Milky Way refer to the major and minor Arms of the Milky Way. Note that, here, Savitā represents the animating, propelling force of Tvaṣṭā.

RV iii.33.6 author संवादो नदीभिर्विश्वामित्रस्य, to इंद्रः, metre त्रिष्टुप् छंदः
इन्द्रो अस्माँ अरदद्वज्रबाहुरपाहन्वृत्रं परिधिं नदीनां ।
देवोऽनयत्सविता सुपाणिस्तस्य वयं प्रसवे याम उर्वीः ॥३.३३.६॥
इंद्रः । अस्मान् । अरदत् । वज्रबाहुः । अप । अहन् । वृत्रं । परिऽधिं । नदीनां ।
देवः । अनयत् । सविता । सुऽपाणिः । तस्य । वयं । प्रऽसवे । यामः । उर्वीः ॥३.३३.६॥
Indra, thunder-armed, smote down Vṛtra, him who enclosed rivers, and opened a path. Deva Savitā, the beautiful-handed, led us, and by his propulsion, we proceed to the six expanses of the Milky Way. (RV iii.33.6)

इंद्रः <m. nom. sg. of इंद्र – Indra>, अस्मान् <m. acc. pl. of 1st person pron. अस्मद् – us>, अरदत् <ipf. 3rd person sg. of √रद् – he cut, opened a road or path>, वज्रबाहुः <m. nom. sg. of adj. वज्रबाहु – thunder-armed>, अप <ind. prefix to verb अहन् (√हन्)>, अप-अहन् <ipf. 3rd person sg. of अप-√हन् – he beat off, destroyed>, वृत्रं <m. acc. sg. of वृत्र – Vṛtra, the plasma sheath>, परिधिं <m. acc. sg. of परिधि – an enclosure, fence, wall>, नदीनां <f. gen. pl. of नदी – of rivers, of flowing waters (cosmic plasmas)>;

देवः <m. nom. sg. of देव – Deva>, अनयत् <ipf. 3rd person sg. of √नी – he led, guided, directed>, सविता <m. nom. sg. of सवितृ – Savitā, the vivifier>, सुपाणिः <m. nom. sg. of adj. सुपाणि – having beautiful hands, dexterous-handed>, तस्य

<m. gen. sg. of 3rd person pron. तद् – his>, वयं <m. nom. pl. of 1st person pron. अस्मद् – we>, प्रसवे <m. loc. sg. of प्रसव – in/at setting or being set in motion or stimulation, being stimulated (by Savitā), i.e. by (Savitā's) propulsion>, याम: <pre. 1st person pl. of √या – we proceed, advance, travel>, उर्वी: <f. acc. pl. of उर्वी – here in the plural, with and without षष्, it refer to the six expanses of the Milky Way (उर्वी in the singular "the earth", refers to the Milky Way)>.

4.4.3 Savitā Impels Our Hymns

Savitā not only propels cosmic plasmas to flow, but also impels 'thoughts' and 'prayers' of cosmic plasmas. Savitā carries the thoughts and prayers of the Devas, as well as ours. He, Savitā, impels our prayers and hymns so 'we' may attain his splendour. It is what this celebrated gāyatrī (गायत्री) stanza RV iii.62.10 tells us. It is also called sāvitrī (सावित्री) stanza. The stanza is addressed to Savitā.

RV iii.62.10 author विश्वामित्रः, to सविता, metre गायत्री छंदः
तत्सवितुर्वरेण्यं भर्गो देवस्य धीमहि । धियो यो नः प्रचोदयात् ॥३.६२.१०॥
तत् । सवितुः । वरेण्यं । भर्गः । देवस्य । धीमहि । धियः । यः । नः । प्रऽचोदयात् ॥३.६२.१०॥
May we attain that marvellous splendour of Savitā the Deva,
He who shall impel our thoughts and prayers. (RV iii.62.10)

तत् <n. acc. sg. of तद् – that>, सवितुः <m. gen. sg. of सवितृ – of Savitā>, वरेण्यं <n. acc. sg. of adj. वरेण्य – excellent, best>, भर्गः <n. acc. sg. of भर्गस् – radiance, splendour, glory>, देवस्य <m. gen. sg. of देव – of the Deva>, धीमहि <benedictive 1st person pl. of √धा – may we obtain, assume>;

धियः <f. acc. pl. of धी – thoughts, prayers, i.e. sounds or Speeches or hymns>, यः <m. nom. sg. of pron. यद् – he who>, नः <m. gen. pl. of अस्मद् – our>, प्रचोदयात् <sub. cau. 3rd person sg. of प्र-√चुद् – he shall drive on, urge, impel>.

Stanza RV iii.62.10, which addresses an aspect of cosmic plasma phenomenon, can be directly applied to the training of one's mind. There are many who recognise the necessity of training one's body, which is the container of one's soul, but not the necessity of training one's mind, which is "the charioteer" of one's soul.

4.4.4 Three Times a Day Savitā Brings Celestial Waters

Savitā brings wealth and riches to the two celestial Bowls, that is, to the two Soma Ponds, and to the three spheres of the Heaven three times a day every day (RV iii.56.6). The treasures and wealth and riches refer to the celestial Waters, the cosmic plasmas.

RV iii.56.6 author प्रजापतिर्वैश्वामित्रो वाच्यो वा, to विश्वे देवाः, metre त्रिष्टुप् छंदः
त्रिरा दिवः सवितर्वार्याणि दिवेदिव आ सुव त्रिर्नो अह्नः ।
त्रिधातु राय आ सुवा वसूनि भग त्रातर्धिषणे सातये धाः ॥३.५६.६॥
त्रिः । आ । दिवः । सवितः । वार्याणि । दिवेऽदिवे । आ । सुव । त्रिः । नः । अह्नः ।
त्रिऽधातु । रायः । आ । सुव । वसूनि । भग । त्रातः । धिषणे । सातये । धाः ॥३.५६.६॥
Thrice a day, bring thou, O Savitā, from the Heaven, thrice a day, treasures to us day by day.
Bring wealth and riches, O gracious Lord, O Protector, to the two celestial Bowls and to the three spheres of the Heaven, O Bestower, to distribute. (RV iii.56.6)

त्रि: <(त्रिस्) ind. – thrice, three times>, आ <preposition – (with abl.) from>, दिव: <m. abl. sg. of दिव् – from the Heaven>, सवित: <m. voc. sg. of सवितृ – O Savitā>, वार्याणि <n. acc. pl. of वार्य – wealth, treasures, i.e. cosmic plasmas>, दिवेदिवे <n. loc. sg. of दिव दिव – day by day, daily>, आ <ind. prefix to verb सुव (√सू)>, आ-सुव <ipv. 2nd person sg. of आ-√सू 6P – you bring quickly>, त्रि: <(त्रिस्) ind. – thrice, three times>, न: <m. dat. pl. of अस्मद् – to us>, अह्न: <अह्नस् (with त्रि:) – thrice a day>;

त्रिधातु <n. acc. sg. of त्रिधातु – to the threefold Heaven, i.e. to the three spheres of the Heaven>, राय: <m. acc. pl. of रै – wealth, riches>, आ <ind. prefix to verb सुव (√सू)>, आ-सुव <ipv. 2nd person sg. of आ-√सू 6P – you bring quickly>, वसूनि <n. acc. pl. of वसु – wealth, riches>, भग <m. voc. sg. of भग – O Bhaga, O gracious Lord>, त्रात: <m. voc. sg. of त्रातृ – O Protector (applied to Bhaga or Savitā)>, धिषणे <f. acc. du. of धिषणा – to the two celestial bowls, i.e. to two Soma Ponds>, सातये <infinitive of √सन् – to bestow, distribute>, धा: <m. voc. sg. of धा – O Placer, Bestower>.

4.5 VṚTRA, THE SERPENT OF THE HEAVEN, THE PLASMA SHEATH

4.5.1 Dāsa and Ārya, the Two Opposite Charges of the Double Layer

According to stanza RV vi.33.3, Dāsa and Ārya are the two opposite charges that constitute the double layer of Vṛtras, or plasma sheaths. Āryas represent the positive charges carried by Agnis; Dāsas, the negative charges carried by Somas. Indra is implored to break them open with sharp-edged flames. In the stanza, Vṛtra is addressed in the plural as Vṛtras. Vṛtras refer to the plasma sheaths of individual plasma blooms. The stanza is addressed to Indra; the author is Śunahotra (शुनहोत्र:), 'the offerer of auspicious sacrifices'.

RV vi.33.3 author शुनहोत्र:, to इंद्र:, metre त्रिष्टुप् छंद:
त्वं ताँ इंद्रोभयाँ अमित्रान्दासा वृत्राण्यार्या च शूर ।
वधीर्वनेव सुधितेभिरत्कैरा पृत्सु दर्षि नृणां नृतम ॥६.३३.३॥
त्वं । तान् । इंद्र । उभयान् । अमित्रान् । दासा । वृत्राणि । आर्या । च । शूर ।
वधीः । वनाऽइव । सुऽधितेभिः । अत्कैः । आ । पृत्ऽसु । दर्षि । नृणां । नृऽतम ॥६.३३.३॥
O thou Indra, Hero, those Vṛtras of Dāsa and Ārya, the foes of two kinds,
Strikest thou and breakest them open in battle, as cutting down trees, with well-arranged and sharp-edged flames, O most manly of men! (RV vi.33.3)

त्वं <m. voc. sg. of 2nd person pron. युष्मद् – O thou>, तान् <m. acc. pl. of pron. तद् – them, those>, इंद्र <m. voc. sg. of इंद्र – O Indra>, उभयान् <m. acc. pl. of adj. उभय – of both kinds, i.e. of the two opposite charges>, अमित्रान् <m. acc. pl. of अमित्र – enemies, foes>, दासा <m. inst. sg. of दास – with Dāsa, "a knower of the universal spirit", the negative charge carried by Soma>, वृत्राणि <n. acc. pl. of वृत्र – Vṛtras (pl.), plasma sheaths>, आर्या <m. inst. sg. of आर्य – with Ārya, "a honourable man; a master, owner", the positive charge carried by Agni>, च <ind. – and>, शूर <m. voc. sg. of शूर – O hero>;

वधीः <(=बधीः) inj. 2nd person sg. of √वध् – you slay, strike down>, वना <n. acc. pl. of वन – trees>, इव <ind. – like, as if>, सुधितेभिः <m. inst. pl. of adj. सुधित – with well-placed, arranged>, अत्कैः <m. inst. pl. of अत्क (=तेजस्) – with sharp-edged, pointed flames>, आ <ind. prefix to verb दर्षि (√दॄ)>, पृत्सु <f. loc. pl. of पृत् – in battle>, आ-दर्षि <pre. 2nd person

sg. of आ-√दृ – you crush, break open>, नृणां <m. gen. pl. of नृ – of men (the Devas), of heroes>, नृतम <m. voc. sg. of नृतम – O most manly man>.

4.5.2 Let Not My Thread Be Severed Before Time

Vṛtra, the plasma sheath, must be slayed to release cosmic plasmas from the Soma Pond for the sustenance of our galactic system. However, Vṛtra must not be slayed before the weaving of the hymn is accomplished, nor before the preparation of the sacrifice is complete (RV ii.28.5), nor before plasma blooms are fully matured (RV viii.67.20).

In RV ii.28.5, the fountain of sacrifice refers to the Soma Pond equipped with the Stone, the celestial ioniser. The celestial ioniser that continuously produces and discharges cosmic plasmas is called the fountain of sacrifice. The stanza is addressed to Varuṇa.

RV ii.28.5 author कूर्मो गार्त्समदो गृत्समदो वा, to वरुणः, metre त्रिष्टुप् छंदः
वि मच्छ्रथाय रशनामिवाग ऋध्याम ते वरुण खामृतस्य ।
मा तंतुश्छेदि वयतो धियं मे मा मात्रा शार्यपसः पुर ऋतोः ॥२.२८.५॥
वि । मत् । श्रथय । रशनांऽइव । आगः । ऋध्याम । ते । वरुण । खां । ऋतस्य ।
मा । तंतुः ।छेदि । वयतः । धियं । मे । मा । मात्रा । शारि । अपसः । पुरा । ऋतोः ॥२.२८.५॥
Unbind thou the sin from me as though unbinding a rope. O Varuṇa, let us make thy fountain of sacrifice swell. Let not my thread be broken from weaving the hymn, nor the measure of my sacrificial work be shattered before time. (RV ii.28.5)

वि <ind. prefix to verb श्रथय (√श्रथ्)>, मत् <m. abl. sg. of 1st person pron. अस्मद् – from me>, वि-श्रथय <ipv. cau. 2nd person sg. of वि-√श्रथ् – you loosen, unbind, untie>, रशनां <f. acc. sg. of रशना – a cord, strap, rope>, इव <ind. – like, as>, आगः <n. acc. sg. of आगस् – offense, sin (the ionic bond)>, ऋध्याम <ipv. or opt. 1st person pl. of √ऋध् – let us make (acc.) grow, increase>, ते <m. gen. sg. of 2nd person pron. युष्मद् – your>, वरुण <m. voc. sg. of वरुण – O Varuṇa>, खां <f. acc. sg. of खा – a well, fountain>, ऋतस्य <m. gen. sg. of ऋत – of sacrifice>;

मा <ind. – not>, तंतुः <m. nom. sg. of तंतु – a thread, cord, filament (of plasma blooms)>, छेदि <inj. passive 3rd person sg. of √छिद् – it (the thread, cord) be cut off, split>, वयतः <mn. abl. sg. of वयत् – from weaving>, धियं <f. acc. sg. of धी – prayer, hymn>, मे <m. gen. sg. of अस्मद् – my>, मा <ind. – not>, मात्रा <f. nom. sg. of मात्रा – the measure or course of action>, शारि <inj. passive 3rd person sg. of √शॄ – it was crushed, broken>, अपसः <n. gen. sg. of अपस् – of sacred work, of sacrificial work>, पुरा <ind. – before>, ऋतोः <m. gen. sg. of ऋतु – of right time, season, time appointed for any action (esp. sacrifices and worship)>.

In stanza RV viii.67.20, Vivasvān refers to the Stone or the fire of the Stone. Vivasvān is "the Brilliant one". In the stanza, plasma blooms, or Soma plants, are compared to Vivasvān's missile or arrow, which is used by Indra as the weapon for slaying Vṛtra. The stanza is addressed to the Ādityas. The authors are Matsya Sāṃmada (मत्स्यः सांमदः), 'King of fish'; Mānya Maitrāvaruṇi (मान्यः मैत्रावरुणिः), 'venerable Maitrāvaruṇi'; and Bahava Matsyā Jālanaddhā (बहवः मत्स्याः जालनद्धाः), 'many fish caught in the net'. The authors' names are the descriptions of blooming cosmic plasmas within the Soma Pond. The plasma bloom swirling in the Ocean of celestial Waters is compared to the King of fish. Note, Christ, the anointed one, is identified in the early Christian church with the Greek word for fish (ἰχθύς ichthýs). Charged embryonic particles, especially Agnis, attracted to and blooming on the surface of a triply-spun filament of Soma plant, are 'fish caught in the net'. They do not want to be destroyed until they are fully-grown and matured.

RV viii.67.20 author मत्स्यः सांमदो मान्यो वा मैत्रावरुणिर्बहवो वा मत्स्या जालनद्धाः, to आदित्याः, metre गायत्री छंदः
मा नो हेतिर्विवस्वत आदित्याः कृत्रिमा शरुः । पुरा नु जरसो वधीत् ॥८.६७.२०॥
मा । नः । हेतिः । विवस्वतः । आदित्याः । कृत्रिमा । शरुः । पुरा । नु । जरसः । वधीत् ॥८.६७.२०॥
Let not Vivasvān's missile, nor the arrow, O Ādityas, wrought with skill,
Slay us before we are fully-grown and matured. (RV viii.67.20)

मा <ind. particle of negation – not>, नः <m. acc. pl. of 1st person pron. अस्मद् – us>, हेतिः <f. nom. sg. of हेति – a missile weapon>, विवस्वतः <m. gen. sg. of विवस्वत् – of Vivasvān (विवस्वान्), of the brilliant one, the celestial ioniser>, आदित्याः <m. voc. pl. of आदित्य – O Ādityas (pl.), O sons of Aditi>, कृत्रिमा <f. nom. sg. of adj. कृत्रिम – made artificially, i.e. engineered, made with skill (cf. कृत्याकृत्)>, शरुः <f. nom. sg. of शरु – a missile, arrow>;

पुरा <ind. – before>, नु <ind. – indeed, surely>, जरसः <m. acc. pl. of adj. जर – becoming old, i.e. fully-grown, matured>, वधीत् <inj. 3rd person sg. of √वध् – let (Vivasvān) destroy, slay>.

4.6 VARUṆA'S NOOSE, THE SIN OF THE IONIC BOND

4.6.1 The Bonds Not Wrought of Ropes

In RV vii.84.2, the terms 'sheath (हेळः heḷaḥ)' and 'bonds (सेतारः setāraḥ)' are used for the sheath of Varuṇa, the plasma sheath. In RV i.24.14-15, the terms 'sheath (हेळः heḷaḥ)', 'noose (पाशः pāśaḥ)', and 'sins (एनांसि enaṃsi)' are used for the ionic bond of the atomic structure. In general, when Varuṇa is implored to release 'them' from the noose, sheath, or sin, it indicates that the ionic bond is involved. When Indra is implored to remove the bond or sheath of Varuṇa, it indicates that Vṛtra, the plasma sheath, is involved.

RV vii.84.2 tells us that the bonds Varuṇa uses to bind men, Agnis (protons) and Somas (electrons), are not constructed with actual cords. In the stanza, Indra is implored to remove the sheath of Varuṇa. The lofty kingdom of Varuṇa refers to the Soma Pond. Note, the stanza, which describes cosmic plasma phenomena, can be interpreted in the context of spiritual discipline.

RV vii.84.2 author वसिष्ठः, to इंद्रावरुणौ, metre त्रिष्टुप् छंदः
युवो राष्ट्रं बृहदिन्वति द्यौर्यौ सेतृभिररज्जुभिः सिनीथः ।
परि नो हेळो वरुणस्य वृज्या उरुं न इंद्रः कृणवदु लोकं ॥७.८४.२॥
युवोः । राष्ट्रं । बृहत् । इन्वति । द्यौः । यौ । सेतृऽभिः । अरज्जुऽभिः । सिनीथः ।
परि । नः । हेळः । वरुणस्य । वृज्याः । उरुं । नः । इंद्रः । कृणवत् । ऊं इति । लोकं ॥७.८४.२॥
The Heaven invigorates the lofty kingdom of yours, ye who bind with bonds not wrought of ropes. May Indra break the sheath of Varuṇa for our sake and grant us great freedom. (RV vii.84.2)

युवोः <(=युवयोः) m. gen. du. of 2nd person pron. युष्मद् – of you both, i.e. of Indra-Varuṇa>, राष्ट्रं <n. acc. sg. of राष्ट्र – a kingdom, empire>, बृहत् <n. acc. sg. of adj. बृहत् – lofty, tall, great, mighty>, इन्वति <pre. 3rd person sg. of √इन्व् – it/he (the Heaven) infuses strength, invigorates>, द्यौः <m. nom. sg. of दिव् – the Heaven>, यौ <m. nom. du. of relative pron. यद् – ye (du.) who>, सेतृभिः <m. inst. pl. of सेतृ – with bonds>, अरज्जुभिः <m. inst. pl. of adj. अरज्जु – not consisting of ropes or cords>, सिनीथः <pre. 2nd person du. of √सि – ye (du.) bind, tie, fetter>;

परि <ind. – on account of, for the sake of>, नः <m. acc. pl. of 1st person pron. अस्मद् – us>, हेळः <(=हेडः) n. acc. sg. of हेडस् (=हेड) – (from √हेड् to surround, clothe) cover, sheath (here, it refers to the plasma sheath)>, वरुणस्य <m. gen. sg. of वरुण – of Varuṇa>, वृज्याः <opt. 2nd person sg. of √वृज् – may you wring off, break>, उरुं <m. acc. sg. of adj. उरु – broad, excellent, great>, नः <m. acc. pl. of 1st person pron. अस्मद् – us>, इंद्रः <m. nom. sg. of इंद्र – Indra>, कृणवत् <sub. 3rd person sg. of √कृ – (with लोकं acc.) may he make room or grant freedom>, उ <ind. particle – and, also>, लोकं <m. acc. sg. of लोक – (with उरु) free or open space, free motion>.

4.6.2 Remove the Noose from Us O Varuṇa

In stanza RV i.24.14, Varuṇa is implored to take his sheath away and untie the sin of the ionic bond. In stanza RV i.24.15, Varuṇa is implored to untie the noose, the noose placed above, the noose placed middlemost, and the noose placed under. Likely the noose placed above refers to the outer shell formed by Somas orbiting the nucleus. The noose placed middlemost refers to the bond between the nucleus and the outer shell of Somas. The noose placed under refers to the ionic bond amongst the Agnis within the nucleus.

RV i.24.14 author शुनःशेपः आजीगर्तिः (कृत्रिमो वैश्वामित्रो देवरातः), to वरुणः, metre त्रिष्टुप् छंदः
अव ते हेळो वरुण नमोभिरव यज्ञेभिरीमहे हविर्भिः ।
क्षयन्नस्मभ्यमसुर प्रचेता राजन्नेनांसि शिश्रथः कृतानि ॥१.२४.१४॥
अव । ते । हेळः । वरुण । नमःऽभिः । अव । यज्ञेभिः । ईमहे । हविःऽभिः ।
क्षयन् । अस्मभ्यं । असुर । प्रचेत इति प्रऽचेतः । राजन् । एनांसि । शिश्रथः । कृतानि ॥१.२४.१४॥
With salutations, with Yajñas, with oblations, O Varuṇa, we implore you to drive your sheath away. O wise Asura, King, Ruler, for us, untie thou the sins committed. (RV i.24.14)

अव <ind. – off, away>, ते <m. gen. sg. of युष्मद् – your>, हेळः <(=हेडः) n. acc. sg. of हेडस् (=हेड) – (from √हेड् to surround, clothe) a cover, cloth (here it refers to the ionic bond)>, वरुण <m. voc. sg. of वरुण – O Varuṇa>, नमोभिः <m. inst. pl. of नमस् – with salutations>, अव <ipv. 2nd person sg. of √अव् – you drive, impel>, यज्ञेभिः <m. inst. pl. of यज्ञ – with Yajñas, sacrifices>, ईमहे <pre. middle 1st person pl. of √इ 2Ā – we ask, request>, हविर्भिः <n. inst. pl. of हविस् – with oblations>;

क्षयन् <m. voc. sg. of pre. active pt. क्षयत् (√क्षि 1P) – he ruling, O ruler>, अस्मभ्यं <m. dat. pl. of 1st person pron. अस्मद् – for us>, असुर <m. voc. sg. of असुर – O Asura (Varuṇa)>, प्रचेतः <m. voc. sg. of adj. प्रचेतस् – wise>, राजन् <m. voc. sg. of राजन् – O King>, एनांसि <n. acc. pl. of एनस् – offences, sins>, शिश्रथः <ipv. 2nd person sg. of √श्रथ् – you loosen, untie>, कृतानि <n. acc. pl. of adj. कृत – done, made, committed (by us)>.

RV i.24.15 author शुनःशेपः आजीगर्तिः (कृत्रिमो वैश्वामित्रो देवरातः), to वरुणः, metre त्रिष्टुप् छंदः
उदुत्तमं वरुण पाशमस्मदवाधमं वि मध्यमं श्रथाय ।
अथा वयमादित्य व्रते तवानागसो अदितये स्याम ॥१.२४.१५॥
उत् । उत्ऽतमं । वरुण । पाशं । अस्मत् । अव । अधमं । वि । मध्यमं । श्रथय ।
अथ । वयं । आदित्य । व्रते । तव । अनागसः । अदितये । स्याम ॥१.२४.१५॥

Untie thou the noose from us, O Varuṇa, the noose placed above, middlemost, and under. Now, in thy dominions, O Āditya, may we be sinless and belong to Aditi. (RV i.24.15)

उत् <उद् ind. particle – upon, on>, उत्तमं <m. acc. sg. of adj. उत्तम – upper, outermost, above>, वरुण <m. voc. sg. of वरुण – O Varuṇa>, पाशं <m. acc. sg. of पाश – the noose (the ionic bond)>, अस्मत् <m. abl. pl. of 1st person pron. अस्मद् – from us>, अव <ind. prefix to verb श्रथय (√श्रथ्)>, अधमं <m. acc. sg. of adj. अधम – below, under>, वि <ind. – apart, asunder, away, off>, मध्यमं <m. acc. sg. of adj. मध्यम – in the middle, middlemost>, अव-श्रथय <ipv. cau. 2nd person sg. of अव-√श्रथ् – you loosen, untie>;

अथ <ind. particle – now>, वयं <m. nom. pl. of 1st person pron. अस्मद् – we>, आदित्य <m. voc. sg. of आदित्य – O Āditya>, व्रते <n. loc. sg. of व्रत – in the dominions, realm (Āditya's dominion is the Soma Pond)>, तव <m. gen. sg. of 2nd person pron. युष्मद् – your>, अनागसः <m. nom. pl. of adj. अनागस् – sinless, i.e. freed from the ionic bond>, अदितये <f. dat. sg. of अदिति – to Aditi>, स्याम <opt. 1st person pl. of √अस् 2P – may we belong to (with dat.)>.

4.7 VĀK, THE SPEECH OF COSMIC PLASMAS

4.7.1 Vāk, the Queen of the Devas

Vāk is the Queen (राष्ट्री) of the Devas. As noted earlier, the Devas utter their own unique Speeches. Vāk is called the milch Cow who yields invigorating libation. Vāk, humming pleasantly, yields cosmic plasmas for the four Arms of the Milky Way (RV viii.100.10-11). The two stanzas presented here indicate that Speeches or sounds break the ionic bonds and the plasma sheaths in the process of cosmic ionisation and cosmic plasma discharge and release. Vāk is the sound, the Speech, of cosmic plasmas personified and deified.

RV viii.100.10 author नेमो भर्गभः, to वाक्, metre त्रिष्टुप् छंदः
यद्वाग्वदंत्यविचेतनानि राष्ट्री देवानां निषसाद मंद्रा ।
चतस्र ऊर्जं दुदुहे पयांसि क्व स्विदस्याः परमं जगाम ॥८.१००.१०॥
यत् । वाक् । वदंती । अविऽचेतनानि । राष्ट्री । देवानां । निऽससाद । मंद्रा ।
चतस्रः । ऊर्जं । दुदुहे । पयांसि । क्व । स्वित् । अस्याः । परमं । जगाम ॥८.१००.१०॥
When they utter intelligent words, Vāk, Queen of the Devas, humming pleasantly, sits down,
And yields invigorating food to the four Arms of the Milky Way. Where do you think is the furthest distance she has travelled? (RV viii.100.10)

यत् <conjunction – when, that>, वाक् <f. nom. sg. of वाच् – Vāk, sound, Speech>, वदंति <pre. 3rd person. pl. of √वद् – they (the Devas) speak, utter>, अविचेतनानि <n. acc. pl. of अविचेतन – (अवि adj. favourable, kindly disposed + चेतन adj. excellent, intelligent) kind and intelligent (words, speeches)>, राष्ट्री <f. nom. sg. of राष्ट्री – a female sovereign, queen>, देवानां <m. gen. pl. of देव – of the Devas>, निषसाद <prf. 3rd person sg. of √निषद् – she sits down>, मंद्रा <f. nom. sg. of adj. मंद्र – charming, (esp.) sounding or speaking pleasantly>;

चतस्रः <f. acc. pl. of चतुर् – the four, i.e. the four spiral Arms of the Milky Way (here प्रदिशः, pradiśaḥ f. acc. pl., is omitted)>, ऊर्जं <f. acc. sg. of ऊर्ज् – vigour, sap, juice, refreshment, food (refers to Soma or cosmic plasmas)>, दुदुहे <prf. middle 3rd person sg. of √दुह् – she (Vāk) yields, gives (acc.)>, क्व <ind. – where? in what place? (connected with the particle स्विद्)>, स्वित् <(स्विद्) ind. particle of enquiry – "do you think?" (क्व स्विद्, where do you think?)>, परमं <mn. acc. sg. of परम – to remotest, most distant>, जगाम <prf. 3rd person sg. of √गं – she (Vāk) has gone to or travelled (see RV x.125.8 for the answer); note that Vāk, Speech, always travels along with cosmic plasmas>.

The Devī Vāk is generated by the Devas, yet Vāk is the milch Cow who produces the Devas. This kind of reciprocal relationship, once a force is produced, then becomes the producer in turn, is a common magneto-plasma phenomenon described in the Ṛgveda, as we have seen in section 4.3.4.

RV viii.100.11 author नेमो भर्गभः, to वाक्, metre त्रिष्टुप् छंदः
देवीं वाचमजनयंत देवास्तां विश्वरूपाः पशवो वदंति ।
सा नो मंद्रेषमूर्जं दुहाना धेनुर्वागस्मानुप सुष्टुतैतु ॥८.१००.११॥
देवीं । वाचं । अजनयंत । देवाः । तां । विश्वऽरूपाः । पशवः । वदंति ।
सा । नः । मंद्र । इषं । ऊर्जं । दुहाना । धेनुः । वाक् । अस्मान् । उप । सुऽस्तुता । आ । एतु ॥८.१००.११॥
The Devas generated the Devī Vāk and the herds of cattle in varied forms speak Vāk.
She, Vāk the milch Cow, well-lauded, sitting beside us, humming pleasantly, yields invigorating libation for us. (RV viii.100.11)

देवीं <f. acc. sg. of देवी – Devī>, वाचं <f. acc. sg. of वाच् – Vāk>, अजनयंत <ipf. cau. middle 3rd person pl. of जन् – they (the Devas) generated, produced>, देवाः <m. nom. pl. of देव – the Devas>, तां <f. acc. sg. of तद् – her, i.e. Vāk, Speech>, विश्वरूपाः <m. nom. pl. of adj. विश्वरूप – wearing varied forms, manyfold>, पशवः <m. nom. pl. of पशु – the herds of cattle, i.e. celestial lights, cosmic plasmas>, वदंति <pre. 3rd person. pl. of √वद् – they speak, utter>;

सा <f. nom. sg. of 3rd person pron. तद् – she>, नः <m. dat. pl. of 1st person pron. अस्मद् – to/for us>, मंद्रा <f. nom. sg. of adj. मंद्र – humming or speaking pleasantly>, इषं <f. acc. sg. of इष् – a draught, libation, refreshment>, ऊर्जं <f. acc. sg. of adj. ऊर्ज् – invigorating, refreshing>, दुहाना <f. nom. sg. of pre. middle pt. दुहाना (√दुह्) – she yields, gives>, धेनुः <f. nom. sg. of धेनु – the milch Cow>, वाक् <f. nom. sg. of वाच् – Vāk>, अस्मान् <m. acc. pl. of 1st person pron. अस्मद् – us>, उप <ind. – near, by the side of>, सुस्तुता <f. nom. sg. of adj. सुस्तुत – well-praised, celebrated>.

4.7.2 A Hymn of Vāk

The eight stanzas of RV x.125, are the self-eulogy of Vāk. The author is Vāgāṃbhṛṇī (वागांभृणी = वाक् Vāk + आंभृणी Āṃbhṛṇī), and the hymn is addressed to herself. Āṃbhṛṇī, Vāk, is a daughter of Aṃbhṛṇa (आंभृणः). Aṃbhṛṇa, the father of Vāk, is "powerful" and "roaring terribly".

Let's hear the hymn of Vāk. Note Vāk is the Speeches, or the sounds, or the oscillations generated by cosmic plasmas.

RV x.125.1 author वागांभृणी (वाक् आम्भृणी), to वागांभृणी, metre त्रिष्टुप् छंदः
अहं रुद्रेभिर्वसुभिश्चराम्यहमादित्यैरुत विश्वदेवैः ।
अहं मित्रावरुणोभा बिभर्म्यहमिन्द्राग्नी अहमश्विनोभा ॥१०.१२५.१॥
अहं । रुद्रेभिः । वसुऽभिः । चरामि । अहं । आदित्यैः । उत । विश्वऽदेवैः ।
अहं । मित्रावरुणा । उभा । बिभर्मि । अहं । इंद्राग्नी इति । अहं । अश्विना । उभा ॥१०.१२५.१॥
I travel with the Rudras and the Vasus, also with the Ādityas and all the Devas.
I hold both Mitra and Varuṇa, I sustain Indra and Agni, and both of the twins Aśvinā.
(RV x.125.1)

अहं <f. nom. sg. of 1st person pron. अस्मद् – I, Vāk>, रुद्रेभिः <m. inst. pl. of रुद्र – with the Rudras, i.e. with the Maruts>, वसुभिः <m. inst. pl. of वसु – with the Vasus (the bright ones)>, चरामि <pre. 1st person sg. of √चर् – I walk,

move, travel>, अहं <mf. nom. sg. of 1st person pron. अस्मद् – I, Vāk>, आदित्यैः <m. inst. pl. of आदित्य – with Ādityas>, उत <ind. – and, also, even>, विश्वदेवैः <m. inst. pl. of विश्वदेव – with all the Devas>;

अहं <mf. nom. sg. of 1st person pron. अस्मद् – I, Vāk>, मित्रावरुणा <m. acc. du. of मित्रावरुण – Mitra and Varuṇa>, उभा <m. acc. du. of pron. उभ – both>, बिभर्मि <pre. 1st person sg. of √भृ – I bear, foster, support, maintain>, अहं <mf. nom. sg. of 1st person pron. अस्मद् – I, Vāk>, इंद्राग्नी <m. acc. du. of इंद्राग्नि – Indra and Agni>, अहं <mf. nom. sg. of 1st person pron. अस्मद् – I, Vāk>, अश्विना <m. acc. du. of अश्विन् – the twins Aśvinā>, उभा <m. acc. du. of adj. उभ – both>.

RV x.125.2 author वागांभृणी (वाक् आम्भृणी), to वागांभृणी, metre जगती छंदः
अहं सोममाहनसं बिभर्म्यहं त्वष्टारमुत पूषणं भगं ।
अहं दधामि द्रविणं हविष्मते सुप्राव्ये३ यजमानाय सुन्वते ॥१०.१२५.२॥
अहं । सोमं । आहनसं । बिभर्मि । अहं । त्वष्टारं । उत । पूषणं । भगं ।
अहं । दधामि । द्रविणं । हविष्मते । सुप्रऽअव्ये । यजमानाय । सुन्वते ॥१०.१२५.२॥
I carry pressed Soma. I bear Tvaṣṭā, even Pūṣa and Bhaga.
I bestow riches to the zealous Soma sacrificer who worships and offers an oblation.
(RV x.125.2)

अहं <mf. nom. sg. of 1st person pron. अस्मद् – I, Vāk>, सोमं <m. acc. sg. of सोम – Soma>, आहनसं <m. acc. sg of adj. आहनस् – pressed out (as Soma)>, बिभर्मि <pre. 1st person sg. of √भृ – I bear, carry>, अहं <mf. nom. sg. of 1st person pron. अस्मद् – I, Vāk>, त्वष्टारं <m. acc. sg. of त्वष्टृ – Tvaṣṭā>, उत <ind. – and, also, even>, पूषणं <m. acc. sg. of पूषन् – Pūṣa>, भगं <m. acc. sg. of भग – Bhaga>;

अहं <mf. nom. sg. of 1st person pron. अस्मद् – I, Vāk>, दधामि <pre. 1st person sg. of √धा – bestow, present>, द्रविणं <n. acc. sg. of द्रविण – wealth, i.e. cosmic plasmas in all forms>, हविष्मते <m. dat. sg. of adj. हविष्मत् – offering an oblation>, सुप्राव्ये <m. dat. sg. of adj. सुप्रावी – attentive, zealous>, यजमानाय <m. dat. sg. of adj. यजमान – sacrificing, worshiping>, सुन्वते <m. dat. sg. of सुन्वत् – to the offerer of a Soma sacrifice>.

RV x.125.3 author वागांभृणी (वाक् आम्भृणी), to वागांभृणी, metre त्रिष्टुप् छंदः
अहं राष्ट्री संगमनी वसूनां चिकितुषी प्रथमा यज्ञियानां ।
तां मा देवा व्यदधुः पुरुत्रा भूरिस्थात्रां भूर्यावेशयंतीं ॥१०.१२५.३॥
अहं । राष्ट्री । संऽगमनी । वसूनां । चिकितुषी । प्रथमा । यज्ञियानां ।
तां । मा । देवाः । वि । अदधुः । पुरुऽत्रा । भूरिऽस्थात्रां । भूरि । आऽवेशयंतीं ॥१०.१२५.३॥
I am the Queen, the gatherer of riches, observant, the foremost among those who are worthy of worship. Devas placed me everywhere and furnished me with many abodes to abide in.
(RV x.125.3)

अहं <mf. nom. sg. of 1st person pron. अस्मद् – I, Vāk>, राष्ट्री <f. nom. sg. of राष्ट्री – a female sovereign, queen>, संगमनी <f. nom. sg. of संगमन – a gatherer>, वसूनां <n. gen. pl. of वसु – of riches, wealths>, चिकितुषी <f. nom. sg. of prf. active pt. चिकितुषी (√चित् 1P) – she being attentive, observant>, प्रथमा <f. nom. sg. of adj. प्रथम – the first, foremost>, यज्ञियानां <f. gen. pl. of यज्ञिय – of those who are worthy of worship>;

तां <f. acc. sg. of 3rd person pron. तद् – her, that very person or self (used to emphasise मा)>, मा <f. acc. sg. of 1st person pron. अस्मद् – me>, देवाः <m. nom. pl. of देव – the Devas>, वि <ind. prefix to verb अदधुः (√धा)>, वि-अदधुः <ipf. 3rd person pl. of वि-√धा – they furnished, established>, पुरुत्रा <ind. – variously, in many ways>, भूरिस्थात्रां <f. acc. sg. of adj. भूरिस्थात्र – being at many places>, भूरि <(=भूरि स्थानं) n. acc. sg. of भूरि स्थान – many dwellings, places, abodes>, आवेशयंतीं <f. acc. sg. of pre. cau. active pt. आवेशयंती (आ-√विश्) – her to enter or abide in (acc.)>.

RV x.125.4 author वागांभृणी (वाक् आम्भृणी), to वागांभृणी, metre त्रिष्टुप् छंदः
मया सो अन्नमत्ति यो विपश्यति यः प्राणिति य ईं शृणोत्युक्तं ।

अमंतवो मां त उप क्षियंति श्रुधि श्रुत श्रद्धिवं ते वदामि ॥१०.१२५.४॥
मया । सः । अन्नं । अत्ति । यः । विऽपश्यति । यः । प्राणिति । यः । ईं । शृणोति । उक्तं ।
अमंतवः । मां । ते । उप । क्षियंति । श्रुधि । श्रुत । श्रद्धिऽवं । ते । वदामि ॥१०.१२५.४॥
Through me he, who observes, breathes, hears what is spoken, through me alone, eats the food.
They who know it not, yet dwell beside me, hear, one and all, the truth I speak to you.
(RV x.125.4)

मया < f. inst. sg. of 1st person pron. अस्मद् – through me>, सः <m. pron. sg. of 3rd person pron. तद् – he (the Deva)>, अन्नं <n. acc. sg. of अन्न – "food in a mystical sense" (in the Ṛgveda Soma is the food)>, अत्ति <pre. 3rd person sg. of √अद् – he eats>, यः <m. nom. sg. of pron. यद् – he who>, विपश्यति <pre. 3rd person sg. of वि-√दृश् or वि-√पश् – he observes>, यः <m. nom. sg. of pron. यद् – he who>, प्राणिति <pre. 3rd person sg. of √प्राण् or √प्रान् (प्र-अन्) – he breathes, blows>, यः <m. nom. sg. of pron. यद् – he who>, ईं <ind. – a particle of affirmation use with यः (यद्)>, शृणोति <pre. 3rd person sg. of श्रु – he who hears (speech, sound)>, उक्तं <n. acc. sg. of उक्त – the word spoken, uttered>;

अमंतवः <mf. nom. pl. of adj. अमंतु (अमन्तु) – they who are ignorant>, मां <mf. acc. sg. of 1st person pron. अस्मद् – me>, ते <m. nom. pl. of 3rd person pron. तद् – they>, उप <ind. – near, by the side of>, क्षियंति <pre. 3rd person pl. of √क्षि – they who stay, dwell, reside>, श्रुधि <ipv. 2nd person sg. of √श्रु – you hear>, श्रुत <ipv. 2nd person pl. of √श्रु – you all hear>, श्रद्धिवं <n. acc. sg. of श्रद्धिव – the truth>, ते <m. dat. sg of युष्मद् – to you>, वदामि <pre. 1st person sg. of √वद् – I speak>.

RV x.125.5 author वागांभृणी (वाक् आम्भृणी), to वागांभृणी, metre त्रिष्टुप् छंदः
अहमेव स्वयमिदं वदामि जुष्टं देवेभिरुत मानुषेभिः ।
यं कामये तंतमुग्रं कृणोमि तं ब्रह्माणं तमृषिं तं सुमेधां ॥१०.१२५.५॥
अहं । एव । स्वयं । इदं । वदामि । जुष्टं । देवेभिः । उत । मानुषेभिः ।
यं । कामये । तंऽतं । उग्रं । कृणोमि । तं । ब्रह्माणं । तं । ऋषिं । तं ।सुऽमेधां ॥१०.१२५.५॥
I, indeed, myself utter the Speech welcomed by both the Devas and men alike.
This or that impetuous one whom I value, I make this one a Brahmā, that one a Ṛṣi, and the other a great Intelligence. (RV x.125.5)

अहं <mf. nom. sg. of 1st person pron. अस्मद् – I>, एव <ind. – indeed, truly>, स्वयं <ind. – myself>, इदं <n. acc. sg. of इदं – this (refers to Speech, sound)>, वदामि <pre. 1st person sg. of √वद् – I speak, utter>, जुष्टं <n. acc. sg. of adj. जुष्ट – liked, loved, welcome>, देवेभिः <m. inst. pl. of देव – by the Devas>, उत <ind. – and, also>, मानुषेभिः <m. inst. pl. of मानुष – by mānuṣas, by men (Agnis)>;

यं <m. acc. sg. of pron. यद् – whom, which>, कामये <pre. middle 1st person sg. of √कं 1Ā – I wish or value>, तंतं <m. acc. sg. of तद् तद् – this or that, various, different>, उग्रं <m. acc. sg. of उग्र – powerful, impetuous (oscillating, spinning rapidly)>, कृणोमि <pre. 1st person sg. of √कृ – I make>, तं <m. acc. sg. of तद् – this man>, ब्रह्माणं <m. acc. sg. of ब्रह्मन् – Brahmā (ब्रह्मा m. nom. sg. of ब्रह्मन्)>, तं <m. acc. sg. of तद् – that man>, ऋषिं <m. acc. sg. of ऋषि – Ṛṣi>, तं <m. acc. sg. of तद् – the other man>, सुमेधां <f. acc. sg. of सु-मेधा (cf. मेधा) – great or virtuous force, intelligence personified>.

RV x.125.6 author वागांभृणी (वाक् आम्भृणी), to वागांभृणी, metre त्रिष्टुप् छंदः
अहं रुद्राय धनुरा तनोमि ब्रह्मद्विषे शरवे हंतवा उ ।
अहं जनाय समदं कृणोम्यहं द्यावापृथिवी आ विवेश ॥१०.१२५.६॥
अहं । रुद्राय । धनुः । आ । तनोमि । ब्रह्मऽद्विषे । शरवे । हंतवै । ऊं इति ।
अहं । जनाय । सऽमदं । कृणोमि । अहं । द्यावापृथिवी इति । आ । विवेश ॥१०.१२५.६॥
I am the bow for Rudra. I stretch so that his arrow may strike and slay the wicked.
I undertake battle for the people, and I settle in the twofold Heaven and the Milky Way.
(RV x.125.6)

अहं <f. nom. sg. of 1st person pron. अस्मद् – I, Vāk>, रुद्राय <m. dat. sg. of रुद्र – for Rudra>, धनुः <m. nom. sg. of धनु – a bow>, आ <ind. prefix to verb तनोमि (√तन्)>, आ-तनोमि <pre. 1st person sg. of आ-√तन् – I extend or stretch>, ब्रह्मद्विषे <m. dat. sg. of ब्रह्मद्विष् – the impious, the wicked, i.e. Vṛtra>, शरवे <m. dat. sg. of शरु – for a missile, for an arrow>, हंतवै <infinitive of √हं (√हन्) – to strike, to slay (acc. dat.)>, उ <ind. – and, also, further>;

अहं <f. nom. sg. of 1st person pron. अस्मद् – I Vāk>, जनाय <m. dat. sg. of जन – for man, person, race, people (Agni and other Devas)>, समदं <f. acc. sg. of समद् – strife, battle>, कृणोमि <pre. 1st person sg. of √कृ – I cause or order, undertake (battle)>, अहं <f. nom. sg. of 1st person pron. अस्मद् – I Vāk>, द्यावापृथिवी <f. acc. dual compound of द्यावा and पृथिवी – Dyāvāpṛthivī, Twofold Heaven and Pṛthivī, the Milky Way>, आ <ind. prefix to verb विवेश (√विश्)>, आ-विवेश <prf. 1st person sg. of आ-√विश् – I enter, settle down>.

RV x.125.7 author वागांभृणी (वाक् आम्भृणी), to वागांभृणी, metre त्रिष्टुप् छंदः
अहं सुवे पितरमस्य मूर्धन्मम योनिरप्स्वं१◌◌तः समुद्रे ।
ततो वि तिष्ठे भुवनानु विश्वोतामूं द्यां वर्ष्मणोप स्पृशामि ॥१०.१२५.७॥
अहं । सुवे । पितरं । अस्य । मूर्धन् । मम । योनिः । अप्ऽसु । अंतरिति । समुद्रे ।
ततः । वि । तिष्ठे । भुवना । अनु । विश्वा । उत । अमूं । द्यां । वर्ष्मणा । उप । स्पृशामि ॥१०.१२५.७॥
In the beginning, I bring forth Father of this World. My seat is midst the Waters in the celestial Ocean. Thence, I dwell in all beings, and I thereupon reach up to the Heaven in auspicious form. (RV x.125.7)

अहं <f. nom. sg. of 1st person pron. अस्मद् – I Vāk>, सुवे <pre. middle 1st person sg. of √सू 2Ā – I bring forth, produce>, पितरं <m. acc. sg. of पितृ – Father (refers to the Heaven, the Heavenly Vault)>, अस्य <m. gen. sg of इदं – of this World (the galactic system, the Milky Way)>, मूर्धन् <(=मूर्धनि) m. loc. sg. of मूर्धन् – at first, in the beginning>, मम <f. gen. sg of 1st person sg. of अस्मद् – my>, योनिः <m. nom. sg. of योनि – the abode, seat>, अप्सु <f. loc. pl. of अप् – among celestial Waters, i.e. among cosmic plasmas>, अंतः <(अन्तर्) ind. – within, midst>, समुद्रे <m. loc. sg. of समुद्र – in the Samudra, i.e. in the Ocean of cosmic plasmas, the Soma Pond>;

ततः <(ततस्) ind. – thence, from there, there>, वि <preposition (with acc.) – through or between>, तिष्ठे <pre. middle 1st person sg. of √स्था – I stand by, abide, adhere to, stay at>, भुवना <n. acc. pl. of भुवन – all beings (here, refer to cosmic plasmas and all the associated phenomena)>, अनु <ind. – afterwards, thereupon>, विश्वा <n. acc. pl. of adj. विश्व – all>, उत <ind. – and, even>, अमूं <f. acc. sg. of pron. अदस् – that>, द्यां <f. acc. sg. of दिव् – the Heaven>, वर्ष्मणा <mn. inst. sg. of वर्ष्मन् – in handsome form or auspicious appearance>, उप <ind. prefix to verb स्पृशामि (√स्पृश्)>, उप-स्पृशामि <pre. 1st person sg. of उप-√स्पृश् – I touch, reach up to>.

RV x.125.8 author वागांभृणी (वाक् आम्भृणी), to वागांभृणी, metre त्रिष्टुप् छंदः
अहमेव वात इव प्र वाम्यारभमाणा भुवनानि विश्वा ।
परो दिवा पर एना पृथिव्यैतावती महिना सं बभूव ॥१०.१२५.८॥
अहं । एव । वातःऽइव । प्र । वामि । आऽरभमाणा । भुवनानि । विश्वा ।
परः । दिवा । परः । एना । पृथिव्या । एतावती । महिना । सं । बभूव ॥१०.१२५.८॥
I indeed blow forth like Vāta, and I hold together all the worlds and beings.
Beyond the Heaven, so far beyond the mighty Milky Way, I have travelled.
(RV x.125.8)

अहं <mf. nom. sg. of 1st person pron. अस्मद् – I, Vāk>, एव <ind. – so, indeed>, वातः <m. nom. sg. of वात – Vāta>, इव <ind. – like, in the same manner>, प्र <ind. prefix to verb वामि (√वा)>, प्र-वामि <pre. 1st person sg. of प्र-√वा – I blow forth>, आरभमाणा <f. nom. sg. of pre. middle pt. आरभमाणा (आ-√रभ्) – I hold together>, भुवनानि <n. acc. pl. of भुवन – all the worlds and beings (all the worlds in the Milky Way and all beings)>, विश्वा <n. acc. pl. of adj. विश्व – all>;

परः <(परस्) ind. – (with inst.) beyond >, दिवा <m. inst. sg. of दिव् – the Heaven>, परः <परस् ind. – (with inst.) beyond>, एना <ind. – (with पर and inst.) beyond>, पृथिव्या <f. inst. sg. of पृथिवि or पृथिवी – Pṛthivī, the Milky Way>,

एतावती <f. nom. sg. of adj. एतावत् – a such a measure, so far>, महिना <ind. – great, mighty>, सं <ind. prefix to verb बभूव (√भू)>, सं-बभूव <prf. 1st. person sg. of सं-√भू – I have been found, I have been>.

4.8 THE WATCHERS

4.8.1 Varuṇa's Watchers

It is known that a plasma sheath, which is a double layer of the two opposite charges, is formed "wherever plasmas with different densities, temperatures, or magnetic field strengths come in contact".[84] It presupposes that they have the ability to discern the differences in these factors in their environments.

According to the Ṛgveda, Varuṇa, Agni, and Soma place their Watchers (स्पशः spaśaḥ pl.) so that they can observe everything and know what is going on in their environments. Varuṇa places Watchers everywhere so that he can observe vigilantly without even winking (RV vii.61.3). They are observant and gifted with insight, and they are the carriers of the hymn (RV vii.87.3). Stanza RV vii.61.3 is addressed to Mitra and Varuṇa, and stanza RV vii.87.3 to Varuṇa alone.

RV vii.61.3 author वसिष्ठः, to मित्रावरुणौ, metre त्रिष्टुप् छंदः
प्रोरोर्मित्रावरुणा पृथिव्याः प्र दिव ऋष्वाद्बृहतः सुदानू ।
स्पशो दधाथे ओषधीषु विक्ष्वृधग्यतो अनिमिषं रक्षमाणा ॥७.६१.३॥
प्र । उरोः । मित्रावरुणा । पृथिव्याः । प्र । दिवः । ऋष्वात् । बृहतः । सुदानू इति सुऽदानू ।
स्पशः । दधाथे इति । ओषधीषु । विक्षु । ऋधक् । यतः । अनिऽमिषं । रक्षमाणा ॥७.६१.३॥
From the spacious Milky Way, O Varuṇa and Mitra, from the lofty, sublime Heaven, O bounteous Two, Ye place the Watchers in the dwellings and in the light-containing plants, wherefrom, one by one, ye watch without even winking. (RV vii.61.3)

प्र <ind. prefix to adj. उरोः – very, much>, उरोः <f. abl. sg. of adj. उरु – wide, broad, spacious>, मित्रावरुणा <m. voc. du. of मित्रावरुण – O Mitra and Varuṇa>, पृथिव्याः <f. abl. sg. of पृथिवी – from Pṛthivī (the Milky Way)>, प्र <ind. prefix to adj. बृहतः and/or ऋष्वात् – very, much>, दिवः <m. abl. sg. of दिव् – from the Heaven>, ऋष्वात् <m. abl. sg. of adj. ऋष्व – sublime>, बृहतः <m. abl. sg. of adj. बृहत् – lofty, high>, सुदानू <m. voc. du. of सुदानु – O bounteous Two>;

स्पशः <m. acc. pl. of स्पश् – watchers, spies>, दधाथे <pre. or prf. middle 2nd person du. of √धा 3Ā – ye (du.) put, place, set, lay in (loc.)>, ओषधीषु <f. loc. pl. of ओषधी (=ओषधि) – in the burning or light-containing plants, i.e. in the plasma blooms or Soma plants>, विक्षु <mf. loc. pl. of विश् – in the houses, dwellings>, ऋधक् <ind. – one by one, separately>, यतः <(यतस्) ind. – wherefrom>, अनिमिषं <ind. – without winking, vigilantly>, रक्षमाणा <m. nom. du. of pre. middle pt. रक्षमाण (√रक्ष्) – ye (du.) watch>.

RV vii.87.3 author वसिष्ठः, to वरुणः, metre त्रिष्टुप् छंदः
परि स्पशो वरुणस्य स्मदिष्टा उभे पश्यंति रोदसी सुमेके ।

84 Anthony L. Peratt, *Physics of the Plasma Universe*, 2nd ed., New York, Springer, 2015, p. 2.

ऋतावानः कवयो यज्ञधीराः प्रचेतसो य इषयंत मन्म ॥७.८७.३॥
परि । स्पशः । वरुणस्य । स्मत्ऽइष्टाः । उभे इति । पश्यंति । रोदसी इति । सुमेके इति सुऽमेके ।
ऋतऽवानः । कवयः । यज्ञऽधीराः । प्रऽचेतसः । ये । इषयंत । मन्म ॥७.८७.३॥
Varuṇa's Watchers, well trained, survey the beautifully constructed pair of Rodasī.
Performers of sacred works, they, who impel the hymn, are gifted with insight, conversant with sacrifice, and observant. (RV vii.87.3)

परि <ind. prefix to verb पश्यंति (√दृश्)>, स्पशः <m. nom. pl. of स्पश् – spies, watchers>, वरुणस्य <m. gen. sg. of वरुण – of Varuṇa>, स्मदिष्टाः <m. nom. pl. of adj. स्मदिष्ट – (cf. स्मद्दिष्टि) well trained or practised>, उभे <n. acc. du. of pron. उभ – both (the pair)>, परि-पश्यंति <pre. 3rd person pl. of परि-√दृश् – they observe, survey>, रोदसी <n. acc. du. of रोदस् – Rodasī (the pair of Soma Pond and the Stone)>, सुमेके <n. acc. du. of adj. सुमेक – well fixed, beautifully constructed>;

ऋतावानः <m. nom. pl. of ऋतावन् – performers of sacred works>, कवयः <m. nom. pl. of adj. कवि – gifted with insight>, यज्ञधीराः <m. nom. pl. of adj. यज्ञधीर – conversant with sacrifice>, प्रचेतसः <m. nom. pl. of adj. प्रचेतस् – attentive, observant>, ये <m. nom. pl. of pron. यद् – they who>, इषयंत <inj. 3rd person pl. of √इष् 4P – they cause to move quickly, impel, incite>, मन्म <n. acc. sg. of मन्मन् – thought, prayer, hymn>.

4.8.2 Agni's Watchers

In stanza RV iv.4.3, Agni is implored not to allow the wicked to go unobserved by his Watchers so that he can stop the wicked from assailing the people. Here, the people or the tribe refers to the charged particles of cosmic plasmas that share common attributes, such as Speech patterns, densities, temperatures, or magnetic field strengths. The stanza is addressed to Agnī Rakṣohā (अग्नी रक्षोहा), 'Agnī the destroyer of the Rakṣa (रक्षः n. nom. sg. of रक्षस्)'. Rakṣa is the demon or the enemy of the Devas. Note that Agnī, the name of the addressee, is the feminine gender form of Agni.

RV iv.4.3 author वामदेवः, to अग्नी रक्षोहा, metre त्रिष्टुप् छंदः
प्रति स्पशो वि सृज तूर्णितमो भवा पायुर्विशो अस्या अदब्धः ।
यो नो दूरे अघशंसो यो अंत्यग्ने माकिष्टे व्यथिरा दधर्षीत् ॥४.४.३॥
प्रति । स्पशः । वि । सृज । तूर्णिऽतमः । भव । पायुः । विशः । अस्याः । अदब्धः ।
यः । नः । दूरे । अघऽशंसः । यः । अंति । अग्ने । माकिः । ते । व्यथिः । आ । दधर्षीत् ॥४.४.३॥
Dispatch thou thy Watchers in all directions and be the most expeditious, never-deceived guard of this tribe. Let not the wicked, O Agni, who is near or in the distance, unobserved by thee, assail us. (RV iv.4.3)

प्रति <ind. prefix to verb सृज (√सृज्)>, स्पशः <m. acc. pl. of स्पश् – watchers, spies>, वि <ind. – away from, in different directions>, प्रति-सृज <ipv. 2nd person sg. प्रति-√सृज् – you send away, dispatch>, तूर्णितमः <m. nom. sg. of superlative adj. तूर्णि – the most expeditious, zealous>, भव <ipv. 2nd person sg. of √भू – be thou>, पायुः <m. nom. sg. of पायु – a guard, protector>, विशः <f. gen. sg. of विश् – of the tribe (i.e. the tribe of people that share common attributes)>, अस्याः <f. gen. sg. of pron. इदं – of this>, अदब्धः <m. nom. sg. of adj. अदब्ध – not deceived or tampered with>;

यः <m. nom. sg. of यद् – who>, नः <m. acc. pl. of the 1st person pron. अस्मद् – us>, दूरे <ind. – in the distance, far away>, अघशंसः <m. nom. sg. of अघशंस – an evil, the wicked (here, it refers to the people who do not share common attributes)>, यः <m. nom. sg. of यद् – who>, अंति <ind. – near, in the proximity of>, अग्ने <m. voc. sg. of अग्नि – O Agni>, माकिः <(माकिस्) ind. – may not, let not>, ते <m. gen. sg. of 2nd person pron. युष्मद् – (with व्यथिः) by you>, व्यथिः <m. nom. sg. of adj. व्यथिस् – unobserved by (gen.)>, आ <ind. prefix to verb दधर्षीत् (√धृष्)>, आ-दधर्षीत् <sub. 3rd person sg. of आ-√धृष् – may he (the wicked) assail, attack>.

4.8.3 Soma's Watchers

Soma's Watchers never close their eyes. They are found everywhere, each carrying a noose and binding. They utter the Speech together. They are honey-tongued and restless (RV ix.73.4). They are vigorous, free from malice, swift-moving. They roar. They are the keen-sighted eyes of the people, that is, the 'people' of cosmic plasmas (RV ix.73.7). The thousand streams refer to the blooming plasmas, or the Soma plants, in the Soma Pond.

RV ix.73.4 author पवित्रः, to पवमानः सोमः, metre जगती छंदः
सहस्रधारेऽव ते समस्वरन्दिवो नाके मधुजिह्वा असश्चतः ।
अस्य स्पशो न नि मिषंति भूर्णयः पदेपदे पाशिनः संति सेतवः ॥९.७३.४॥
सहस्रऽधारे । अव । ते । सं । अस्वरन् । दिवः । नाके । मधुऽजिह्वाः । असश्चतः ।
अस्य । स्पशः । न । नि । मिषंति । भूर्णयः । पदेऽपदे । पाशिनः । संति । सेतवः ॥९.७३.४॥
In the vault of the Heaven, they, honey-tongued and exhaustless, flowing in a thousand streams, uttered a sound together. His Watchers never close their eyes. They are everywhere, restless, each carrying a noose, binding. (RV ix.73.4)

सहस्रधारे <m. loc. sg. of adj. सहस्रधार – thousand-streamed, flowing in a thousand streams>, अव <ind. prefix to verb अस्वरन् (√स्वृ)>, ते <m. nom. pl. of 3rd person pron. तद् – they (Soma's watchers)>, सं <ind. – with, together>, अव-अस्वरन् <ipf. 3rd person pl. of अव-√स्वृ – they sang, uttered a sound>, दिवः <m. gen. sg. of दिव् – of the Heaven>, नाके <m. loc. sg. of नाक – in the vault (of the Heaven)>, मधुजिह्वाः <m. nom. pl. of adj. मधुजिह्व – honey-tongued, sweet-tongued>, असश्चतः <m. nom. pl. of adj. असश्चत् – not ceasing>;

अस्य <m. gen. sg. of pron. इदं – his (Soma's)>, स्पशः <m. nom. pl. of स्पश् – spies, watchers>, न <ind. – not, no, never>, नि <ind. prefix to verb मिषंति (√मिष्)>, नि-मिषंति <pre. 3rd person pl. of नि-√मिष् – they shut the eyelid, close the eyes>, भूर्णयः <m. nom. pl. of adj. भूर्णि – active, excited, restless>, पदेपदे <(पदे n. loc. sg. of पद) – पदेपदे at every step, everywhere>, पाशिनः <m. nom. pl. of adj. पाशिन् – having a net or noose, carrying a noose>, संति <pre. 3rd person pl. of √अस् – they are>, सेतवः <m. nom. pl. of adj. सेतु – binding>.

RV ix.73.7 author पवित्रः, to पवमानः सोमः, metre जगती छंदः
सहस्रधारे वितते पवित्र आ वाचं पुनंति कवयो मनीषिणः ।
रुद्रास एषामिषिरासो अद्रुहः स्पशः स्वंचः सुदृशो नृचक्षसः ॥९.७३.७॥
सहस्रऽधारे । विऽतते । पवित्रे । आ । वाचं । पुनंति । कवयः । मनीषिणः ।
रुद्रासः । एषां । इषिरासः । अद्रुहः । स्पशः । सुऽअंचः । सुऽदृशः । नृऽचक्षसः ॥९.७३.७॥
The wise singers, flowing in a thousand streams, pure, far-reaching, carrying the Speech in its course, Roaring, vigorous, free from malice, swiftly moving are their Watchers, the keen-sighted eyes of the people. (RV ix.73.7)

सहस्रधारे <m. loc. sg. of adj. सहस्रधार – thousand-streamed, flowing in a thousand streams>, वितते <m. loc. sg. of adj. वितत – spread out, far-spreading>, पवित्रे <m. loc. sg. of adj. पवित्र – purifying, pure, holy>, आ <ind. prefix to verb पुनंति (√पू)>, वाचं <f. acc. sg. of वाच् – Speech, sound>, आ-पुनंति <pre. 3rd person pl. of आ-√पू – they carry in its course>, कवयः <m. nom. pl. of कवि – the singers>, मनीषिणः <m. nom. pl. of adj. मनीषिन् – intelligent, wise>;

रुद्रासः <m. nom. pl. of adj. रुद्र – howling, roaring>, एषां <m. gen. pl. of इदं – their>, इषिरासः <m. nom. pl. of adj. इषिर – vigorous, active>, अद्रुहः <m. nom. pl. of adj. अद्रुह् – free from malice>, स्पशः <m. nom. pl. of स्पश् – spies, watchers>, स्वंचः <m. nom. pl. of adj. स्वंच् (स्वञ्च्) – moving swiftly, rapid>, सुदृशः <m. nom. pl. of adj. सुदृश् – keen-sighted>, नृचक्षसः <m. nom. pl. of नृचक्षस् – eyes of men or people>.

4.9 MĀYĀ

4.9.1 Māyā, the Magic Force of Varuṇa

Māyā is the magic force of Varuṇa and Mitra. Māyā (माया) is "extraordinary or supernatural power" and "sorcery, witchcraft magic" according to Monier-Williams. In RV x.147.5 and RV x.99.10, Varuṇa (and Mitra) is called the magician (मायी māyī m. nom. sg. of मायिन् māyin). The Ṛgveda and Śatapathabrāhmaṇaṃ tell us that māyā, the magic force of Varuṇa and Mitra, is not a single, homogeneous force. In the extant text of the Ṛgveda, we find one short hymn (RV x.177), with three stanzas, addressed to 'the division of māyā (मायाभेदः māyābhedaḥ)', which indicates that māyā has the division of forces associated with opposite charges or polarities. The word 'bheda (भेदः bhedaḥ)' means 'division', 'partition', or 'duality'.

According to the Śatapathabrāhmaṇaṃ (शतपथब्राह्मणं, ŚB iii.6.2.2), there are 'two māyās (माये māye du.)': Suparṇī (सुपर्णी) and Kadru (कद्रुः). We do not find the exact paired names 'Suparṇī and Kadru' in the extant text of the Ṛgveda. However, we find many similarly paired names, such as Dāsa (दासः) and Ārya (आर्यः), Yadu (यदुः) and Turvaśa (तुर्वशः), and so on, which show that these are the dual, polaric forces of the opposite charges associated with magnetism and cosmic plasmas.

Stanza RV v.85.5 states that māyā, the magic force, is of Varuṇa who, standing in the Midheaven, measures out the Milky Way with Sūrya.

RV v.85.5 author अत्रिः, to वरुणः, metre त्रिष्टुप् छंदः
इमामू ष्वासुरस्य श्रुतस्य महीं मायां वरुणस्य प्र वोचं ।
मानेनेव तस्थिवाँ अंतरिक्षे वि यो ममे पृथिवीं सूर्येण ॥५.८५.५॥
इमां । ऊं इति । सु । आसुरस्य । श्रुतस्य । महीं । मायां । वरुणस्य । प्र । वोचं ।
मानेनऽइव । तस्थिऽवान् । अंतरिक्षे । वि । यः । ममे । पृथिवीं । सूर्येण ॥५.८५.५॥
I laud the mighty māyā of the much celebrated Varuṇa, the Divine,
He who, standing in the Midheaven, measures out the Milky Way with Sūrya as if with a measuring-cord. (RV v.85.5)

इमां <f. acc. sg. of इदं – this>, उ <ind. – and, also, further>, सु <ind. prefix to adj. श्रुतस्य (श्रुत)>, आसुरस्य <m. gen. sg. of adj. आसुर – spiritual, divine>, सु-श्रुतस्य <m. gen. sg. of adj. सु-श्रुत – very well known, famous, celebrated>, महीं <f. acc. sg. of adj. मह् – great, powerful>, मायां <f. acc. sg. of माया – māyā, "extraordinary or supernatural power", i.e. magic power>, वरुणस्य <m. gen. sg. of वरुण – of Varuṇa>, प्र <ind. prefix to verb वोचं (√वच्)>, प्र-वोचं <inj. 1st person sg. of प्र-√वच् – I proclaim, announce, praise>;

मानेन <n. inst. sg. of मान – with a measuring-cord>, इव <ind. – as, as it were>, तस्थिवान् <m. nom. sg. of adj. तस्थिवस् – standing in, being in (loc.)>, अंतरिक्षे <m. loc. sg. of अंतरिक्ष – in the Midheaven or the Innerfield>, वि <ind. prefix to verb ममे (√मा)>, यः <m. nom. sg. of pron. यद् – he who>, वि-ममे <prf. middle 3rd person sg. of वि-√मा – he measures out>, पृथिवीं <f. acc. sg. of पृथिवी – Pṛthivī, the Milky Way>, सूर्येण <m. inst. sg. of सूर्य – with Sūrya>.

4.9.2 Māyā Drives the Flying Horse

As noted, the hymn RV x.177 is addressed to Māyābheda (मायाभेदः māyābhedaḥ), 'the division of māyā', the twofold force associated with opposite charges or polarities. The author is Pataṃga Prājāpatya (पतंगः प्राजापत्यः), 'Pataṃga, the son of Prajāpati (प्रजापतिः)', the Creator. Pataṃga (पतंगः), the flying horse, is driven by māyā, Asura's magic force. The definitions provided by Monier-Williams for Pataṃga include "flying" as an adjective; "a horse", "the Flier, name of Kṛṣṇa", "name of tree", "name of a mythical river", and so on as a noun. All these definitions describe the different aspects of plasma blooms, or Soma plants. Thus, one can see Pataṃga refers to the blooming cosmic plasmas in the Soma Pond. The wise men and the singers behold him in the Samudra (समुद्रः), the Ocean of celestial Waters (RV x.177.1), the Soma Pond. Māyā is the herdsman who never rests, running to and fro on his paths, turning fast within all beings (RV x.177.3).

RV x.177.1 author पतंगः प्राजापत्यः, to मायाभेदः, metre जगती छंदः
पतंगमक्तमसुरस्य मायया हृदा पश्यंति मनसा विपश्चितः ।
समुद्रे अन्तः कवयो वि चक्षते मरीचीनां पदमिच्छंति वेधसः ॥१०.१७७.१॥
पतंगं । अक्तं । असुरस्य । मायया । हृदा । पश्यंति । मनसा । विपःऽचितः ।
समुद्रे । अंतरिति । कवयः । वि । चक्षते । मरीचीनां । पदं । इच्छंति । वेधसः ॥१०.१७७.१॥
The wise men behold the flying horse driven by māyā of Asura with their heart and mind.
The pious singers appear midst the celestial Ocean and seek the abode of celestial lights.
(RV x.177.1)

पतंगं <m. acc. sg. of पतंग – a flying horse>, अक्तं <m. acc. sg. of adj. अक्त – driven>, असुरस्य <m. gen. sg. of असुर – of Asura (Varuṇa)>, मायया <f. inst. sg. of माया – by māyā, the magic force>, हृदा <n. inst. sg. of हृद् – <n. inst. sg. of हृद् – in the heart>, पश्यंति <pre. 3rd person pl. of √दृश् – they look at, behold>, मनसा <n. inst. sg. of मनस् – in the mind, (with हृदा) with all the heart>, विपश्चितः <m. nom. pl. of विपश्चित् – wise, learned men>;

समुद्रे <m. loc. sg. of समुद्र – in the Samudra, the Ocean of cosmic plasmas>, अंतः <(अंतर्) ind. – (with loc.) in the middle>, कवयः <mf. nom. pl. of कवि – singers>, वि <ind. prefix to verb चक्षते (√चक्ष्)>, वि-चक्षते <pre. middle 3rd person pl. of वि-√चक्ष् – they appear, become visible>, मरीचीनां <mf. gen. pl. of मरीचि – of the particles of light, of the rays of light (cosmic plasmas)>, पदं <n. acc. sg. of पद – station, abode, home>, इच्छंति <pre. 3rd person pl. of √इष् 6P – they endeavour to obtain, seek for>, वेधसः <m. nom. pl. of adj. वेधस् – virtuous, pious>.

RV x.177.2 author पतंगः प्राजापत्यः, to मायाभेदः, metre त्रिष्टुप् छंदः
पतंगो वाचं मनसा बिभर्ति तां गंधर्वोऽवदद्गर्भे अंतः ।
तां द्योतमानां स्वर्यं मनीषामृतस्य पदे कवयो नि पांति ॥१०.१७७.२॥
पतंगः । वाचं । मनसा । बिभर्ति । तां । गंधर्वः । अवदत् । गर्भे । अंतरिति ।
तां । द्योतमानां । स्वर्यं । मनीषां । ऋतस्य । पदे । कवयः । नि । पांति ॥१०.१७७.२॥
The flying horse bears Speech in his mind, the Speech Gandharva uttered in the womb.
At the seat of sacrifice, the singers observe the radiant hymn reverberating. (RV x.177.2)

पतंगः <m. nom. sg. of पतंग – a flying horse>, वाचं <f. acc. sg. of वाच् – Speech>, मनसा <ind. – in the mind>, बिभर्ति <pre. 3rd person sg. of √भृ – he carries, bears>, तां <f. acc. sg. of 3rd person pron. तद् – that/her (Speech)>, गंधर्वः <m. nom. sg. of गंधर्व – Gandharva, the celestial singer>, अवदत् <ipf. 3rd person sg. of √वद् – he spoke, uttered>, गर्भे <m. loc. sg. of गर्भ – in the womb, i.e. in the Soma Pond>, अंतः <(अंतर्) ind. – in, in the middle>;

तां <f. acc. sg. of 3rd person pron. तद् – that/her (prayer, hymn)>, द्योतमानां <f. acc. sg. of pre. middle pt. द्योतमाना (√द्युत्) – she/it being bright or brilliant>, स्वर्यं <(स्वर्यां?) f. acc. sg. of adj. स्वर्य – resounding, roaring>, मनीषां <f. acc. sg. of मनीषा – prayer, hymn>, ऋतस्य <n. gen. sg. of ऋत – of sacrifice>, पदे <n. loc. sg. of पद – in the abode, at the seat>, कवयः <mf. nom. pl. of कवि – singers>, नि <ind. prefix to verb पांति (√पा)>, पांति <pre. 3rd person pl. of नि-√पा 2P – they observe, watch>.

RV x.177.3 author पतंगः प्राजापत्यः, to मायाभेदः, metre त्रिष्टुप् छंदः
अपश्यं गोपामनिपद्यमानमा च परा च पथिभिश्चरंतं ।
स सध्रीचीः स विषूचीर्वसान आ वरीवर्ति भुवनेष्वंतः ॥१०.१७७.३॥
अपश्यं । गोपां । अनिऽपद्यमानं । आ । च । परा । च । पथिऽभिः । चरंतं ।
सः । सध्रीचीः । सः । विषूचीः । वसानः । आ । वरीवर्ति । भुवनेषु । अंतरिति ॥१०.१७७.३॥
I beheld the herdsman, who never rests, running to and fro on his paths.
He, converging, parting asunder, he, entering into all beings, revolves fast within them.
(RV x.177.3)

अपश्यं <ipf. 1st person sg. of √दृश् – I saw, beheld, looked at>, गोपां <m. acc. sg. of गोपा – a cowherd, herdsman>, अनिपद्यमानं <m. acc. sg. of adj. अनिपद्यमान – not falling down, untiring>, आ <ind. – to, towards>, च <ind. – and, also>, परा <ind. – away, off>, च <ind. – and, also>, पथिभिः <m. inst. pl. of पथिन् – on the paths>, चरंतं <m. acc. sg. of pre. active pt. चरत् (√चर्) – him running, moving>;

सः <m. nom. sg. of 3rd person pron. तद् – he>, सध्रीचीः <m. nom. sg. of adj. सध्रीची (=सध्र्यञ्च्) – turn to the centre, converging>, सः <m. nom. sg. of 3rd person pron. तद् – he>, विषूचीः <m. nom. sg. of adj. विषूची (=विष्वञ्च्) – going asunder or apart>, वसानः <m. nom. sg. of pre. middle pt. वसान (√वस् 2Ā) – he entering into>, आ <ind. prefix to verb वरीवर्ति (√वृत्)>, आ-वरीवर्ति <(=आ-वरीवर्त्ति) int. 3rd person sg. of आ-√वृत् – he turns or revolves quickly or repeatedly>, भुवनेषु <n. loc. pl. of भुवन – into all beings>, अंतः <(अंतर्) ind. – within, in the middle or interior>.

4.10 Ātmā

4.10.1 Ātmā, the Soul of Cosmic Yajña

Ātmā is the soul of cosmic Yajña (RV ix.6.8) and the soul of the Devas (RV x.168.4). Mostly, in the Ṛgveda, Vāta and Soma are called ātmā, the individuated soul. According to RV ix.6.8, ātmā is also the pressed Soma juice. The Ṛgveda tells us that ātmā, the individuated soul, is released through the cosmic ionisation process along with Agnis and Somas.

Stanza RV ix.6.8 is addressed to Pavamāna Soma (पवमानः सोमः), bright or flowing-clear Soma. For the stanza RV x.168.4, see section 2.5.24.

RV ix.6.8 author असितः काश्यपो देवलो वा, to पवमानः सोमः, metre गायत्री छंदः
आत्मा यज्ञस्य रंह्या सुष्वाणः पवते सुतः । प्रत्नं नि पाति काव्यं ॥९.६.८॥
आत्मा । यज्ञस्य । रंह्या । सुस्वानः । पवते । सुतः । प्रत्नं । नि । पाति । काव्यं ॥९.६.८॥
Ātmā, the soul of Yajña, the pressed Soma juice, becomes bright as he runs swiftly.
He guards the ancient wisdom. (RV ix.6.8)

आत्मा <m. nom. sg. of आत्मन् – ātmā, the individuated soul>, यज्ञस्य <m. gen. sg. of यज्ञ – of yajña>, रंह्या <f. inst. sg. of रंहि – by/through running, flowing, speed>, सुष्वाणः <m. nom. sg. of pre. middle pt. सुष्वाण (√सु) – he being pressed, extracted>, पवते <pre. middle 3rd person sg. of √पू – he becomes clear or bright>, सुतः <m. nom. sg. of सुत – the pressed Soma juice>;

प्रत्नं <n. acc. sg. of adj. प्रत्न – ancient, traditional>, नि <ind. prefix to verb पाति (√पा)>, नि-पाति <pre. 3rd person sg. of नि-√पा 2P – he protects, guards>, काव्यं <n. acc. sg. of काव्य – intelligence, wisdom>.

4.10.2 Have I Indeed Drunk Soma?

Hymn RV x.119 is addressed to Ātmastuti (आत्मस्तुतिः), which means a 'eulogy (स्तुतिः) of Ātmā (आत्मा)'. The author is Laba Aindra (लब ऐंद्रः), 'Laba the lunar mansion'. According to Monier-Williams, laba (लबः) is "a quail" and Aindra, as a neuter noun, is "the lunar mansion", which is the Soma Pond. Probably the lunar mansion, the Soma Pond, is called 'quail', for the variegated feathers of a quail resemble cosmic plasmas blooming in the Soma Pond.

The hymn is a soliloquy of Ātmā. Through this enchanting soliloquy, Ātmā reveals who he is and what he does.

RV x.119.1 author लब ऐंद्रः, to आत्मस्तुतिः, metre गायत्री छंदः
इति वा इति मे मनो गामश्वं सनुयामिति । कुवित्सोमस्यापामिति ॥१०.११९.१॥
इति । वै । इति । मे । मनः । गां । अश्वं । सनुयां । इति । कुवित् । सोमस्य । अपां । इति ॥१०.११९.१॥
I must certainly procure the cow and the steed. Thus was my resolve.
Have I indeed drunk Soma? (RV x.119.1)

इति <ind. – in this manner, thus>, वै <ind. – (a particle of emphasis and affirmation) indeed, truly, certainly>, इति <ind. – in this manner, thus>, मे <m. gen. sg. of 1st person pron. अस्मद् – my>, मनः <n. nom. sg. of मनस् – intention, will, resolve>, गां <m. acc. sg. of गो – the cow (celestial light, cosmic plasmas)>, अश्वं <m. acc. sg. of अश्व – the steed (force behind the plasma current)>, सनुयां <opt. 1st person sg of √सन् 8P – I should or must obtain, acquire, procure>, इति <ind. – thus (repeated for emphasis or for the governing of metre)>;

कुवित् <(कुविद्) ind. – if, whether (a particle of interrogation used in direct and indirect questions)>, सोमस्य <m. gen. sg. of सोम – of Soma>, अपां <ipf. 1st person sg. of √पा – I have drunk, I drank (with gen.)>, इति <ind. – thus>.

RV x.119.2 author लब ऐंद्रः, to आत्मस्तुतिः, metre गायत्री छंदः
प्र वाता इव दोधत उन्मा पीता अयंसत । कुवित्सोमस्यापामिति ॥१०.११९.२॥
प्र । वाताःऽइव । दोधतः । उत् । मा । पीताः । अयंसत । कुवित् । सोमस्य । अपां । इति ॥१०.११९.२॥
Like impetuous gusts of wind, the draughts that I have drunk lifted me up.
Have I indeed drunk Soma? (RV x.119.2)

प्र <ind. prefix to verb अयंसत (√यं)>, वाताः <m. nom. pl. of वात – Vātas, gusts>, इव <ind. – like, as>, दोधतः <m. gen. sg. of दोधत् (originally pre. active pt. of √दुध् to stir up) – of the impetuous, wild wind>, उत् <(उद्) ind. – up, above>, मा <m. acc. sg. of 1st person pron. अस्मद् – me>, पीताः <m. nom. pl. of past passive pt. पीत (√पा) – (the draughts) being drunk (by me)>, प्र-अयंसत <aor. middle 3rd person pl. of प्र-√यं – they held up, raised>;

कुवित् सोमस्य अपां इति <see RV x.119.1 – Have I indeed drunk Soma?>.

RV x.119.3 author लब ऐंद्रः, to आत्मस्तुतिः, metre गायत्री छंदः
उन्मा पीता अयंसत रथमश्वा इवाशवः । कुवित्सोमस्यापामिति ॥१.११९.३॥
उत् । मा । पीताः । अयंसत । रथं । अश्वाःऽइव । आशवः । कुवित् । सोमस्य । अपां । इति ॥१०.११९.३॥
The draughts I have drunk raised me up, as swift horses draw a chariot.
Have I indeed drunk Soma? (RV x.119.3)

उत् <(उद्) ind. – up, above>, मा <m. acc. sg. of 1st person pron. अस्मद् – me>, पीताः <m. nom. pl. of past passive pt. पीत (√पा) – (the draughts) being drunk (by me)>, अयंसत <aor. middle 3rd person pl. of √यं – they have raised, lifted>, रथं <m. acc. sg. of रथ – a chariot>, अश्वाः <m. nom. pl. of अश्व – horses>, इव <ind. – like, as>, आशवः <m. nom. pl. of adj. आशु – fast, quick>;
कुवित् सोमस्य अपां इति <see RV x.119.1 – Have I indeed drunk Soma?>.

RV x.119.4 author लब ऐंद्रः, to आत्मस्तुतिः, metre गायत्री छंदः
उप मा मतिरस्थित वाश्रा पुत्रमिव प्रियं । कुवित्सोमस्यापामिति ॥१०.११९.४॥
उप । मा । मतिः । अस्थित । वाश्रा । पुत्रंऽइव । प्रियं । कुवित् । सोमस्य । अपां । इति ॥१०.११९.४॥
The sacred hymn reached me, as a mother reaches for her beloved son.
Have I indeed drunk Soma? (RV x.119.4)

उप <ind. prefix to verb अस्थित (√स्था)>, मा <m. acc. sg. of 1st person pron. अस्मद् – me>, मतिः <f. nom. sg. of मति – prayer, hymn, sacred utterance>, उप-अस्थित <aor. middle 3rd person sg. of उप-√स्था – it (hymn) placed one's self, approached>, वाश्रा <f. nom. sg. of वाश्रा – a cow, mother>, पुत्रं <m. acc. sg. of पुत्र – a son>, इव <ind. – like, in the same manner as>, प्रियं <m. acc. sg. of adj. प्रिय – beloved, dear>;
कुवित् सोमस्य अपां इति <see RV x.119.1 – Have I indeed drunk Soma?>.

RV x.119.5 author लब ऐंद्रः, to आत्मस्तुतिः, metre गायत्री छंदः
अहं तष्टेव वंधुरं पर्यचामि हृदा मतिं । कुवित्सोमस्यापामिति ॥१.११९.५॥
अहं । तष्टाऽइव । वंधुरं । परि । अचामि । हृदा । मतिं । कुवित् । सोमस्य । अपां । इति ॥१०.११९.५॥
As a builder of chariots tends to the seat of a charioteer, I tend the sacred hymn in my heart.
Have I indeed drunk Soma? (RV x.119.5)

अहं <m. nom. sg. of 1st person pron. अस्मद् – I>, तष्टा <m. nom. sg. of तष्टृ – a carpenter, builder of chariots>, इव <ind. – like, in the same manner as>, वंधुरं <n. acc. sg. of वंधुर – the seat of a charioteer>, परि <ind. – fully, richly>, अचामि <pre. 1st person sg. of √अच् – I tend, honour>, हृदा <n. inst. sg. of हृद् – in one's heart>, मतिं <f. acc. sg. of मति – prayer, hymn, sacred utterance>;
कुवित् सोमस्य अपां इति <see RV x.119.1 – Have I indeed drunk Soma?>.

RV x.119.6 author लब ऐंद्रः, to आत्मस्तुतिः, metre गायत्री छंदः
नहि मे अक्षिपच्चनाच्छांत्सुः पंच कृष्टयः । कुवित्सोमस्यापामिति ॥१.११९.६॥
नहि । मे । अक्षिऽपत् । चन । अच्छांत्सुः । पंच । कृष्टयः । कुवित् । सोमस्य । अपां । इति ॥१०.११९.६॥
Surely, not even the five tribes of the Devas eluded the glance of my eye.
Have I indeed drunk Soma? (RV x.119.6)

नहि <ind. – surely not, by no means>, मे <m. gen. sg. of अस्मद् – my>, अक्षिपत् <ind. (from √क्षिप् to throw a glance) – throwing a glance, the glance of the eye>, चन <ind. – not even>, अच्छांत्सुः <aor. 3rd person pl. of √छंद् (=√छद्) – they hid, concealed>, पंच <f. nom. pl. of पंचन् (पञ्चन्) – five>, कृष्टयः <f. nom. pl. of कृष्टि – races of men, races of the Devas, (original meaning of कृष्टयः kṛṣṭayaḥ (pl.) are 'the ploughers of the field', who are the men or the Devas who plough the field (the Soma Pond) with discharged cosmic plasmas.)>;
कुवित् सोमस्य अपां इति <see RV x.119.1 – Have I indeed drunk Soma?>.

RV x.119.7 author लब ऐंद्रः, to आत्मस्तुतिः, metre गायत्री छंदः
नहि मे रोदसी उभे अन्यं पक्षं चन प्रति । कुवित्सोमस्यापामिति ॥१.११९.७॥
नहि । मे । रोदसी इति । उभे इति । अन्यं । पक्षं । चन । प्रति । कुवित् । सोमस्य । अपां । इति ॥१०.११९.७॥
Both of the Rodasī, surely, are not even comparable to one of my wings.
Have I indeed drunk Soma? (RV x.119.7)

नहि <ind. – surely not, by no means>, मे <m. gen. sg. of अस्मद् – my>, रोदसी <n. nom. du. of रोदस् – Rodasī, the pair of the the Stone and the Soma Pond>, उभे <n. nom. du. of pron. उभ – both>, अन्यं <m. acc. sg. of अन्य – other, one of>, पक्षं <m. acc. sg. sg. of पक्ष – a flank, wing, pinion>, चन <ind. – not even>, प्रति <ind. preposition – in comparision, on a par with>;

कुवित् सोमस्य अपां इति <see RV x.119.1 – Have I indeed drunk Soma?>.

RV x.119.8 author लब ऐंद्रः, to आत्मस्तुतिः, metre गायत्री छंदः
अभि द्यां महिना भुवमभी३◌◌मां पृथिवीं महीं । कुवित्सोमस्यापामिति ॥१.११९.८॥
अभि । द्यां । महिना । भुवं । अभि । इमां । पृथिवीं । महीं । कुवित् । सोमस्य । अपां । इति ॥१०.११९.८॥
I, by my supreme dominion, surpass the Heaven, surpass this great World, the Milky Way, the region of life. Have I indeed drunk Soma? (RV x.119.8)

अभि <ind. – over, superior, surpass>, द्यां <m. acc. sg. of दिव् – the Heaven>, महिना <n. inst. sg. of महिन – through/by sovereignty, dominion>, भुवं <f. acc. sg. of भू – the place of being, i.e. the region of life>, अभि <ind. – over, superior, surpass>, इमां <f. acc. sg. of pron. इदं – this>, पृथिवीं <f. acc. sg. of पृथिवी – Pṛthivī, the Milky Way>, महीं <f. acc. sg. of मही – the great world>;

कुवित् सोमस्य अपां इति <see RV x.119.1 – Have I indeed drunk Soma?>.

RV x.119.9 author लब ऐंद्रः, to आत्मस्तुतिः, metre गायत्री छंदः
हंताहं पृथिवीमिमां नि दधानीह वेह वा । कुवित्सोमस्यापामिति ॥१.११९.९॥
हंत । अहं । पृथिवीं । इमां । नि । दधानि । इह । वा । इह । वा । कुवित् । सोमस्य । अपां । इति ॥१.११९.९॥
Ah! this Milky Way, should I lay myself down here in this place or there.
Have I indeed drunk Soma? (RV x.119.9)

हंत <(हन्त) ind. a particle of exclamation or inceptive – ah! oh!>, अहं <m. nom. sg. of 1st person pron. अस्मद् – I>, पृथिवीं <f. acc. sg. of पृथिवी – Pṛthivī, “the broad and extended One”, the Milky Way>, इमां <f. acc. sg. of pron. इदं – this>, नि <ind. prefix to verb दधानि (√धा)>, नि-दधानि <ipv. 1st person sg. of नि-√धा – should I put, lay down, deposit>, इह <ind. – in this place, here>, वा <ind. – or>, इह <ind. – in that place>, वा <ind. – or>;

कुवित् सोमस्य अपां इति <see RV x.119.1 – Have I indeed drunk Soma?>.

RV x.119.10 author लब ऐंद्रः, to आत्मस्तुतिः, metre गायत्री छंदः
ओषमित्पृथिवीमहं जंघनानीह वेह वा । कुवित्सोमस्यापामिति ॥१.११९.१०॥
ओषं । इत् । पृथिवीं । अहं । जंघनानि । इह । वा । इह । वा । कुवित् । सोमस्य । अपां । इति ॥१.११९.१०॥
With burning ardour, shall I smite the Milky Way here in this place or in that place.
Have I indeed drunk Soma? (RV x.119.10)

ओषं <ind. – with burning ardour, fervour>, इत् <(इद्) ind. – (a particle of affirmation) even, just>, पृथिवीं <f. acc. sg. of पृथिवी – Pṛthivī, the Milky Way>, अहं <m. nom. sg. of 1st person pron. अस्मद् – I>, जंघनानि <(जङ्घनानि) sub. 1st person sg. of √हन् – will I strike, beat, hammer>, इह <ind. – in this place, here, to this place>, वा <ind. – or>, इह <ind. – in that place>, वा <ind. – or>;

कुवित् सोमस्य अपां इति <see RV x.119.1 – Have I indeed drunk Soma?>.

RV x.119.11 author लब ऐंद्रः, to आत्मस्तुतिः, metre गायत्री छंदः
दिवि मे अन्यः पक्षो३ऽधो अन्यमचीकृषं । कुवित्सोमस्यापामिति ॥१.११९.११॥
दिवि । मे । अन्यः । पक्षः । अधः । अन्यं । अचीकृषं । कुवित् । सोमस्य । अपां । इति ॥१.११९.११॥
One of my wings is in the Heaven, I have dragged the other down.
Have I indeed drunk Soma? (RV x.119.11)

दिवि <m. loc. sg. of दिव् – in the Heaven>, मे <m. gen. sg. of अस्मद् – my>, अन्यः <m. nom. sg. of अन्य – the one>, पक्षः <m. nom. sg. of पक्ष – a wing, pinion, flank>, अधः <(अधस्) ind. – below, down (to the Milky Way)>, अन्यं <m. acc. sg. of अन्य – the other>, अचीकृषं <aor. 1st person sg. of √कृष् – I have dragged, pulled>;
कुवित् सोमस्य अपां इति <see RV x.119.1 – Have I indeed drunk Soma?>.

RV x.119.12 author लब ऐंद्रः, to आत्मस्तुतिः, metre गायत्री छंदः
अहमस्मि महामहोऽभिनभ्यमुदीषितः । कुवित्सोमस्यापामिति ॥१.११९.१२॥
अहं । अस्मि । महाऽमहः । अभिऽनभ्यं । उत्ऽईषितः । कुवित् । सोमस्य । अपां । इति ॥१.११९.१२॥
I, the mightiest, have risen above the nave of the Milky Way.
Have I indeed drunk Soma? (RV x.119.12)

अहं <m. nom. sg. of 1st person pron. अस्मद् – I>, अस्मि <pre. 1st person sg. of √अस् – I am>, महामहः <m. nom. sg. of adj. महामह – very mighty>, अभिनभ्यं <ind. (अभि above + नभ्यं the nave) – above the nave (Here, the nave refers to the celestial ioniser located at the base of the Vault of the Heaven, which is anchored at the centre of the Milky Way. see Figures 4 and 6)>, उदीषितः <m. nom. sg. of adj. उदीषित – elevated, risen>;
कुवित् सोमस्य अपां इति <see RV x.119.1 – Have I indeed drunk Soma?>.

RV x.119.13 author लब ऐंद्रः, to आत्मस्तुतिः, metre गायत्री छंदः
गृहो याम्यरंकृतो देवेभ्यो हव्यवाहनः । कुवित्सोमस्यापामिति ॥१.११९.१३॥
गृहः । यामि । अरंऽकृतः । देवेभ्यः । हव्यऽवाहनः । कुवित् । सोमस्य । अपां । इति ॥१.११९.१३॥
I, the servant, travel prepared carrying the oblation to the Devas.
Have I indeed drunk Soma? (RV x.119.13)

गृहः <m. nom. sg. of गृह – an assistant, servant>, यामि <pre. 1st person sg. of √या – I travel, journey>, अरंकृतः <m. nom. sg. of adj. अरंकृत – prepared, ready>, देवेभ्यः <m. dat. pl. of देव – to the Devas>, हव्यवाहनः <m. nom. sg. of adj. हव्यवाहन (=हव्यवह्) – bearing, carrying, bringing the oblation (to the Devas)>;
कुवित् सोमस्य अपां इति <see RV x.119.1 – Have I indeed drunk Soma?>.

4.11 MANA, THE COSMIC MIND

Mana (मनः manaḥ) is the intelligent cosmic Mind or spirit. It is inherent in aṃbha (अंभः), the celestial hydrogen, and arises following the mighty cosmic force, which turns the wheel of cosmic ionisation. As hydrogen atoms are pulled into the Stone, the celestial ioniser, the ionic bond is broken, the charges are separated, and Agnis and Somas, the embryonic particles of cosmic plasmas, along with ātmā, are discharged into the Soma Pond. During the celestial plasma discharge and release processes, Agni and Soma recover their old body, the body of hydrogen, through recombination. When more than one Agnis are combined in the nucleus, new elements will be formed. Along with the recovered body of hydrogen, the newly formed elements, such as helium or lithium, are pulled back into the house of Yama, the Stone, the celestial ioniser, the house of death (the death of hydrogens) and rebirth (as cosmic plasmas). To maintain the unbroken chain of cosmic ionisation to sustain the Milky Way, wherein lie the worlds of life, they need to be brought back to the house of Yama for reionisation. Yama (यमः) is the Deva who presides over the deceased forefathers and rules the spirits of the dead (read ionised). In the hymn, mana (मनः manaḥ) is used almost synonymously with Agni.

The twelve stanzas of hymn RV x.58 describe this cyclical journey of the cosmic Mind or spirit (मनः manaḥ) as an unbroken chain of cosmic ionisation. Hymn RV x.58 is the only hymn found in the extant text of the Ṛgveda addressed to Mana Āvartanaṃ (मनः आवर्तनं), 'turning round or returning mana'. The authors are 'the sons of Gopa (गोपः) beginning with Bandhu (बंध्वादयो गौपायनाः)'. The sons of Gopa are Bandhu (बंधुः), Śrutabandhu (श्रुतबंधुः), and Viprabandhu (विप्रबंधुः). Gopa is a cowherd or herdsman.

RV x.58.1 author बंध्वादयो गौपायनाः, to मनआवर्तनं, metre अनुष्टुप् छंदः
यत्ते॑ य॒मं वै॑वस्व॒तं मनो॑ ज॒गाम॑ दू॒र॒कं । तत्त॒ आ व॑र्तयामसी॒ह क्षया॑य जी॒वसे॑ ॥१०.५८.१॥
यत् । ते॒ । य॒मं । वै॒व॒स्व॒तं । मनः॑ । ज॒गाम॑ । दू॒र॒कं । तत् । ते॒ । आ । व॒र्त॒या॒म॒सि॒ । इ॒ह । क्षया॑य । जी॒वसे॑
॥१०.५८.१॥
When your spirit goes far away to Yama, the son of Vivasvān,
We make it turn back to the house of Yama to bring it back to life. (RV x.58.1)

यत् <conjunction – that, when>, ते <m. gen. sg. of 2nd person pron. युष्मद् – your>, यमं <m. acc. sg. of यम – to Yama>, वैवस्वतं <m. acc. sg. of वैवस्वत – to the son of विवस्वान् (Vivasvān, the Stone, the celestial ioniser)>, मनः <n. nom. sg. of मनस् – mind, intelligence, spirit>, जगाम <prf. 3rd person sg. of √गं – it goes>, दूरकं <(=दूरं) ind. – far away>;
तत् <n. acc. sg. of तद् – it/that (mind, spirit)>, ते <m. gen. sg. of 2nd person pron. युष्मद् – your>, आ <ind. prefix to verb वर्तयामसि (√वृत्)>, आ-वर्तयामसि <pre. causative 1st person pl. of आ-√वृत् – we cause to turn back>, इह <ind. adverb – here, to this place>, क्षयाय <mn. dat. sg. of क्षय – to the seat, "the house of Yama" (here, Yama's house refers to the Stone, celestial ioniser)>, जीवसे <infinitive of √जीव् – to revive, vivify, restore to life>.

RV x.58.2 author बंध्वादयो गौपायनाः, to मनआवर्तनं, metre अनुष्टुप् छंदः
यत्ते॒ दिवं॒ यत्पृ॑थि॒वीं मनो॑ ज॒गाम॑ दू॒र॒कं । तत्त॒ आ व॑र्तयामसी॒ह क्षया॑य जी॒वसे॑ ॥१०.५८.२॥
यत् । ते॒ । दिव॑ं । यत् । पृ॒थि॒वीं । मनः॑ । ज॒गाम॑ । दू॒र॒कं । तत् । ते॒ । आ । व॒र्त॒या॒म॒सि॒ । इ॒ह । क्षया॑य । जी॒वसे॑
॥१०.५८.२॥
When that spirit of yours goes to the Heaven and far away to the Milky Way,
We make it turn back to the house of Yama to bring it back to life. (RV x.58.2)

यत् <conjunction – that, when>, ते <m. gen. sg. of 2nd person pron. युष्मद् – your>, दिवं <m. acc. sg. of दिव् – to the Heaven>, यत् <n. nom. sg. of यद् – that>, पृथिवीं <f. acc. sg. of पृथिवी – to Pṛthivī, i.e. to the Milky Way>, मनः <n. nom. sg. of मनस् – mind, spirit>, जगाम <prf. 3rd person sg. of √गं – it goes>, दूरकं <(=दूरं) ind. – far away>;
तत् ते आ वर्तयामसि इह क्षयाय जीवसे <see RV x.58.1 – We make it turn back to the house of Yama to bring it back to life.>.

RV x.58.3 author बंध्वादयो गौपायनाः, to मनआवर्तनं, metre अनुष्टुप् छंदः
यत्ते भूमिं चतुर्भृष्टिं मनो जगाम दूरकं । तत्त आ वर्तयामसीह क्षयाय जीवसे ॥१०.५८.३॥
यत् । ते । भूमिं । चतुःभृष्टिं । मनः । जगाम । दूरकं । तत् । ते । आ । वर्तयामसि । इह । क्षयाय । जीवसे ॥१०.५८.३॥
When your spirit goes to the Soma Pond, and far away to the four spiral Arms of the Milky Way,
We make it turn back to the house of Yama to bring it back to life. (RV x.58.3)

यत् <conjunction – that, when>, ते <m. gen. sg. of 2nd person pron. युष्मद् – your>, भूमिं <f. acc. sg. of भूमि – to the ground, field (refers to the Soma Pond)>, चतुर्भृष्टिं <f. acc. sg. of चतुर्भृष्टि – to the four spiked-regions i.e. four spiral Arms of the Milky Way>, मनः <n. nom. sg. of मनस् – mind, spirit>, जगाम <prf. 3rd person sg. of √गं – it goes>, दूरकं <(=दूरं) ind. – far away>;

तत् ते आ वर्तयामसि इह क्षयाय जीवसे <see RV x.58.1 – We make it turn back to the house of Yama to bring it back to life.>.

RV x.58.4 author बंध्वादयो गौपायनाः, to मनआवर्तनं, metre अनुष्टुप् छंदः
यत्ते चतस्रः प्रदिशो मनो जगाम दूरकं । तत्त आ वर्तयामसीह क्षयाय जीवसे ॥१०.५८.४॥
यत् । ते । चतस्रः । प्रऽदिशः । मनः । जगाम । दूरकं । तत् । ते । आ । वर्तयामसि । इह । क्षयाय । जीवसे ॥१०.५८.४॥
When your spirit goes far away to the four Arms of the Milky Way,
We make it turn back to the house of Yama to bring it back to life. (RV x.58.4)

यत् <conjunction – that, when>, ते <m. gen. sg. of 2nd person pron. युष्मद् – your>, चतस्रः <f. acc. pl. of चतुर् – four>, प्रदिशः <f. acc. pl. of प्रदिश् – to the pointed-out regions, i.e. to the four Arms of the Milky Way>, मनः <n. nom. sg. of मनस् – mind, spirit>, जगाम <prf. 3rd person sg. of √गं – it goes>, दूरकं <(=दूरं) ind. – far away>;

तत् ते आ वर्तयामसि इह क्षयाय जीवसे <see RV x.58.1 – We make it turn back to the house of Yama to bring it back to life.>.

RV x.58.5 author बंध्वादयो गौपायनाः, to मनआवर्तनं, metre अनुष्टुप् छंदः
यत्ते समुद्रमर्णवं मनो जगाम दूरकं । तत्त आ वर्तयामसीह क्षयाय जीवसे ॥१०.५८.५॥
यत् । ते । समुद्रं । अर्णवं । मनः । जगाम । दूरकं । तत् । ते । आ । वर्तयामसि । इह । क्षयाय । जीवसे ॥१०.५८.५॥
When your spirit goes far away to the billowy Ocean of celestial Waters,
We make it turn back to the house of Yama to bring it back to life. (RV x.58.5)

यत् <conjunction – that, when>, ते <m. gen. sg. of 2nd person pron. युष्मद् – your>, समुद्रं <m. acc. sg. of समुद्र – to the Ocean of celestial Waters>, अर्णवं <m. acc. sg. of adj. अर्णव – billowy, foaming>, मनः <n. nom. sg. of मनस् – mind, spirit>, जगाम <prf. 3rd person sg. of √गं – it goes>, दूरकं <(=दूरं) ind. – far away>;

तत् ते आ वर्तयामसि इह क्षयाय जीवसे <see RV x.58.1 – We make it turn back to the house of Yama to bring it back to life.>.

RV x.58.6 author बंध्वादयो गौपायनाः, to मनआवर्तनं, metre अनुष्टुप् छंदः
यत्ते मरीचीः प्रवतो मनो जगाम दूरकं । तत्त आ वर्तयामसीह क्षयाय जीवसे ॥१०.५८.६॥
यत् । ते । मरीचीः । प्रऽवतः । मनः । जगाम । दूरकं । तत् । ते । आ । वर्तयामसि । इह । क्षयाय । जीवसे ॥१०.५८.६॥
When your spirit goes far away to the beams of lights that are blazing forth,
We make it turn back to the house of Yama to bring it back to life. (RV x.28.6)

यत् <conjunction – that, when>, ते <m. gen. sg. of 2nd person pron. युष्मद् – your>, मरीचीः <f. acc. pl. of मरीचि – to the particles of light, the rays of light>, प्रवतः <f. acc. pl. of adj. प्रवत् – blazing forth>, मनः <n. nom. sg. of मनस् – mind, spirit>, जगाम <prf. 3rd person sg. of √गं – it goes>, दूरकं <(=दूरं) ind. – far away>;

तत् ते आ वर्तयामसि इह क्षयाय जीवसे <see RV x.58.1 – We make it turn back to the house of Yama to bring it back to life.>.

RV x.58.7 author बंध्वादयो गौपायनाः, to मनआवर्तनं, metre अनुष्टुप् छंदः
यत्ते अपो यदोषधीर्मनो जगाम दूरकं । तत्त आ वर्तयामसीह क्षयाय जीवसे ॥१०.५८.७॥
यत् । ते । अपः । यत् । ओषधीः । मनः । जगाम । दूरकं । तत् । ते । आ । वर्तयामसि । इह । क्षयाय । जीवसे ॥१०.५८.७॥
When that spirit of yours goes to the celestial Waters, far away to the light-containing plants,
We make it turn back to the house of Yama to bring it back to life. (RV x.58.7)

यत् <conjunction – that, when>, ते <m. gen. sg. of 2nd person pron. युष्मद् – your>, अपः <f. acc. pl. of अप् – to the celestial Waters, i.e. to cosmic plasmas>, यत् <n. nom. sg. of यद् – that>, ओषधीः <f. acc. pl. of ओषधि – to the light-containing or burning plants, i.e. to the Soma plants or the plasma blooms>, मनः <n. nom. sg. of मनस् – mind, spirit>, जगाम <prf. 3rd person sg. of √गं – it goes>, दूरकं <(=दूरं) ind. – far away>;

तत् ते आ वर्तयामसि इह क्षयाय जीवसे <see RV x.58.1 – We make it turn back to the house of Yama to bring it back to life.>.

RV x.58.8 author बंध्वादयो गौपायनाः, to मनआवर्तनं, metre अनुष्टुप् छंदः
यत्ते सूर्यं यदुषसं मनो जगाम दूरकं । तत्त आ वर्तयामसीह क्षयाय जीवसे ॥१०.५८.८॥
यत् । ते । सूर्यं । यत् । उषसं । मनः । जगाम । दूरकं । तत् । ते । आ । वर्तयामसि । इह । क्षयाय । जीवसे ॥१०.५८.८॥
When that spirit of yours goes to Sūrya and far away to Dawn,
We make it turn back to the house of Yama to bring it back to life. (RV x.58.8)

यत् <conjunction – that, when>, ते <m. gen. sg. of 2nd person pron. युष्मद् – your>, सूर्यं <m. acc. sg. of सूर्य – to Sūrya>, यत् <n. nom. sg. of यद् – that>, उषसं <f. acc. sg. of उषस् – to Uṣā, Dawn>, मनः <n. nom. sg. of मनस् – mind, spirit>, जगाम <prf. 3rd person sg. of √गं – it goes>, दूरकं <(=दूरं) ind. – far away>;

तत् ते आ वर्तयामसि इह क्षयाय जीवसे <see RV x.58.1 – We make it turn back to the house of Yama to bring it back to life.>.

RV x.58.9 author बंध्वादयो गौपायनाः, to मनआवर्तनं, metre अनुष्टुप् छंदः
यत्ते पर्वतान्बृहतो मनो जगाम दूरकं । तत्त आ वर्तयामसीह क्षयाय जीवसे ॥१०.५८.९॥
यत् । ते । पर्वतान् । बृहतः । मनः । जगाम । दूरकं । तत् । ते । आ । वर्तयामसि । इह । क्षयाय । जीवसे ॥१०.५८.९॥
When your spirit goes far away to the lotfy mountains,
We make it turn back to the house of Yama to bring it back to life. (RV x.58.9)

यत् <conjunction – that, when>, ते <m. gen. sg. of 2nd person pron. युष्मद् – your>, पर्वतान् <m. acc. pl. of पर्वत – to the mountains (mountains refer to individual plasma blooms)>, बृहतः <m. acc. pl. of adj. बृहत् – high, lofty>, मनः <n. nom. sg. of मनस् – mind, spirit>, जगाम <prf. 3rd person sg. of √गं – it goes>, दूरकं <(=दूरं) ind. – far away>;

तत् ते आ वर्तयामसि इह क्षयाय जीवसे <see RV x.58.1 – We make it turn back to the house of Yama to bring it back to life.>.

RV x.58.10 author बंध्वादयो गौपायनाः, to मनआवर्तनं, metre अनुष्टुप् छंदः
यत्ते विश्वमिदं जगन्मनो जगाम दूरकं । तत्त आ वर्तयामसीह क्षयाय जीवसे ॥१०.५८.१०॥
यत् । ते । विश्वं । इदं । जगत् । मनः । जगाम । दूरकं । तत् । ते । आ । वर्तयामसि । इह । क्षयाय । जीवसे ॥१०.५८.१०॥
When your spirit goes far away to the whole of this World,
We make it turn back to the house of Yama to bring it back to life. (RV x.58.10)

यत् <conjunction – that, when>, ते <m. gen. sg. of 2nd person pron. युष्मद् – your>, विश्वं <n. acc. sg. of adj. विश्व – every, whole, entire>, इदं <n. acc. sg. of pron. इदं – this>, जगत् <n. acc. sg. of जगत् – to the whole World, i.e. to the whole of the galactic system>, मनः <n. nom. sg. of मनस् – mind, spirit>, जगाम <prf. 3rd person sg. of √गं – it goes>, दूरकं <(=दूरं) ind. – far away>;

तत् ते आ वर्तयामसि इह क्षयाय जीवसे <see RV x.58.1 – We make it turn back to the house of Yama to bring it back to life.>.

RV x.58.11 author बंध्वादयो गौपायनाः, to मनआवर्तनं, metre अनुष्टुप् छंदः
यत्ते पराः परावतो मनो जगाम दूरकं । तत्त आ वर्तयामसीह क्षयाय जीवसे ॥१०.५८.११॥
यत् । ते । पराः । पराऽवतः । मनः । जगाम । दूरकं । तत् । ते । आ । वर्तयामसि । इह । क्षयाय । जीवसे ॥१०.५८.११॥
When your spirit goes far away to distant places and far beyond,
We make it turn back to the house of Yama to bring it back to life. (RV x.58.11)

यत् <conjunction – that, when>, ते <m. gen. sg. of 2nd person pron. युष्मद् – your>, पराः <f. acc. pl. of adj. पर – far distant, farther than>, परावतः <f. acc. pl. of परावत् – to distances>, मनः <n. nom. sg. of मनस् – mind, spirit>, जगाम <prf. 3rd person sg. of √गं – it goes>, दूरकं <(=दूरं) ind. – far away>;

तत् ते आ वर्तयामसि इह क्षयाय जीवसे <see RV x.58.1 – We make it turn back to the house of Yama to bring it back to life.>.

RV x.58.12 author बंध्वादयो गौपायनाः, to मनआवर्तनं, metre अनुष्टुप् छंदः
यत्ते भूतं च भव्यं च मनो जगाम दूरकं । तत्त आ वर्तयामसीह क्षयाय जीवसे ॥१०.५८.१२॥
यत् । ते । भूतं । च । भव्यं । च । मनः । जगाम । दूरकं । तत् । ते । आ । वर्तयामसि । इह । क्षयाय । जीवसे ॥१०.५८.१२॥
When your spirit goes far away to that which is, and to that which is to be,
We make it turn back to the house of Yama to bring it back to life. (RV x.58.12)

यत् <conjunction – that, when>, ते <m. gen. sg. of 2nd person pron. युष्मद् – your>, भूतं <n. acc. sg. भूत – to that which is>, च <ind. – and>, भव्यं <n. acc. sg. of भव्य – to that which is to be>, च <ind. – and>, मनः <n. nom. sg. of मनस् – mind, spirit>, जगाम <prf. 3rd person sg. of √गं – it goes>, दूरकं <(=दूरं) ind. – far away>;

तत् ते आ वर्तयामसि इह क्षयाय जीवसे <see RV x.58.1 – We make it turn back to the house of Yama to bring it back to life.>.

V. In the Beginning

5.1 A Hymn of the Creation

In the extant text of the Ṛgveda, we find only three hymns (RV x.129, RV x.130, RV x.190) that are addressed to Bhāvavṛttaṃ (भाववृत्तं), 'the creation of the World'. Bhāvavṛttaṃ is a compound word of bhāva (भावः) and vṛttaṃ (वृत्तं). The World here refers to the whole of the galactic system that includes the Milky Way (see Figures 4, 6). Of the three hymns addressed to Bhāvavṛttaṃ, hymns RV x.129 and RV x.130 are substantial.

Hymn RV x.129 explains how the state of Real (सत् sat), the existence, has evolved from the state of Non-Real (असत् asat), the non-existence. The hymn covers the very beginning of cosmic creation: what the condition was before the start of cosmic creation and how cosmic creation is initiated. In the beginning, there was only aṃbha (अंभः aṃbhaḥ), pervading throughout the vast emptiness, concealed in darkness. It was the vast ocean of aṃbha, celestial hydrogen, with no beginning and no end. It is a given. It is the ultimate God's domain. The ultimate God, the unknowable, is the One who provided the aṃbha and the boundless space. How did he prepare the aṃbha? How did he establish the vast domain? Who truly knows?

In the Ṛgveda, the 'aṃbha', though its definition provided by Monier-Williams is "the celestial waters", is never used to represent the celestial Waters, the cosmic plasmas. In the extant text of the Ṛgveda, 'aṃbha' appears only once in RV x.129.1, which is before the start of the first ionisation event. The word 'aṃbha' is derived from √ambh (√अम्भ् to sound), which implies that 'sounding' or 'giving forth sound' is a key attribute of hydrogen. Aṃbha is the original 'water-bearer', hydrogen.

In the void of cosmic wind, aṃbha, the celestial hydrogen, starts to breathe of its own accord. Then, the inherent force of aṃbha arises through the arduous penance (तपः tapaḥ), that is, through long, arduous spins and oscillations of hydrogen atoms and the subsequent outward projection of their innate topology. Thereafter, the desire (कामः kāmaḥ), the primaeval flow of the cosmic Mind, arises. Once risen, the inherent force becomes the mighty exerting force. The mighty exerting forces (महिमानः pl.), the magnetic field and magnetic forces, which are represented by Tvaṣṭā and the Ṛbhus, build the Heavenly structures. Tvaṣṭā (त्वष्टा) is called 'the first born (अग्रियः agriyaḥ RV i.13.10)'. Where there is Tvaṣṭā, the magnetic field, there will be the Ṛbhus (ऋभवः ṛbhavaḥ pl.), the magnetic forces. Tvaṣṭā and the Ṛbhus, after building the heavenly structures, turn the wheel of cosmic Yajña, which is the origin of the World and all created beings.

The author of the hymn is Prajāpati Parameṣṭhī (प्रजापतिः परमेष्ठी), 'Creator the Supreme'. Prajāpati is another name for Brahmā. In hymn RV x.129, the Creator himself explains about the beginning of the creation and raises ultimate philosophical enquiries. Whence comes this creation? Who shall here proclaim whence it has originated? Who truly knows?

RV x.129.1 author प्रजापतिः परमेष्ठी, to भाववृत्तं, metre त्रिष्टुप् छंदः
नासदासीन्नो सदासीत् तदानीं नासीद्रजो नो व्योमा परो यत् ।
किमावरीवः कुह कस्य शर्मन्नंभः किमासीद्गहनं गभीरं ॥१०.१२९.१॥
न । असत् । आसीत् । नो इति । सत् । आसीत् । तदानीं । न । आसीत् । रजः । नो इति । विऽओम । परः । यत् ।
किं । आ । अवरीवरिति । कुह । कस्य । शर्मन् । अंभः । किं । आसीत् । गहनं । गभीरं ॥१०.१२९.१॥
Then there was not Non-Real nor Real. There was no Sphere of the Heaven nor that ancient Vault of the Heaven. Who concealed it? Where? Under whose shelter? Was there aṃbha deep-sounding and unfathomed? (RV x.129.1)

न <ind. – not, no, nor, neither>, असत् <n. nom. sg. of असत् – non-existence, the state of non-real>, आसीत् <ipf. 3rd person sg. of √अस् – it was>, नो <ind. – and not>, सत् <n. nom. sg. of सत् – that which really is, existence, the truth, the state of real>, आसीत् <ipf. 3rd person sg. of √अस् – it was>, तदानीं <ind. – at that time, then>, न <ind. – not, no, nor, neither>, आसीत् <ipf. 3rd person sg. of √अस् – it was>, रजः <n. nom. sg. of रजस् – the Heavenly Sphere>, नो <ind. – and not>, व्योमा <m. nom. sg. of व्योमन् – Vyomā, the Vault of the Heaven>, परः <m. nom. sg. of adj. पर – prior, ancient>, यत् <m. nom. sg. of यद् – that>;

किं <n. nom. of interrogative pron. क – who? which? what?>, आ <ind. prefix to verb अवरीवः (√वृ)>, आ-अवरीवः <int. 3rd person sg. of आ-√वृ – who covered, concealed>, कुह <ind. – where?>, कस्य <mn. gen. sg. of क – whose?>, शर्मन् <(=शर्म?) n. nom. sg. of शर्मन् – (with कस्य) whose protection or shelter?>, अंभः <n. nom. sg. of अंभस् (from √अम्भ् to sound) – aṃbha, the original water-bearer, i.e. hydrogen>, किं <n. nom. of interrogative pron. क – who? which? what?>, आसीत् <ipf. 3rd person sg. of √अस् – it was there>, गहनं <n. nom. sg. of adj. गहन – inexplicable, unfathomed>, गभीरं <n. nom. sg. of adj. गभीर – deep in sound, deep-sounding, hollow-toned, i.e. giving forth deep sound>.

RV x.129.2 author प्रजापतिः परमेष्ठी, to भाववृत्तं, metre त्रिष्टुप् छंदः
न मृत्युरासीदमृतं न तर्हि न रात्र्या अह्न आसीत्प्रकेतः ।
आनीदवातं स्वधया तदेकं तस्माद्धान्यन्न परः किं चनास ॥१०.१२९.२॥
न । मृत्युः । आसीत् । अमृतं । न । तर्हि । न । रात्र्याः । अह्नः । आसीत् । प्रऽकेतः ।

आनीत् । अवातं । स्वधया । तत् । एकं । तस्मात् । ह । अन्यत् । न । परः । किं । चन । आस ॥१०.१२९.२॥
Then there was no death, nor yet the immortal. And there was no sight of Night and Day.
That One alone, in the windless space, breathed of its own accord. Apart from that, nothing at all existed. (RV x.129.2)

न <ind. – not, no, nor, neither>, मृत्युः <m. nom. sg. of मृत्यु – death (destruction of the ionic bond of hydrogen atoms), i.e. cosmic ionisation>, आसीत् <ipf. 3rd person sg. of √अस् – it was>, अमृतं <n. nom. sg. of अमृत – the immortal being, i.e. cosmic plasmas, which are described as immortals>, न <ind. – not, no, nor, neither>, तर्हि <ind. – at that time, then>, न <ind. – not, no, nor, neither>, रात्र्याः <f. gen. sg. of रात्री – of night, i.e. of Varuṇa (plasma dark discharge, hence Night)>, अह्नः <n. gen. sg. of अहन् (=अहर्) – of day, i.e. of Mitra (plasma light discharge, hence Day)>, आसीत् <ipf. 3rd person sg. of √अस् – it was>, प्रकेतः <m. nom. sg. of प्रकेत – appearance, sight>;

आनीत् <ipf. 3rd person sg. of √अन् – he/it breathed, respired>, अवातं <n. nom. sg. of adj. अवात – "windless, the windless atmosphere" i.e. in the windless space (devoid of 'cosmic wind')>, स्वधया <f. inst. sg. of स्वधा – by self-power, of one's own accord>, तत् <n. nom. sg. of pron. तद् – that, it>, एकं <n. nom. sg. of pron. एक – one, that one only>, तस्मात् <mn. abl. sg. of pron. तद् – from that>, ह <ind. – a particle for emphasising a preceeding word, indeed>, अन्यत् <n. nom. sg. of अन्य – other than, other>, न <ind. – not, no>, परः <(परस्) ind. – (with abl.) beyond>, किं-चन <ind. – (with a negation) not at all>, आस <prf. 3rd person sg. of √अस् – (nothing) was, existed>.

RV x.129.3 author प्रजापतिः परमेष्ठी, to भाववृत्तं, metre त्रिष्टुप् छंदः
तम आसीत्तमसा गूळ्हमग्रेऽप्रकेतं सलिलं सर्वमा इदं ।
तुच्छ्येनाभ्वपिहितं यदासीत् तपसस्तन्महिनाजायतैकं ॥१०.१२९.३॥
तमः । आसीत् । तमसा । गूळ्हं । अग्रे । अप्रऽकेतं । सलिलं । सर्वं । आः । इदं ।
तुच्छ्येन । आभु । अपिऽहितं । यत् । आसीत् । तपसः । तत् । महिना । अजायत । एकं ॥१०.१२९.३॥
Darkness was there, in the beginning, concealed by darkness. What arose was an imperceptible surge. That One alone, which is pervading throughout the emptiness, mightily arose from arduous penance. (RV x.129.3)

तमः <n. nom. sg. of तमस् – darkness>, आसीत् <ipf. 3rd person sg. of √अस् – it was there>, तमसा <n. inst. sg. of तमस् – by darkness>, गूळ्हं <(=गूढं) n. nom. sg. of adj. गूळ्ह (=गूढ) – hidden, concealed>, अग्रे <ind. – first, in the beginning>, अप्रकेतं <n. nom. sg. of adj. अप्रकेत – indiscriminate, unrecognisable>, सलिलं <n. nom. sg. of सलिल – waves, surge>, सर्वं <n. nom. sg. of adj. सर्व – every, all>, आः <ipf. 3rd person sg. of √अस् – it was>, इदं <n. nom. sg. of इदं – this>;

तुच्छ्येन <n. inst. sg. of तुच्छ्य – through/with emptiness (of space)>, आभु <n. nom. sg. of adj. आभु – pervading, reaching>, अपिहितं <n. nom. sg. of adj. अपिहित – put to, placed into>, यत् <n. nom. sg. of pron. यद् – that, which>, आसीत् <ipf. 3rd person sg. of √अस् – it was>, तपसः <n. abl. sg. of तपस् – from observance, penance>, तत् <n. nom. sg. of pron. तद् – that>, महिना <ind. – mightily, forcibly>, अजायत <ipf. middle 3rd person sg. of √जन् – it came into being, arose>, एकं <n. nom. sg. of pron. एक – one, that one only>.

RV x.129.4 author प्रजापतिः परमेष्ठी, to भाववृत्तं, metre त्रिष्टुप् छंदः
कामस्तदग्रे समवर्तताधि मनसो रेतः प्रथमं यदासीत् ।
सतो बंधुमसति निरविंदन्हृदि प्रतीष्या कवयो मनीषा ॥१०.१२९.४॥
कामः । तत् । अग्रे । सं । अवर्तत । अधि । मनसः । रेतः । प्रथमं । यत् । आसीत् ।
सतः । बंधु । असति । निः । अविंदन् । हृदि । प्रतिऽइष्य । कवयः । मनीषा ॥१०.१२९.४॥
Thereafter, the desire arose, the desire, which was the primaeval flow of cosmic Mind.
With hymn, the singers, seeking to find a kinsman of Real in Non-Real, found him in the nucleus of aṃbha. (RV x.129.4)

कामः <m. nom. sg. of काम – kāma, wish, desire>, तत् <mn. nom. sg. of pron. तद् – that>, अग्रे <ind. – in the beginning, first, further on, subsequently>, सं <ind. prefix to verb अवर्तत (√वृत्)>, सं-अवर्तत <ipf. middle 3rd person sg. of सं-√वृत् – it took shape, came into being, arose>, अधि <ind. (as a separable adverb or preposition) – from, after, for, on, at>, मनसः <n. gen. sg of मनस् – of cosmic mind, intelligence, spirit>, रेतः <n. nom. sg. रेतस् – a flow, current, seed>, प्रथमं <n. nom. sg. of adj. प्रथम – foremost, primary, original>, यत् <n. nom. sg. of relative pron. यद् – that, which>, आसीत् <ipf. 3rd person sg. of √अस् – it was>;

सतः <n. gen. sg. of सत् – of Sat, of that which is real, of real>, बंधुं <m. acc. sg. of बंधु – a kinsman, relative>, असति <n. loc. sg. of असत् – in Asat, non-real)>, निः <(=निस्) prefix to verb अविंदन् (√विद्)>, निर्-अविंदन् <ipf. 3rd person pl. of निर्-√विद् – they found out>, हृदि <n. loc. sg. of हृद् – (in older language) the interior of the body, i.e. in the nucleus of aṃbha, hydrogen>, प्रतीष्य <ind. pt. of √प्रतीष् (प्रति-√इष्) – striving after, seeking>, कवयः <m. nom. pl. of कवि – the seers, singers (refer to the firestones of the Stone)>, मनीषा <f. inst. sg. of मनीषा – with prayer, hymn>.

RV x.129.5 author प्रजापतिः परमेष्ठी, to भाववृत्तं, metre त्रिष्टुप् छंदः
तिरश्चीनो विततो रश्मिरेषामधः स्विदासी३दुपरि स्विदासी३त् ।
रेतोधा आसन्महिमान आसन्त्स्वधा अवस्तात्प्रयतिः परस्तात् ॥१०.१२९.५॥
तिरश्चीनः । विऽततः । रश्मिः । एषां । अधः । स्वित् । आसी३त् । उपरि । स्वित् । आसी३त् । रेतःऽधाः । आसन् । महिमानः । आसन् । स्वधा । अवस्तात् । प्रऽयतिः । परस्तात् ॥१०.१२९.५॥
Their beam of light extended across. Indeed, it was extended upwards and down.
Bearers of seeds were there and mighty forces were there. At first, it was the inherent force, later it became the exerting force. (RV x.129.5)

तिरश्चीनः <m. nom. sg. of adj. तिरश्चीन – transverse, across>, विततः <m. nom. sg. of adj. वितत – extended, diffused, drawn>, रश्मिः <m. nom. sg. of रश्मि – a ray of light, a beam of light>, एषां <m. gen. pl. of pron. इदं – their (singers')>, अधः <(अधस्) ind. – below, beneath, down>, स्वित् <(स्विद्) ind. – (a particle of interrogation or enquiry or doubt) do you think? perhaps, indeed, any>, आसीत् <ipf. 3rd person sg. of √अस् – it was>, उपरि <ind. – above, upwards>, स्वित् <(स्विद्) ind. – see above>, आसीत् <ipf. 3rd person sg. of √अस् – it was>;

रेतोधाः <m. nom. pl. of रेतोधा – bearers of seeds (aṃbhas, hydrogens, are the bearers of seeds)>, आसन् <ipf. 3rd person pl. of √अस् – they were>, महिमानः <m. nom. pl. of महिमन् – might, powers, i.e. mighty forces>, आसन् <ipf. 3rd person pl. of √अस् – there were>, स्वधा <f. nom. sg. of स्वधा – self-power, inherent power>, अवस्तात् <ind. – before (in time)>, प्रयतिः <f. nom. sg. of प्रयति – will, exertion, i.e. the exerting force>, परस्तात् <ind. – afterwards, later>.

RV x.129.6 author प्रजापतिः परमेष्ठी, to भाववृत्तं, metre त्रिष्टुप् छंदः
को अद्धा वेद क इह प्र वोचत्कुत आजाता कुत इयं विसृष्टिः ।
अर्वाग्देवा अस्य विसर्जनेनाथा को वेद यत आबभूव ॥१०.१२९.६॥
कः । अद्धा । वेद । कः । इह । प्र । वोचत् । कुतः । आऽजाता । कुतः । इयं । विऽसृष्टिः । अर्वाक् । देवाः । अस्य । विऽसर्जनेन । अथ । कः । वेद । यतः । आऽबभूव ॥१०.१२९.६॥
Who truly knows? Who shall here proclaim whence was it born? Whence comes this creation? Even the Devas came after the emergence of this creation. Then, who knows whence has it originated? (RV x.129.6)

कः <m. nom. sg. of pron. क – who>, अद्धा <ind. – certainly, truly>, वेद <pre. 3rd person sg. of √विद् – who knows>, कः <m. nom. sg. of pron. क – who>, इह <ind. – in this place, here>, प्र <ind. prefix to verb वोचत् (√वच्)>, प्र-वोचत् <inj. 3rd person sg. of प्र-√वच् – who shall proclaim>, कुतः <(कुतस्) ind. – from where? whence?>, आजाता <f. nom. sg. of adj. आजात – born>, कुतः <(कुतस्) ind. – whence?>, इयं <f. nom. sg. of pron. इदं – this>, विसृष्टिः <f. nom. sg. of विसृष्टि – creation>;

अर्वाक् <ind. – (with inst.) from a certain point, after>, देवाः <m. nom. pl. of देव – the Devas>, अस्य <n. gen. sg. of इदं – of this, its>, विसर्जनेन <n. inst. sg. of विसर्जन – (with अर्वाक्) the act of creation>, अथ <ind. – now, then>, कः <m. nom. sg.

of pron. क – who>, वेद <pre. 3rd person sg. of √विद् – who knows>, यतः <(यतस्) ind. – from which or what, wherefrom, whence>, आबभूव <prf. 3rd person sg. of √आभू – it has originated>.

RV x.129.7 author प्रजापतिः परमेष्ठी, to भाववृत्तं, metre त्रिष्टुप् छंदः
इयं विसृष्टिर्यत आबभूव यदि वा दधे यदि वा न ।
यो अस्याध्यक्षः परमे व्योमन्त्सो अंग वेद यदि वा न वेद ॥१०.१२९.७॥
इयं । विऽसृष्टिः । यतः । आऽबभूव । यदि । वा । दधे । यदि । वा । न ।
यः । अस्य । अधिऽअक्षः । परमे । विऽओमन् । सः । अंग । वेद । यदि । वा । न । वेद ॥१०.१२९.७॥
Whence has this creation originated, whether he has caused it or not,
He who supervises at the highest point in the Vault of the Heaven verily knows, or perhaps he knows not. (RV x.129.7)

इयं <f. nom. sg. of इदं – this>, विसृष्टिः <f. nom. sg. of विसृष्टि – production, creation>, यतः <(यतस्) ind. – from which or whom, whence>, आबभूव <prf. 3rd sg. of √आभू – it has originated>, यदि वा <ind. – whether or>, दधे <prf. middle 3rd person sg. of √धा – he has caused, executed>, यदि वा न <<ind. – or not>;

यः <m. nom. sg. of relative pron. यद् – he who>, अस्य <m. gen. sg. of इदं – of this, it>, अध्यक्षः <m. nom. sg. of adj. अध्यक्ष – exercising supervision>, परमे <n. loc. sg. of परम – at the highest point>, व्योमन् <(=व्योमनि) m. loc. sg. of व्योमन् – in the Vault of the Heaven>, सः <m. nom. sg. of 3rd person pron. तद् – he>, अंग <(अङ्ग) ind. particle of assent or desire – indeed, true>, वेद <pre. 3rd person sg. of √विद् – he knows (gen.)>, यदि वा न <ind. – or rather not>, वेद <pre. 3rd person sg. of √विद् – he knows (gen.)>.

5.2 WEAVING THE NEBULOUS FABRIC WITH SĀMA HYMNS

In the Ṛgveda, chandas (छंदांसि chaṃdāṃsi pl. metres) are associated with Speeches (वाचः vācaḥ pl.), the oscillations generated by cosmic plasmas across a broad frequency band. The celestial beings involved in the processes of cosmic ionisation and plasma discharge and release sing and make Speeches. In the Ṛgveda, these Speeches, or hymns, are numerically coded in metres of stanzas (RV x.130.4-5, RV viii.7.1). The units of Vedic metres are the verse (पदं n. padaṃ or पादः m. pādaḥ), the stanza (ऋक् ṛk), and the hymn (सूक्तं sūktaṃ). A verse or pāda mostly contains eight, eleven, or twelve syllables each but may occasionally consist of fewer or more syllables.[85]

According to Vedic scholars, Vedic metres are regulated by the number of syllables in each verse of a stanza. Each stanza generally has three, four, or five pādas with a marked interval at the end of the second pāda, commonly with a vertical bar. In the printed text, the first and second pādas form one line and the third, third and fourth, or third and fourth and fifth, complete the stanza. However, considering the importance the Ṛgveda places

---

[85] E. Vernon Arnold, *Vedic Metre in Its Historical Development*, Cambridge, Cambridge University Press, 1905, p. 7.

on Speeches, or sounds, or hymns, originally, the cadences of intonation might have been considered in the metrical structure of the stanzas, not just the number of syllables. It is possible that the original cadences of intonation have been lost and the current ones have deviated from the original ones. For in-depth discussions of the Vedic metres, refer to *Vedic Metre in its Historical Development* by E. Vernon Arnold (1905) and *Rig Veda: A Metrically Restored Text with an Introduction and Notes* by Barend A. Van Nooten and Gary B. Holland (1994).

In stanzas 1 and 2 of RV x.130, the discharge of cosmic plasmas from the Stone, the celestial ioniser, into the Soma Pond and the subsequent blooming of cosmic plasmas in the Soma Pond are compared to weaving. Stanzas 4 and 5 of RV x.130 explain which metre is associated with what particular cosmic plasma phenomenon. Chandas or metres listed in hymn RV x.130 include Gāyatrī (गायत्री), Uṣṇihā (उष्णिहा), Anuṣṭup (अनुष्टुप्), Virāṭ (विराट्), Bṛhatī (बृहती), Triṣṭup (त्रिष्टुप्), and Jagatī (जगती), which are the seven most common metres of the Ṛgveda. Listed below are the common forms of metrical structures of these metres.[86] The vertical bar indicates the line separation. Uṣṇihā and Virāṭ metres show many more variations in their forms when compared to the metrical structures of other metres. Note, only common forms are listed and the variations in their metrical structures are not included here.

Gāyatrī (गायत्री): three pādas of eight syllables each (8 8 | 8) = 24 syllables
Uṣṇihā (उष्णिहा): two pādas of eight syllables and one pāda of twelve syllables (8 8 | 12) = 28 syllables
Anuṣṭup (अनुष्टुप्): four pādas of eight syllables each (8 8 | 8 8) = 32 syllables
Virāṭ (विराट्): three pādas of eleven syllables each (11 11 | 11) = 33 syllables
Bṛhatī (बृहती): three pādas of eight syllables and one pāda of twelve syllables (8 8 | 12 8) = 36 syllables
Triṣṭup (त्रिष्टुप्): four pādas of eleven syllables each (11 11 | 11 11) = 44 syllables
Jagatī (जगती): four pādas of twelve syllables each (12 12 | 12 12) = 48 syllables

What hymn RV x.130 implies is that Speeches, or sounds, are inherent in physical reality and are intimately associated with the creation, manifestation, and sustenance of physical reality. However, exactly how the metrical structures of these metres are related to the frequencies of the Speeches of cosmic plasmas have not yet been deciphered.

The author of hymn RV x.130 is Yajña Prājāpatya (यज्ञः प्राजापत्यः), ‘Yajña, son of Prajāpati (प्रजापतिः)’. Prajāpati is another name for Brahmā, the Creator.

---

86 Barend A. Van Nooten and Gary B. Holland (ed.), *Rig Veda: A Metrically Restored Text with an Introduction and Notes*, Department of Sanskrit and Indian Studies, Harvard University, 1994, p. xiv-xvi. Havard Oriental Series 50.

RV x.130.1 author यज्ञः प्राजापत्यः, to भाववृत्तं, metre जगती छंदः
यो यज्ञो विश्वतस्तंतुभिस्तत एकशतं देवकर्मेभिरायतः ।
इमे वयंति पितरो य आययुः प्र वयाप वयेत्यासते तते ॥१०.१३०.१॥
यः । यज्ञः । विश्वतः । तंतुऽभिः । ततः । एकऽशतं । देवऽकर्मेभिः । आऽयतः ।
इमे । वयंति । पितरः । ये । आऽययुः । प्र । वय । अप । वय । इति । आसते । तते ॥१०.१३०.१॥
That Yajña is performed with filaments extended on all sides by the one hundred and one masters of sacred work who arrived there. These forefathers, who arrived, sit by the warp and weave singing, Weave forth, weave back! (RV x.130.1)

यः <m. nom. sg. of relative pron. यद् – what, that, which>, यज्ञः <m. nom. sg. of यज्ञ – cosmic Yajña>, विश्वतः <(विश्वतस्) ind. – from or on all sides>, तंतुभिः <m. inst. pl. of तंतु – threads or filaments (of cosmic plasmas)>, ततः <m. nom. sg. of adj. तत – extended, stretched, stretching, extending>, एकशतं <n. nom. sg. of एकशत – one hundred and one>, देवकर्मेभिः <m. inst. pl. of देवकर्म – by the masters of sacred work (refer to the Devas)>, आयतः <m. nom. sg. of adj. आयत् – arriving, entering>;

इमे <m. nom. pl. of pron. इदं – these>, वयंति <pre. 3rd person pl. of √वे – they weave, braid>, पितरः <m. nom. pl. of पितृ – forefathers>, ये <m. nom. pl. of pron. यद् – who>, आययुः <prf. 3rd person pl. of आ-√या – they have arrived>, प्र <ind. prefix to verb वय (√वे)>, प्र-वय <ipv. 2nd person sg. of प्र-√वे – you weave forth>, अप <ind. prefix to verb वय (√वे)>, अप-वय <ipv. 2nd person sg. of अप-√वे – you weave back>, इति <ind. – in this manner, thus>, आसते <pre. middle 3rd person pl. of √आस् – they sit>, तते <n. loc. sg. of तत – on/by the warp>.

RV x.130.2 author यज्ञः प्राजापत्यः, to भाववृत्तं, metre त्रिष्टुप् छंदः
पुमाँ एनं तनुत उत्कृणत्ति पुमान्वि तत्ने अधि नाके अस्मिन् ।
इमे मयूखा उप सेदुरू सदः सामानि चक्रुस्तसराण्योतवे ॥१०.१३०.२॥
पुमान् । एनं । तनुते । उत् । कृणत्ति । पुमान् । वि । तत्ने । अधि । नाके । अस्मिन् ।
इमे । मयूखाः । उप । सेदुः । ऊं इति । सदः । सामानि । चक्रुः । तसराणि । ओतवे ॥१०.१३०.२॥
The man weaves and continues spinning, spreading it out over the Vault of the Heaven. These pegs appeared from the seat of fire and made Sāma hymns weaving shuttles for the cross-threads. (RV x.130.2)

पुमान् <m. nom. sg. of पुंस् – a man, a male being, i.e. Agni or the Deva>, एनं <m. acc. sg. of pron. इदं – this, it>, तनुते <pre. middle 3rd person sg. of √तन् – he extends, spreads, weaves>, उत् <(उद्) ind. prefix to verb कृणत्ति (√कृत्)>, उत्-कृणत्ति <pre. 3rd person sg. of उत्-√कृत् 7P – he continues spinning>, पुमान् <m. nom. sg. of पुंस् – a man>, वि <ind. prefix to verb तत्ने (√तन्)>, वि-तत्ने <prf. middle 3rd person sg. of वि-√तन् – he spread out, stretched, extended (a net, snare, cord)>, अधि <ind. – (with loc.) over, on, at>, नाके <m. loc. sg. of नाक – over the Vault of the Heaven>, अस्मिन् <m. loc. sg. of इदं – this>;

इमे <m. nom. pl. of pron. इदं – these>, मयूखाः <m. nom. pl. of मयूख – pegs (of the loom), beams of light, flames (plasma blooms, or Soma plants, are compared to the pegs of the loom)>, उप <ind. prefix to verb सेदुः (√सद्)>, उप-सेदुः <prf. 3rd person pl. of उप-√सद् – they have approached>, ऊं <(उ) ind. – and, also, besides>, सदः <m. abl. sg of सद् – from the seat (of fire); here, the seat refers to the Stone, the celestial ioniser>, सामानि <n. acc. pl. of सामन् – Sāma hymns>, चक्रुः <prf. 3rd person pl. of √कृ – they have made>, तसराणि <n. acc. pl. of तसर – weaving shuttles>, ओतवे <m. dat. sg. of ओतु – for the woof (weft) or cross-threads of a web>.

RV x.130.3 author यज्ञः प्राजापत्यः, to भाववृत्तं, metre त्रिष्टुप् छंदः
कासीत्प्रमा प्रतिमा किं निदानमाज्यं किमासीत्परिधिः क आसीत् ।
छंदः किमासीत्प्रउगं किमुक्थं यद्देवा देवमयजंत विश्वे ॥१०.१३०.३॥
का । आसीत् । प्रऽमा । प्रतिऽमा । किं । निऽदानं । आज्यं । किं । आसीत् । परिऽधिः । कः । आसीत्
। छंदः । किं । आसीत् । प्रउगं । किं । उक्थं । यत् । देवाः । देवं । अयजंत । विश्वे ॥१०.१३०.३॥

What was the foundation? What was the image? What was the primary cause? What was the fuel for the sacrificial fire? What was the enclosure laid round the sacrificial fire to keep it together? When all the Devas worshipped that one Deva, what was the Chanda? Which Praugaṃ? What was the Ukthaṃ? (RV x.130.3)

का <f. nom. sg. of interrogative pron. क – what, which>, आसीत् <ipf. 3rd person sg. of √अस् – it was>, प्रमा <f. nom. sg. of प्रमा – basis, foundation, measure>, प्रतिमा <f. nom. sg. of प्रतिमा – picture, image, i.e. the plan or design>, किं <n. nom. sg. of क – what, which>, निदानं <n. nom. sg. of निदान – a first or primary cause>, आज्यं <n. nom. sg. of आज्य – "melted or clarified butter (used for oblations, or for pouring into the holy fire at the sacrifice)", i.e. oil or fuel to keep the sacrificial fire burning>, किं <n. nom. sg. of क – what, which>, आसीत् <ipf. 3rd person sg. of √अस् – it was>, परिधिः <m. nom. sg. of परिधि – an enclosure, fence; sticks laid round a sacrificial fire to keep it together>, कः <m. nom. sg. of interrogative pron. क – what, which>, आसीत् <ipf. 3rd person sg. of √अस् – it was>;

छंदः <n. nom. sg. of छंदस् – chanda, metre>, किं <n. nom. sg. of क – what, which>, आसीत् <ipf. 3rd person sg. of √अस् – it was>, प्रउगं <n. nom. sg. of प्रउग (=प्रउगशस्त्र) – praugaṃ, "name of the second Śastra or hymn at the morning libation", i.e. hymn or Speech>, किं <n. nom. sg. of क – what, which>, उक्थं <n. nom. sg. of उक्थ – ukthaṃ, "a certain recited verse forming a subdivision of the śastras", i.e. a verse or stanza>, यत् <conjunction – when>, देवाः <m. nom. pl. of देव – the Devas>, देवं <m. acc. sg. of देव – the Deva>, अयजंत <ipf. 3rd person pl. of √यज् – they (the Devas) worshipped, honoured>, विश्वे <m. nom. pl. of of pron. विश्व – every, all>.

RV x.130.4 author यज्ञः प्राजापत्यः, to भाववृत्तं, metre त्रिष्टुप् छंदः
अग्नेर्गायत्र्यभवत्सयुग्वोष्णिहया सविता सं बभूव ।
अनुष्टुभा सोम उक्थैर्महस्वान्बृहस्पतेर्बृहती वाचमावत् ॥१०.१३०.४॥
अग्नेः । गायत्री । अभवत् । सऽयुग्वा । उष्णिहया । सविता । सं । बभूव ।
अनुऽस्तुभा । सोमः । उक्थैः । महस्वान् । बृहस्पतेः । बृहती । वाचं । आवत् ॥१०.१३०.४॥
From Agni Gāyatrī arose. By Uṣṇihā associated Savitā was born.
By Anuṣṭup, delightful Soma. With Ukthas, Bṛhatī impelled the Speech of Bṛhaspati.
(RV x.130.4)

अग्नेः <m. abl. sg. of अग्नि – from Agni (proton)>, गायत्री <f. nom. sg. of गायत्री – Gāyatrī metre>, अभवत् <ipf. 3rd person sg. of √भू – it arose, came into being>, सयुग्वा <m. nom. sg. of adj. सयुग्वन् – united or associated>, उष्णिहया <f. inst. sg. of उष्णिहा – by Uṣṇihā metre>, सविता <m. nom. sg. of सवितृ – Savitā>, सं <ind. prefix to verb बभूव √भू>, सं-बभूव <prf. 3rd person sg. of सं-√भू – he has been born, has risen>;

अनुष्टुभा <f. inst. sg. of अनुष्टुभ् – by Anuṣṭup metre>, सोमः <m. nom. sg. of सोम – Soma (electron)>, उक्थैः <n. inst. pl. of उक्थ – with Ukthas, i.e. with a series of stanzas recited or sung>, महस्वान् <m. nom. sg. of adj. महस्वत् – gladdening, delightful>, बृहस्पतेः <m. gen. sg. of बृहस्पति – of Bṛhaspati, of the lord of prayer (Bṛhaspati refers to the Stone, the celestial ioniser)>, बृहती <f. nom. sg. of बृहती – Bṛhatī metre>, वाचं <f. acc. sg. of वाच् – sound, Speech>, आवत् <ipf. 3rd person sg. of √अव् – it impelled, animated>.

RV x.130.5 author यज्ञः प्राजापत्यः, to भाववृत्तं, metre त्रिष्टुप् छंदः
विराण्मित्रावरुणयोरभिश्रीरिंद्रस्य त्रिष्टुबिह भागो अह्नः ।
विश्वान्देवाञ्जगत्या विवेश तेन चाक्लृप्र ऋषयो मनुष्याः ॥१०.१३०.५॥
विऽराट् । मित्रावरुणयोः । अभिऽश्रीः । इंद्रस्य । त्रिऽस्तुप् । इह । भागः । अह्नः ।
विश्वान् । देवान् । जगती । आ । विवेश । तेन । चाक्लृप्रे । ऋषयः । मनुष्याः ॥१०.१३०.५॥
Virāṭ the radiant light of Mitra-Varuṇa, Triṣṭup Indra's fortunate lot of the day of sacrifice.
Jagatī takes possession of all the Devas. Through this process, the Manuṣyas become the Ṛṣis.
(RV x.130.5)

विराट् <f. nom. sg. of विराज् – Virāṭ metre>, मित्रावरुणयोः <m. gen. du. of मित्रावरुण – of Mitra-Varuṇa>, अभिश्रीः <mf. nom. sg. of अभिश्री (अभि ind. prefix that expresses superiority, intensity + श्री f. light, radiance, splendour) – radiant light>, इंद्रस्य <m. gen. sg. of इंद्र – of Indra>, त्रिष्टुप् <f. nom. sg. of त्रिष्टुभ् – Triṣṭup metre>, इह <ind. – in this place, here>, भागः <m. nom. sg. of भाग – portion, share, fortunate lot>, अह्नः <n. gen. sg. of अहन् (=अहर्) – of a sacrificial day>;

विश्वान् <m. acc. pl. of pron. विश्व – all, every>, देवान् <m. acc. pl. of देवा – the Devas>, जगती <f. nom. sg. of जगती (=जगत्) – Jagatī metre>, आ <ind. prefix of verb विवेश (√विश्)>, आ-विवेश <prf. 3rd person sg. of आ-√विश् – it enters, reaches, takes possession of>, तेन <mn. inst. sg. of pron तद् – by/through this>, चाक्लृप्रे <prf. middle 3rd person pl. of √कॢप् – they have become (with nom.)>, ऋषयः <m. nom. pl. of ऋषि – Ṛṣis, the singers of sacred hymns>, मनुष्याः <m. nom. pl. of मनुष्य – Manuṣyas, men, a class of deceased (ionised) ancestors>.

RV x.130.6 author यज्ञः प्राजापत्यः, to भाववृत्तं, metre त्रिष्टुप् छंदः
चाक्लृप्रे तेन ऋषयो मनुष्या यज्ञे जाते पितरो नः पुराणे ।
पश्यन्मन्ये मनसा चक्षसा तान्य इमं यज्ञमयजंत पूर्वे ॥१०.१३०.६॥
चाक्लृप्रे । तेन । ऋषयः । मनुष्याः । यज्ञे । जाते । पितरः । नः । पुराणे ।
पश्यन् । मन्ये । मनसा । चक्षसा । तान् । ये । इमं । यज्ञं । अयजंत । पूर्वे ॥१०.१३०.६॥
By this process, Manuṣyas, our forefathers, became Ṛṣis at the ancient Yajña that has emerged. Beholding them with the mind and the eye, I honour those who first performed this Yajña. (RV x.130.6)

चाक्लृप्रे <prf. middle 3rd person pl. of √कॢप् – they have become (with nom.)>, तेन <mn. inst. sg. of pron तद् – by/through this (process)>, ऋषयः <m. nom. pl. of ऋषि – Ṛṣis, the singers of sacred hymns>, मनुष्याः <m. nom. pl. of मनुष्य – Manuṣyas, men, a class of deceased ancestors>, यज्ञे <m. loc. sg. of यज्ञ – at Yajña>, जाते <m. loc. sg. of adj. जात – born, brought into existence, arisen>, पितरः <m. nom. pl. of पितृ – fathers, forefathers>, नः <m. gen. pl. of 1st person pron. अस्मद् – our>, पुराणे <m. loc. sg. of adj. पुराण – old, ancient>;

पश्यन् <m. nom. sg. of pre. active pt. पश्यत् (√पश्) – I seeing, observing>, मन्ये <pre. middle 1st person sg. of √मन् – I regard, honour, esteem>, मनसा <n. inst. sg. of मनस् – with the mind>, चक्षसा <n. inst. sg. of चक्षस् – with the eye>, तान् <m. acc. pl. of 3rd person pron. तद् – them>, ये <m. nom. pl. of relative pron. यद् – those who>, इमं <m. acc. sg. of pron. इदं – this>, यज्ञं <m. acc. sg. of यज्ञ – Yajña>, अयजंत <ipf. 3rd person pl. of √यज् – they offered, performed>, पूर्वे <m. nom. pl. of adj. पूर्व – first>.

RV x.130.7 author यज्ञः प्राजापत्यः, to भाववृत्तं, metre त्रिष्टुप् छंदः
सहस्तोमाः सहछंदस आवृतः सहप्रमा ऋषयः सप्त दैव्याः ।
पूर्वेषां पंथामनुदृश्य धीरा अन्वालेभिरे रथ्यो३ न रश्मीन् ॥१०.१३०.७॥
सहऽस्तोमाः । सहऽछंदसः । आऽवृतः । सहऽप्रमाः । ऋषयः । सप्त । दैव्याः ।
पूर्वेषां । पंथां । अनुऽदृश्य । धीराः । अनुऽआलेभिरे । रथ्यः । न । रश्मीन् ॥१०.१३०.७॥
The seven celestial Ṛṣis abound with the hymns and measures of the mighty Chanda. Beholding the ancient path, the wise ones have taken up the reins as charioteers. (RV x.130.7)

सहस्तोमाः <m. nom. pl. of सहस्तोम – with eulogia, hymns>, सहछंदसः <n. gen. sg. of सहछंदस् – of mighty chanda, metre>, आवृतः <m. nom. sg. of adj. आवृत – filled with, abounding with>, सहप्रमाः <m. nom. pl. of adj. सहप्रम – with the measures (rhythms or metrical units)>, ऋषयः <m. nom. pl. of ऋषि – Ṛṣis>, सप्त <m. nom. pl. of सप्तन् – seven>, दैव्याः <m. nom. pl. of adj. दैव्य – celestial, divine>;

पूर्वेषां <m. gen. pl. of adj. पूर्व – old, ancient, traditional>, पंथां <m. acc. sg. of पथिन् – path>, अनुदृश्य <ind. pt. of अनु-√दृश् – surveying, beholding>, धीराः <m. nom. pl. of धीर – the wise ones>, अन्वालेभिरे <prf. middle 3rd person pl. of अन्व्-आ-√लभ् – they have taken in the hand, they have taken up>, रथ्यः <m. nom. pl. of रथी – carriage-drivers, charioteers>, न <ind. – like, as>, रश्मीन् <m. acc. pl. of रश्मि – reins>.

The names of the Seven Ṛṣis (सप्त ऋषयः pl. sapta ṛṣayaḥ), as provided by Monier-Williams, are Gotama (गोतमः the best of oxen), Bharadvāja (भरद्वाजः the bearer of speed), Viśvā-Mitra (विश्वा-मित्रः all-containing Mitra), Jamadagni (जमदग्निः Agni the blazing fire), Vasiṣṭha (वसिष्ठः the owner of the cow of plenty, the richest), Kaśyapa (कश्यपः one who has black teeth), and Atri (अत्रिः devourer).

Gotama, the best of oxen, is Agni, proton. Bharadvāja, the bearer of speed, is Soma, electron. Viśvā-Mitra represents Mitra, the plasma light discharge. Jamadagni, Agni the blazing fire, is Aryamā, the plasma fire discharge. Kaśyapa, the one who has black teeth, represents Varuṇa, the plasma dark discharge in particular and cosmic plasmas in general. Vasiṣṭha is Rudra or the Soma Pond. Atri, the devourer, is the Stone, the celestial ioniser.

Thus, the seven Ṛṣis are the ones who perform cosmic sacrifices, through the processes of cosmic ionisation and cosmic plasma discharge and release, and sustain our galactic system. Now that the seven Ṛṣis have taken up the reins as charioteers, they are ready to release the cosmic plasmas, the secret lights and fires of cosmic creation that sustain our galactic system. Let us watch the spectacles of the upward and downward releases of cosmic plasmas.

VI. The Events of Cosmic Plasma Release

The plasma blooms, or Soma plants (सोमाः somāḥ pl. RV i.137.1), or light-containing plants (ओषधयः oṣadhayaḥ pl.), growing in the Soma Pond, when matured, are released. According to the Ṛgveda, they are released in two different modes: Aśvinā (अश्विना du.) the twin Jets and the celestial river Gaṅgā (गंगा). Aśvinā, the twin Jets, one from the Upper Vault of the Heaven and the other from the Lower Vault of the Heaven, fly upwards with winged steeds, travel around the Vault of the Heaven, and down towards the Milky Way (see Figure 13).

As the twin Jets shoot upwards over the summits of the Upper and the Lower Vaults of the Heaven, the two Heavenly Vaults simultaneously release the downward flow, the majestic river Gaṅgā (गंगा). Gaṅgā flows downwards from the base of the Soma Pond or the Samudra, the Ocean of celestial Waters, and falls upon Mt. Meru (मेरुः), the Galactic Bar, at the centre of the Milky Way. Then, from the summit of Mt. Meru, it divides itself into four parts and flows down along the four Arms of the Milky Way. There are two branches of the river Gaṅgā: the upper Gaṅgā and the lower Gaṅgā. The upper Gaṅgā falls from the base of the Soma Pond in the Upper Vault of the Heaven and the lower Gaṅgā falls from the base of the Soma Pond in the Lower Vault of the Heaven. Both branches of Gaṅgā fall upon Mt. Meru.

These upward and downward releases of celestial Waters, or cosmic plasmas, from the Upper and Lower Vaults of the Heaven are the great floods that inundate the Milky Way with celestial Waters, which sustain the Milky Way, three times a day every day.

6.1 Aśvinā, the Twin Jets

Aśvinā represent the cosmic twin Jets. The word Aśvinā (du.) literally means "the two charioteers". The chariot of Aśvinā is 'drawn by eagles (श्येनपत्वा RV i.118.1)' or flies 'with winged steeds (विभिः RV i.46.3)' and travels around the Heavenly Vault (RV iv.45.1). They are called Nāsatyā (नासत्या du. RV i.46.5). They are two performers of wondrous deeds (दस्रा dasrā du. RV i.46.2, RV vi.62.5). Their two paths are circumambulatory (RV i.46.14); one circumambulates the Upper Vault of the Heaven, the other circumambulates the Lower Vault of the Heaven (see Figure 13). They are the two lords of splendour (शुभस्पती du. RV x.93.6).

It appears that the stanzas that explain the cosmic plasma discharge and release events are notoriously misinterpreted. Stanzas RV x.61.5-9 are good examples of this notorious misinterpretation and mistranslation. Griffith passed over these stanzas in his English translation and instead provided Latin translations of stanzas RV x.61.5-8. For the stanza RV x.61.9, even a Latin translation was omitted and Griffith chose to provide Wilson's translation of the stanza. It turns out that stanzas RV x.61.5-9, presented in section 6.1.2, actually provide information about how cosmic plasmas are released as cosmic twin Jets.

<Figure 12> "Artist's Illustration of Galaxy with Jets from a Supermassive Black Hole".
Image credit: ESA/Hubble, L. Calçada (ESO)

Figure 12 is an illustration of the cosmic twin Jets from a "supermassive Black Hole" created by Calçada at the European Space Agency (ESA). According to the Ṛgveda, however, the cosmic twin Jets are released from the two Soma Ponds. They are released three times a day every day, by which, along with the downward release of the river Gaṅgā, the whole galactic system is sustained. One Jet is released from the Soma Pond in the Upper Vault of the Heaven and the other Jet from the Soma Pond in the Lower Vault of the Heaven. The evidence for the cosmic twin Jets' existence in the Milky Way has been presented by astrophysicists. For example, see Su and Finkbeiner (2012).[87]

6.1.1 Aśvinā Transport the Plasma Sheaths Filled with Soma Mead

According to hymn RV iv.45, the chariot of Aśvinā carries three swift steeds, which refer to the triply-spun magnetic field-aligned filament, and a sheath filled full with sweet mead as a fourth part (RV iv.45.1). What stanza RV iv.45.1 reveals is that each plasma bloom is a plasma sheath, which contains a triply-spun filament and cosmic plasmas blooming on the surface of this filament.

Tearing away the veil of darkness, Aśvinā's chariots, with the swift steeds and celestial lights, ascend to the Heavenly Sphere (RV iv.45.2). Note, here, Aśvinā's chariots refer to plasma blooms, or plasma sheaths. In RV iv.45.3, Aśvinā are implored to transport the plasma sheaths filled full with Soma mead. Imagine a fleet of long, serpentine plasma sheaths sailing through the three Heavenly Spheres and beyond.

The first three stanzas of RV iv.45 are addressed to Aśvinā. The author is Vāmadeva (वामदेवः). For the word 'vāma', as an adjective, definitions such as "beautiful, splendid, noble" and as a noun, "the female breast", "name of Rudra", and so on are provided by Monier-Williams. These definitions provide clues that Vāmadeva refers to the Soma Pond.

RV iv.45.1 author वामदेवः, to अश्विनौ, metre जगती छंदः
ए॒ष स्य भा॒नुरुदि॑यर्ति यु॒ज्यते॒ रथः॒ परि॑ज्मा दि॒वो अ॒स्य सान॑वि ।
पृ॒क्षासो॑ अ॒स्मिन्मि॑थु॒ना अधि॒ त्रयो॒ दृति॑स्तु॒रीयो॒ मधु॑नो॒ वि र॑प्शते ॥४.४५.१॥
ए॒षः । स्यः । भा॒नुः । उत् । इ॒य॒र्ति॒ । यु॒ज्यते॑ । रथः॑ । परि॑ऽज्मा । दि॒वः । अ॒स्य । सान॑वि ।
पृ॒क्षासः॑ । अ॒स्मि॒न । मि॒थु॒नाः । अधि॑ । त्रयः॑ । दृतिः॑ । तु॒रीयः॑ । मधु॑नः । वि । र॒प्श॒ते॒ ॥४.४५.१॥
Here rises the light, to which your chariot is harnessed, driving over the summit of the Heaven. In this chariot are a set of three swift steeds and a sheath filled full with sweet mead, the fourth part. (RV iv.45.1)

एषः <m. nom. sg. of pron. एतद् – this, this here, here>, स्यः <m. nom. sg. of pron. त्यद् – that>, भानुः <m. nom. sg. of भानु – light, a ray of light>, उत् <(उद्) ind. – prefix to verb इयर्ति (√ऋ)>, उद्-इयर्ति <pre. 3rd person sg. of उद्-√ऋ – it rises>, युज्यते <pre. passive 3rd person sg. of √युज् – it (the chariot) is yoked or harnessed>, रथः <m. nom. sg. of रथ – a chariot>, परिज्मा <m. nom. sg. of adj. परिज्मन् – running or driving round>, दिवः <m. gen. sg. of दिव् – of the Heaven>, अस्य <m. gen. sg. of इदं – of this>, सानवि <n. loc. sg. of सानु – around the summit, ridge>;

पृक्षासः <(=पृक्षाः) m. nom. pl. of पृक्ष – swift horses>, अस्मिन <m. loc. sg. of इदं – in/on this (the chariot of Aśvinā)>, मिथुनाः <m. nom. pl. of adj. मिथुन – forming a pair, a set>, अधि <ind. – (with loc.) over, on, at>, त्रयः < m. nom. sg. of adj. त्रय – threefold, consisting of three>, दृतिः <m. nom. sg. of दृति – skin, hide, i.e. a plasma sheath>, तुरीयः <m. nom.

---

87 Meng Su and Douglas P. Finkbeiner, 'Evidence for Gama-Ray Jets in the Milky Way', *The Astrophysical Journal* 753:61(13pp), 01 July 2012. doi:10.1088/0004-637X/753/1/61.

sg. of adj. तुरीय – fourth, the 4th part>, मधुनः <n. gen. sg. of मधु – of mead (refers to cosmic plasmas)>, वि <ind. prefix to verb रप्शते (√रप्श्)>, वि-रप्शते <pre. middle 3rd person sg. of वि-√रप्श् – it is full, abound in (gen.)>.

RV iv.45.2 author वामदेवः, to अश्विनौ, metre जगती छंदः
उद्वां पृक्षासो मधुमंत ईरते रथा अश्वास उषसो व्युष्टिषु ।
अपोर्णुवंतस्तम आ परीवृतं स्व१र्ण शुक्रं तन्वंत आ रजः ॥४.४५.२॥
उत् । वां । पृक्षासः । मधुऽमंतः । ईरते । रथाः । अश्वासः । उषसः । विऽउष्टिषु ।
अपऽऊर्णुवंतः । तमः । आ । परिऽवृतं । स्वः । न । शुक्रं । तन्वंतः । आ । रजः ॥४.४५.२॥
Your chariots, fleet horses, and sweet mead rise up at the first gleam of dawn.
Tearing away the veil of darkness, as light, they spread towards the resplendent Heavenly sphere. (RV iv.45.2)

उत् <(उद्) ind. prefix to verb ईरते (√ईर्)>, वां <m. gen. du. of युष्मद् – your (du.)>, पृक्षासः <m. nom. pl. of adj. पृक्ष – swift, fleet>, मधुमंतः <m. nom. pl. of मधुमत् – sweets, sweet mead>, उद्-ईरते <pre. middle 3rd person pl. of उद्-√ईर् – they rise, ascend>, रथाः <m. nom. pl. of रथ – chariots>, अश्वासः <m. nom. pl. of अश्व – horses>, उषसः <f. gen. sg. of उषस् – of the dawn>, व्युष्टिषु <f. loc. pl. of व्युष्टि – at the first gleam or breaking of dawn>;

अपोर्णुवंतः <m. nom. pl. of pre. active pt. अपोर्णुवत् (√अपोर्णु=अप-√ऊर्णु) – they uncovering, unveiling>, तमः <n. acc. sg. of तमस् – darkness>, आ <ind. prefix to pt. तन्वंतः (√तन्)>, परिवृतं <n. acc. sg. of adj. परिवृत (=परिवारित) – covered with, veiled in>, स्वः <(स्वर्) ind. – light, i.e. light of Sūrya>, न <ind. – like, as>, शुक्रं <n. acc. sg. of adj. शुक्र – bright, resplendent>, आ-तन्वंतः <m. nom. pl. of pre. active pt. आ-तन्वत् (आ-√तन्) – they stretch, spread>, आ <ind. preposition – (with a following acc.) to, up to>, रजः <n. acc. sg. of रजस् – to the Heavenly Sphere>.

RV iv.45.3 author वामदेवः, to अश्विनौ, metre जगती छंदः
मध्वः पिबतं मधुपेभिरासभिरुत प्रियं मधुने युंजाथां रथं ।
आ वर्तनिं मधुना जिन्वथस्पथो दृतिं वहेथे मधुमंतमश्विना ॥४.४५.३॥
मध्वः । पिबतं । मधुऽपेभिः । आसऽभिः । उत । प्रियं । मधुने । युंजाथां । रथं ।
आ । वर्तनिं । मधुना । जिन्वथः । पथः । दृतिं । वहेथे इति । मधुऽमंतं । अश्विना ॥४.४५.३॥
Drink ye the Soma mead with jaws accustomed to drinking madhu. Fasten ye the beloved chariot for the sweet mead's sake. With sweet mead, O Aśvinā, refresh the wheel track of the path and bring the sheath that holds the sweet mead. (RV iv.45.3)

मध्वः <n. gen. sg. of मधु – of madhu, sweet mead, Soma mead>, पिबतं <ipv. 2nd person du. of √पा – ye (du.) draw in, drink (with gen.)>, मधुपेभिः <(=मधुपैः) m. inst. pl. of adj. मधुप – with drinking sweetness, drinking madhu>, आसभिः <n. inst. pl. of आसन् – with mouths, jaws>, उत <ind. – and, slso, even>, प्रियं <m. acc. sg. of adj. प्रिय – beloved, liked, favourite>, मधुने <n. dat. sg. of मधु – for madhu, the sweet mead>, युंजाथां <(युञ्जाथाम्) ipv. 2nd person du. of √युज् – ye (du.) yoke, fasten, make ready>, रथं <m. acc. sg. of रथ – chariot>;

आ <ind. prefix to verb जिन्वथः (√जिन्व्)>, वर्तनिं <f. acc. sg. of वर्तनि – a wheel track, rut>, मधुना <n. inst. sg. of मधु – with sweet mead>, आ-जिन्वथः <pre. 2nd person du. of आ-√जिन्व् – ye (du.) refresh>, पथः <m. gen. sg. of पथिन् – of the path, road>, दृतिं <m. acc. sg. of दृति – a skin, a leather bag (for holding water), i.e. a plasma sheath that holds the Soma mead>, वहेथे <pre. middle 2nd person du. of √वह् – ye (du.) carry, transport>, मधुमंतं <m. acc. sg. of adj. मधुमत् – containing sweet mead>, अश्विना <m. voc. du. of अश्विन् – O Aśvinā (twin)>.

6.1.2 The Spectacle of Cosmic Eruption, the Dappled Cows Are Released

Stanzas RV x.61.4-9 describe the events of cosmic plasma discharge from the Stone, the celestial ioniser, into the Soma Pond and the subsequent release of cosmic plasmas from the Soma Pond, the Ocean of celestial Waters. The releasing of the matured plasma blooms from the Soma Pond is called the Soma Yajña, the Soma sacrifice.

In stanza RV x.61.4, the Aśvinā are invoked to sit among the herd of ruddy cattle and lead the Soma Yajña. They are called 'two black sons of the Heaven'. Note that the colour 'black' or 'dark-blue' is the colour of King Varuṇa (वरुणः) who represents cosmic plasma in general and cosmic plasma dark discharge in particular. The enemy refers to Vṛtra, the serpent of the Heaven, the plasma sheath.

RV x.61.4 author नाभानेदिष्ठ मानवः, to विश्वे देवाः, metre त्रिष्टुप् छंदः
कृष्णा यद्गोष्वरुणीषु सीदद्दिवो नपाताश्विना हुवे वां ।
वीतं मे यज्ञमा गतं मे अन्नं ववन्वांसा नेषमस्मृतध्रू ॥१०.६१.४॥
कृष्णा । यत् । गोषु । अरुणीषु । सीदत् । दिवः । नपाता । अश्विना । हुवे । वां ।
वीतं । मे । यज्ञं । आ । गतं । मे । अन्नं । ववन्वांसा । न । इषं । अस्मृतध्रू इत्यस्मृतऽध्रू ॥१०.६१.४॥
When I call on you, O Aśvinā, the two black sons of the Heaven, to sit among the herd of ruddy cattle, Once ye procured my juicy food, lead my Yajña now taking place, and forget not to slay the enemy. (RV x.61.4)

कृष्णा <m. acc. du. of adj. कृष्ण – black, dark blue>, यत् <(यद्) conjunction – that, when>, गोषु <f. loc. pl. of गो – among the herd of cattle>, अरुणीषु <f. loc. pl. of adj. अरुण – red, ruddy>, सीदत् <infinitive? inj.? of √सद् – to sit>, दिवः <m. gen. sg. of दिव् – of the Heaven>, नपाता <m. acc. du. of नपात् – two descendants, two sons>, अश्विना <m. voc. du. of अश्विन् – O Aśvinā>, हुवे <pre. 1st person sg. of √हू – I call, invoke>, वां <m. acc. du. of युष्मद् – you (du.)>;

वीतं <ipv. 2nd person du. of √वी – ye (du.) seize, impel, lead>, मे <m. gen. sg. of 1st person pron. अस्मद् – my>, यज्ञं <m. acc. sg. of यज्ञ – Yajña>, आ <ind. prefix to गतं (गत)>, आ-गतं <m. acc. sg. of adj. आ-गत – occurred, happened, anything that has taken place>, मे <m. gen. sg. of 1st person pron. अस्मद् – my>, अन्नं <n. acc. sg. अन्न – food in a mystical sense, i.e. Soma>, ववन्वांसा <m. nom. du. of prf. active pt. ववन्वांस् (√वन्) – (the two) having acquired, procured>, न <ind. – not, no, nor>, इषं <n. acc. sg. of adj. इष – possessing sap, sappy, juicy>, अस्मृतध्रू <m. nom. du. of adj. अस्मृतध्रु (अस्मृत forgotten, forgetting + √ध्रु (=√ध्वृ) to cause to fall, hurt, injure) – (with न above) not forgetting to slay (the enemy)>.

Stanza RV x.61.5 explains how plasma blooms tear away to release cosmic plasmas. Once released, the remaining stump thrusts back. The word 'once more' indicates that it is a repeated event. Note that the maidens refer to the firestones of the Stone, the celestial ioniser.

RV x.61.5 author नाभानेदिष्ठ मानवः, to विश्वे देवाः, metre त्रिष्टुप् छंदः
प्रथिष्ट यस्य वीरकर्ममिष्णदनुष्ठितं नु नर्यो अपौहत् ।
पुनस्तदा वृहति यत्कनाया दुहितुरा अनुभृतमनर्वा ॥१०.६१.५॥

प्रथिष्ट । यस्य । वीरऽकर्मं । इष्णात् । अनुऽस्थितं । नु । नर्यः । अप । औहत् ।
पुनरिति । तत् । आ । वृहति । यत् । कनायाः । दुहितुः । आः । अनुऽभृतं । अनर्वा ॥१०.६१.५॥
The man has stretched. Now, the man, whose heroic act of releasing accomplished, thrusts back. Ah! that, which ye procured abundantly from the maidens and the Soma extractor, tears away once more. (RV x.61.5)

प्रथिष्ट <aor. middle 3rd person sg. of √प्रथ् – he (the man, a plasma bloom) extended, stretched>, यस्य <m. gen. sg. of pron. यद् – whose>, वीरकर्मं <n. nom. sg. of वीरकर्म – a heroic or manly act>, इष्णत् <n. nom. sg. of pre. active pt. इष्णत् (√इष्) – pouring out, streaming out, releasing (of cosmic plasmas)>, अनुष्ठितं <n. nom. sg. of adj. अनुष्ठित – done, executed, accomplished>, नु <ind. – so now, now>, नर्यः <m. nom. sg. of नर्य – the man (refers to a plasma bloom or Soma Plant)>, अपौहत् <ipf. 3rd person sg. of √अपोह् (अप away, back + √ऊह् to push, thrust) – he pushed back, thrust back>;

पुनः <(पुनर्) ind. – again, once more>, तत् <m. nom. sg. of तद् – that, i.e. plasma bloom or Some plant>, आ <ind. prefix of verb वृहति (√वृह्)>, आ-वृहति <(=आ-बृहति) pre. 3rd person sg. of आ-√वृह् (=आ-√बृह्) – it tears away>, यत् <n. nom. sg. of यद् – what, which>, कनायाः <f. abl. sg. of कना – from maidens (the firestones of the Stone, the celestial ioniser)>, दुहितुः <f. abl. sg. of दुहितृ – from the Soma extractor (the Soma Pond)>, आः <(आस्) ind. – Ah! Oh!>, अनुभृतं <ipv. 2nd person du. of √अनुभृ (अनु-√भृ) – ye (the twins Aśvinā) brought or procured (here, ipv. 2nd person du. corresponds to injunctive, which is an unaugmented past tense)>, अनर्वा <m. nom. du. of adj. अनर्व – not to be limited, not to be obstructed, i.e. abundantly>.

Stanza RV x.61.6 explains that, as the matured plasma blooms are released, the Heaven and the maiden of the celestial ioniser produce and discharge cosmic plasmas again. And the discharged cosmic plasmas spread along the ridge of the mountain. The ridge of the mountain refers to the ridge of an individual plasma bloom. The 'virtuous deed' (सुकृतं sukṛtaṃ) refers to the cosmic Yajña, the act of producing and discharging cosmic plasmas. The seat of virtuous deed refers to the Soma Pond, into which cosmic plasmas are discharged and, when fully bloomed and matured, from which they are released.

RV x.61.6 author नाभानेदिष्ठ मानवः, to विश्वे देवाः, metre त्रिष्टुप् छंदः
मध्या यत्कर्त्वमभवदभीके कामं कृण्वाने पितरि युवत्यां ।
मनानग्रेतो जहतुर्वियंता सानौ निषिक्तं सुकृतस्य योनौ ॥१०.६१.६॥
मध्या । यत् । कर्त्वं । अभवत् । अभीके । कामं । कृण्वाने । पितरि । युवत्यां ।
मनानक् । रेतः । जहतुः । विऽयंता । सानौ । निऽसिक्तं । सुऽकृतस्य । योनौ ॥१०.६१.६॥
In the meantime, as the Heaven and the Maiden accomplished that desired task together,
At once, they discharged a stream of celestial Water. And the discharged stream spread out along the mountain ridge at the seat of virtuous deed. (RV x.61.6)

मध्या <ind. – meanwhile>, यत् <conjunction – that, when, as>, कर्त्वं <n. acc. sg. of कर्त्व – duty, task (of pressing Soma, i.e. ionisation)>, अभवत् <ipf. 3rd person sg. of √भू – it (the task) was, became, accomplished>, अभीके <ind. – in meeting together>, कामं <n. acc. sg. of adj. काम – wished, desired>, कृण्वाने <m. loc. sg. of pre. middle pt. कृण्वान (√कृ) – (with locative pre. pt. + with young maiden and the father Heaven in locative, it expresses the simultaneous action with another action) as the Maiden and the Heaven accomplished>, पितरि <m. loc. sg. of पितृ – Father the Heaven>, युवत्यां <f. loc. sg. of युवति – the young maiden, i.e. the Stone, the celestial ioniser>;

मनानक् <(=मनाक्) ind. – immediately, at once>, रेतः <n. acc. sg. of रेतस् – a flow, stream, current, a flow of Water (a stream of cosmic plasmas)>, जहतुः <prf. 3rd person du. of √हा 3P – they (the Heaven and the maiden) relinquished,

discharged>, वियंता <(वियन्ता) prf. 3rd person sg. of वि-√यं – it (the discharged stream of Water) spread out>, सानौ <m. loc. sg. of सानु – along the mountain ridge (refers to the ridge of a plasma bloom)>, निषिक्तं <n. nom. sg. of निषिक्त – the sprinkled, irrigated, i.e. the discharged stream of Water>, सुकृतस्य <n. gen. sg. of सुकृत – of good or righteous deed, meritorious act (refers to the cosmic Yajña)>, योनौ <m. loc. sg. of योनि – in the womb, abode, at the seat>.

RV x.61.7 explains how a plasma sheath is fashioned to form a new Soma Pond. Note that when the matured cosmic plasmas are released, the previously formed Soma Pond is destroyed. As newly produced cosmic plasmas are discharged, a new sheath is fashioned by their Speeches to form a new Soma Pond. In the stanza, the plasma sheath is called 'the protector of the house (वास्तोष्पतिः vāstoṣpatiḥ)' indicating the positive aspect of Vṛtra as a protective sheath of discharged cosmic plasmas until they are matured and ready to be released. The parched field refers to the destroyed, dry field of the Soma Pond after the release of the matured plasma blooms.

RV x.61.7 author नाभानेदिष्ठ मानवः, to विश्वे देवाः, metre त्रिष्टुप् छंदः
पिता यत्स्वां दुहितरमधिष्कन्क्ष्मया रेतः संजग्मानो नि षिंचत् ।
स्वाध्योऽजनयन्ब्रह्म देवा वास्तोष्पतिं व्रतपां निरतक्षन् ॥१०.६१.७॥
पिता । यत् । स्वां । दुहितरं । अधिऽस्कन् । क्ष्मया । रेतः । संऽजग्मानः । नि । सिंचत् ।
सुऽआध्यः । अजनयन् । ब्रह्म । देवाः । वास्तोः । पतिं । व्रतऽपां । निः । अतक्षन् ॥१०.६१.७॥
When the Heaven, united with his own Soma milker, poured out a stream of Water over the parched field, The devout Devas produced the sacred Speech and fashioned the protector of the house who upholds the ordinances. (RV x.61.7)

पिता <m. nom. sg. of पितृ – Father the Heaven>, यत् <conjunction – that, when>, स्वां <f. acc. sg. of स्व – one's own>, दुहितरं <f. acc. sg. of दुहितृ – the milker, extractor (of Soma), i.e. daughter milch Cow (refers to the Soma Pond)>, अधिष्कन्क्ष्मया <अधि ind. over + स्कन्न = शुष्क adj. dried, parched + क्ष्मया ind. in the field – over the dried or parched field (refers to the empty Soma Pond after the release of matured cosmic plasmas)>, रेतः <n. acc. sg. of रेतस् – a stream of Water or seed (cosmic plasmas)>, संजग्मानः <m. nom. sg. of prf. middle pt. संजग्मान (सं-√गम्) – he joining or united with (acc.)>, नि <ind. – in, into>, षिंचत् <(सिश्चत्) inj. 3rd person sg. of √सिच् – he poured out, discharged>;

स्वाध्यः <m. nom. pl. of adj. स्वाधी – thoughtful, devout>, अजनयन् <ipf. cau. 3rd person pl. of √जन् – they produced>, ब्रह्म <n. acc. sg. of ब्रह्मन् – pious utterance, prayer, i.e. Speech, hymn>, देवाः <m. nom. pl. of देव – Devas>, वास्तोष्पतिं <m. acc. sg. of वास्तोष्पति – the protector of the house, the name of Rudra (indicates the positive aspect of Vṛtra, the plasma sheath, as the protector of discharged cosmic plasmas)>, व्रतपां <m. acc. sg. of adj. व्रतपा – upholding ordinances>, निः <(निस्) ind. prefix to verb अतक्षन् (√तक्ष्)>, निरतक्षन् <(निस्-अतक्षन्) ipf. 3rd person pl. of निस्-√तक्ष् – they fashioned, formed>.

Stanza RV x.61.8 explains the event of cosmic plasma release from the Soma Pond, which is described as the ancient Ocean of splendour. Here, plasma blooms are described as 'foam'. The milch Cow refers to the Soma Pond and plasma blooms growing in it; the dappled cows, to the celestial lights, the cosmic plasmas, Agnis in particular.

RV x.61.8 author नाभानेदिष्ठ मानवः, to विश्वे देवाः, metre त्रिष्टुप् छंदः
स ईं वृषा न फेनमस्यदाजौ स्मदा परैदप दभ्रचेताः ।
सरत्पदा न दक्षिणा परावृङ्न ता नु मे पृशन्यो जगृभ्रे ॥१०.६१.८॥

सः । ई । वृषा । न । फेनं । अस्यत् । आजौ । स्मत् । आ । परा । एत् । अप । दभ्रऽचेताः ।
सरत् । पदा । न । दक्षिणा । पराऽवृक् । न । ताः । नु । मे । पृशन्यः । जगृभ्रे ॥१०.६१.८॥
He, like a bull, destroyed the foam at once in battle, which arose from the ancient Ocean of splendour. The milch Cow was plucked at once, and attached to the foot of flow, these dappled cows of mine were seized. (RV x.61.8)

सः <m. nom. sg. of 3rd person pron. तद् – he>, ईं <ind. – now>, वृषा <m. nom. sg. of वृषन् – a bull>, न <ind. – as, like>, फेनं <m. acc. sg. of फेन – foam (refers to plasma blooms that resemble the foam (see Fig. 7)>, अस्यत् <ipf. 3rd person sg. of √सो – he destroyed, killed>, आजौ <m. loc. sg. of आजि – in battle>, स्मत् <ind. – at once>, आ <ind. – (with a preceding loc.) in, at>, परा <f. nom. sg. of पर – past, ancient>, ऐत् <ipf. 3rd person sg. of √इ – it arose from, came from>, अप <ind. prefix to noun दभ्रचेताः – away, off>, दभ्रचेताः <mf. nom. sg. of दभ्र-चेतस् (दभ्र ocean + चेतस् splendour) – the Ocean of splendour (the Ocean of celestial Waters or cosmic plasmas), i.e. the Soma Pond>;

सरत्पदा <mfn. inst. sg. of सरत्पद् (सरत् flow + पदा inst. of पद् (at the end of a compound) "sticking to the feet of") – sticking to the feet of flow, at the feet of flow>, न <ind. – (with another न, it forms a strong affirmation) – certainly, surely>, दक्षिणा <f. nom. sg. of दक्षिणा – the milch Cow, i.e. the daughter milch Cow, the Soma Pond>, परावृक् <f. nom. sg. of adj. परावृज् (परा away, off + वृज् twist off, pull up, pluck) – twisted off, plucked, pulled up>, न <ind. – see न above>, ताः <f. nom. pl. of 3rd person pron. तद् – these>, नु <ind. – now, at once>, मे <m. gen. sg. of अस्मद् – my>, पृशन्यः <f. nom. pl. of पृशनी – dappled cows, rays of light, i.e. plasma blooms or cosmic plasmas>, जगृभ्रे <prf. middle 3rd person pl. of √ग्रभ् (=√ग्रह्) – they were seized, they relinquished>.

In stanza RV x.61.9, the fire (वह्निः vahniḥ) is the god of fire, Trita (त्रितः), or tritium, of the Stone. As soon as the milch Cow was plucked and the dappled cows were released, the Stone immediately produces a new batch of freed Agnis and brings them to the udder, the newly created Soma Pond. Agnis are born again, eager to fight. In the Stanza, Agni is described as the procurer of fuel for the sacred fire, the fire of the Stone.

RV x.61.9 author नाभानेदिष्ठ मानवः, to विश्वे देवाः, metre त्रिष्टुप् छंदः
मक्षू न वह्निः प्रजाया उपब्दिरग्निं न नग्न उप सीददूधः ।
सनितेध्मं सनितोत वाजं स धर्ता जज्ञे सहसा यवीयुत् ॥१०.६१.९॥
मक्षु । न । वह्निः । प्रऽजायाः । उपब्दिः । अग्निं । न । नग्नः । उप । सीदत् । ऊधः ।
सनिता । इध्मं । सनिता । उत । वाजं । सः । धर्ता । जज्ञे । सहसा । यविऽयुत् ॥१०.६१.९॥
The Stone, sounding like a bard, quickly brings forth subsequently produced Agni to the udder. The procurer of fuel for the sacred fire, bearer and bestower of wealth, he is born mightily, eager to fight. (RV x.61.9)

मक्षू <ind. – quickly, rapidly>, न <ind. – like, as>, वह्निः <m. nom. sg. of वह्नि – "fire (in general or 'the god of fire')", i.e. the fire of the Stone, the celestial ioniser>, प्रजायाः <f. gen. sg. of प्रजा – of descendants or aftergrowth, i.e. of subsequently produced cosmic plasmas>, उपब्दिः <m. nom. sg. of adj. उपब्दि (=उपब्द) – sounding, rattling>, अग्निं <m. acc. sg. of अग्नि – Agni>, न <ind. – like, as>, नग्नः <m. nom. sg. of नग्न – a bard accompanying an army>, उप <ind. prefix to verb सीदत् (√सद्)>, उप-सीदत् <inj. 3rd person sg. of उप-√सद् – it (the sacred fire) places, brings forth>, ऊधः <n. acc. sg. of ऊधस् – to the udder, i.e. to the Soma Pond>;

सनिता < m. nom. sg. of सनितृ – a procurer, bestower (with acc.)>, इध्मं <m. acc. sg. of इध्म – fuel as used for the sacred fire>, सनिता < m. nom. sg. of adj. सनितृ – a procurer, bestower (with acc.)>, उत <ind. – and, also, even>, वाजं <m. acc. sg. of वाज – vigour, wealth, treasure>, सः <m. nom. sg. of 3rd person pron. तद् – he>, धर्ता <m. nom. sg. of धर्तृ – a bearer, supporter>, जज्ञे <prf. middle 3rd person sg. of √जन् – he is born, produced>, सहसा <m. inst. sg. of सहस् – with strength or might, mightily>, यवीयुत् <m. nom. sg. of adj. यवीयुध् – eager to fight, fond of war>.

6.1.3 The Mighty Aśvinā Perform Wondrous Deeds

Hymn RV i.46 is addressed to Aśvinā (du.). Of the fifteen stanzas of the hymn, RV i.46.1-7 and 14 provide characteristic attributes of Aśvinā. Stanza RV i.46.1 invokes Aśvinā by praising them.

RV i.46.1 author प्रस्कण्वः काण्वः, अश्विनौ, गायत्री छंदः
एषो उषा अपूर्व्या व्युच्छति प्रिया दिवः । स्तुषे वामश्विना बृहत् ॥१.४६.१॥
एषो इति । उषाः । अपूर्व्या । वि । उच्छति । प्रिया । दिवः । स्तुषे । वां । अश्विना । बृहत् ॥१.४६.१॥
This Uṣā, beloved daughter of the Heaven, shines forth with her first light.
She lauds you two, O mighty Aśvinā. (RV i.46.1)

एषा <f. nom. sg. of pron. एतद् – this, she>, उषाः <f. nom. sg. of उषस् – Uṣā, Dawn (here, Uṣā refers to the Soma Pond)>, अपूर्व्या <f. nom. sg. of adj. अपूर्व्य – unpreceded, first>, व्युच्छति <pre. 3rd person sg. of वि-√वस् – she shines forth>, प्रिया <f. nom. sg. of प्रिय – the beloved (daughter)>, दिवः <m. gen. sg. of दिव् – of the Heaven>;

स्तुषे <pre. middle 3rd person sg. of √स्तु – she (Uṣā) praises, lauds, celebrates in song or hymns>, वां <m. acc. du. of 2nd person pron. युष्मद् – you (du.)>, अश्विना <m. voc. du. of अश्विन् – O Aśvinā>, बृहत् <ind. – lofty, tall, mighty>.

In RV i.46.2, Aśvinā are called Dasrā (दस्रा du.) who perform wondrous deeds of releasing celestial lights. Sindhu (सिंधुः siṃdhuḥ), the river of celestial Waters in the Soma Pond, is their mother.

RV i.46.2 author प्रस्कण्वः काण्वः, अश्विनौ, गायत्री छंदः
या दस्रा सिंधुमातरा मनोतरा रयीणां । धिया देवा वसुविदा ॥१.४६.२॥
या । दस्रा । सिंधुऽमातरा । मनोतरा । रयीणां । धिया । देवा । वसुऽविदा ॥१.४६.२॥
The two, the performers of wondrous deeds, the offerers of riches, who have Sindhu as their mother, With prayer the twin Devas bestow riches. (RV i.46.2)

या <m. nom. du. of यद् – who (du.)>, दस्रा <m. nom. du. of दस्र – two performers of wonderful deeds>, सिंधुमातरा <m. nom. du. of adj. सिंधुमातृ – having Sindhu for mother>, मनोतरा <m. nom. du. of मनोतृ – the two offerers, presenters>, रयीणां <m. gen. pl. of रयि – of riches>;

धिया <f. inst. sg of धी – with thought, prayer, i.e. hymn, Speech>, देवा <m. nom. du of देव – the twin Devas, i.e. Aśvinā>, वसुविदा <m. nom. du. of adj. वसुविद् – bestowing wealth>.

6.1.4 The Coursers of Aśvinā Rush Over the Blazing Summit of the Heaven

In RV i.46.3, the summit of the Heaven is described as 'blazing', for blooming plasmas look as if they are set ablaze (see Figure 7).

RV i.46.3 author प्रस्कण्वः काण्वः, अश्विनौ, गायत्री छंदः
वच्यंते वां ककुहासो जूर्णायामधि विष्टपि । यद्वां रथो विभिष्पतात् ॥१.४६.३॥
वच्यंते । वां । ककुहासः । जूर्णायां । अधि । विष्टपि । यत् । वां । रथः । विऽभिः । पतात् ॥१.४६.३॥
Your coursers hasten on over the blazing summit of the Heaven,
When your chariot flies with winged coursers. (RV i.46.3)

वच्यंते <pre. middle 3rd person pl. of √वञ्च् – they hasten, rush>, वां <m. gen. du. of 2nd person pron. युष्मद् – your (du.)>, ककुहासः <m. nom. pl. of ककुह – horses of Aśvinā>, जूर्णायां <f. loc. sg. of adj. जूर्ण (cf.जूर्णि and जूर्णिन्) – blazing, glowing>, अधि <ind. – above, over>, विष्टपि <f. loc. sg. of विष्टप् – on/over top, summit, height (of the Heaven)>;

यत् <conjunction – that, when>, वां <m. gen. du. of 2nd person pron. युष्मद् – your (du.)>, रथः <m. nom. sg. of रथ – a chariot>, विभिः <m. inst. pl. of वि – with birds (also applied to horses, arrows and the Maruts), with flying or winged horses; note that, here, birds and flying or winged horses refer to the plasma blooms or Soma plants>, पतात् <sub. 3rd person sg. of √पत् 1P – it (the chariot) flies>.

In RV i.46.4, the lord of the fortress and the cultivator refer to the Stone, the celestial ioniser. The Stone presents Aśvinā with abundant oblation. The fortress refers to the fortress of fire, that is, the 'Ring of Fire' of the celestial ioniser.

RV i.46.4 author प्रस्कण्वः काण्वः, अश्विनौ, गायत्री छंदः
हविषा जारो अपां पिपर्ति पपुरिर्नरा । पिता कुटस्य चर्षणिः ॥१.४६.४॥
हविषा । जारः । अपां । पिपर्ति । पपुरिः । नरा । पिता । कुटस्य । चर्षणिः ॥१.४६.४॥
He, the bountiful, the beloved of celestial Waters, grants the twin heroes with abundant oblation.
Lord of the fortress, the cultivator. (RV i.46.4)

हविषा <n. inst. sg. of हविस् – with oblation>, जारः <m. nom. sg. of जार – a lover, beloved>, अपां <f. gen. pl. of अप् – of celestial Waters, i.e. of cosmic plasmas>, पिपर्ति <pre. 3rd person. sg. of √पॄ – he grants abundantly, presents with (inst.)>, पपुरिः <m. nom. sg. of adj. पपुरि – liberal, bountiful>, नरा <m. acc. du. of नृ or नर – the twin heroes>;

पिता <m. nom. sg. of पितृ – a father, lord>, कुटस्य <m. gen. sg. of कुट – of the fort, stronghold (here, it refers to the Stone, the celestial ioniser)>, चर्षणिः <m. nom. sg. of चर्षणि – the cultivator (of Soma plants or plasma blooms)>.

In RV i.46.5, Aśvinā (du.) are called Nāsatyā (नासत्या m. nom. du. of नासत्य), 'two heroes of satyaṃ (सत्यं)'. Nāsatya (नासत्यः sg.) is a compound word of Nā (ना m. nom. sg. of नृ a man, hero) + satya (सत्यः m. Agni or Sūrya). In the Ṛgveda, Agni and Sūrya are called satyaṃ or satyā (सत्यं satyaṃ n. RV x.170.2, सत्या satyā f. RV iv.1.7), the truth, the real. Thus, Nāsatyā (du.) are 'the two heroes of Agni'. Aśvinā are called to bust open the Soma Pond with the command of their sacred hymns. The stanza indicates that the 'hymns' are the cause of the eruption of Vṛtra, the serpent of the Heaven, the plasma sheath.

RV i.46.5 author प्रस्कण्वः काण्वः, अश्विनौ, गायत्री छंदः
आदारो वां मतीनां नासत्या मतवचसा । पातं सोमस्य धृष्णुया ॥१.४६.५॥
आऽदारः । वां । मतीनां । नासत्या । मतऽवचसा । पातं । सोमस्य । धृष्णुऽया ॥१.४६.५॥
O Nāsatyā, tearing up the Soma Pond with the honoured command of your hymns,
Drink ye boldly of the Soma juice. (RV i.46.5)

आदारः <(आदारस्) ind(?). (from आ-√दॄ to crush, split open) – crushing, tearing up (the Soma Pond)>, वां <m. gen. du. of 2nd person pron. युष्मद् – your (du.)>, मतीनां <f. gen pl. of मति – of prayers, hymns, sacred utterances (Speeches)>, नासत्या <m. voc. du. of नासत्य (ना a man, hero + सत्य truth, reality, a name of Agni) – O Nāsatyā, O two heroes of Agni>, मतवचसा <mf. inst. sg. of मतवचस् (मत honoured, esteemed + वचस् speech, voice, command, order) – with the honoured command>;

पातं <ipv. 2nd person du. of √पा – ye (du.) drink (with acc. or gen.)>, सोमस्य <m. gen. sg. of सोम – of Soma juice>, धृष्णुया <ind. – boldly, strongly>.

6.1.5 Aśvinā Bring the Two Streams of Light

Stanza RV i.46.6 explains that Aśvinā bring and bestow two streams of light (ज्योतिष्मती jyotiṣmatī du.).

RV i.46.6 author प्रस्कण्वः काण्वः, अश्विनौ, गायत्री छंदः
या नः पीपरदश्विना ज्योतिष्मती तमस्तिरः । तामस्मे रासाथामिषं ॥१.४६.६॥
या । नः । पीपरत् । अश्विना । ज्योतिष्मती । तमः । तिरः । तां । अस्मे इति । रासाथां । इषं ॥१.४६.६॥
O Aśvinā, ye twins who, through darkness, brought two streams of celestial light to us,
Bestow upon us that refreshing draught of celestial Waters. (RV i.46.6)

या <m. voc. du. of pron. यद् – the two who>, नः <m. dat. pl. of 1st person pron. अस्मद् – to us>, पीपरत् <inj. 2nd person du. of √पृ or √पॄ – ye (du.) brought, delivered>, अश्विना <m. voc. du. of अश्विन् – O Aśvinā>, ज्योतिष्मती <n. acc. du. of ज्योतिष्मत् – two (streams or jets), two streames of celestial light>, तमः <n. acc. sg. of तमस् – darkness>, तिरः <(तिरस्) ind. – through, across>;

तां <f. acc. sg. of तद् – that>, अस्मे <m. loc. pl. of अस्मद् – upon us>, रासाथां <ipv. 2nd person du. of √रा – ye (du.) grant, bestow>, इषं <f. acc. sg. of इष् – libation, the refreshing waters of the Heaven (cosmic plasmas)>.

In RV i.46.7, Aśvinā are asked to journey to the other shore, with the boat of hymns, where the Arms of the Milky Way swirl in the ocean of celestial hydrogen and cosmic plasmas. The boat of hymns refers to the plasma sheaths, carrying hymns, that sail over the summit of the Heaven down to the Milky Way and beyond.

RV i.46.7 author प्रस्कण्वः काण्वः, अश्विनौ, गायत्री छंदः
आ नो नावा मतीनां यातं पाराय गंतवे । युंजाथामश्विना रथं ॥१.४६.७॥
आ । नः । नावा । मतीनां । यातं । पाराय । गंतवे । युंजाथां । अश्विना । रथं ॥१.४६.७॥
To set out on your journey, with the boat of hymns, to us on the other shore,
O Aśvinā, fasten ye the chariot. (RV i.46.7)

आ <ind. prefix to verb यातं (√या)>, नः <m. dat. pl. of 1st person pron. अस्मद् – to us>, नावा <m. inst. sg. of. नाव – with the boat>, मतीनां <f. gen pl. of मति – of prayers, hymns>, आ-यातं <ipv. 2nd person du. of आ-√या – ye (du.) travel, advance, journey>, पाराय <n. dat. sg. of पार – to the further shore or to the other shore>, गंतवे <infinitive of √गम् – to come, to set out>;

युंजाथां <(युञ्जाथाम्) ipv. 2nd person du. of युज् – ye (du.) yoke, harness, fasten, make ready>, अश्विना <m. voc. du. of अश्विन् – O Aśvinā>, रथं <m. acc. sg. of रथ – the chariot>.

6.1.6 The Two Circumambulatory Paths of Aśvinā

Stanza RV i.46.14 explains that the two paths of Aśvinā circumambulate the perimeters of the Upper and the Lower Vaults of the Heaven down towards the Milky Way in the Innerfield, the Midheaven (see Figure 13).

RV i.46.14 author प्रस्कण्वः काण्वः, अश्विनौ, गायत्री छंदः
युवोरुषा अनु श्रियं परिज्मनोरुपाचरत् । ऋता वनथो अक्तुभिः ॥१.४६.१४॥
युवोः । उषाः । अनु । श्रियं । परिऽज्मनोः । उपऽआचरत् । ऋता । वनथः । अक्तुऽभिः ॥१.४६.१४॥

Dawn travelled together with the diffusing light on your two circumambulatory paths.
Prepare ye the sacirifices with beams of light. (RV i.46.14)

युवोः <m. gen. du. of 2nd person pron. युष्मद् – your, of you two>, उषा <f. nom. sg. of उषा – Uṣā, morning light, Dawn>, अनु <ind. – after, along>, श्रियं <mf. acc. sg. of श्री – diffusing light, light>, परिज्मनोः <m. loc. du. of परिज्मन् – on the two paths that go around, i.e. on the two circumambulatory paths>, उपाचरत् <ipf. 3rd person sg. of उप-√चर् (उप by the side of, together with + √चर् to move, travel through – she (Dawn) travelled through>;

ऋता <n. acc. pl. of ऋत – sacrifices (Soma sacrifices of releasing celestial lights)>, वनथः <pre. 2nd person du. of √वन् – ye (du.) prepare>, अक्तुभिः <m. inst. pl. of अक्तु – with rays or beams of light (beams or rays of cosmic plasmas)>.

6.2 THE CELESTIAL RIVERS

In the Ṛgveda, streaming currents of cosmic plasmas are called celestial rivers. Though Sarasvatī (सरस्वती) is praised as the most divine (असुर्या) of the rivers (RV vii.96.1), Sindhu (सिंधुः) is the mightiest and most rapid of the rapids (RV x.75.7). Sindhu surpasses in its grandeur all the rivers that flow (RV x.75.1). In the hymn of RV x.75, in addition to Sindhu, sixteen celestial rivers are listed. These rivers all travel with Sindhu. In the hymn, Sindhu is addressed in feminine and masculine nouns and adjectives in turn.

The different names of the celestial rivers appear to represent the different aspects of cosmic plasmas. For example, Sindhu is "a river, stream (esp. of the Indus, ...)". Indus (इंदवः iṃdavaḥ pl.) are bright drops of Soma. Gaṅgā (गंगा) is a "swift-goer". Sarasvatī (सरस्वती) is "a region abounding in pools and lakes". Yamunā (यमुना), "identified with Yamī (यमी)", is the river of Soma, for Yamī represents Soma. Asiknī (असिक्नी) is "the dark one" or "the night". Both 'the dark one' and 'the Night' represent Varuṇa, the cosmic plasma dark discharge. Paruṣṇī (परुष्णी) means "knotty (as reed)" or "spotted, variegated" which are the descriptions of Soma plants growing in the Soma Pond. Marudvṛdhā (मरुद्वृधा) means "rejoicing in the wind or in the Maruts". Suṣomā (सुषोमा) means "containing good sap" and is the "name of a particular Soma vessel". Rasā (रसा) means "sap or juice of plants", "elixir". All these names of the celestial rivers are the characteristic descriptions of cosmic plasmas blooming in the Soma Pond. Though the river Gaṅgā (गंगा) is mentioned in several stanzas, no hymn addressed to Gaṅgā is found in the extant text of the Ṛgveda.

According to the Ṛgveda, there are two different categories of celestial rivers. The rivers that flow into the Soma Pond, or Samudra, the Ocean of celestial Waters, and the rivers that flow out of the Soma Pond, or Samudra. When the cosmic plasmas are discharged from the Stone, the celestial ioniser, the rivers flow upwards into the Soma Pond or the Samudra, the Ocean of celestial Waters. When the cosmic plasmas are released, the celestial rivers flow out from the Soma Pond and fall upon Mt. Meru, the Galactic Bar, at the centre of the Milky Way.

The nine stanzas of hymn RV x.75 describe the key attributes of these celestial rivers. The author of the hymn is Sindhukṣit Praiyamedha (सिंधुक्षित्प्रैयमेधः). Sindhukṣit means 'dwelling in Sindhu' or 'inhabitant of Sindhu'. Praiyamedha, patronymic from priyamedha (प्रियमेधः), means 'beloved nourishing drink'. The author's name literally means 'nourishing drink dwelling in the river Sindhu'. The hymn is addressed to the celestial rivers (नद्यः nadyaḥ pl.). Nourishing drink refers to cosmic plasmas, Soma in particular.

6.2.1 Sindhu Flows Upwards Roaring Like a Charging Bull

According to RV x.75.1-3, Varuṇa cuts the channels for Sindhu. Sindhu surpasses all other rivers in its majesty and, roaring like a charging bull, flows over the precipitous ridge of the plasma bloom as he leads the celestial Waters. Note that the rivers set out 'triply', indicating that the rivers are the triply-spun filaments of plasma blooms, or Soma plants.

RV x.75.1 author सिंधुक्षित्प्रैयमेधः, to नद्यः, metre जगती छंदः
प्र सु वं आपो महिमानमुत्तमं कारुर्वोचाति सदने विवस्वतः ।
प्र सप्तसप्त त्रेधा हि चक्रमुः प्र सृत्वरीणामति सिंधुरोजसा ॥१०.७५.१॥
प्र । सु । वः । आपः । महिमानं । उत्ऽतमं । कारुः । वोचाति । सदने । विवस्वतः ।
प्र । सप्तऽसप्त । त्रेधा । हि । चक्रमुः । प्र । सृत्वरीणां । अति । सिंधुः । ओजसा ॥१०.७५.१॥
The architect, at the seat of fire, O Waters, shall commend of your majestic grandeur.
The rivers set out triply, seven and seven. Sindhu in its splendour surpasses all the rivers that flow. (RV x.75.1)

प्र <ind. prefix to verb वोचाति (√वच्)>, सु <ind. adverb – greatly, highly>, वः <m. gen. pl. of 2nd person pron. युष्मद् – your (pl.)>, आपः <f. voc. pl. of अप् – O Waters (cosmic plasmas)>, महिमानं <m. acc. sg. of महिमन् – greatness, might, majesty, glory>, उत्तमं <m. acc. sg. of adj. उत्तम – highest, most elevated, excellent>, कारुः <m. nom. sg. of कारु – a maker, "architect of the gods", "one who sings or praises" (the one who sings and produces the Devas is the Stone, the Soma Press)>, प्र-वोचाति <sub. 3rd person sg. of प्र-√वच् 3P – he shall praise, commend>, सदने <n. loc. sg. of सदन – at the seat, (with विवस्वतः) "at the seat of Fire">, विवस्वतः <m. gen. sg. of विवस्वत् – of Vivasvān (विवस्वान्), of the Brilliant one, of Fire (i.e. the fire of the Stone, the celestial ioniser)>;

प्र <ind. – a simple repetition (sometimes repeated before the verb), or "used in the sense of गति" (going, flowing)>, सप्तसप्त <सप्त mf. nom. pl of सप्तन् – seven and seven (often used to express an indefinite plurality)>, त्रेधा <(=त्रिधा) ind. – in three parts, triply, i.e. a triply-spun filament of plasma bloom>, हि <ind. – indeed, surely>, प्र-चक्रमुः <prf. 3rd person pl. of प्र-√क्रम् – they (the rivers) have set out, advanced>, प्र <ind. prefix to verb चक्रमुः (√क्रम्)>, सृत्वरीणां <f. gen. pl. of adj. सृत्वन् – running, swift (rivers)>, अति <ind. adverb or preposition – (with gen.) beyond>, सिंधुः <mf. nom. sg. of सिंधु – Sindhu, the name of a river in the Soma Pond>, ओजसा <n. inst. sg. of ओजस् – by/with vigour, splendour>.

RV x.75.2 author सिंधुक्षित्प्रैयमेधः, to नद्यः, metre जगती छंदः
प्र तेऽरदद्वरुणो यातवे पथः सिंधो यद्वाजाँ अभ्यद्रवस्त्वं ।
भूम्या अधि प्रवता यासि सानुना यदेषामग्रं जगतामिरज्यसि ॥१०.७५.२॥
प्र । ते । अरदत् । वरुणः । यातवे । पथः । सिंधो इति । यत् । वाजान् । अभि । अद्रवः । त्वं ।
भूम्याः । अधि । प्रऽवता । यासि । सानुना । यत् । एषां । अग्रं । जगतां । इरज्यसि ॥१०.७५.२॥
Varuṇa cut the channels, O Sindhu, for you to travel, when you hastened to win the battles.
You march on over the precipitous ridge of the field, when you lead these living Waters to the summit. (RV x.75.2)

प्र <ind. prefix to verb अरदत् (√रद्)>, ते <m. dat. sg. of युष्मद् – for you>, प्र-अरदत् <ipf. 3rd person sg. of प्र-√रद् – he cuts in, digs, marks out>, वरुणः <m. nom. sg. of वरुण – Varuṇa>, यातवे <infinitive of √या – to travel>, पथः <m. acc. pl. of पथिन् – paths, channels>, सिंधो <mf. voc. sg. of सिंधु – O Sindhu>, यत् <conjunction – that, when>, वाजान् <m. acc. pl. of वाज – races, battles, wars>, अभि <ind. prefix to verb अद्रवः (√द्रु)>, अभि-अद्रवः <ipf. 2nd person sg. of अभि-√द्रु – you hastened to>, त्वं <mf. nom. sg. of युष्मद् – you>;

भूम्याः <f. gen. sg. of भूमि – of the field, i.e. of the Soma Pond>, अधि <ind. prefix to verb यासि (√या)>, प्रवता <mn. inst. sg. of adj. प्रवत् – precipitous, sloping>, अधि-यासि <pre. 2nd person sg. of अधि-√या – you proceed, travel, march on or over>, सानुना <mn. inst. sg. of सानु – through/over ridge>, यत् <conjunction – that, when>, एषां <mn. gen. pl. of pron. इदं – of these>, अग्रं <n. acc. sg. of अग्र – to surface, top, summit>, जगतां <mn. gen. pl. of adj. जगत् – the moving, living, that which moves or is alive, i.e. living or live Waters or cosmic plasmas>, इरज्यसि <pre. 2nd person sg. of √इरज्य – you lead (with gen.)>.

RV x.75.3 author सिंधुक्षित्प्रैयमेधः, to नद्यः, metre जगती छंदः
दिवि स्वनो यतते भूम्योपर्यनंतं शुष्ममुदियर्ति भानुना ।
अभ्रादिव प्र स्तनयंति वृष्टयः सिंधुर्यदेति वृषभो न रोरुवत् ॥१०.७५.३॥
दिवि । स्वनः । यतते । भूम्या । उपरि । अनंतं । शुष्मं । उत् । इयर्ति । भानुना ।
अभ्रात्ऽइव । प्र । स्तनयंति । वृष्टयः । सिंधुः । यत् । एति । वृषभः । न । रोरुवत् ॥१०.७५.३॥
In the Heaven, the roaring Water marches upwards through the field, and the eternal flame rises up in its splendour. As floods of rain thunder forth from the water-bearer, roaring like a charging bull, Sindhu advances. (RV x.75.3)

दिवि <m. loc. sg. of दिव् – in the Heaven>, स्वनः <m. nom. sg. of स्वन – sound, the roar of thunder ot water, roaring water>, यतते <pre. middle 3rd person sg. of √यत् – it (the roaring Water) marches, flies together or in line>, भूम्या <f. inst. sg. of भूमि – through the field, i.e. through the Soma Pond>, उपरि <ind. – upwards, towards the upper side of>, अनंतं <n. nom. sg. of adj. अनंत – boundless, eternal>, शुष्मं <n. nom. sg. of शुष्म – fire, flame>, उत् <(उद्) ind. prefix to verb इयर्ति (√ऋ)>, उद्-इयर्ति <pre. 3rd person sg. of उद्-√ऋ – it (the eternal flame) comes up, rises>, भानुना <m. inst. sg. of भानु – with light, a ray of light, splendour>;

अभ्रात् <n. abl. sg. of अभ्र – from the "water-bearer" (it refers to the Soma Pond)>, इव <ind. – like, as>, प्र <ind. prefix to verb स्तनयंति (√स्तन्)>, प्र-स्तनयंति <pre. cau. 3rd person pl. of प्र-√स्तन् – they (the floods of rain) thunder forth>, वृष्टयः <f. nom. pl. of वृष्टि – rains, floods of rain>, सिंधुः <mf. nom. sg. of सिंधु – Sindhu>, यत् <m. nom. sg. of adj. यत् (pre. pt. of √इ) – advancing, charging>, एति <pre. 3rd person. sg. of √इ – it (Sindhu) flows, advances>, वृषभः <m. nom. sg. of वृषभ – a bull>, न <ind. – like, as>, रोरुवत् <int. pt. of √रु – bellowing, roaring>.

6.2.2 Sindhu Leads the Two Wings of Army Like a Warrior King

In RV x.75.4, two wings of army can refer either to Agnis and Somas or to the twin Jets, the Aśvinā. Either way makes sense, though, in the context of this stanza, Agnis and Somas are more likely candidates, for each plasma current carries Agnis (protons) and Somas (electrons).

RV x.75.4 author सिंधुक्षित्प्रैयमेधः, to नद्यः, metre जगती छंदः
अभि त्वा सिंधो शिशुमिन्न मातरो वाश्रा अर्षंति पयसेव धेनवः ।
राजेव युध्वा नयसि त्वमित्सिचौ यदासामग्रं प्रवतामिनक्षसि ॥१०.७५.४॥

अभि । त्वा । सिंधो इति । शिशुं । इत् । न । मातरः । वाश्राः । अर्षंति । पयसाऽइव । धेनवः ।
राजाऽइव । युध्वा । नयसि । त्वं । इत् । सिचौ । यत् । आसां । अग्रं । प्रऽवतां । इनक्षसि ॥१०.७५.४॥
Like bellowing mother Cows yielding milk move quickly to their calves, O Sindhu, you are full of juice. You lead the two wings of army, like a warrior king, in your desire to reach the summit of these mountain slopes. (RV x.75.4)

अभि <ind. separate preposition – to, towards>, त्वा <f. nom. sg. of 2nd person pron. त्व – you (Sindhu)>, सिंधो <f. voc. sg. of सिंधु – O Sindhu>, शिशुं <m. acc. sg. of शिशु – child, infant, the young of any animal (as a calf)>, इत् <(इद्) ind. particle – even, just, indeed>, न <ind. – like, as>, मातरः <f. nom. pl. of मातृ – mother Cows>, वाश्राः <f. nom. pl. of adj. वाश्र – roaring, lowing, howling>, अर्षंति <pre. 3rd person. pl. of √ऋष् – they (the mother Cows) move with quick motion>, पयसा <f. nom sg. of adj. पयस – full of juice or sap>, इव <ind. – like, in the same manner as>, धेनवः <f. nom. pl. of adj. धेनु – milch, yielding or giving milk>;

राजा <m. nom. sg. of राजन् – a king, sovereign>, इव <ind. – like, in the same manner as>, युध्वा <m. nom. sg. of adj. युध्वन् – warlike, a warrior>, नयसि <pre. 2nd person sg. of √नी – you lead, guide, govern>, त्वं <m. nom. sg. of 2nd person pron. युष्मद् – you>, इत् <(इद्) ind. – even, just, indeed>, सिचौ <f. acc. du. of सिच् – the two wings of army (refer to Agnis and Somas)>, यत् <conjunction – when, so that, in order that>, आसां <f. gen. pl. of pron. इदं – of these>, अग्रं <n. acc. sg. of अग्र – uppermost part, top, summit>, प्रवतां <f. gen. pl. of प्रवत् – of the slopes of a mountain (refer to the steep upward paths of plasma blooms or Soma plants)>, इनक्षसि <desiderative 2nd person sg. of √नश् – you desire to reach, attain>.

6.2.3 Seek My Hymn O Gaṅgā, Yamunā, Sarasvatī, Śutudrī, and Paruṣṇī

In stanzas RV x.75.5-6, the names of sixteen celestial rivers that travel with Sindhu are enumerated. Gomatī (गोमती) and Krumu (क्रुमुः) are the destinations to where Sindhu, with other rivers, travels. Gomatī means “a place abounding in herds of cattle”. Krumu (क्रुमुः) is probably the woodland or forest of kṛmuka (कृमुकः) trees and associated with kārmuka (कार्मुकः as an adjective “consisting of the wood कृमुक”, as a noun “bamboo” or “the white Khadira tree”). These are the descriptions of the Soma Pond where plasma blooms, or Soma plants grow.

The celestial rivers are asked to seek out or follow ‘my’ hymn, or Speech (RV x.75.5). This is the author’s call to the celestial rivers. Note that the author’s name is Sindhukṣit Praiyamedha, ‘nourishing drink dwelling in the river Sindhu’, which is Soma. And Soma is the Lord of Speech. Sindhu, together with these celestial rivers, is asked to come to the Soma Pond to journey over the summit of the Heaven and down to the Milky Way in the Innerfield, the Midheaven (RV x.75.6).

RV x.75.5 author सिंधुक्षित्प्रैयमेधः, to नद्यः, metre जगती छंदः
इमं मे गंगे यमुने सरस्वति शुतुद्रि स्तोमं सचता परुष्ण्या ।
असिक्न्या मरुद्वृधे वितस्तयार्जीकीये शृणुह्या सुषोमया ॥१०.७५.५॥
इमं । मे । गंगे । यमुने । सरस्वति । शुतुद्रि । स्तोमं । सचत । परुष्णि । आ ।
असिक्न्या । मरुत्ऽवृधे । वितस्तया । आर्जीकीये । शृणुहि । आ । सुऽसोमया ॥१०.७५.५॥

Seek ye this my hymn, O Gaṅgā, O Yamunā, O Sarasvatī, O Śutudrī, O Paruṣṇī,
With Asiknī, O Marudvṛdhā, with Vitastā, O Ārjīkīyā, with Suṣomā, hear my call. (RV x.75.5)

इमं <m. acc. sg. of pron. इदं – this>, मे <m. gen. sg. of 1st person pron. अस्मद् – my>, गंगे <f. voc. sg. of गंगा – O Gaṅgā>, यमुने <f. voc. sg. of यमुना – O Yamunā>, सरस्वति <f. voc. sg. of सरस्वती – O Sarasvatī>, शुतुद्रि <f. voc. sg. of शुतुद्री – O Śutudrī>, स्तोमं <m. acc. sg. of स्तोम – praise, eulogium, hymn>, आ-सचत <ipv. 2nd person pl. of आ-√सच् – ye seek, pursue, follow>, परुष्णि <f. voc. sg. of परुष्णी – O Paruṣṇī>, आ <ind. prefix to verb सचत (√सच्)>;

असिक्न्या <f. inst. sg. of असिक्नी – with Asiknī>, मरुद्वृधे <f. voc. sg. of मरुद्वृधा – O Marudvṛdhā>, वितस्तया <f. inst. sg. of वितस्ता – with Vitastā>, आर्जीकीये <f. voc. sg. of आर्जीकीया – O Ārjīkīyā>, आ-शृणुहि <ipv. 2nd person sg. of आ-√श्रु – you hear, listen to>, आ <ind. prefix to verb शृणुहि (√श्रु)>, सुषोमया <f. inst. sg. of सुषोमा – with Suṣomā>.

RV x.75.6 author सिंधुक्षित्प्रैयमेधः, to नद्यः, metre जगती छंदः
तृष्टामया प्रथमं यातवे सजूः सुसर्त्वा रसया श्वेत्या त्या ।
त्वं सिंधो कुभया गोमतीं क्रुमुं मेहत्न्वा सरथं याभिरीयसे ॥१०.७५.६॥
तृष्टऽअमया । प्रथमं । यातवे । सऽजूः । सुऽसर्त्वा । रसया । श्वेत्या । त्या ।
त्वं । सिंधो इति । कुभया । गोऽमतीं । क्रुमुं । मेहत्न्वा । सऽरथं । याभिः । ईयसे ॥१०.७५.६॥
First, to journey with Tṛṣṭāmā, and with Susartu, with Rasā, with Śvetī,
In the same chariot with Kubhā, with Mehatnū, with all of them, O Sindhu, you are asked to come to Gomatī and to Krumu. (RV x.75.6)

तृष्टामया <f. inst. sg. of तृष्टामा – with Tṛṣṭāmā>, प्रथमं <ind. adverb – foremost, first>, यातवे <infinitive of √या – to set out, travel, journey>, सजूः <सजूस् ind. – at the same time, besides, moreover>, सुसर्त्वा <f. inst. sg. of सुसर्तु – with Susartu>, रसया <f. inst. sg. of रसा – with Rasā>, श्वेत्या <f. inst. sg. of श्वेती – with Śvetī>, त्या <f. inst. sg. of adj. त्य – with this>;

त्वं <m. nom. sg. of युष्मद् – you>, सिंधो <m. voc. sg. of सिंधु – O Sindhu>, कुभया <f. inst. sg. of कुभा – with Kubhā>, गोमतीं <f. acc. sg. of गोमती – to Gomatī, "a place abounding in herds of cattle" (refers to the Soma Pond where herds of cattle, celestial lights, abound)>, क्रुमुं <f. acc. sg. of क्रुमु – to Krumu, to the woodland or forest of white Khadira trees or bamboos (refers to the forest of Soma plants, which is the Soma Pond)>, मेहत्न्वा <f. inst. sg. of मेहत्नू – with Mehatnū>, सरथं <ind. – in the same chariot>, याभिः <f. inst. pl. of pron. यद् – with which or who (pl.)>, ईयसे <pre. passive 2nd person sg. of √इ – you are asked to come to>.

6.2.4 Sindhu, Most Rapid of the Rapids, Traverses the Heavenly Spheres

In stanzas RV x.75.7 and 9, Sindhu is described as the most rapid of the rapids and beautiful to behold. To get a rough idea of how rapidly Sindhu flows, let's assume that the fully grown and matured height of plasma blooms is, conservatively, half the height of the Heavenly Vault. Since the height of the Heavenly Vault is 28,710.8 light-years according to the Vedic source, we calculate about 14,355 light-years for the height of the fully matured plasma blooms (see Figure 4 and section 2.3).

According to the Ṛgveda, cosmic plasmas are discharged and released three times a day. If we assume the discharges of cosmic plasmas occur more or less in equal intervals, Sindhu has to travel 14,355 light-years of distance in eight hours to reach the summit of the Soma Pond. This implies that cosmic plasmas travel, at the very least, 1,794 light-years per hour when they are discharged from the Stone, the celestial ioniser, into the Soma Pond. This is not an accurate scientific measurement or calculation. However, it gives one an idea of how rapidly cosmic plasmas travel in the spheres of the Heaven. It seems unplausible, but this is what is implied. Note that the Heavenly Vault and the celestial ioniser are magnetic field structures. If this is plausible, very likely, the impelling forces of these stupendously large magnetic field structures cause this enormous speed of streaming cosmic plasmas.

In the following two stanzas, Sindhu is used almost synonymously as Aśvinā by stating that Sindhu traverses beyond the Heavenly Spheres (RV x.75.7) and that he shall procure wealth in battle (RV x.75.9).

RV x.75.7 author सिंधुक्षित्प्रैयमेधः, to नद्यः, metre जगती छंदः
ऋजीत्येनी रुशती महित्वा परि ज्रयांसि भरते रजांसि ।
अदब्धा सिंधुरपसामपस्तमाश्वा न चित्रा वपुषीव दर्शता ॥१०.७५.७॥
ऋजीती । एनी । रुशती । महिऽत्वा । परि । ज्रयांसि । भरते । रजांसि ।
अदब्धा । सिंधुः । अपसां । अपःऽतमा । अश्वा । न । चित्रा । वपुषीऽइव । दर्शता ॥१०.७५.७॥
Rushing upwards, Sindhu, bright, shining, exalted, traverses beyond the wide expanses of the Heavenly Spheres. Like a magnificent, unbroken mare, Sindhu, most rapid of the rapids, is beautiful to behold. (RV x.75.7)

ऋजीती <f. nom. sg. of adj. ऋजीति (ऋजीती f.) – going or tending upwards>, एनी <f. nom. sg. of adj. एनी (= एत) – of a variegated colour, shining, brilliant>, रुशती <f. nom. sg. of adj. रुशत् – bright, white>, महित्वा <ind. pt. of √मह् – being elated, exalted>, परि <ind. prefix to verb भरते (√भृ)>, ज्रयांसि <n. acc. pl. of ज्रयस् – wide expanses, spaces>, परि-भरते <pre. middle 3rd person sg. of परि-√भृ – she (Sindhu) extends, passes beyond>, रजांसि <n. acc. pl. of रजस् – the Heavenly Spheres>;

अदब्धा <f. nom. sg. of adj. अदब्ध – not tampered, unbroken>, सिंधुः <f. nom. sg. of सिंधु – Sindhu>, अपसां <f. gen. pl. of अपस् – of the rapids>, अपस्तमा <f. nom. sg. of adj. अपस्तम – most rapid>, अश्वा <f. nom. sg. of अश्वा – a mare>, न <ind. – like, as>, चित्रा <f. nom. sg. of adj. चित्र – excellent, distinguished>, वपुषी <f. nom. sg. of adj. वपुषी (=वपुस्) – wonderfully beautiful, having beautiful form, handsome>, इव <ind. – like, in the same manner as>, दर्शता <f. nom. sg. of adj. दर्शत – striking the eye, beautiful>.

RV x.75.9 author सिंधुक्षित्प्रैयमेधः, to नद्यः, metre जगती छंदः
सुखं रथं युयुजे सिंधुरश्विनं तेन वाजं सनिषदस्मिन्नाजौ ।
महान्ह्यस्य महिमा पनस्यतेऽदब्धस्य स्वयशसो विरप्शिनः ॥१०.७५.९॥
सुऽखं । रथं । युयुजे । सिंधुः । अश्विनं । तेन । वाजं । सनिषत् । अस्मिन् । आजौ ।
महान् । हि । अस्य । महिमा । पनस्यते । अदब्धस्य । स्वऽयशसः । विऽरप्शिनः ॥१०.७५.९॥
Sindhu fastened the swift-rolling chariot drawn by steeds. With that chariot shall he procure wealth in battle. Mighty Sindhu is praised for his unimpaired, overflowing, sublime majesty. (RV x.75.9)

सुखं <m. acc. sg. of adj. सुख – running swiftly, easily (only applied to cars or chariots)>, रथं <m. acc. sg. of रथ – a chariot>, युयुजे <prf. middle 3rd person sg. of √युज् – he (Sindhu) yoked, harnessed, fastened>, सिंधुः <m. nom sg. of सिंधु – Sindhu>, अश्विनं <m. acc. sg. of adj. अश्विन् – possessed of horses, consisting of horses, i.e. drawn by horses>, तेन <m. inst. sg. of तद् – with it (chariot)>, वाजं <m. acc. sg. of वाज – wealth, treasure>, सनिषत् <sub. 3rd person sg. of

√सन् – he (Sindhu) shall gain, procure, bestow, distribute>, अस्मिन् <m. loc. sg. of इदं – in this>, आजौ <m. loc. sg. of आजि – in battle (of destroying Vṛtra)>;

महान् <m. nom. sg. of adj. महत् – eminent, great>, हि <ind. – for, because, on account of>, अस्य <m. gen. sg. of इदं – of this, his>, महिमा <m. nom. sg. of महिमन् – glory, majesty>, पनस्यते <pre. middle 3rd person sg. of nominal verb √पनस्य – he (Sindhu) excites admiration or praise>, अदब्धस्य <m. gen. sg. of adj. अदब्ध – unimpaired, unbroken>, स्वयशसः <m. gen. sg. of adj. स्वयशस् – glorious or illustrious through one's own (acts)>, विरप्शिनः <m. gen. sg. of adj. विरप्शिन् (from वि-√रप्श्) – copious, full, overflowing>.

6.2.5 Exalted by Soma Mead Sindhu Grows Bright

RV x.75.8 states that Sindhu grows bright upon being exhilarated by Soma mead. Sindhu is described as woolly or rich in wool (ऊर्णावती). In the Ṛgveda, plasma blooms or Soma plants are often called sheep's golden fleece (अव्यः वारः हरिः RV ix.7.6) or sheep's fleeces (रोमाणि अव्यया RV i.135.6). Sindhu is the river channel through which 'woolly' cosmic plasmas run and bloom. Thus the river Sindhu is described as such (see Figure 7).

RV x.75.8 author सिंधुक्षित्प्रैयमेधः, to नद्यः, metre जगती छंदः
स्वश्वा सिंधुः सुरथा सुवासा हिरण्ययी सुकृता वाजिनीवती ।
ऊर्णावती युवतिः सीलमावत्युताधि वस्ते सुभगा मधुवृधं ॥१०.७५.८॥
सुऽअश्वा । सिंधुः । सुऽरथा । सुऽवासाः । हिरण्ययी । सुऽकृता । वाजिनीऽवती ।
ऊर्णाऽवती । युवतिः । सीलमाऽवती । उत । अधि । वस्ते । सुऽभगा । मधुऽवृधं ॥१०.७५.८॥
Sindhu has excellent horses and a chariot, wears beautiful garments. Well adorned, spirited, rich in gold is she. Prosperous, youthful, rich in wool, rich in celestial Water, running upwards, upon being exhilarated by Soma mead, she grows bright. (RV x.75.8)

स्वश्वा <f. nom. sg. of adj. स्वश्व – having excellent horses>, सिंधुः <f. nom. sg. of सिंधु – Sindhu>, सुरथा <f. nom. sg. of adj. सुरथ – having a good chariot>, सुवासाः <f. nom. sg. of adj. सुवासस् – having beautiful garments>, हिरण्ययी <f. nom. sg. of adj. हिरण्य – abounding in gold>, सुकृता <f. nom. sg. of adj. सुकृत – well arranged, adorned>, वाजिनीवती <f. nom. sg. of adj. वाजिनीवत् – impetuous, spirited>;

ऊर्णावती <f. nom. sg. of adj. ऊर्णावत् – abounding in wool, having wool, woolly>, युवतिः <f. nom. sg. of युवति – a youth, youthful>, सीलमावती <f. nom. sg. of सीलमावती – rich in plants or rich in Water>, उत् <(उद्) ind. prefix to verb – up, upwards, i.e she flows, runs, rushes upwards (when appears uncompounded with a verb, the verb has to be supplied from the context)>, अधि <ind. – as a separable adverb or preposition (with acc.) over, upon>, वस्ते <pre. middle 3rd person sg. of √वस् 6Ā – she (Sindhu) shines, grows bright>, सुभगा <f. nom. sg. of adj. सुभग – blessed, prosperous>, मधुवृधं <f. acc. sg. of adj. – exhilarated by madhu (i.e. Soma mead)>.

6.2.6 Mighty Gaṅgā Falls Upon Mt. Meru

In the extant text of the Ṛgveda, there is no stanza that connects the downward release of cosmic plasmas to the river Gaṅgā, though the downward release of cosmic plasmas is explained (see section 2.4.6). However, we find that stanzas 37 and 38 of Bhuvanakośa of Siddhāntaśiromaṇi explicitly call this downward flow the river Gaṅgā, and explain how the downward release of cosmic plasmas occurs. The Gaṅgā emerges from the foot of the Heavenly Vault and falls upon Mt. Meru, the Galactic Bar, at the centre of the Milky

Way, divides itself into four parts, and flows down along the four buttress mountain ranges of Mt. Meru, that is, the four major Arms of the Milky Way (stanza 37).

Out of four parts of Gaṅgā, the first part called Sītā flows into Bhadrāśva Varṣa (भद्राश्ववर्षः), the second part Alakanandā to Bhārata Varṣa (भारतवर्षः), the third part Cakṣu to Ketumāla Varṣa (केतुमालावर्षः), and the fourth part Bhadrā enters into Uttara Kurus (उत्तरकुरुवर्षः), or Upper Kurus (stanza 38). The Bhadrāśva Varṣa is likely the innermost Arm and the Uttara Kurus, or Upper Kurus, the outermost Arm of the Milky Way.

For the word Varṣa (m. वर्षः varṣaḥ or n. वर्षं varṣam), Monier-Williams provides two distinct categories of definitions: "rain, a shower (either of flowers, arrows, dust etc.)" and "a division of the earth as separated off by certain mountain ranges". In the Ṛgveda, rain or shower represents the discharged or released cosmic plasmas. Arrows (of Agni, of Rudra, of the Maruts) and flowers represent the plasma blooms, the Soma plants. In the second definition, as noted previously, the 'earth' refers to Prthivī, the Milky Way, not the planet earth as has been mistakenly translated. When we consider that the Arms of the Milky Way are the concentric 'peninsulas' or 'mountain ranges' that project out from the Galactic Bar at centre of the Milky Way and separated off by the 'invisible' celestial oceans, the symbolically expressed definitions provided by Monier-Williams are right on the mark, though no one has deciphered the meanings of these definitions suitably.

Stanza 37 भुवनकोशः सिद्धान्तशिरोमणिः (Bhuvanakośa, Siddhāntaśiromaṇi)
विष्णुपदी विष्णुपदात् पतिता मेरौ चतुर्धास्मात् ।
विष्कम्भाचलमस्तकशस्तसरः संगतागता वियता ॥३७ भुवनकोशः सिद्धान्तशिरोमणिः ॥
Gaṅgā, from the foot of Viṣṇu, fell upon Mt. Meru, and from there, dividing itself into four,
Like a cast shuttle, flowing along the immovable buttress, arrived fitted together, stretched out separately. (37 Bhuvanakośa, Siddhāntaśiromaṇi)

विष्णुपदी <f. nom. sg. of विष्णुपदी – Viṣṇupadī, the name of Gaṅgā (गंगा)>, विष्णुपदात् <nm. abl. sg. of विष्णुपद – from the foot of Viṣṇu, from "the sea of milk", i.e. from the foot of the Samudra, the Ocean of celestial Waters (the Soma Pond)>, पतिता <prf. 3rd person sg. of √पत् 1P – it (Gaṅgā) fell on>, मेरौ <m. loc. sg. of मेरु – on Mt. Meru>, चतुर्धा <ind. – in four parts>, अस्मात् <mn. abl. sg. of pron. इदं – from this>;

विष्कम्भाचलमस्तकशस्तसरः <f. nom. sg. of adj. विष्कम्भाचलमस्तकशस्तसर (विष्कम्भाचलं n. an immovable buttress or mountain range + अस्त adj. thrown, cast + कशस् n. moving, flowing + तसरः m. a shuttle) – flowing on the immovable buttress like a cast shuttle>, संगतागता <f. nom. sg. of adj. संगतागता (संगत adj. coming together, fitted together + आगता adj. come, arrived) – arrived fitted together>, वियता <f. nom. sg. of adj. वियत – kept apart, stretched out separately>.

Stanza 38 भुवनकोशः सिद्धान्तशिरोमणिः (Bhuvanakośa, Siddhāntaśiromaṇi)
सीताख्या भद्राश्वं सालकनन्दा च भारतं वर्षं ।
चक्षुश्च केतुमालं भद्राख्या चोत्तरान् कुरून् याता ॥३८ भुवनकोशः सिद्धान्तशिरोमणिः ॥
(Out of the four parts of Gaṅgā) The first part called Sītā flowed into Bhadrāśva Varṣa, this Alakanandā to Bhārata Varṣa, And Cakṣu to Ketumāla Varṣa, Bhadrā to Uttara Kurus Varṣa. (38 Bhuvanakośa, Siddhāntaśiromaṇi)

सीता <f. nom. sg. of सीता – Sītā (1st part of Gaṅgā)>, आख्या <f. nom. sg. of आख्या – (at the end of a compound) named, called>, भद्राश्वं <n. acc.sg. of भद्राश्व – to Bhadrāśva Varṣa (the innermost Arm of the Milky Way)>, सा <f. nom. sg. of pron. तद् – that, this>, अलकनन्दा <f. nom. sg. of अलकनन्दा – Alakanandā, (2nd part of Gaṅgā)>, च <ind. – and, also>, भारतं <n. acc. sg. of भारतं – to Bhārata (2nd Arm of the Milky Way)>, वर्षम् <n. acc. sg. of वर्ष – to Varṣa, a division or region of "the earth", i.e. to a division (Arm) of the Milky Way>;

चक्षुः <f. nom. sg. of चक्षुस् – Cakṣu (3rd part of Gaṅgā)>, च <ind. – and, also>, केतुमालं <mn. acc. sg. of केतुमाल – to Ketumāla Varṣa (3rd Arm of the Milky Way)>, भद्रा <f. nom. sg. of भद्रा – Bhadrā (4th part of Gaṅgā)>, आख्या <f. nom. sg. of आख्या – named, called>, च <ind. – and, also>, उत्तरान् कुरून् <m. acc. pl. of उत्तर कुरु – to Uttara Kurus or Northern Kurus or Upper Kurus (the outermost Arm of the Milky Way)>, याता <prf. 3rd person sg. of √या – it entered, proceeded, entered>.

VII. Agni and Soma

7.1 AGNI, PROTON

Agni, proton, and Soma, electron, are the two embryonic particles (गर्भा garbhā du.) of cosmic plasmas and of all beings, sentient and insentient (RV i.70.3/4). Agni, proton, is the lord of the house (विश्पतिः RV x.135.1). The 'house' refers either to the individual Soma plants or the individual atomic structures of elements. The number of protons in the nucleus determines the identity of an element and the number of electrons in the atomic structure; thus, Agni, proton, is the lord of the house. Agni is also the lord of light (स्वर्णरः svarṇaraḥ or स्वर्णा svarṇā RV viii.19.1). No neutrons, quarks, hardrons, leptons, force carrier particles, photons, antimatter, antiproton, antiparticles, gluons, nutrinos, and so on are necessary in the magneto-plasma universe of the Ṛgveda. Agni, proton, and Soma, electron, are the two 'fundamentals'. Agni and Soma are literally 'god particles' according to the Ṛgveda. They are the Devas, the gods, and they are the two embryonic particles of celestial Waters, that is, the embryonic particles of cosmic plasmas. Agnis and Somas interact, attract and bind, or repel one another based on their Speech patterns, that is, the patterns of their oscillations and directions of their spins. When Agnis are kindled, they become lights or fires. Agnis and Somas themselves carry lights, forces, and Speeches.

7.1.1 Agni, an Embryonic Particle of Celestial Waters and All Beings

Stanza RV i.70.3/4 is addressed to Agni. According to the stanza, Agni, who is a charged embryonic particle of cosmic plasmas, is the omnipresent, imperishable, and dedicated one in the house of the atomic structure. The author is Parāśara Śāktyaḥ (पराशरः शाक्त्यः),

the 'crusher or destroyer (पराशरः), the son of Śakti (शक्तिः energy, might, regal power)'. Parāśara Śāktyaḥ refers to the Stone, the celestial ioniser.

RV i.70.3/4 author पराशरः शाक्त्यः, to अग्निः, metre द्विपदा विराट् छंदः
गर्भो यो अपां गर्भो वनानां गर्भश्च स्थातां गर्भश्चरथां ।
अद्री चिदस्मा अंतर्दुरोणे विशां न विश्वो अमृतः स्वाधीः ॥१.७०.३।४॥
गर्भः । यः । अपां । गर्भः । वनानां । गर्भः । च । स्थातां । गर्भः । चरथां ।
अद्री । चित् । अस्मै । अंतः । दुरोणे । विशां । न । विश्वः । अमृतः । सुऽआधीः ॥१.७०.३।४॥
He, who is an embryonic particle of celestial Waters, of trees, of beings that move not and that move, In the house of the tribes, built for him on the mountain, indeed, is omnipresent, imperishable, and devout. (RV i.70.3/4)

गर्भः <m. nom. sg of गर्भ – "a foetus or embryo", "offspring of the sky", i.e. an embryonic particle produced by the Heaven>, यः <m. nom. sg. of यद् – he who>, अपां <f. gen. pl. of अप् – of celestial Waters, i.e. of cosmic plasmas>, गर्भः <m. nom. sg of गर्भ – an embryonic particle>, वनानां <n. gen. pl. of वन – of trees (refer to plasma blooms or Soma plants)>, गर्भः <m. nom. sg of गर्भ – an embryonic particle>, च <ind. – and, also>, स्थातां <(=स्थानां) mfn. gen. pl. of स्थ – of beings stationary, immovable>, गर्भः <m. nom. sg of गर्भ – an embryonic particle>, चरथां <(=चरथानां) mfn. gen. pl. of चरथ – of beings moving, of living beings>;

अद्रौ <m. loc. sg. of अद्रि – on the mountain (refers to the Soma Pond)>, चित् <(चिद्) ind. particle – even, indeed>, अस्मै <m. dat. sg. of pron. इदं – (built or prepared) for him>, अंतः <(अन्तर्) ind. – in, into (with loc.)>, दुरोणे <n. loc. sg. of दुरोण – in the residence, dwelling, house; refers to the individual plasma blooms in the Soma Pond>, विशां <f. gen. pl. of विश् – of tribes, races, the people (refer to the tribes of cosmic plasmas)>, न <ind. – like, as, as it were>, विश्वः <m. nom. sg. of adj. विश्व – omnipresent>, अमृतः <m. nom. sg. of adj. अमृत – immortal, imperishable>, स्वाधीः <m. nom. sg. of adj. स्वाधी – well-minded, thoughtful, devout>.

7.1.2 Agni Comes Out of the Stone as Celestial Waters

Stanza RV ii.1.1 explains the transformation of Agni as he is ionised and kindled. Agni comes out of the Stone as cosmic plasmas and, gleaming forth through the Heavens, he becomes the forests of Soma plants. Again, he, glowing, changes into the light-containing plants (RV ii.1.1).

RV ii.1.1 author गृत्समद आंगिरसः शौनहोत्रः पश्चाद्भार्गवः शौनकः, to अग्निः, metre जगती छंदः
त्वमंग्ने द्युभिस्त्वमाशुशुक्षणिस्त्वमुद्भ्यस्त्वमश्मनस्परि ।
त्वं वनेभ्यस्त्वमोषधीभ्यस्त्वं नृणां नृपते जायसे शुचिः ॥२.१.१॥
त्वं । अग्ने । द्युऽभिः । त्वं । आऽशुशुक्षणिः । त्वं । अत्ऽभ्यः । त्वं । अश्मनः । परि ।
त्वं । वनेभ्यः । त्वं । ओषधीभ्यः । त्वं । नृणां । नृऽपते । जायसे । शुचिः ॥२.१.१॥
O thou Agni, out of the Stone, thou becomest celestial Waters. O thou, gleaming forth through the Heavens, Thou becomest the forests. O thou, glowing, O lord of men, thou art changed into the light-containing plants. (RV ii.1.1)

त्वं <m. voc. sg. of 2nd person pron. युष्मद् – O thou>, अग्ने <m. voc. sg. of अग्नि – O Agni>, द्युभिः <m. inst. pl. of दिव् – through the Heavens>, त्वं <m. nom. sg. of 2nd person pron. युष्मद् – you>, आशुशुक्षणिः <m. nom. sg. of adj. आशुशुक्षणि – gleaming or shining forth>, त्वं <m. nom. sg. of 2nd person pron. युष्मद् – you>, अद्भ्यः <f. dat. pl. of अप् – to Waters, i.e. become cosmic plasmas>, त्वं <m. voc. sg. of 2nd person pron. युष्मद् – O thou>, अश्मनः <m. abl. sg. अश्मन् – from the Stone, i.e. from the celestial ioniser>, परि <ind. – (with abl.) from, out of>;

त्वं <m. nom. sg. of 2nd person pron. युष्मद् – thou>, वनेभ्यः <n. dat. pl. of वन – to the forests, woods, trees (of Soma plants)>, त्वं <m. nom. sg. of 2nd person pron. युष्मद् – thou>, ओषधीभ्यः <f. dat. pl. of ओषधी – to the flaming or light-containing plants>, त्वं <m. voc. sg. of 2nd person pron. युष्मद् – O thou>, नृणां <m. gen. pl. of नृ – of men, heroes>, नृपते <m. voc. sg. of नृपति – O lord of men>, जायसे <pre. middle 2nd person sg. of √जन् 4Ā – you become, you are changed into (dat.)>, शुचिः <m. nom. sg. of adj. शुचि – shining, glowing>.

7.1.3 The Three Forms of Agni

Stanza RV iii.20.2 emphasises that there are three forms or fires of Agni, which are termed three isotopes of hydrogen in modern physics. Agni's three forms are described as three coursers, three abodes, or three ancient tongues of flame. Agni is born of the fire sacrifice, that is, through ionisation. The author is Gāthī (गाथी), the 'singer'. Kauśika (कौशिकः) is the patronymic of Gāthī, so he is a descendant of Kuśika, the ploughshare. All the Devas are singers, but the most prominent singer is the Stone who drives the cutting blade to plough the field (Soma Pond) to grow plasma blooms, or Soma plants (refer to section 2.7.11).

RV iii.20.2 author गाथी, to अग्निः, metre त्रिष्टुप् छंदः
अग्ने त्री ते वाजिना त्री षधस्था तिस्रस्ते जिह्वा ऋतजात पूर्वीः ।
तिस्र उ ते तन्वो देववातास्ताभिर्नः पाहि गिरो अप्रयुच्छन् ॥३.२०.२॥
अग्ने । त्री । ते । वाजिना । त्री । सधऽस्था । तिस्रः । ते । जिह्वाः । ऋतऽजात । पूर्वीः ।
तिस्रः । ऊं इति । ते । तन्वः । देवऽवाताः । ताभिः । नः । पाहि । गिरः । अप्रऽयुच्छन् ॥३.२०.२॥
O Agni, born of sacrifice, three are thy coursers, three are thine abodes, three are thine ancient tongues of flame. Three are thy manifestations, desired by the Devas. With these three forms, attend thou to our invocations, O thou heedful. (RV iii.20.2)

अग्ने <m. voc. sg. of अग्नि – O Agni>, त्री <(=त्रीणि) n. nom. pl. of त्रि – three>, ते <m. gen. sg. of 2nd person pron. युष्मद् – your>, वाजिना <(=वाजिनानि) n. nom. pl. of वाजिन (=वाजिन्) – the steeds of a war-chariot, horses>, त्री <त्रीणि n. nom. pl. of त्रि – three>, सधस्था <(=सधस्थानि) n. nom. pl. of सधस्थ – abodes, homes>, तिस्रः <f. nom. pl. of त्रि – three>, ते <m. gen. sg. of 2nd person pron. युष्मद् – your>, जिह्वाः <f. nom. pl. of जिह्वा – tongues of Agni's flame>, ऋतजात <m. voc. sg. of adj. ऋतजात – O born of sacrifice>, पूर्वीः <(=पूर्वाः) f. nom. pl. of adj. पूर्व – ancient, old>;

तिस्रः <f. nom. pl. of त्रि – three>, उ <ind. – and, also, further>, ते <m. gen. sg. of 2nd person pron. युष्मद् – your>, तन्वः <f. nom. pl. of तनू – bodies, forms, manifestations>, देववाताः <f. nom. pl. of adj. देववात – wished for, desired by the Devas>, ताभिः <f. inst. pl. of 3rd person pron. तद् – with these (i.e. with these three forms)>, नः <m. gen. pl. of 1st person pron. अस्मद् – our>, पाहि <ipv. 2nd person sg. of √पा 2P – you observe, attend to>, गिरः <f. acc. pl. of गिर् – Speeches, invocations, praises>, अप्रयुच्छन् <m. voc. sg. of अप्रयुच्छत् – O thou attentive, heedful, mindful>.

7.1.4 Ekata, Dvita, Trita: Protium, Deuterium, Tritium

7.1.4.1 Ekata

The three forms of Agni's fires, according to the Ṛgveda, are Ekata (एकतः), Dvita (द्वितः), and Trita (त्रितः). Ekata, protium (^1H), does not appear in the extant text of the Ṛgveda. Very likely, it is because Ekata is a general form of Agni, thus it is represented by Agni.

Dvita, deuterium (^2H), and Trita, tritium (^3H), two particular forms of Agni's fires, appear in RV v. 9.5, RV v.18.2, RV viii.47.14-16, and in several other stanzas. We find Trita, Dvita, and Ekata listed together as three Āptyas (आप्त्याः āptyāḥ, pl.) in Śatapathabrāhmaṇaṃ (शतपथब्राह्मणं) i.2.3.1 and 5. However, in the extant text of the Ṛgveda, Āptya is mostly associated with Trita as in 'Trita Āptya' (RV viii.47.13-15). Monier-Williams defines Āptya as "name of Trita". It seems that Āptya is derived from āpta (आप्त), which means "reached", "full", "complete", "respected", "true". Of the three fires of Agni, Trita is the most powerful fire that attained its full potential. Thus, Āptya befits the name of Trita.

7.1.4.2 Dvita Brings Slayed Offerings

Dvita (द्वितः), deuterium (^2H), appears in stanza RV v.18.2. The author of the hymn RV v.18 is Dvita himself. His full name is Dvita Mṛktavāhā Ātreya (द्वितः मृक्तवाहाः आत्रेयः), 'Dvita who brings the slayed offerings, the descendant of Atri (अत्रिः)'. Atri is the devourer, the Stone. The entire hymn is addressed to Agni. The first three stanzas of the hymn, RV v.18.1-3, are presented here.

In stanza RV v.18.1, Agni is called Atithi (अतिथिः), which has been translated as a 'guest' by Vedic students. However, note that, in the Ṛgveda, Agni is the lord of the house, never a guest. The word atithi (अतिथि) must have been derived from 'ati' (अति ind. beyond, surpassing) or 'at' (अत् ind. extraordinary). Agni is the beloved lord of the house. The deceased men (मर्ताः pl.) refer to the men reborn as cosmic plasmas after death, that is, the destruction of atomic structures through the ionisation process.

RV v.18.1 author द्वितो मृक्तवाहा आत्रेयः, to अग्निः, metre अनुष्टुप् छंदः
प्रातरग्निः पुरुप्रियो विशः स्तवेतातिथिः । विश्वानि यो अमर्त्यो हव्या मर्तेषु रण्यति ॥५.१८.१॥
प्रातः । अग्निः । पुरुऽप्रियः । विशः । स्तवेत । अतिथिः ।
विश्वानि । यः । अमर्त्यः । हव्या । मर्तेषु । रण्यति ॥५.१८.१॥
At dawn, may Agni, the beloved lord of the house, be celebrated in hymns,
He, the immortal among the deceased men, who delights in all oblations. (RV v.18.1)

प्रातः <(प्रातर्) ind. – at daybreak, at dawn>, अग्निः <m. nom. sg. of अग्नि – Agni>, पुरुप्रियः <m. nom. sg. of adj. पुरुप्रिय – much beloved, liked, wanted>, विशः <f. gen. sg. of विश् – of the house, dwelling, settlement, i.e. of each plasma bloom or the Soma Pond>, स्तवेत <opt. middle 3rd person sg. of √स्तु – may he be praised, celebrated in song or hymns>, अतिथिः <m. nom. sg. of अतिथि (derived from अति beyond, surpassing or from अत् ind. extraordinary) – the extraordinary, the lord (Traditionally it has been translated as a 'guest'. However, note that, in the Ṛgveda, Agni is the lord of the house, never a guest.)>;

विश्वानि <n. acc. pl. of adj. विश्व – all>, यः <m. nom. sg. of pron. यद् – he who>, अमर्त्यः <m. nom. sg. of अमर्त्य – the immortal, imperishable, divine>, हव्या <n. acc. pl. of हव्य – oblations, sacrificial foods>, मर्तेषु <m. loc. pl. of मर्त (from √मृ to die, decease) – amongst deceased men, among mortals (refers to the men reborn as cosmic plasmas after death, i.e. the destruction of the atomic structures of hydrogen)>, रण्यति <pre. 3rd person sg. of √रण् – he rejoices, delights, take pleasure in (loc.)>.

According to stanza RV v.18.2, the 'slayed offerings' which Dvita brings are the gifts from his own Dakṣa (दक्षः). The implication is that Dakṣa, the energy or force of the celestial ioniser, is Dvita's own. He, Trita, the praiser of Agni, bestows the bright drop of Soma upon Divta who delivers the slayed offerings.

RV v.18.2 author द्वितो मृक्तवाहा आत्रेयः, to अग्निः, metre अनुष्टुप् छंदः
द्विताय मृक्तवाहसे स्वस्य दक्षस्य मंहना । इंदुं स धत्त आनुषक्स्तोता चित्ते अमर्त्य ॥५.१८.२॥
द्विताय । मृक्तऽवाहसे । स्वस्य । दक्षस्य । मंहना ।
इंदुं । सः । धत्ते । आनुषक् । स्तोता । चित् । ते । अमर्त्य ॥५.१८.२॥
Upon Dvita who brings slayed offerings, the gifts from his own Dakṣa,
He, thy praiser, bestows the bright drop of Soma uninterrupted, O immortal One!
(RV v.18.2)

द्विताय <m. dat. sg. of द्वित – to Dvita, Deuterium (^{2}H)>, मृक्तवाहसे <m. dat. sg. of मृक्तवाहस् (मृक्त the pressed, crushed, slayed (from √मृच्) + वाहस् offering) – to Mṛktavāhā (मृक्तवाहाः), the name of Dvita, 'one who brings slayed offerings'>, स्वस्य <m. gen. sg. of adj. स्व – of own, one's own>, दक्षस्य <m. gen. sg. of दक्ष – of Dakṣa (mental power, strength of will, energy)>, मंहना <(मंहनानि) n. acc. pl. of मंहन – presents, gifts>;

इंदुं <m. acc. sg. of इंदु – Indu, the bright drop (of Soma)>, सः <m. nom. sg. of pron. तद् – he (Trita)>, धत्ते <pre. middle 3rd person sg. of √धा – he bestows on, presents to (dat.)>, आनुषक् <ind. – in continuous order, uninterruptedly>, स्तोता <m. nom. sg. of स्तोतृ – a praiser (refers to Trita)>, चित् <(चिद् ind. – even, indeed>, ते <m. gen. sg. of युष्मद् – your (of Agni)>, अमर्त्य <m. voc. sg. of अमर्त्य – O immortal One (refers to Agni)>.

In stanza RV v.18.3, Maghavās (मघवानः maghavānaḥ pl.), the institutors of sacrifices, and Aśvadāvā (अश्वदावा), the provider of coursers, refer to Agnis and Agni respectively.

RV v.18.3 द्वितो मृक्तवाहा आत्रेयः, to अग्निः, metre अनुष्टुप् छंदः
तं वो दीर्घायुशोचिषं गिरा हुवे मघोनां । अरिष्टो येषां रथो व्यश्वदावन्नीयते ॥५.१८.३॥
तं । वः । दीर्घायुऽशोचिषं । गिरा । हुवे । मघोनां ।
अरिष्टः । येषां । रथः । वि । अश्वऽदावन् । ईयते ॥५.१८.३॥
With hymn I invoke that lofty flame of Maghavās,
Whose chariot traverses unimpeded, O Aśvadāvā, the provider of coursers.
(RV v.18.3)

तं <m. acc. sg. of pron. तद् – that>, वः <m. gen. pl. युष्मद् – yours (pl.)>, दीर्घायुशोचिषं <m. acc. sg. of दीर्घायुशोचिस् (दीर्घायु long + शोचिस् = शोचि flame) – a lofty or tall flame, i.e. the plasma bloom>, गिरा <mf. inst. sg. गिर् – with song, hymn>, हुवे <pre. 1st person sg. of √हू – I (Dvita) call, invoke>, मघोनां <m. gen. pl. of मघवत् (=मघवन्) – of Maghavās (मघवानः maghavānaḥ pl.), of the Institutors of sacrifices (here Maghavās refer to Agnis, not Indra)>;

अरिष्टः <m. nom. sg. of adj. अरिष्ट – unhurt, secure, not torn, unbroken>, येषां <m. gen. pl. of pron. यद् – whose (pl.)>, रथः <m. nom. sg. of रथ – a chariot>, वि <ind. prefix to verb ईयते (√इ)>, अश्वदावन् <m. voc. sg. of अश्वदावन् (=अश्वद) – O Aśvadāvā (अश्वदावा m. nom. sg. of अश्वदावन् the provider of horses)>, वीयते <(वि-ईयते) int. middle 3rd person sg. of √वी (वि-√इ) 4Ā – it (the chariot) traverses, goes through>.

7.1.4.3 Trita Blasts and Whets Like a Blower in the Smelting Furnace

When Trita, tritium (^3H), blasts and whets like a blower in the smelting furnace, Agni's smoking flames rise together in one direction (RV v.9.5). Plasma blooms, the light-containing plants, are called Agni's smoking flames. Note the charged particles of cosmic plasmas that are blown out of the Stone, the celestial ioniser, are described as smoke. The stanza shows that Trita is the fire that is responsible for the ionisation of hydrogen by breaking up the atomic structures. This process is called Soma pressing in the Ṛgveda, for Soma (electron) is pressed out from atomic structures as Soma juice.

Stanza RV v.9.5 is addressed to Agni. The author is Gaya Ātreya (गयः आत्रेयः), which means 'what has been acquired (गयः) by the descendant of Atri'. Atri (अत्रिः) is the devourer, the Stone. So, the author's name is 'what is acquired by the Stone, the celestial ioniser'. Of course, what is acquired are the flames of Agni, the blooming cosmic plasmas, the light-containing plants.

RV v.9.5 author गय आत्रेयः, to अग्निः, metre अनुष्टुप् छंदः
अध स्म यस्यार्चयः सम्यक्संयंति धूमिनः ।
यदीमह त्रितो दिव्युप ध्मातेव धमति शिशीते ध्मातरी यथा ॥५.९.५॥
अध । स्म । यस्य । अर्चयः । सम्यक् । संऽयंति । धूमिनः ।
यत् । ईं । अह । त्रितः । दिवि । उप । ध्माताऽइव । धमति । शिशीते । ध्मातरि । यथा ॥५.९.५॥
His smoking flames rise together turned in one direction,
When Trita, in the Heaven, blasts and whets like a smelter in the smelting furnace.
(RV v.9.5)

अध <ind. – now, then, moreover>, स्म <ind. particle – indeed, verily>, यस्य <m. gen. sg. of pron. यद् – whose, of which, his (Agni's)>, अर्चयः <m. nom. pl. of अर्चि – rays, flames>, सम्यक् <ind. – turned together or in one direction, forming one line>, संयंति <pre. 3rd person pl. of सं-√इ – they (the flames) go or come together, rise together>, धूमिनः <m. nom. pl. of adj. धूमिन् – smoking, steaming>;

यत् <conjunction – that, when>, ईं <ind. particle of affirmation – (used with the conjunction यद्)>, अह <ind. – certainly, indeed>, त्रितः <m. nom. sg. of त्रित – Trita (Tritium)>, दिवि <m. loc. sg. of दिव् – in the Heaven>, उप <ind. prefix to verb धमति (√ध्मा)>, ध्माता <m. nom. sg. of ध्मातृ – a blower, smelter>, इव <ind. – like, in the same manner as>, उप-धमति <pre. 3rd person sg. of उप-√ध्मा – it (Trita) blows upon, melts by blowing>, शिशीते <pre. middle 3rd person sg. of √शो – it (Trita) whets or sharpens>, ध्मातरि <m. loc. sg. of ध्मातृ – in the smelter, in the smelting furnace>, यथा <ind. – as, like>.

Stanzas RV i.52.5 and RV viii.47.13-16, presented below, also provide information that is helpful for understanding Dvita and Trita. Stanza RV i.52.5 states that Indra cleft Vṛtra as Trita cleaves Vala's fences. Vala's fences refer to the ionic bonds of hydrogens. The stanza confirms that Trita is the fire that destroys the ionic bonds. The stanza is addressed to Indra. The author is Savya Āṅgirasa (सव्यः आंगिरसः). Savya (सव्यः), according to Monier-Williams, is "a fire lighted at a person's death". Āṅgirasa (आंगिरसः) is a descendant of

Aṅgirā (अङ्गिराः aṅgirāḥ, m. nom. sg. of अङ्गिरस्), the celestial light, the messenger of the Heaven. The fire ignited at the time of the death of hydrogen, that is, at the time of ionisation, are the celestial lights, the cosmic plasmas, that grow into the flaming light-containing plants, the plasma blooms.

RV i.52.5 author सव्य आंगिरसः, to इंद्रः, metre जगती छंदः
अभि स्ववृष्टिं मदे अस्य युध्यतो रघ्वीरिव प्रवणे सस्रुरूतयः ।
इंद्रो यद्वज्री धृषमाणो अंधसा भिनद्वलस्य परिधीँरिव त्रितः ॥१.५२.५॥
अभि । स्वऽवृष्टिं । मदे । अस्य । युध्यतः । रघ्वीःऽइव । प्रवणे । सस्रुः । ऊतयः ।
इंद्रः । यत् । वज्री । धृषमाणः । अंधसा । भिनत् । वलस्य । परिधीन्ऽइव । त्रितः ॥१.५२.५॥
Refreshing draughts advanced in rapture along a precipitous course as fleet coursers led to war, When Indra, emboldened by Soma juice, wielding a thunderbolt, cleft the rain-retainer as Trita cut asunder Vala's fences. (RV i.52.5)

अभि <ind. prefix to verb सस्रुः (√सृ)>, स्ववृष्टिं <mf. acc. sg. of स्ववृष्टि – the rain-retainer, i.e. the retainer of cosmic plasmas (refers to Vṛtra, the plasma sheath)>, मदे <m. loc. sg. of मद) – in rapture>, अस्य <m. gen. sg. of इदं – of this, his>, युध्यतः <f. nom. pl. of adj. युध्यत् – led to war, engaged in battle>, रघ्वीः <f. nom. pl. of रघ्वि (?) – racers, fleet coursers>, इव <ind. – like, in the same manner as>, प्रवणे <ind. – in a precipitous course (note that plasma blooms grow upwards vertically)>, अभि-सस्रुः <prf. 3rd person pl. of अभि-√सृ – they (refreshing draughts) flowed towards, advanced>, ऊतयः <f. nom. pl. of ऊति – riches, refreshments, i.e. refreshing draughts>;

इंद्रः <m. nom. sg. of इंद्र – Indra>, यत् <conjunction – that, when>, वज्री <m. nom. sg. of adj. वज्रिन् – wielding a thunderbolt>, धृषमाणः <m. nom. sg. of pre. middle pt. धृषमाण (√धृष्) – he (Indra) being bold or courageous>, अंधसा <n. inst. sg. of अंधस् – by Soma juice>, भिनत् <inj. or ipf. 3rd person sg. of √भिद् – he split, cleft, cut asunder>, वलस्य <m. gen. sg. of वल – of Vala (refers to an atom or atomic structure)>, परिधीन् <m. acc. pl. of परिधि – enclosures, walls, fences (refer to the ionic bonds of hydrogens)>, इव <<ind. – like, in the same manner as>, त्रितः <m. nom. sg. of त्रित – Trita (Tritium)>.

Trita Āptya also appears in RV viii.47.13-16. Dvita is invoked with Trita in RV viii.47.16. The author of these stanzas is Trita Āptya (त्रितः आप्त्यः) himself. Stanza RV viii.47.13 is addressed to the Ādityas (आदित्याः pl.). Stanzas RV viii.47.14-16 are addressed to the Ādityas and the Dawns (आदित्याः उषाः च). The wicked evil-doer (दुष्कृतं duṣkṛtaṃ) refers to the ionic bond that binds Agnis and Somas within the atomic structure.

Note, in RV viii.47.13-16, a refrain repeated at the end of each stanza (अनेहसः वः ऊतयः सुऊतयः वः ऊतयः), can be interpreted and translated differently depending on how the words 'ūtayaḥ (ऊतयः mf. nom. pl. of ऊति)' and 'suūti (सुऊति suūti = सूति sūti)' are intepreted. The word ūti (ऊति), according to Monier-Williams, can be interpreted as "refreshment" or "means of helping or promoting or refreshing goods, riches (also pl.)". Here, "refreshment" refers to Soma juice, "refreshing goods, riches" to Soma draughts. The "means" of procuring "refreshing goods, riches" refers to the ionisation process that produces refreshing goods, that is, Soma draughts. Thus, the same refrain can be interpreted differently between RV viii.47.13 and RV viii.47.14-16 based on the contents of the stanzas. One cannot help but believe the ancient poet was masterful in choosing the words ‘ūtayaḥ (ऊतयः mf. nom. pl. of ऊति)’ and ‘suūti (सुऊति = सूति)’ in the refrain.

RV viii.47.13 author त्रित आप्त्यः, to आदित्याः, metre महापंक्तिः छंदः
यदाविर्यदपीच्यं१◌◌ देवासो अस्ति दुष्कृतं ।
त्रिते तद्विश्वमाप्त्य आरे अस्मद्दधातनानेहसो व ऊतयः सुऊतयो व ऊतयः ॥८.४७.१३॥
यत् । आविः । यत् । अपीच्यं । देवासः । अस्ति । दुःऽकृतं ।
त्रिते । तत् । विश्वं । आप्त्ये । आरे । अस्मत् । दधातन । अनेहसः । वः । ऊतयः । सुऽऊतयः । वः । ऊतयः
॥८.४७.१३॥
Evidently, there is that wicked evil-doer, which is hidden, O Devas,
In the fire of Trita Āptya, remove ye every one of that evil-doer from us. Your unrivalled means of procuring riches, the pressing out of Soma draughts, your measures of procuring riches. (RV viii.47.13)

यत् <n. nom. sg. of यद् – that which>, आविः <आविस् ind. – before the eyes, openly, evidently>, यत् <n. nom. sg. of यद् – that which>, अपीच्यं <n. nom. sg. of adj. अपीच्य – secret, hidden>, देवासः <m. voc. pl. of देव – O Devas>, अस्ति <pre. 3rd person sg. of √अस् – it is, there is>, दुष्कृतं <n. nom. sg. of दुष्कृत् – the wicked, evil-doer (refers to the ionic bond of the atomic structure)>;

त्रिते <m. loc. sg. of त्रित – in Trita (Tritium)>, तत् <n. acc. sg. of तद् – that (the wicked)>, विश्वं <n. acc. sg. of विश्व – all, every>, आप्त्ये <m. loc. sg. of आप्त्य – in Āptya (the name of Trita)>, आरे <ind. – far>, अस्मत् <m. abl. pl. of 1st person pron. अस्मद् – from us>, दधातन <ipv. 2nd person pl. of √धा – (with आरे) ye (the firestones of Trita) remove>, अनेहसः <f. nom. pl. of adj. अनेहस् – without a rival, incomparable>, वः <m. gen. pl. of युष्मद् – your (pl.)>, ऊतयः <f. nom. pl. of ऊति – means of procuring or promoting riches, i.e. means of procuring cosmic plasmas>, सुऊतयः <(=सूतयः) f. nom. pl. of सुऊति (=सूति) – pressing out the Soma juice>, वः <m. gen. pl. of युष्मद् – your (pl.)>, ऊतयः <f. nom. pl. of ऊति – means of procuring or promoting riches, i.e. cosmic plasmas>.

RV viii.47.14 author त्रित आप्त्यः, to आदित्या उषाश्च, metre महापंक्तिः छंदः
यच्च गोषु दुष्ष्वप्न्यं यच्चास्मे दुहितर्दिवः ।
त्रिताय तद्विभावर्याप्त्याय परा वहानेहसो व ऊतयः सुऊतयो व ऊतयः ॥८.४७.१४॥
यत् । च । गोषु । दुःऽस्वप्न्यं । यत् । च । अस्मे इति । दुहितः । दिवः ।
त्रिताय । तत् । विभाऽवरि । आप्त्याय । परा । वह । अनेहसः । वः । ऊतयः । सुऽऊतयः । वः । ऊतयः
॥८.४७.१४॥
O milch Cow of the Heaven, cast the pressed Soma upon us and upon the herds of cattle.
For Trita Āptya, O bright stream, carry off that pressed Soma. Your unrivalled Soma draughts, pressed Soma juice, your refreshing drinks. (RV viii.47.14)

यच्च...यच्च <(यद् च...यद् च) – both...and>, गोषु <m. loc. pl. of गो – upon the herds of cattle>, दुष्ष्वप्न्यं <n. acc. sg. of दुष्ष्वप्न्य (दुः evil, bad + स्वप् dead, killed + न्य = न्यस् = न्य्-√अस् to throw, cast) – the pressed Soma (when the evil, i.e. the ionic bond, is killed and thrown off, Soma juice or cosmic plasmas are produced)>, यच्च <see above>, अस्मे <m. loc. pl. of 1st person pron. अस्मद् – upon us>, दुहितः <f. voc. sg. of दुहितृ – O milch Cow, the extractor, the Soma Pond>, दिवः <m. gen. sg. of दिव् – of the Heaven>;

त्रिताय <m. dat. sg. of त्रित – for Trita>, तत् <n. acc. sg. of तद् – that (the pressed Soma)>, विभावरि <f. voc. sg. of विभावरी (विभा brilliant, bright, light + वरी stream) – O bright stream (refers to उषाः Uṣā)>, आप्त्याय <m. dat. sg. of आप्त्य – for Āptya>, परा <ind. prefix to verb वह् (√वह्)>, परा-वह <ipv. 2nd person sg. of परा-√वह् – you (the bright stream or Uṣā) carry off, take away>, अनेहसः <f. nom. pl. of adj. अनेहस् – without a rival, incomparable>, वः <m. gen. pl. of युष्मद् – your (pl.)>, ऊतयः <f. nom. pl. of ऊति – refreshing Soma draughts (draughts of cosmic plasmas)>, सुऊतयः <(=सूतयः) f. nom.

pl. of सुऊति (=सूति) – pressed out Soma juice>, वः <m. gen. pl. of युष्मद् – your (pl.)>, ऊतयः <f. nom. pl. of ऊति – refreshing drinks>.

RV viii.47.15 author त्रित आप्त्यः, to आदित्या उषाश्च, metre महापंक्तिः छंदः
निष्कं वा घा कृणवते स्रजं वा दुहितर्दिवः ।
त्रिते दुष्ष्वप्न्यं सर्वमाप्त्ये परि दद्मस्यनेहसो व ऊतयः सुऊतयो व ऊतयः ॥८.४७.१५॥
निष्कं । वा । घ । कृणवते । स्रजं । वा । दुहितः । दिवः ।
त्रिते । दुःऽस्वप्न्यं । सर्वं । आप्त्ये । परि । दद्मसि । अनेहसः । वः । ऊतयः । सुऽऊतयः । वः । ऊतयः
॥८.४७.१५॥
O milch Cow of the Heaven, whether he shall make a golden vessel or a garland,
Upon Trita Āptya, we entrust all the pressed Soma. Your unrivalled Soma draughts, pressed Soma juice, your refreshing drinks. (RV viii.47.15)

निष्कं <m. acc. sg. निष्क – a golden ornament for the neck or breast, golden vessel>, वा...वा <ind. – either...or>, घ <ind. – surely, verily, indeed>, कृणवते <sub. middle 3rd person sg. of √कृ – he (Trita) shall make>, स्रजं <f. acc. sg. of स्रज् – a wreath of flowers, garland>, वा <ind. – see above>, दुहितः <f. voc. sg. of दुहितृ – O daughter milch Cow, the extractor>, दिवः <m. gen. sg. of दिव् – of the Heaven>;

त्रिते <m. loc. sg. of त्रित – upon Trita>, दुष्ष्वप्न्यं <n. acc. sg. of दुष्ष्वप्न्य (दुः evil, bad + स्वप् dead, killed + न्य = न्यस् = न्य्-√अस् to throw, take off) – the pressed Soma (when the evil, i.e. the ionic bond, is killed and thrown off, Soma juice or cosmic plasmas are produced)>, सर्वं <n. acc. sg. of adj. सर्व – every, all>, आप्त्ये <m. loc. sg. of आप्त्य – upon Āptya>, परि <ind. prefix to verb दद्मसि (√दा)>, परि-दद्मसि <pre. 1st person pl. of परि-√दा – we entrust upon (loc.)>, अनेहसः वः ऊतयः सुऊतयः वः ऊतयः <see RV viii.47.14 – Your unrivalled Soma draughts, pressed Soma juice, your refreshing drinks.>.

RV viii.47.16 author त्रित आप्त्यः, to आदित्या उषाश्च, metre महापंक्तिः छंदः
तदन्नाय तदपसे तं भागमुपसेदुषे ।
त्रिताय च द्विताय चोषो दुष्ष्वप्न्यं वहानेहसो व ऊतयः सुऊतयो व ऊतयः ॥८.४७.१६॥
तत्ऽअन्नाय । तत्ऽअपसे । तं । भागं । उपऽसेदुषे ।
त्रिताय । च । द्विताय । च । उषः । दुःऽस्वप्न्यं । वह । अनेहसः । वः । ऊतयः । सुऽऊतयः । वः । ऊतयः
॥८.४७.१६॥
To that region, which is accustomed to having that food and to that streaming celestial Waters, to that region, For Trita, and for Dvita, O Dawn, carry off the pressed Soma. Your unrivalled Soma draughts, pressed Soma juice, your refreshing drinks. (RV viii.47.16)

तदन्नाय <m. dat. sg. of adj. तदन्न (तद् + अन्न food) – to/for the one (the region) accustomed to that food (in the Ṛgveda, Soma is food)>, तदपसे <m. dat. sg. of adj. तदपस् (तद् + अपस् running Waters) – to/for the one (i.e. the region) accustomed to that running Waters (streaming celestial Waters)>, तं <m. acc. sg. of तद् – to this, that>, भागं <m. acc. sg. of भाग – to the place, region (refers to the Innerfield where the Milky Way is formed and sustained)>, उपसेदुषे <m. dat. sg. of prf. active pt. उपसेदुष् (उप-√सद्) – (for the region) having possessed of>;

त्रिताय <m. dat. sg. of त्रित – for Trita>, च <ind. – and>, द्विताय <m. dat. sg. of द्वित – for Dvita>, च <ind. – and>, उषः <f. voc. sg. of उषस् – O Dawn>, दुष्ष्वप्न्यं <n. acc. sg. of दुष्ष्वप्न्य (दुः evil, bad + स्वप् dead, killed + न्य = न्यस् = न्य्-√अस् to throw, take off) – the pressed Soma (when the evil, i.e. the ionic bond, is killed and thrown off, Soma juice or cosmic plasmas are produced)>, वह <ipv. 2nd person sg. of √वह् – you carry off>, अनेहसः वः ऊतयः सुऊतयः वः ऊतयः <see RV viii.47.14 – Your unrivalled Soma draughts, pressed Soma juice, your refreshing drinks.>.

7.1.5 Agni's Path: Black, White, Yellow, Red, Purple, and Glorious

In stanza RV x.20.9, the path upon which Agni treads, as he is born, grows, matures, and is kindled, is described in colours. Agni's path is also glorious and fleet. The stanza is addressed to Agni.

RV x.20.9 author विमद ऐंद्रः प्राजापत्यो वा वसुकृद्वा वासुक्रः, to अग्निः, metre विराट् छंदः
कृष्णः श्वेतोऽरुषो यामो अस्य ब्रध्न ऋज्र उत शोणो यशस्वान् ।
हिरण्यरूपं जनिता जजान ॥१०.२०.९॥
कृष्णः । श्वेतः । अरुषः । यामः । अस्य । ब्रध्नः । ऋज्रः । उत । शोणः । यशस्वान् ।
हिरण्यऽरूपं । जनिता । जजान ॥१०.२०.९॥
The path Agni treads is black, white, yellow, red, purple, and glorious and fleet.
His sire has begotten him in hues of gold. (RV x.20.9)

कृष्णः <m. nom. sg. of adj. कृष्ण – dark-blue, black>, श्वेतः <m. nom. sg. of adj. श्वेत – white>, अरुषः <m. nom. sg. of adj. अरुष – red>, यामः <m. nom. sg. of याम – a road, way, path>, अस्य <m. gen. sg. of इदं – of this, his (Agni's)>, ब्रध्नः <m. nom. sg. of adj. ब्रध्न – bay, yellow>, ऋज्रः <m. nom. sg. of adj. ऋज्र – quick (as horses)>, उत <ind. – and, also, even>, शोणः <m. nom. sg. of adj. शोण – crimson, purple>, यशस्वान् <m. nom. sg. of adj. यशस्वत् – splendid, magnificent, glorious>;

हिरण्यरूपं <m. acc. sg. of adj. हिरण्यरूप – gold-coloured, hues of gold>, जनिता <m. nom. sg. of जनितृ – a progenitor, father>, जजान <prf. 3rd person sg. of √जन् – has begotten>.

7.1.6 Frogs Capture One Another Based on Their Speech Patterns

According to the Ṛgveda, there are two different classes of Brāhmaṇas: Agnis (RV ii.1.2, iii.13.6, iv.9.4, vii.103.1, viii.17.3) and Somas (RV ix.96.6). These Brāhmaṇas have divine knowledge and know the intention of the Creator.

The ten stanzas of hymn RV vii.103 show how frogs, that is, Agnis, capture and bind with one another (RV vii.103.3-4) and how they modulate their Speech (RV vii.103.5-6). According to the hymn RV vii.103, frogs, or Agnis, capture one another based on their Speech patterns, indicating that protons attract and bind with one another based on the frequencies of their Speeches and directions of their spins. In these stanzas, the rain and the celestial Waters refer to discharged cosmic plasmas. Once matured plasma blooms are released by Indra, the subsequent cycle of rainy season or year begins. Ionisation continues, cosmic plasmas are discharged, and the Soma plants, or plasma blooms, will grow and mature in the Soma Pond as in the previous cycle. When fully matured, they are released again.

All ten stanzas of the hymn are addressed to frogs (मंडूकाः maṇḍukāḥ, pl.), or Agnis. The first stanza is in anuṣṭup metre (अनुष्टुप् छंदः) and the rest are in triṣṭup metre (त्रिष्टुप् छंदः). The author is Vasiṣṭha (वसिष्ठः), the owner of the cow of plenty, that is, Rudra, the Soma Pond.

RV vii.103.1 author वसिष्ठः, to मंडूकाः, metre अनुष्टुप् छंदः
संवत्सरं शशयाना ब्राह्मणा व्रतचारिणः । वाचं पर्जन्यजिन्वितां प्र मंडूका अवादिषुः ॥७.१०३.१॥
संवत्सरं । शशयानाः । ब्राह्मणाः । व्रतऽचारिणः ।
वाचं । पर्जन्यऽजिन्वतां । प्र । मंडूकाः । अवादिषुः ॥७.१०३.१॥
Brāhmaṇas, the frogs, performing penance in repose,
Raised their voice urging for rain to the Soma Pond. (RV vii.103.1)

संवत्सरं <m. acc. sg. of संवत्सर – to Saṃvatsara, i.e. to the Soma Pond>, शशयानाः <m. nom. pl. of adj. शशयान – lying, reposing>, ब्राह्मनाः <m. nom. pl. of ब्राह्मण - Brāhmaṇas (pl.) (refer to the frogs or Agnis)>, व्रतचारिणः <m. nom. pl. of adj. व्रतचारिन् – vow-performing, engaged in religious observance>;

वाचं <f. acc. sg. of वाच् – speech, voice>, पर्जन्यजिन्वितां <f. acc. sg. of adj. पर्जन्यजिन्वित (पर्जन्य rain + जिन्वित (from √जिन्व् to urge)) – urging for rain>, प्र <ind. prefix to verb अवादिषुः (√वद्)>, मंडूकाः <m. nom. pl. of मंडूक – maṇḍūkas, frogs>, प्र-अवादिषुः <aor. 3rd person pl. of प्र-√वद् – (with वाचं) they raised their voice, spoke>.

RV vii.103.2 author वसिष्ठः, to मंडूकाः, metre त्रिष्टुप् छंदः
दिव्या आपो अभि यदेनमायन्दृतिं न शुष्कं सरसी शयानं ।
गवामह न मायुर्वत्सिनीनां मंडूकानां वग्नुरत्रा समेति ॥७.१०३.२॥
दिव्याः । आपः । अभि । यत् । एनं । आयन् । दृतिं । न । शुष्कं । सरसी इति । शयानं ।
गवां । अह । न । मायुः । वत्सिनीनां । मंडूकानां । वग्नुः । अत्र । सं । एति ॥७.१०३.२॥
When the celestial Waters fell upon the troop of frogs lying like parched skin on the lake bed,
The sound of the frogs arises in concert as cows bellowing with their calves beside them.
(RV vii.103.2)

दिव्याः <f. nom. pl. of adj. दिव्य – heavenly, celestial>, आपः <f. nom. pl. of अप् – Waters>, अभि <ind. prefix to verb आयन् (√इ)>, यत् <conjunction – that, when>,एनं <m. acc. sg. of इदं – to this (to the troop of frogs)>, अभि-आयन् <ipf. 3rd person. pl. of अभि-√इ – they (Waters) fell upon (with acc.)>, दृतिं <m. acc. sg. of दृति – skin, hide>, न <ind. – like, as>, शुष्कं <m. acc. sg. of adj. शुष्क – dried, parched>, सरसी <(सरसि?) n. loc. sg. of सरस् – in the lake, i.e. in the Soma Pond>, शयानं <m. acc. sg. of adj. शयान – lying>;

गवां <m. gen. pl. of गो – of cows>, अह <ind. – (a particle of affirmation) surely, certainly>, न <ind. – like, as>, मायुः <m. nom. sg. of मायु – bleating, bellowing>, वत्सिनीनां <f. gen. pl. of adj. वत्सिन् – having calves>, मंडूकानां <m. gen. pl. of मंडूक – of maṇḍūkas, of frogs>, वग्नुः <m. nom. sg. of वग्नु – a cry, call, sound (esp. of animals)>, अत्र <ind. – here, then>, सं <ind. prefix to verb एति (√इ)>, सं-एति <pre. 3rd person sg. of सं-√इ – it (sound) arises, comes together>.

RV vii.103.3 author वसिष्ठः, to मंडूकाः, metre त्रिष्टुप् छंदः
यदीमेनाँ उशतो अभ्यवर्षीत्तृष्यावतः प्रावृष्यागतायां ।
अख्खलीकृत्या पितरं न पुत्रो अन्यो अन्यमुप वदंतमेति ॥७.१०३.३॥
यत् । ईं । एनान् । उशतः । अभि । अवर्षीत् । तृष्याऽवतः । प्रावृषि । आऽगतायां ।
अख्खलीकृत्य । पितरं । न । पुत्रः । अन्यः । अन्यं । उप । वदंतं । एति ॥७.१०३.३॥
When the rainy season arrived and the rain fell upon them, who were thirsty and longing for rain, One frog approaches the other who utters an exclamation of joy as a son who greets his father. (RV vii.103.3)

यत् <conjunction – that, when>, ईं <ind. – a particle of affirmation and restriction, now>, एनान् <m. acc. pl. of इदं – them>, उशतः <m. acc. pl. of adj. उशत् – wishing, desiring>, अभि <ind. prefix to verb अवर्षीत् (√वृष्)>, अभि-अवर्षीत् <aor. 3rd person sg. of अभि-√वृष् – the rain fell upon>, तृष्यावतः <m. acc. pl. of adj. तृष्यावत् – thirsty>, प्रावृषि <f. loc. sg. of प्रावृष् – the rainy season>, आगतायां <f. loc. sg. of adj. आगत – come, arrived>;

अख्खलीकृत्य <ind. – (cf. अख्खल ind. an exclamation of joy) uttering the exclamation of joy>, पितरं <m. acc. sg. of पितृ – father>, न <ind. – like, as>, पुत्रः <m. nom. sg. of पुत्र – a son>, अन्यः <m. nom. sg. of pron. अन्य – the one (frog)>, अन्यं <m. acc. sg. of pron. अन्य – the other (frog)>, उप <ind. – near to, towards>, वदंतं <m. acc. sg. of pre. active pt. वदंत् (√वद्) – him speaking, uttering>, एति <pre. 3rd person sg. of √इ – he approaches>.

RV vii.103.4 author वसिष्ठः, to मंडूकाः, metre त्रिष्टुप् छंदः

अन्यो अन्यमनु गृभ्णात्येनोरपां प्रसर्गे यदमंदिषातां ।

मंडूको यदभिवृष्टः कनिष्कन्पृश्निः संपृंक्ते हरितेन वाचं ॥७.१०३.४॥

अन्यः । अन्यं । अनु । गृभ्णाति । एनोः । अपां । प्रऽसर्गे । यत् । अमंदिषातां ।

मंडूकः । यत् । अभिऽवृष्टः । कनिस्कन् । पृश्निः । संऽपृंक्ते । हरितेन । वाचं ॥७.१०३.४॥

One of the two frogs seizes the other as the two are revelling in the flow of celestial Waters,
The dappled frog, when rained upon, leaps and joins his Speech with that of the green frog.
(RV vii.103.4)

अन्यः <m. nom. sg. of pron. अन्य – the one (frog)>, अन्यं <m. acc. sg. of pron. अन्य – the other>, अनु <ind. preposition – after, along, over>, गृभ्णाति <pre. 3rd person sg. of √ग्रभ् (=√ग्रह्) – he seizes, takes captive>, एनोः <m. gen. du. of एन – of the two>, अपां<f. gen. pl. of अप् – of Waters>, प्रसर्गे <m. loc. sg. of प्रसर्ग – in pouring, in the flow>, यत् <conjunction – that, when, as>, अमंदिषातां <aor. middle 3rd person du. of √मंद् (√मन्द्) – the two are delighted, rejoiced>;

मंडूकः <m. nom. sg. of मंडूक – a frog, Agni>, यत् <conjunction – that, when>, अभिवृष्टः <m. nom. sg. of adj. अभिवृष्ट – rained upon>, कनिष्कन् <ipf. 3rd person sg. of √स्कन्द् – jumped, leapt>, पृश्निः <m. nom. sg. of adj. पृश्नि – dappled, spotted>, संपृंक्ते <(संपृङ्क्ते) pre. middle 3rd person sg. of सं-√पृच् – he mixes, unites, joins>, हरितेन <m. inst. sg. of हरित – with the green or verdant (as opposed to शुष्क 'dry') frog>, वाचं <f. acc. sg. of वाच् – Speech, voice>.

RV vii.103.5 author वसिष्ठः, to मंडूकाः, metre त्रिष्टुप् छंदः

यदेषामन्यो अन्यस्य वाचं शाक्तस्येव वदति शिक्षमाणः ।

सर्वं तदेषां समृधेव पर्व यत्सुवाचो वदथनाध्यप्सु ॥७.१०३.५॥

यत् । एषां । अन्यः । अन्यस्य । वाचं । शाक्तस्यऽइव । वदति । शिक्षमाणः ।

सर्वं । तत् । एषां । समृधाऽइव । पर्व । यत् । सुऽवाचः । वदथन । अधि । अप्ऽसु ॥७.१०३.५॥

When one of the frogs learns the other's Speech as a student learns that of the teacher,
And as ye all utter eloquent Speeches in the celestial Waters, every member of the frogs becomes perfect. (RV vii.103.5)

यत् <conjunction – that, when>, एषां <m. gen pl. of इदं – of them (frogs)>, अन्यः <m. nom. sg. of pron. अन्य – the one>, अन्यस्य <m. gen. sg. of अन्य – of the other>, वाचं <f. acc. sg. of वाच् – Speech>, शाक्तस्य <m. gen. sg. of शाक्त – of the teacher>, इव <ind. – like, in the same manner as, as>, वदति <pre. 3rd person sg. of √वद् – speaks>, शिक्षमाणः <m. nom. sg. of pre. middle pt. शिक्षमाण (√शिक्ष्) – he learning, learns>;

सर्वं <n. nom. sg. of adj. or pron. सर्व – whole, entire, all, every>, तत् <n. nom. sg. of तद् – that, the>, एषां <n. gen pl. of इदं – of them, of these frogs>, समृधा <n. nom. pl. of adj. समृध – full, complete, perfect>, इव <ind. – like, just so, indeed, very>, पर्व <n. nom. sg. of पर्वन् – a member>, यत् <conjunction – that, when, as>, सुवाचः <f. acc. pl. of सुवाच् – beautiful, eloquent Speeches>, अधि-वदथन <(=वदथ) pre. 2nd person pl. of अधि-√वद् – ye speak, utter>, अधि <ind. prefix to verb वदथन (√वद्)>, अप्सु <f. loc. pl. of अप् – in the celestial Waters, i.e. in the flow of cosmic plasmas>.

RV vii.103.6 author वसिष्ठः, to मंडूकाः, metre त्रिष्टुप् छंदः
गोमायुरेको अजमायुरेकः पृश्निरेको हरित एक एषां ।
समानं नाम बिभ्रतो विरूपाः पुरुत्रा वाचं पिपिशुर्वदंतः ॥७.१०३.६॥
गोऽमायुः । एकः । अजऽमायुः । एकः । पृश्निः । एकः । हरितः । एकः । एषां ।
समानं । नाम । बिभ्रतः । विऽरूपाः । पुरुऽत्रा । वाचं । पिपिशुः । वदंतः ॥७.१०३.६॥
One of the frogs bellows like a cow, the other bleats like a goat, one is dappled, the other green. Bearing the same name, yet varying in form, speaking in many different ways, they modulate their Speech. (RV vii.103.6)

गोमायुः <m. nom. sg. of adj. गोमायु – bellowing like a cow>, एकः <m. nom. sg. of pron. एक – one>, अजमायुः <m. nom. sg. of adj. अजमायु – bleating like a goat>, एकः <m. nom. sg. of pron. एक – one>, पृश्निः <m. nom. sg. of adj. पृश्नि – spotted, dappled>, एकः <m. nom. sg. of pron. एक – one>, हरितः <m. nom. sg. of adj. हरित – verdant, green>, एकः <m. nom. sg. of pron. एक – one>, एषां <m. gen. pl. of इदं – of these (frogs)>;

समानं <n. acc. sg. of adj. समान – same, identical, common>, नाम <n. acc. sg. of नामन् – name>, बिभ्रतः <m. nom. pl. of adj. बिभ्रत् – they bearing, carrying>, विरूपाः <m. nom. pl. of adj. विरूप – vary in form, nature, character>, पुरुत्रा <ind. – variously, in many ways>, वाचं <f. acc. sg. of वाच् – Speech>, पिपिशुः <prf. 3rd person pl. of √पिश् – they form, mould>, वदंतः <m. nom. pl. of pre. active pt. वदत् (√वद्) – they speaking, speak, uttering>.

RV vii.103.7 author वसिष्ठः, to मंडूकाः, metre त्रिष्टुप् छंदः
ब्राह्मणासो अतिरात्रे न सोमे सरो न पूर्णमभितो वदंतः ।
संवत्सरस्य तदहः परि ष्ठ यन्मंडूकाः प्रावृषीणं बभूव ॥७.१०३.७॥
ब्राह्मणासः । अतिऽरात्रे । न । सोमे । सरः । न । पूर्णं । अभितः । वदंतः ।
संवत्सरस्य । तत् । अहरिति । परि । स्थ । यत् । मंडूकाः । प्रावृषीणं । बभूव ॥७.१०३.७॥
Brāhmaṇas, sitting around in the Soma Pond filled full, speak at the preparation of Soma sacrifice. On that very day, at the beginning of the rainy season, O frogs, ye are here in the Soma Pond. (RV vii.103.7)

ब्राह्मणासः <m. nom. pl. of ब्राह्मण – Brāhmaṇas (refers to frogs, i.e. Agnis)>, अतिरात्रे <m. loc. sg. of adj. अतिरात्र – at the preparation, in preparing>, न <ind. – not, no (with another न) forms a strong affirmation>, सोमे <m. loc. sg. of सोम – at the Soma sacrifice>, सरः <n. acc. sg. of सरस् – a pond, Lake, i.e. the Soma Pond>, न <see above>, पूर्णं <n. acc. sg. of adj. पूर्ण – filled, full>, अभितः <(अभितस्) ind. – (with acc.) everywhere, around>, वदंतः <m. nom. pl. of pre. active pt. वदत् (√वद्) – they speaking, uttering>;

संवत्सरस्य <m. gen. sg. of संवत्सर – of Saṃvatsara (of the Soma Pond)>, तत् अहः <तद्-अहर् ind. – on that very day>, परि <ind. prefix to verb बभूव (√भू)>, ष्ठ <(स्थ) pre. 2nd person pl. of √अस् – ye are around, belong to (gen.)>, यत् <conjunction – that, when>, मंडूकाः <m. voc. pl. of मंडूक – O frogs>, प्रावृषीणं <n. nom. sg. of adj. प्रावृषीण – (day) beginning the rainy season>, परि-बभूव <prf. 2nd person pl. of परि-√भू – ye are around>.

RV vii.103.8 author वसिष्ठः, to मंडूकाः, metre त्रिष्टुप् छंदः
ब्राह्मणासः सोमिनो वाचमक्रत ब्रह्म कृण्वंतः परिवत्सरीणं ।
अध्वर्यवो घर्मिणः सिष्विदाना आविर्भवंति गुह्या न के चित् ॥७.१०३.८॥
ब्राह्मणासः । सोमिनः । वाचं । अक्रत । ब्रह्म । कृण्वंतः । परिवत्सरीणं ।
अध्वर्यवः । धर्मिणः । सिस्विदानाः । आविः । भवंति । गुह्याः । न । के । चित् ॥७.१०३.८॥
These Brāhmaṇas, performing the Soma sacrifice, made Speech offering their year-long prayer. Adhvaryus, perspiring, engaged in operating the Cauldron, are manifest, and none are concealed. (RV vii.103.8)

ब्राह्मणासः <m. nom. pl. of ब्राह्मण – Brāhmaṇas, ones who have divine knowledge, i.e. frogs, Agnis>, सोमिनः <m. nom. pl. of adj. सोमिन् – performing the Soma sacrifice>, वाचं <f. acc. sg. of वाच् – Speech, voice>, अक्रत <ipf. middle 3rd person pl. of √कृ – they made>, ब्रह्म <n. acc. sg. of ब्रह्मन् – prayer, pious utterance>, कृण्वंतः <m. nom. pl. of pre. active pt. कृण्वत् (√कृ) – they offering>, परिवत्सरीणं <n. acc. sg. of adj. परिवत्सरीण – relating to a full year, lasting a whole year (a cycle of plasma discharge and release is described as a year, a season, or a rainy season)>;

अध्वर्यवः <m. nom. pl. of अध्वर्यु – Adhvaryus (pl.)>, घर्मिणः <m. nom. pl. of adj. घर्मिन् – "engaged in preparing the Gharma offering" (cf. घर्मः gharma is a cauldron, boiler and refers to the Stone, the celestial ioniser)>, सिष्वदानाः <m. nom. pl. of pre. middle pt. सिष्वदान (√स्विद्) – they (Adhvaryus) sweating, perspiring>, आविः <(आविस्) ind. – before the eyes, openly, manifestly, evidently>, भवंति <pre. 3rd person pl. of √भू – they become, be>, गुह्याः <m. nom. pl. of adj. गुह्य – concealed, hidden>, न केचित् <no one, none of them>.

RV vii.103.9 author वसिष्ठः, to मंडूकाः, metre त्रिष्टुप् छंदः
देवहितिं जुगुपुर्द्वादशस्य ऋतुं नरो न प्र मिनंत्येते ।
संवत्सरे प्रावृष्यागतायां तप्ता घर्मा अश्नुवते विसर्गं ॥७.१०३.९॥
देवऽहितिं । जुगुपुः । द्वादशस्य । ऋतुं । नरः । न । प्र । मिनंति । एते ।
संवत्सरे । प्रावृषि । आऽगतायां । तप्ताः । घर्माः । अश्नुवते । विऽसर्गं ॥७.१०३.९॥
These men guard the devine ordinance of the Twelve and never neglect the season.
In the Soma Pond, when the rainy season arrives, the red-hot cauldrons discharge Soma draughts. (RV vii.103.9)

देवहितिं <f. acc. sg. of देवहिति – divine ordinance>, जुगुपुः <prf. 3rd person pl. of √गुप् – they (the men) guard, defend>, द्वादशस्य <mn. gen. sg. of द्वादश - of the twelve (the twelve tribes of the Devas or twelve Ādityas)>, ऋतुं <m. acc. sg. of ऋतु – the order, rule, the season>, नरः <m. nom. pl. of नृ – the men, the Devas (refers to the Brāhmaṇas or frogs or Agnis)>, न <ind. – not, no, never>, प्र <ind. prefix to verb मिनंति (√मी)>, प्र-मिनंति <pre. 3rd person pl. of प्र-√मी – they destroy, violate, neglect>, एते <m. nom. pl. of एतद् – these>;

संवत्सरे <m. loc. sg. of संवत्सर – in the Soma Pond>, प्रावृषि <f. loc. sg. of प्रावृष् – in the rainy season>, आगतायां <f. loc. sg. of adj. आगत – arrived, come to (loc.)>, तप्ताः <m. nom. pl. of adj. तप्त – heated, inflamed, made red-hot>, घर्माः <m. nom. pl. of घर्म – cauldrons, boilers (cauldrons in the plural refer to the firestones of the Stone, the celestial ioniser)>, अश्नुवते <pre. middle 3rd person pl. of √अश् – they (red-hot cauldrons) offer>, विसर्गं <m. acc. sg. of विसर्ग – sending forth, emission, discharge (of Soma draughts or cosmic plasmas)>.

RV vii.103.10 author वसिष्ठः, to मंडूकाः, metre त्रिष्टुप् छंदः
गोमायुरदादजमायुरदात्पृश्निरदाद्धरितो नो वसूनि ।
गवां मंडूका ददतः शतानि सहस्रसावे प्र तिरंत आयुः ॥७.१०३.१०॥
गोऽमायुः । अदात् । अजऽमायुः । अदात् । पृश्निः । अदात् । हरितः । नः । वसूनि ।
गवां । मंडूकाः । ददतः । शतानि । सहस्रऽसावे । प्र । तिरंते । आयुः ॥७.१०३.१०॥
The frog who bellows like a cow offered us riches, the frog who bleats like a goat offered us riches, the dappled frog and the green frog offered us riches. May the frogs bestow upon us cows in hundreds and raise our food in a thousand Soma pressings. (RV vii.103.10)

गोमायुः <m. nom. sg. of गोमायु – the frog who bellows like a cow>, अदात् <ipf. 3rd person sg. of √दाश् – the frog offered, granted>, अजमायुः <m. nom. sg. of अजमायु – the frog who bleats like a goat>, अदात् <ipf. 3rd person sg. of √दाश् – he offered, granted>, पृश्निः <m. nom. sg. of पृश्नि – the dappled frog>, अदात् <ipf. 3rd person sg. of √दाश् – he offered, granted>, हरितः <m. nom. sg. of हरित –the green frog>, नः <m. acc. pl. of अस्मद् – us>, वसूनि <n. acc. pl. of वसु – treasures, riches>;

गवां <m. gen. pl. of गो – cows (celestial lights)>, मंडूकाः <m. nom. pl. of मंडूक – maṇḍūkas, frogs>, ददतः <sub. 3rd person pl. of √दा – may they (frogs) give, bestow, offer (dat., gen., loc.)>, शतानि <n. acc. pl. of शत – to hundreds (शत

is neuter in gender and declined as neuter)>, सहस्रसावे <m. loc. sg. of सहस्रसाव – in thousandfold Soma pressing, in a thousand Soma pressings>, प्र <ind. prefix to verb तिरंते (√तॄ)>, प्र-तिरंते <pre. middle 3rd person pl. of प्र-√तॄ – they raise, increase>, आयुः <n. acc. sg. of आयुस् – food (Soma)>.

7.1.7 The Bird That Is but One, Singers with Hymns Shape It into Many Forms

Two stanzas RV x.114.5 and RV i.164.46 inform us that Agni is shaped into many different forms by the Speeches, or hymns, of the wise singers. These stanzas confirm that the Speeches, or hymns, the oscillations generated by cosmic plasmas across a broad band of the electromagnetic spectrum, are responsible for cosmic ionisation, for the manifested cosmic plasma phenomena we observe in the cosmic field, and for the constitution of the atomic structures of elements.

RV x.114.5 सध्रिर्वैरूपो घर्मो वा तापसः, to विश्वे देवाः, metre त्रिष्टुप् छंदः
सुपर्णं विप्राः कवयो वचोभिरेकं संतं बहुधा कल्पयंति ।
छंदांसि च दधतो अध्वरेषु ग्रहान्त्सोमस्य मिमते द्वादश ॥१०.११४.५॥
सुऽपर्णं । विप्राः । कवयः । वचःऽभिः । एकं । संतं । बहुधा । कल्पयंति ।
छंदांसि । च । दधतः । अध्वरेषु । ग्रहान् । सोमस्य । मिमते । द्वादश ॥१०.११४.५॥
The bird that is but one, wise singers, with their Speeches, shape it into many forms.
And while producing hymns, they measure out twelve ladles of Soma at sacrifices.
(RV x.114.5)

सुपर्णं <m. acc. sg. of सुपर्ण – the mythical or supernatural bird (the bird represents individual plasma blooms, Soma or Agni)>, विप्राः <m. nom. pl. of adj. विप्र – wise>, कवयः <m. nom. pl. of कवि – singers, bards>, वचोभिः <n. inst. pl. of वचस् – with Speeches, songs>, एकं <m. acc. sg. of adj. एक – one, one and the same, identical>, संतं <m. acc. sg. of pre. active pt. सत् (√अस्) – it being>, बहुधा <ind. – in many forms, manifold>, कल्पयंति <pre. cau. 3rd person pl. of √कॢप् (√kḷp) – they make, arrange, bring about, form>;

छंदांसि <n. acc. pl. of छंदस् – metres, hymns>, च <ind. – and, also, moreover>, दधतः <m. nom. pl. of adj. दधत् (interpreted as adj., not as pt. of √धा) – making, producing, generating>, अध्वरेषु <m. loc. pl. of अध्वर (from √ध्वृ cause to fall, hurt) – at sacrifices (refer to Agni Yajñas, fire sacrifices, the events of cosmic ionisation)>, ग्रहान् <m. acc. pl. of ग्रह – ladles, vessels>, सोमस्य <m. gen. sg. of सोम – of Soma>, मिमते <pre. middle 3rd person pl. of √मा 4Ā – they apportion, measure out>, द्वादश <m. acc. pl. of द्वादशन् – twelve>.

RV i.164.46 author दीर्घतमा औचथ्यः, to सूर्यः, metre त्रिष्टुप् छंदः
इंद्रं मित्रं वरुणमग्निमाहुरथो दिव्यः स सुपर्णो गरुत्मान् ।
एकं सद्विप्रा बहुधा वदंत्यग्निं यमं मातरिश्वानमाहुः ॥१.१६४.४६॥
इंद्रं । मित्रं । वरुणं । अग्निं । आहुः । अथो इति । दिव्यः । सः । सुऽपर्णः । गरुत्मान् ।
एकं । सत् । विप्राः । बहुधा । वदंति । अग्निं । यमं । मातरिश्वानं । आहुः ॥१.१६४.४६॥
They call him Indra, Mitra, Varuṇa, Agni, and he is the beautifully winged bird Garuḍa.
To what is one, singers give many a name, and they call it Agni, Yama, Mātariśvā.
(RV i.164.46)

इंद्रं <m. acc. sg. of इंद्र – Indra>, मित्रं <m. acc. sg. of मित्र – Mitra>, वरुणं <m. acc. sg. of वरुण – Varuṇa>, अग्निं <m. acc. sg. of अग्नि – Agni>, आहुः <prf. 3rd person pl. of √अह् – they call (by name)>, अथो <ind. – now, likewise, moreover, certainly>, दिव्यः <m. nom. sg. of adj. दिव्य – divine, heavenly, celestial>, सः <m. nom. sg. of 3rd person pron. तद् –

he>, सुपर्णः <m. nom. sg. of adj. सुपर्ण – having beautiful wings>, गरुत्मान् <m. nom. sg. of गरुत्मत् – Garutmān, the bird Garuḍa (गरुडः); it corresponds to the mythical bird Phoenix>;

एकं <n. acc. sg. of adj. or pron. एक – one, alone, that one only>, सत् <n. acc. sg. of adj. सत् – being, existing, occurring>, विप्राः <mf. nom. pl. of विप्र – singers>, बहुधा <ind. – in many ways, variously>, वदंति <pre. 3rd person. pl. of √वद् – they speak, address>, अग्निं <m. acc. sg. of अग्नि – Agni>, यमं <m. acc. sg. of यम – Yama>, मातरिश्वानं <m. acc. sg. of मातरिश्वन् – Mātariśvā (मातरिश्वा), a name of Agni>, आहुः <prf. 3rd person pl. of √अह् – they call (by name)>.

7.2 SOMA, ELECTRON

7.2.1 Soma, an Embryonic Particle of Celestial Waters

Soma is one of the two charged embryonic particles of celestial Waters, that is, cosmic plasmas. Soma encloses and restrains Agnis in the nucleus, forming the shell of the atomic structure. This act of enclosing Agnis in the nucleus is described as an important task of Soma (RV ix.97.41). Soma is called water buffalo, Pavamāna, and Indu, a bright drop (RV ix.97.41). Soma is a bull, falcon, and a spark of light (RV ix.96.19). Soma is the lord of the mind (RV ix.28.1) and the lord of Speech (RV ix.26.4).

Stanza RV ix.97.41 is addressed to Pavamāna Soma (पवमानः सोमः) and the author is Parāśara (पराशरः), 'the crusher or destroyer'.

RV ix.97.41 author पराशरः, to पवमानः सोमः, metre त्रिष्टुप् छंदः
महत्तत्सोमो महिषश्चकारापां यद्गर्भोऽवृणीत देवान् ।
अदधादिंद्रे पवमान ओजोऽजनयत्सूर्ये ज्योतिरिंदुः ॥९.९७.४१॥
महत् । तत् । सोमः । महिषः । चकार । अपां । यत् । गर्भः । अवृणीत । देवान् ।
अदधात् । इंद्रे । पवमानः । ओजः । अजनयत् । सूर्ये । ज्योतिः । इंदुः ॥९.९७.४१॥
Soma, the water buffalo, an embryonic particle of celestial Waters, performed an important task when he enclosed the Devas. Pavamāna granted power upon Indra, and Indu brought fire to Sūrya. (RV ix.97.41)

महत् <n. acc. sg. of महत् – a great thing, important matter>, तत् <n. acc. sg. of pron. तद् – that>, सोमः <m. nom. sg. of सोम – Soma>, महिषः <m. nom. sg. of महिष – water buffalo>, चकार <prf. 3rd person sg. of √कृ – he made, performed>, अपां <f. gen. pl. of अप् – of celestial Waters>, यत् <conjunction – that, when>, गर्भः <m. nom. sg. of गर्भ – an embryonic particle>, अवृणीत <ipf. middle 3rd person sg. of √वृ – he (Soma) veiled, surrounded, restrained>, देवान् <m. acc. pl. of देव – the Devas (here, the Devas refer to Agnis in the nucleus)>;

अददात् <ipf. 3rd person sg. of √दा – he (Pavamāna) granted, imparted, offered to (loc.)>, इंद्रे <m. loc. sg. of इंद्र – on/upon Indra>, पवमानः <m. nom. sg. of पवमान – Pavamāna, the name of Soma (it literally means 'being purified or flowing clear')>, ओजः <n. acc. sg. of ओजस् – strength, power>, अजनयत् <ipf. cau. 3rd person sg. of √जन् – he (Indu) generated, caused>, सूर्ये <m. loc. sg. of सूर्य – in/unto Sūrya>, ज्योतिः <n. acc. sg. of ज्योतिस् – light, fire>, इंदुः <m. nom. sg. of इंदु – Indu, a bright drop of Soma>.

7.2.2 The Thunderbolt of Indra

As noted earlier, Soma is the thunderbolt (वज्रः vajraḥ) of Indra. The triply-spun filament of Soma plant, or plasma bloom, is called the thunderbolt. In stanza RV ix.72.7, Pavamāna Soma, which is poured into the wave of celestial Waters and into the rivers (सिंधवः siṃdhavaḥ pl. sindhus), is called the nave of the Milky Way and the mighty prop of the Heaven. Exhilarating Soma, the bull, with mighty treasures, flows clear to make the heart rejoice. The stanza is addressed to Pavamāna Soma. The author is Harimanta (हरिमंतः harimaṃtaḥ), 'a son of Aṅgirā (अङ्गिराः m. nom. sg. of अङ्गिरस्)', an angel, the courier of celestial light.

RV ix.72.7 author हरिमंतः, to पवमानः सोमः, metre जगती छंदः
नाभा पृथिव्या धरुणो महो दिवो३ऽपामूर्मौ सिंधुष्वंरुक्षितः ।
इंद्रस्य वज्रो वृषभो विभूवसुः सोमो हृदे पवते चारु मत्सरः ॥९.७२.७॥
नाभा । पृथिव्याः । धरुणः । महः । दिवः । अपां । ऊर्मौ । सिंधुषु । अंतः । उक्षितः ।
इंद्रस्य । वज्रः । वृषभः । विभुऽवसुः । सोमः । हृदे । पवते । चारु । मत्सरः ॥९.७२.७॥
The nave of the Milky Way, the mighty prop of the Heaven, which is sprinkled into the wave of Waters, into the rivers, The exhilarating Soma, the thunderbolt of Indra, the bull who possesses mighty treasures, flows clear to make the heart rejoice. (RV ix.72.7)

नाभा <f. nom. sg. of नाभ – nave, central point>, पृथिव्याः <f. gen. sg. of पृथिवी – of Pṛthivī, of the Milky Way>, धरुणः <m. nom. sg. of धरुण – support, stay, prop>, महः <m. nom. sg. of adj. मह – mighty, strong>, दिवः <m. gen. sg. of दिव् – of the Heaven>, अपां <f. gen. pl. of अप् – of celestial Waters>, ऊर्मौ <m. loc. sg. of ऊर्मि – in the wave>, सिंधुषु <m. loc. pl. of सिंधु – among/into the Sindhus or rivers>, अंतः <m. nom. sg. of अंत – within, inside>, उक्षितः <m. nom. sg. of adj. उक्षित – sprinkled, poured>;

इंद्रस्य <m. gen. sg. of इंद्र – of Indra>, वज्रः <m. nom. sg. of वज्र – the thunderbolt>, वृषभः <m. nom. sg. of वृषभ – a bull>, विभूवसुः <m. nom. sg. of adj. विभूवसु – possessing mighty treasures or wealth>, सोमः <m. nom. sg. of सोम – Soma>, हृदे <n. dat. sg. of हृद् – for the heart>, पवते <pre. middle 3rd person sg. of √पू – he (Soma) flows clear>, चारु <ind. – so as to please, agreeably (with dat.)>, मत्सरः <m. nom. sg. of adj. मत्सर – exhilarating, intoxicating>.

Indra's thunderbolt, the vajra (वज्रः), is represented in "the form of two transverse bolts crossing each other thus X". This 'symbol X' is also "applied to similar weapons used by the various gods" such as arrows, lances or "to any mythical weapon destructive of spells or charms" according to Monier-Williams. The two flaming swords, crossed in the form of the symbol X, represent Indra's celestial weapon used for slaying Vṛtra, the serpent of the Heaven, to release the celestial Waters, the celestial lights, the cosmic plasmas.

7.2.3 Soma Spins a Triple Filament

According to stanza RV ix.86.32, Soma spins a triple filament, indicating that the Birkeland currents, or magnetic ropes, are triply-braided filaments. The stanza tells us that Soma knows how to spin it. Soma is the lord of the sacrifice and the lord of the maidens, the firestones of the celestial ioniser. He guides the newly atoned 'people' of cosmic plasmas to the Place of the Atoned, that is, to the Soma Pond, where ionised Agnis and Somas gather and bloom.

The stanza is addressed to Pavamāna Soma, the author is Traya Ṛṣigaṇā (त्रयः ऋषिगणाः), 'the three tribes of Ṛṣis'.

RV ix.86.32 author त्रय ऋषिगणाः, to पवमानः सोमः, metre जगती छंदः
स सूर्यस्य रश्मिभिः परि व्यत तंतुं तन्वानस्त्रिवृतं यथा विदे ।
नयन्नृतस्य प्रशिषो नवीयसीः पतिर्जनीनामुप याति निष्कृतं ॥९.८६.३२॥
सः । सूर्यस्य । रश्मिऽभिः । परि । व्यत । तंतुं । तन्वानः । त्रिऽवृतं । यथा । विदे ।
नयन् । ऋतस्य । प्रऽशिषः । नवीयसीः । पतिः । जनीनां । उप । याति । निःऽकृतं ॥९.८६.३२॥
He wraps himself with Sūrya's rays of light, spinning, as he knows how, a triple filament.
The lord of sacrifice, lord of the maidens, guiding the newly atoned, goes to the Place of the Atoned. (RV ix.86.32)

सः <m. nom. sg. of 3rd person pron. तद् – he (Soma)>, सूर्यस्य <m. gen. sg. of सूर्य – of Sūrya>, रश्मिभिः <m. inst. pl. of रश्मि – with rays of light>, परि <ind. prefix to verb व्यत (√व्ये)>, परि-व्यत <pre. middle 3rd person sg. of परि-√व्ये (or √व्या) – he covers, wraps, clothes himself>, तंतुं <m. acc. sg. of तंतु – a thread, cord, filament>, तन्वानः <m. nom. sg. of pre. middle pt. तन्वान (√तन्) – he spinning out, weaving>, त्रिवृतं <m. acc. sg. of adj. त्रिवृत् – threefold, consisting of three parts or folds>, यथा <ind. – as, according as>, विदे <prf. middle 3rd person sg. of √विद् – he knows>;

नयन् <m. nom. sg. of pre. active pt. नयत् (√नी) – he (Soma) guiding, directing>, ऋतस्य <n. gen. sg. of ऋत – of sacrifice>, प्रशिषः <f. acc. pl. of प्रशिष् (from √शिष् 1P to hurt, kill) – the killed or atoned (for the sin of the ionic bond), i.e. the ionised 'people'>, नवीयसीः <f. acc. pl. of adj. नवीयसी (f. form of नवीयस्) – recent, being appearing lately>, पतिः <m. nom. sg. of पति – lord, master>, जनीनां <f. gen. pl. of जनि – of the maidens or fingers (i.e. of the firestones of the celestial ioniser)>, उप <ind. prefix to verb याति (√या)>, उप-याति <pre. 3rd person sg. of उप-√या – he goes towards, arrives at>, निष्कृतं <n. acc. sg. of निष्कृत – to the Place of the Atoned (for the sin of ionic bond), i.e. to the Soma Pond where ionised people, cosmic plasmas, gather>.

7.2.4 Soma, the Brahmā of the Devas

Soma is the progenitor of the sacred hymns and of the Heaven (RV ix.96.5). He is the Brahmā of the Devas. Brahmā is the Creator, "the Absolute", "the self-existent impersonal Spirit", "the one universal soul", "one divine essence and source from which all created things emanate or with which they are identified and to which they return". Soma is a falcon midst the vultures, a bull midst the wild beasts, and a lofty tree in the thick of groves (RV ix.96.6). The 'purifying sieve' in RV ix.96.6 refers to the Soma Pond filled full with filaments of blooming plasmas. In the Ṛgveda, Soma stalks or plasma blooms are called the woollen Soma sieve, the fleece, the golden fleece, purifying sieve, and so on (see Figure 7).

Stanzas RV ix.96.5 and 6 need to be read with an understanding of the reciprocal relationship between magnetism and cosmic plasmas. Tvaṣṭā and the Ṛbhus create the Heavenly Vaults, which are magnetic structures. The Heavenly Vaults in turn produce cosmic plasmas and the Milky Way. Now Soma draughts, the currents of electrons, which were begotten by Tvaṣṭā and the Stone, become the progenitor. Once created, Soma draughts, the currents of electrons, become the progenitor of the Heaven, of the Milky Way, of Agni, and of other cosmic plasma phenomena.

The entire hymn RV ix.96 is addressed to Pavamāna Soma, the author is Pratardana Daivodāsi (प्रतर्दनः दैवोदासिः), 'the slayer of the Heaven's demon'. Daivodāsi (दैवोदासिः) is a descendant of Divodāsa (दिवोदासः), the demon of the Heaven. Of course, the Heaven's demon (दिवोदासः divodāsaḥ) is Vṛtra, the serpent of the Heaven, which is withholding the release of the celestial lights, the cosmic plasmas.

RV ix.96.5 author प्रतर्दनो दैवोदासिः, to पवमानः सोमः, metre त्रिष्टुप् छंदः
सोमः पवते जनिता मतीनां जनिता दिवो जनिता पृथिव्याः ।
जनिताग्नेर्जनिता सूर्यस्य जनितेंद्रस्य जनितोत विष्णोः ॥९.९६.५॥
सोमः । पवते । जनिता । मतीनां । जनिता । दिवः । जनिता । पृथिव्याः ।
जनिता । अग्नेः । जनिता । सूर्यस्य । जनिता । इंद्रस्य । जनिता । उत । विष्णोः ॥९.९६.५॥
Soma, the begetter of sacred hymns, flows clearly. The begetter of the Heaven and of the Milky Way, The begetter of Agni, the begetter of Sūrya, the begetter of Indra and even of Viṣṇu. (RV ix.96.5)

सोमः <m. nom. sg. of सोम – Soma>, पवते <pre. middle 3rd person sg. of √पू – he flows off clearly>, जनिता <m. nom. sg. of जनितृ – progenitor, father>, मतीनां <f. gen. pl. of मति – of hymns, sacred utterances>, दिवः <m. gen. sg. of दिव् – of the Heaven>, पृथिव्याः <f. gen. sg. of पृथिवी – of Pṛthivī, of the Milky Way>;

जनिता <m. nom. sg. of जनितृ – progenitor, father>, अग्नेः <m. gen. sg. of अग्नि – of Agni>, सूर्यस्य <m. gen. sg. of सूर्य – of Sūrya>, इंद्रस्य <m. gen. sg. of इंद्र – of Indra>, उत <ind. – and, also, even>, विष्णोः <m. gen. sg. of विष्णु – of Viṣṇu>.

RV ix.96.6 author प्रतर्दनो दैवोदासिः, to पवमानः सोमः, metre त्रिष्टुप् छंदः
ब्रह्मा देवानां पदवीः कवीनामृषिर्विप्राणां महिषो मृगाणां ।
श्येनो गृध्राणां स्वधितिर्वनानां सोमः पवित्रमत्येति रेभन् ॥९.९६.६॥
ब्रह्मा । देवानां । पदऽवीः । कवीनां । ऋषिः । विप्राणां । महिषः । मृगाणां ।
श्येनः । गृध्राणां । स्वऽधितिः । वनानां । सोमः । पवित्रं । अति । एति । रेभन् ॥९.९६.६॥
Brahmā of the Devas, the leader of the bards, Ṛṣi of the wise, bull midst the wild beasts, Falcon among the vultures, lofty tree in the thick of groves, through the cleansing sieve goes Soma singing. (RV ix.96.6)

ब्रह्मा <m. nom. sg. of ब्रह्मन् – the Brahmā, the Creator, the self-existent impersonal Spirit, the one universal Soul>, देवानां <m. gen. pl. of देव – of the Devas>, पदवीः <m. nom. sg. of पदवी – a leader, forerunner>, कवीनां <m. gen. pl. of कवि – of singers, of bards>, ऋषिः <m. nom. sg. of ऋषि – Ṛṣi, the singer of sacred hymns>, विप्राणां <m. gen. pl. विप्र – of the wise ones, seers>, महिषः <m. nom. sg. of महिष – a buffalo, bull>, मृगाणां <m. gen. pl. of मृग – of mṛgas, of wild games, of wild beasts>;

श्येनः <m. nom. sg. of श्येन – a falcon>, गृध्राणां <m. gen. pl. of गृध्र – of vultures>, स्वधितिः <mf. nom. sg. of स्वधिति – a lofty tree>, वनानां <n. gen. pl. of वन – of the woods, groves, forests>, सोमः <m. nom. sg. of सोम – Soma>, पवित्रं <n. acc. sg. of पवित्र – a strainer, cleansing sieve>, अति <ind. prefix to verb एति (√इ)>, अति-एति <pre. 3rd person sg. of अति-√इ – he passes through>, रेभन् <m. nom. sg. of pre. active pt. रेभत् (√रेभ् or √रिभ्) – he singing>.

7.2.5 The Bestower of Light Skilled in a Thousand Arts

Soma, of enlightened mind, the bestower of light, is skilled in a thousand arts. Soma, wishing to acquire the third mode, uttering joyful sounds, shines in accordance with Virāṭ metre (विराट् f. nom. sg. of विराज् RV ix.96.18). In the Soma Pond, Soma, the spark of light, following the wave of cosmic plasmas to the Samudra, the Ocean of celestial Waters, declares his fourth mode (RV ix.96.19). According to RV ix.96.19, the Virāṭ metre is associated with making Soma (electron) shine. It appears that the third and fourth modes are the modes of electrons' oscillations and spins associated with certain cosmic plasma phenomena that occur at certain frequency bands.

RV ix.96.18 author प्रतर्दनो दैवोदासिः, to पवमानः सोमः, metre त्रिष्टुप् छंदः
ऋषिमना य ऋषिकृत्स्वर्षाः सहस्रणीथः पदवीः कवीनां ।
तृतीयं धाम महिषः सिषासन्त्सोमो विराजमनु राजति ष्टुप् ॥९.९६.१८॥
ऋषिऽमनाः । यः । ऋषिऽकृत् । स्वःऽसाः । सहस्रऽनीथः । पदऽवीः । कवीनां ।
तृतीयं । धाम । महिषः । सिसासन् । सोमः । विऽराजं । अनु । राजति । स्तुप् ॥९.९६.१८॥
He, who is of enlightened mind, the Ṛṣi-maker, bestower of light, who is skilled in a thousand arts, the leader of the bards, The bull, uttering joyful sounds, wishing to acquire the third mode, shines in accordance with Virāṭ metre. (RV ix.96.18)

ऋषिमनाः <m. nom. sg. of adj. ऋषिमनस् – of far-seeing or of enlightened mind>, यः <m. nom. sg. of pron. यद् – he who (Soma)>, ऋषिकृत् <m. nom. sg. of ऋषिकृत् – a Ṛṣi-maker>, स्वर्षाः <m. nom. sg. of स्वर्षा (=स्वर्विद्) – a bestower of light>, सहस्रणीथः <m. nom. sg. of adj. सहस्रणीथ – praised in a thousand hymns, skilled in a thousand sciences>, पदविः <m. nom. sg. of पदवी – a leader, forerunner>, कवीनां <m. gen. pl. of कवि – of the bards>;

तृतीयं <n. acc. sg. of तृतीय – the third>, धाम <n. acc. sg. of धामन् – mode, tone, form>, महिषः <m. nom. sg. of महिष – a buffalo, bull>, सिषासन् <m. nom. sg. of pre. desiderative active pt. सिषासत् (√सन्) – he wishing to acquire or obtain>, सोमः <m. nom. sg. of सोम – Soma>, विराजं <f. acc. sg. of विराज् – Virāṭ (विराट् f. nom. sg. of विराज्) metre>, अनु <ind. prefix to verb राजति (√राज्)>, अनु-राजति <pre. 3rd person sg. of अनु-√राज् – he shines in accordance with (said of corresponding metres)>, ष्टुप् <(स्तुप्) m. nom. sg. of adj. स्तुभ् – uttering joyful sounds>.

RV ix.96.19 author प्रतर्दनो दैवोदासिः, to पवमानः सोमः, metre त्रिष्टुप् छंदः
चमूषच्छ्येनः शकुनो विभृत्वा गोविंदुर्द्रप्स आयुधानि बिभ्रत् ।
अपामूर्मिं सचमानः समुद्रं तुरीयं धाम महिषो विवक्ति ॥९.९६.१९॥
चमूऽसत् । श्येनः । शकुनः । विऽभृत्वा । गोऽविंदुः । द्रप्सः । आयुधानि । बिभ्रत् ।
अपां । ऊर्मिं । सचमानः । समुद्रं । तुरीयं । धाम । महिषः । विवक्ति ॥९.९६.१९॥
In the Camū Vessel, Falcon, the mighty bird, the spark of fire, procuring cattle and carrying weapons, spreads out. Following the wave of celestial Waters to the Ocean, the bull declares his fourth mode. (RV ix.96.19)

चमूषत् <m. nom. sg. of adj. चमूषद् –"lying on the Camū Vessel" (refers to the Soma Pond)>, श्येनः <m. nom. sg. of श्येन – a falcon>, शकुनः <m. nom. sg. of शकुन – a mighty bird>, विभृत्वा <m. nom. sg. of adj. विभृत्वन् (from वि-√भृ) – spreading out>, गोविंदुः <m. nom. sg. of adj. गोविंदु (गो cow + विंदु adj. procuring) – procuring cattle (celestial lights)>, द्रप्सः <m. nom. sg. of द्रप्स – a drop, a spark of fire>, आयुधानि <n. acc. pl. of आयुध – waters, weapons>, बिभ्रत् <m. nom. sg. of adj. बिभ्रत् – bearing, carrying>;

अपां <f. gen. pl. of अप् – of celestial Waters>, ऊर्मिं <mf. acc. sg. of ऊर्मि – a wave>, सचमानः <m. nom. sg. of pre. middle pt. सचमान (√सच्) – he pursuing, following>, समुद्रं <m. acc. sg. of समुद्र – to Samudra (the Ocean of cosmic

plasmas, the Soma Pond)>, तुरीयं <n. acc. sg. of तुरीय – fourth>, धाम <n. acc. sg. of धामन् – mode, form>, महिषः <m. nom. sg. of महिष – a buffalo, bull (Soma)>, विवक्ति <pre. 3rd person sg. of √विवच् – he declares, announces>.

7.2.6 Soma, the Lord of the Mind

In stanza RV ix.28.1, Soma is called the fleet steed of a war-chariot and the lord of the mind (मनसः पतिः manasaḥ patiḥ). The war-chariot refers to a plasma bloom, or Soma plant. Soma is omniscient as Agni is. They both have the sacred knowledge and know all beings sentient and insentient.

RV ix.28.1 author प्रियमेधः, to पवमानः सोमः, metre गायत्री छंदः
ए॒ष वा॒जी हि॒तो नृभि॑र्विश्व॒विन्मन॑स॒स्पति॑ः । अव्यो॒ वारं॒ वि धा॑वति ॥९.२८.१॥
ए॒षः । वा॒जी । हि॒तः । नृऽभि॑ः । वि॒श्व॒ऽवित् । मन॑सः । पति॑ः । अव्य॑ः । वार॑ं । वि । धा॒व॒ति॒ ॥९.२८.१॥
Urged by men, this fleet steed of a war-chariot, the lord of the mind, all-knowing,
Runs through the sieve of the woollen Soma strainer. (RV ix.28.1)

एषः <m. nom. sg. of एतद् – this>, वाजी <m. nom. sg. of वाजिन् – the steed of a war-chariot, swift horse>, हितः <m. nom. sg. of adj. हित – impelled, urged on, set in motion>, नृभिः <m. inst. pl. of नृ – by men or heroes (refers to Agnis)>, विश्ववित् <m. nom. sg. of adj. विश्वविद् – knowing everything, omniscient>, मनसः <n. gen. sg. of मनस् – of the mind>, पतिः <m. nom. sg. of पति – a master, lord>;

अव्यः <m. gen. sg. of अवि – of the woollen Soma strainer>, वारं <mn. acc. sg. वार – a hair-sieve, woollen sieve (the filaments of Soma plants or plasma blooms are called the woollen sieve)>, वि <ind. prefix to verb धावति (√धाव्)>, वि-धावति <pre. 3rd person sg. of वि-√धाव् – he flows through, runs through>.

7.2.7 Soma, the Lord of Speech

"Real plasmas" are known to be "noisy".[88] Cosmic plasmas sing hymns and make Speeches through their oscillations across a band of the electromagnetic spectrum. Vāk (वाक्) is the Speech of the Devas, the cosmic plasmas. The Devas sing hymns and make Speeches while performing cosmic Yajñas. Though all the Devas sing, according to the Ṛgveda, the most prominent singer is the Stone, the celestial ioniser. Soma, electron, is the lord of Speech (पतिः वाचः RV ix.26.4) and the begetter (जंतुः) of the singers' Speeches (वाचः कवीनां RV ix.67.13). Soma, himself the child of the singers' Speeches, becomes the begetter and causes the Stone and cosmic plasmas to sing.

Both stanzas RV ix.26.1 and 4, presented below, are addressed to Pavamāna Soma (पवमानः सोमः), the purified or cleansed (read ionised) Soma. The author is Idhmavāha Dārḷhacyuta (इध्मवाहः दाळर्हच्युतः), 'the carrier of sacrificial fire (इध्मवाहः) expelled from the stronghold or fortress (दाळर्हच्युतः)'. The carrier of sacrificial fire refers to Agni freed from

---

88 Hannes Alfvén, *Cosmic Plasma*, Dordrecht, Holland, D. Reidel Publishing Company, 1981, p. 78. Astrophysics and Space Science Library Series, 82.

the stronghold of the ionic bond. Note that Dārḷhacyutaḥ (दाळ्हच्युतः) can be written as Dārḍhacyutaḥ (दार्ढच्युतः patronymic from दृढच्युतः).

Soma, the steed of a war-chariot, is cleansed, that is, freed from the ionic bond, upon the lap of Aditi by the singers with hymn and with fingers (RV ix.26.1). Once freed, Soma is discharged into the Soma Pond by the singers with 'the hymn of two arms' (RV ix.26.4). The two arms refer to the upper and lower discs of the Stone, the celestial ioniser. Soma is called the fellow-dweller of Vivasvān. Vivasvān is the fire of the Stone, which is Trita. And Trita is one of the three fires of Agni, proton.

RV ix.26.1 author इध्मवाहो दार्ढच्युतः, to पवमानः सोमः, metre गायत्री छंदः
तममृक्षंत वाजिनमुपस्थे अदितेरधि । विप्रासो अण्व्या धिया ॥९.२६.१॥
तं । अमृक्षंत । वाजिनं । उपऽस्थे । अदितेः । अधि । विप्रासः । अण्व्या । धिया ॥९.२६.१॥
Him, the steed of a war-chariot, upon the lap of Aditi, the singers cleansed,
The singers with hymn and with fingers. (RV ix.26.1)

तं <m. acc. sg. of तद् – him (Soma)>, अमृक्षंत <aor. middle 3rd person pl. of √मृज् – they (singers) wiped off, rubbed off, cleansed (read ionised, freed from the ionic bond)>, वाजिनं <m. acc. sg. of वाजिन् – the steed of a war-chariot, swift horse (refers to Pavamāna Soma)>, उपस्थे <m. loc. sg. of उपस्थ – (with अधि) upon the lap>, अदितेः <f. gen. sg. of अदिति – of Aditi, the Stone, the celestial ioniser>, अधि <ind. preposition – (with loc.) at, on>;

विप्रासः <m. nom. pl. of विप्र – sages, singers (refers to the firestones of the celestial ioniser)>, अण्व्या <f. inst. sg. of अण्वी – with the fingers (अण्वी is "the name of the fingers preparing the Soma juice", i.e. the firestones of the Stone who prepare Soma juice)>, धिया <f. inst. sg. of धी – with thought, prayer, i.e. with hymn, Speech>.

RV ix.26.4 author इध्मवाहो दार्ढच्युतः, to पवमानः सोमः, metre गायत्री छंदः
तमह्यन्भुरिजोर्धिया संवसानं विवस्वतः । पतिं वाचो अदाभ्यं ॥९.२६.४॥
तं । अह्यन् । भुरिजोः । धिया । संऽवसानं । विवस्वतः । पतिं । वाचः । अदाभ्यं ॥९.२६.४॥
Him, the fellow-dweller of Vivasvān, they discharged with the hymn of two arms,
The lord of Speech infallible. (RV ix.26.4)

तं <m. acc. sg. of तद् – him (Pavamāna Soma)>, अह्यन् <aor. 3rd person pl. of √हि – they (the singers) discharged, hurled>, भुरिजोः <f. gen. du. of भुरिज् – of two arms (two discs of the Stone, the celestial ioniser)>, धिया <f. inst. sg. of धी – with prayer, i.e. with hymn, Speech>, संवसानं <m. acc. sg. of संवसान (=संवस्) – one who dwells with, the fellow-dweller>, विवस्वतः <m. gen. sg. of विवस्वत् – of Vivasvān (विवस्वान्), here, Vivasvān refers to Trita (fire of the Stone), one of the three forms of Agni>;

पतिं <m. acc. sg. of पति – a lord>, वाचः <f. gen. sg of वाच् – of Speech>, अदाभ्यं <m. acc. sg. of adj. अदाभ्य – free from deceit, trusty, infallible>.

7.2.8 Flow Indu, the Frog is Eager for Water

The four stanzas of RV ix.112, addressed to Pavamāna Soma, explain the relationships between Somas and Agnis and the associated cosmic plasma phenomena. In stanzas RV ix.112.1-4, the herds of cattle and the frog represent Agnis, and the steed represents Soma. The author of the hymn is Śiśu (शिशुः). Śiśu is "a child, infant" and "name of a descendant of Aṅgirās (अङ्गिराः)". In the Ṛgveda, mostly Agni is called Śiśu and it has been translated as an 'infant' or a 'child'. However, as it is noted by Monier-Williams, Śiśu also means "young plants", that is, the new shoots of Soma plants, or plasma blooms.

RV ix.112.1 author शिशुः, to पवमानः सोमः, metre पंक्तिः छंदः
नानानं वा उ नो धियो वि व्रतानि जनानां ।
तक्षा रिष्टं रुतं भिषग्ब्रह्मा सुन्वंतमिच्छतींद्रायेंदो परि स्रव ॥९.११२.१॥
नानानं । वै । ऊं इति । नः । धियः । वि । व्रतानि । जनानां ।
तक्षा । रिष्टं । रुतं । भिषक् । ब्रह्मा । सुन्वंतं । इच्छति । इंद्राय । इंदो इति । परि । स्रव ॥९.११२.१॥
Our hymns are indeed diverse and men's spheres of action also vary.
The carpenter seeks the singer, the physician seeks the shattered, Brahmā seeks the offerer of Soma sacrifice. Flow, O Indu, flow thou for Indra. (RV ix.112.1)

नानानं <ind. – differently, in various ways>, वै <ind. – indeed, certainly>, ऊं <उ ind. – and, also, further>, नः <m. gen. pl. of 1st person pron. अस्मद् – our>, धियः <f. nom. pl. of धी – thoughts, prayers (i.e. Speeches, hymns)>, वि <ind. – apart, in different directions>, व्रतानि <n. nom. pl. of व्रत – spheres of action, functions, modes>, जनानां <m. gen. pl. of जन – of races, men (the Devas)>;

तक्षा <m. nom. sg. of तक्षन् – the wood-cutter, carpenter (refers to the Stone, the celestial ioniser)>, रिष्टं <mn. acc. sg. of रिष्ट (from √रिष् to be hurt, injured) – the torn off, broken, shattered (the ionised, refer to cosmic plasmas)>, रुतं <n. acc. sg. of रुत (from √रु to roar, sing, to make a sound) – "sounded, made to resound", a singer, roarer (i.e. the one who breaks up the ionic bonds with sounds or Speeches)>, भिषक् <m. nom. sg. of भिषज् – a healer, physician (here, refers to the Soma Pond)>, ब्रह्मा <m. nom. sg. of ब्रह्मन् – Brahmā>, सुन्वंतं <(सुन्वन्तं) m. acc. sg. of सुन्वत् – the offerer of Soma sacrifice>, इच्छति <pre. 3rd person sg. of √इष् 6P– he seeks, endeavours to obtain>, इंद्राय <m. dat. sg. of इंद्र – for Indra>, इंदो <m. voc. sg. of इंदु – O Indu, O Bright Drop of Soma>, परि <ind. prefix to verb स्रव (√स्रु)>, परि-स्रव <ipv. 2nd person sg. of परि-√स्रु – you stream, flow>.

RV ix.112.2 author शिशुः, to पवमानः सोमः, metre पंक्तिः छंदः
जरंतीभिरोषधीभिः पर्णेभिः शकुनानां ।
कार्मारो अश्मभिर्द्युभिर्हिरण्यवंतमिच्छतींद्रायेंदो परि स्रव ॥९.११२.२॥
जरंतीभिः । ओषधीभिः । पर्णेभिः । शकुनानां ।
कार्मारः । अश्मऽभिः । द्युऽभिः । हिरण्यऽवंतं । इच्छति । इंद्राय । इंदो इति । परि । स्रव ॥९.११२.२॥
With mature light-containing plants and with wings of the mighty birds,
The smith, with the stones and with fires, endeavours to obtain him who has a store of gold.
Flow, O Indu, flow thou for Indra. (RV ix.112.2)

जरतीभिः <f. inst. pl. of adj. जरत् (f. जरती) – old, mature>, ओषधीभिः <f. inst. pl. of ओषधी (=ओषधि) – with light-containing plants (i.e. Soma plants, plasma blooms)>, पर्णेभिः <(=पर्णैः) n. inst. pl. of पर्ण – with wings or feathers>, शकुनानां <m. gen. pl. of शकुन – of the large birds>;

कार्मारः <m. nom. sg. of कार्मार – a mechanic, smith, i.e. the Stone, the celestial ioniser>, अश्मभिः <m. inst. pl. of अश्मन् – with the stones, i.e. the firestones of the Stone>, द्युभिः <m. inst. pl. of द्यु – with fires, i.e. the fires of Trita, tritium>, हिरण्यवंतं <m. acc. sg. of हिरण्यवत् – the possessor of gold (the name of Agni)>, इच्छति <pre. 3rd person sg. of √इष् 6P – he seeks, strives to obtain>, इंद्राय इंदो परि स्रव <see RV ix.112.1 – Flow, O Indu, flow thou for Indra.>.

RV ix.112.3 author शिशुः, to पवमानः सोमः, metre पंक्तिः छंदः
कारुरहं ततो भिषगुपलप्रक्षिणी नना ।
नानाधियो वसूयवोऽनु गा इव तस्थिमेंद्रायेंदो परि स्रव ॥९.११२.३॥

कारुः । अहं । ततः । भिषक् । उपलऽप्रक्षिणी । नना ।
नानाऽधियः । वसुऽयवः । अनु । गाःऽइव । तस्थिम । इंद्राय । इंदो इति । परि । स्रव ॥९.११२.३॥
A bard I am, my daddy is a physician, my mummy grinds grain upon mill-stones.
With varying hymns, desiring wealth, indeed we have stood with the herds of cattle. Flow,
O Indu, flow thou for Indra. (RV ix.112.3)

कारुः <m. nom. sg. of कारु – one who sings or praises, a bard>, अहं <m. nom. sg. of 1st person pron. अस्मद् – I, I am>, ततः <m. nom. sg. of तत – a daddy (familiar expression corresponding to नना)>, भिषक् <m. nom. sg. of भिषज् – a healer, physician>, उपलप्रक्षिणी <f. nom. sg. of उपलप्रक्षिन् (f. उपलप्रक्षिणी) – one who grinds grain upon mill-stones (mill-stones refer to the firestones of the Stone, the celestial ioniser; grain refers to hydrogen atoms)>, नना <f. nom. sg. of नना – familiar expression of one's mother, mammy, mummy>;

नानाधियः <mf. nom. pl. of adj. नानाधी – having different thoughts, prayers, i.e. with hymns, Speeches)>, वसूयवः <mf. nom. pl. of adj. वसूयु – desiring wealth>, अनु <ind. – near to, alongside, with>, गाः <m. acc. pl. of गो – cattle, herds of cattle>, इव <ind. – about, so, indeed>, तस्थिम <prf. 1st person pl. of √स्था – we have stood, stayed, remained>, इंद्राय इंदो परि स्रव <see RV ix.112.1 – Flow, O Indu, flow thou for Indra.>.

In RV ix.112.4, the two cleavers refer to the upper and the lower discs of the Stone, the celestial ioniser. Here, in the context of this stanza, the tail refers to the Soma Pond though it can represent the individual plasma blooms. Discharged cosmic plasmas are compared to feathers, thus the two cleavers, the two discs of the Stone, are described as covered with feathers.

RV ix.112.4 author शिशुः, to पवमानः सोमः, metre पंक्तिः छंदः
अश्वो वोळ्हा सुखं रथं हसनामुपमंत्रिणः ।
शेपो रोमण्वंतौ भेदौ वारिन्मंडूक इच्छतींद्रायेंदो परि स्रव ॥९.११२.४॥
अश्वः । वोळ्हा । सुऽखं । रथं । हसनां । उपऽमंत्रिणः ।
शेपः । रोमण्ऽवंतौ । भेदौ । वाः । इत् । मंडूकः । इच्छति । इंद्राय । इंदो इति । परि । स्रव ॥९.११२.४॥
The steed draws a swiftly rolling chariot and rumbling thunder of the impeller.
The frog is eager for Water and wishes the two cleavers adorned with feathers to procure a tail.
Flow, O Indu, flow thou for Indra. (RV ix.112.4)

अश्वः <m. nom. sg. of अश्व – a horse (refers to Soma)>, वोळ्हा <(वोढा) m. nom. sg. of adj. वोढृ – bearing, drawing>, सुखं <m. acc. sg. of adj. सुख – running swiftly or easily>, रथं <m. acc. sg. of रथ – a chariot>, हसनां <f. acc. sg. of हसना – an encouraging shout, lightning>, उपमंत्रिणः <m. gen. sg. of उपमंत्रिन् (उपमन्त्रिन्) – of inciter, impeller>;

शेपः <n. acc. sg. of शेपस् (cf. शेप a tail) – a tail (here, refers to individual plasma blooms)>, रोमण्वंतौ <m. acc. du. of adj. रोमण्वत् – covered with wool or feathers (wool or feathers refer to plasma blooms that appear woolly)>, भेदौ <m. acc. du. of भेद – the two breakers, cleavers i.e. the upper and lower discs of the Stone>, वाः <n. acc. sg. of वार् – celestial Water>, इत् <इद् ind. – even, just, indeed>, मंडूकः <m. nom. sg. of मंडूक (मण्डूक) – a frog, i.e. Agni>, इच्छति <pre. 3rd person sg. of √इष् – he wishes>, इंद्राय इंदो परि स्रव <see RV ix.112.1 – Flow, O Indu, flow thou for Indra.>.

VIII. The Cosmology of the Ṛgveda: Summary

8.1 What the Ṛgveda Reveals

What the extant text of the Ṛgveda reveals include: (i) what the condition was before the process of cosmic creation is set in motion, (ii) how the twofold Heavenly Vault and its structural components are built, (iii) how the cosmic ionisation and plasma discharge and release processes, the breaking up of the ionic bonds of celestial hydrogen and the discharging and releasing of cosmic plasmas, are initiated by the twofold Heavenly Vault and the Stone, the celestial ioniser, and (iv) how the nebulous fabric of the Milky Way is woven and sustained.

Three hundred twenty eight select stanzas from the ten maṇḍalas of the Ṛgveda are interpreted and translated to draw a complete picture of the cosmology of the Ṛgveda. In addition, to supplement what are revealed in the extant text of the Ṛgveda, six other stanzas from the Kaṭha Upaniṣad, Sūryasiddhānta, and Siddhāntaśiromaṇi are translated and included in this book.

Though the vital and detailed information about cosmic plasma phenomena and cosmic plasma properties are presented in this book, only the core cosmological principles of the Ṛgveda and the major structural components of the galactic system are included in this summary. As each maṇḍalaṃ of the Ṛgveda is interpreted and translated, it is expected that additional information about the cosmic plasma properties and about the events of cosmic ionisation and cosmic plasma discharge and release will be revealed.

8.1.1 The Universe, the Infinite Ocean of Celestial Hydrogen

According to the Ṛgveda, the universe is the ocean of prima materia, aṃbha (अंभः n. nom. sg. of अंभस्), celestial hydrogen. The universe is in a steady state with no beginning and no end. The universe is not to be created, nor to be destroyed. It is a given. It is the ultimate God's domain. What is created and has evolved are the galactic systems that create and harbour life within the universe. The creation of a galactic system starts with prima materia, aṃbha, the original 'water-bearer'. Refer to Chapter V to see how the wheel of cosmic creation is turned.

8.1.2 A Single Source and a Single Cause

The mighty cosmic force, the magnetic field and magnetic forces, which sets the cosmic creation in motion, the subsequently built Heavenly Vault and its substructures, and the event of cosmic Yajña, which is the event of celestial ionisation and cosmic plasma discharge and release, are born of a single source and of a single cause. Aṃbha, hydrogen, is the single source. The mighty cosmic force (महिमा), emanated from the vast ocean of celestial hydrogen, is the single cause. According to the Ṛgveda, the mighty cosmic force rises first from the ocean of celestial hydrogen. Thereafter, the desire (कामः kāmaḥ), the primaeval flow of cosmic Mind (मनः manaḥ), arises. Enchanting hymns of praise for and supplication to the celestial beings involved in cosmic Yajñas will have direct impact on the cosmic events, on the sustenance of the galactic structure, and on the prosperity of our galactic system, for our mind (मनः manaḥ) and the intelligent cosmic Mind (मनः manaḥ) are one and the same, and our soul (आत्मा ātmā) and the Soul (आत्मा ātmā) of cosmic Yajña are one and the same. This is the basic tenet of the Vedic yajñas and recitations and chantings of the hymns of praise and supplication. The original Vedic yajñas are the ceremonial enactments of cosmic Yajñas.

8.1.3 The Creation and Life are the Magneto-Plasma Phenomena

The cosmic creation and life are magneto-plasma phenomena. According to the Ṛgveda, Tvaṣṭā, the magnetic field, which is emanated from the ocean of celestial aṃbha, hydrogen, is the first born. When born, Tvaṣṭā, along with the Ṛbhus, the magnetic forces, builds the twofold Heavenly Vaults and its substructures, the marvellous Creator's machine, the machine of celestial ionisation and cosmic plasma discharge and release. This Creator's marvellous machine produces and discharges and releases cosmic plasmas, and subsequently builds the Milky Way. Through these cosmic ionisation and cosmic plasma discharge and release events, Tvaṣṭā and the Ṛbhus strengthen and sustain the galactic structure and the worlds within it. And they, along with the Speeches of cosmic plasmas, create and shape all forms according to the image (प्रतिमा pratimā) of Brahmā (ब्रह्मा), the Creator, and his intent (कामः kāmaḥ).

8.1.4 The Twofold Heavenly Vault and Its Substructures are the Magneto-Plasma Structures

The whole of our galactic system is a magneto-plasma structure, which is defined by the magnetic fields arranged in the shape of the Ḍamaru (डमरुः), the sacred drum of Śiva (शिवः), or in the shape of an hourglass. And this galactic structure is sustained by its own unbroken chain of ionisation and plasma discharge and release events, the events of cosmic Yajñas. This galactic structure is the altar of cosmic creation where cosmic Yajñas are performed three times a day every day.

Our galactic structure consists of the three spheres of the Heaven, the two Vaults of the Heaven, the two pairs of the Stones and the Soma Ponds, and the Milky Way. The three spheres of the Heaven are the Upper Heaven, the Midheaven, or the Innerfield, and the Lower Heaven. The two Vaults of the Heaven include the Upper and the Lower Vaults, which are vertically aligned, each with its own pair of the Stone and the Soma Pond. The galactic structure is defined and upheld by Tvaṣṭā and the Ṛbhus, the magnetic field and the magnetic forces.

The Ṛgveda and the Bhūgolādhyāya of Sūryasiddhānta state that the Upper and Lower Vaults of the Heaven revolve in opposite directions of each other. The Sūryasiddhānta provides more specific information and states that the Upper Vault of the Heaven revolves westward, or clockwise; the Lower Vault of the Heaven, eastward, or anticlockwise. And at the zero latitude, that is, at the latitude where the disc of the Milky Way is located, it turns always in a westerly direction. This indicates that the disc of the Milky Way Galaxy always revolves clockwise (refer to section 2.4.5).

According to the cosmology chapters of Sūryasiddhānta and the Siddhāntaśiromaṇi, the height of a Vault of the Heaven is approximately eighteen quadrillion seven hundred and twelve trillion (18,712,000,000,000,000) yojanas or 28,710.8 light-years. This measure falls right in the middle of NASA's estimate of 25,000 light-years and the estimate of 10 kpc, which is about 32,000 light-years, provided by Meng Su and Douglas P. Finkbeiner (2012). When we multiply these measures by two to get the total height of the vertically aligned Upper and Lower Vaults of the Heaven, we acquire 57,421 light-years by the Vedic source, 50,000 light-years by NASA, and 64,000 light-years by Meng Su and Douglas P. Finkbeiner (2012) respectively (see Figures 3 and 4, section 2.3).

8.1.5 The Stone, the Celestial Ioniser, the Cold Fire

The Stone, the celestial ioniser, is the celestial machine of ionisation and cosmic plasma discharge and release. In our galactic system, there are two pairs of the Stones and Soma Ponds. Of the two pairs of the Stones and Soma Ponds, one pair is in the Upper Vault of the Heaven, the other pair in the Lower Vault of the Heaven (see Figures 6 and 13). These two pairs of the Stones and the Soma Ponds sustain the Milky Way and power billions of suns, like ours, in the Milky Way by producing, discharging, and releasing cosmic plasmas three times a day every day. In the beginning, once the wheel of the Stone, the celestial ioniser, is turned by Tvaṣṭā and the Ṛbhus, it is then powered and sustained by its own unbroken chain of ionisation and cosmic plasma discharge and release.

Each Stone, one at the base of the Soma Pond in the Upper Vault of the Heaven and the other at the base of the Soma Pond in the Lower Vault of the Heaven, consists of two discs formed by the firestones laid in a circular formation with the Keystone at the center (see Figure 8).

According to the Ṛgveda, the firestones of the Stone are 'cold fire'. These firestones are described as 'cold charcoals' in RV x.34.9 (section 4.1.1). Though they are cold, they destroy the ionic bonds and the atomic structures completely and produce cosmic plasmas. The extremely bright and fiery appearance of these firestones are caused by Speeches, that is, the frequencies of the sound, not by heat. Note Speeches are also called hymns, prayers, or sounds in the Ṛgveda.

Though all the Devas sing and make Speeches, the Stone is the most prominent singer and speaker.

8.1.6 Magnetic Field and Magnetic Forces

According to the Ṛgveda, Tvaṣṭā (त्वष्टा), who represents magnetism in general and the magnetic field in particular, is the first born. Tvaṣṭā is emanated from the vast celestial ocean of hydrogen through arduous penance (तप: tapaḥ), that is, through long, arduous spins and oscillations of hydrogen atoms and the subsequent outward projection of their innate topology of the magnetic field structure. Magnetic forces are represented by the Ṛbhus (ऋभव: ṛbhavaḥ, pl.).

Ṛbhu (ऋभु:) is mostly addressed in the plural as the Ṛbhus, for there are three divisions of magnetic forces according to the Ṛgveda. The three divisions of the Ṛbhus are Ṛbhu (ऋभु:) or Ṛbhukṣā (ऋभुक्षा:), Vāja (वाज:), and Vibhvā (विभ्वा). Ṛbhu is a force that takes hold and embraces. Vāja is a force that impels and flows like a fleet steed. Vibhvā is a force that is all pervading and omnipresent. Thus the embracing and binding force, the repelling and impelling force, and the pervading and omnipresent force that balances the two opposing forces are the three divisions of magnetic forces. It appears that gravity can be explained as an interactive exertion of these magnetic forces between two celestial objects or between any physical objects that have mass. Along the gradient of exerting forces between two celestial objects, there will be a neutral zone where the exerting forces are equalised, creating a neutral or zero gravity zone.

In the Ṛgveda, electric phenomena, such as Savitā, Indra's thunderbolt, or draughts of Soma are treated as different aspects of magneto-plasma phenomena. For example, Savitā, who is closely associated with and often treated as identical to Tvaṣṭā (RV x.10.5), the magnetic field, impels the chariot of Aśvinā, the twin Jets (RV i.34.10), and cosmic plasma currents with his propulsion (RV iii.33.6). The thunderbolt of Indra (वज्रः vajraḥ) is a fiery, magnetic-field aligned, triply-spun filament with plasmas blooming along the surface. According to the Ṛgveda, the electric phenomena are the kindled state of cosmic plasmas and the currents of flowing charges, the draughts of electrons and protons. In the cosmic field, electric phenomena are born of magnetic fields and forces through the cosmic ionisation process, not the other way around as it is generally believed to be. For this reason, the usage of the term 'electric' is limited in this book. In the cosmic field, according to the Ṛgveda, what we observe are the magneto-plasma phenomena.

8.1.7 Māyā

Māyā is the magic force of Varuṇa, the cosmic plasmas in general and cosmic plasma dark discharge in particular, and of Mitra, the cosmic plasma light discharge. Māyā consists of two forces of opposite charges or polarities. Māyā (माया) is "extraordinary or supernatural power" and "sorcery, witchcraft magic" according to Monier-Williams. Varuṇa posseses this magic force, thus he is called the magician (मायी māyī, m. nom. sg. of मायिन् māyin). All celestial beings such as Varuṇa, Mitra, Aryamā, Agni, Soma, the Maruts (pl.), Indra, and so on are different aspects or manifestations of the same cosmic plasmas. Thus they all possess māyā. Māyā sees everything and knows everything. Māyā knows all beings and never rests, running to and fro on his paths, turning fast within all beings (RV x.177.3). Cosmic plasmas perform magical deeds with māyā, the magic force.

8.1.8 Cosmic Yajñas, the Sacred Cosmic Sacrifices

The event of cosmic ionisation and cosmic plasma discharge and release is called cosmic Yajña (यज्ञः), the sacred cosmic sacrifice. A cycle of cosmic Yajña occurs in two major stages. The first stage is the ionisation and discharge of cosmic plasmas produced by the Stone, the celestial ioniser. The Stone destroys the atomic structures of hydrogen and produces cosmic plasmas. The ionic bonds are broken in the fire of Trita (tritium), the charges are separated, and the produced cosmic plasmas are discharged into the Soma Pond, the Samudra, the Ocean of celestial Waters. This is the Agni Yajña, the fire sacrifice. The cosmic plasmas, which are discharged into and gathered in the Soma Pond, grow, bloom, and mature into Soma plants, or light-containing plants, or plasma blooms.

The second stage is the release of the matured Soma plants, or light-containing plants, or plasma blooms. This event of releasing the matured Soma plants, or plasma blooms, is the Soma Yajña. During the Soma Yajña, which is the second and the last stage of a cycle of cosmic Yajña, Indra slays Vṛtra, the serpent of the Heaven, with his thunderbolt to release the plasma blooms that have matured in the Soma Pond. When Indra slays Vṛtra, the plasma blooms, which are also called sacrificial posts or pillars of the Heaven, upon which Agnis and Somas are attached blooming, are plucked and released as Aśvinā (अश्विना, du.), the cosmic twin Jets. This is the upward release of cosmic plasmas, one from the Upper Vault of the Heaven and the other from the Lower Vault of the Heaven. As soon as the matured cosmic plasmas are released as the twin Jets, the subsequent production of cosmic plasmas starts immediately. As the newly produced plasmas are discharged, the discharged cosmic plasmas build a new plasma sheath to form a new Soma Pond. Note the plasma sheath is Vṛtra, the serpent of the Heaven.

The downward release of cosmic plasmas occurs simultaneously as Aśvinā, the twin Jets, fly over the summits of the two Vaults of the Heaven. This downward release is called the river Gaṅgā (गंगा). There are two branches of Gaṅgā. One falls upon Mt. Meru, the Galactic Bar, at the center of the Milky Way, through the base of the Soma Pond in the Upper Vault of the Heaven, the other through the base of the Soma Pond in the Lower Vault of the Heaven. From there, each branch of Gaṅgā divides itself into four parts and flows down along the four Arms of the Milky Way energising the 'terrestrial fields' established along the Arms of the Milky Way to create and sustain life.

This cycle of Agni and Soma Yajñas is repeated three times a day every day.

8.1.9 The Rains of Cosmic Plasmas and the Watchers

Cosmic plasmas rise up day after day and are discharged and released again and again three times a day. The rains of cosmic plasmas sustain all the worlds in the Milky Way and the fires of Agni give renewed vigour to the Heavenly structures. Varuṇa, Mitra, and Aryamā are the three cosmic plasma discharge modes. Varuṇa represents cosmic plasmas in general and plasma dark discharge in particular; Mitra, the plasma light discharge; and Aryamā, the plasma fire discharge. Aryamā is the fire of the Stone, the celestial ioniser.

According to the Ṛgveda, Varuṇa (cosmic plasma), Agni (proton), and Soma (electron) dispatch their Watchers (स्पशः spaśaḥ pl.) everywhere so that they can observe their territory far and wide. The well-trained and practiced Watchers of Varuṇa are observant and gifted with insight. They survey everything without even winking. Agni's Watchers are never-deceived guards of his tribe. Soma's Watchers roar and utter sounds together. They are the carriers of the hymn. They never close their eyes. Soma's Watchers carry nooses and bind to create ionic bonds and plasma sheaths. They are the keen-sighted eyes of the 'people', the 'people' of cosmic plasmas.

8.1.10 The Soma Pond and Plasma Blooms

When the cosmic plasmas, produced by the celestial ioniser, are discharged into the Soma Pond, they grow, bloom, and mature within it (see Figure 7). These blooming cosmic plasmas are called the Soma plants, light-containing plants, forests, trees, sheep's golden fleece, horns of bulls, and so on. These blooming cosmic plasmas are named 'plasma blooms' in this book. When matured, these plasma blooms can reach the height of about 20,000 light-years or more. Each stem of plasma bloom consists of a triply-spun filament and ionised and charged particles that are blooming on the surface of the filament. The Ṛgveda indicates that each Soma plant, or plasma bloom, is a plasma sheath. In the Ṛgveda, the cosmic plasmas are called the celestial Waters, sacred Waters, living Waters, milk, Soma mead, and so on. Agni (अग्निः) and Soma (सोमः) are the two embryonic particles of celestial Waters (आपः āpaḥ pl.).

Indra releases these plasma blooms by slaying Vṛtra, the serpent of the Heaven, the plasma sheath. The released cosmic plasmas sustain the Milky Way and power the hundreds of billions of solar systems within it like ours. The cosmic plasmas produced and discharged by the Stone, the celestial ioniser, into the Soma Pond, that grow and mature into plasma blooms, are the thunderbolts of Indra, the arrows of Agni, and the lances of the Maruts. The lances of the Maruts, that are used to pierce Vṛtra, the serpent of the Heaven, to release celestial lights, are the spears of destiny. The celestial lights that are released by Indra determine the destiny of our galactic system and the hundreds of billions of stars within. The plasma sheaths, filled full with cosmic plasmas, are the serpent people, the Nāgas (नागाः nāgāḥ pl.).

Most phenomenal facts revealed about cosmic plasmas are threefold. First of all, cosmic plasmas discharged into the Soma Pond rapidly grow and bloom at a speed of at least 1,794 light-years per hour (refer to section 6.2.4). Secondly, when Soma plants or plasma blooms are shattered and released, the stumps of these plasma blooms are left behind. These remaining stumps are 'the remnants of sacrifice'. From these remaining stumps, the aftergrowth of plasma shoots emerge and bloom again in the Soma Pond.

And thirdly, cosmic plasmas, as they perform cosmic Yajñas, constantly sing sacred hymns, say prayers, and make Speeches. The Speeches, which are also called hymns, prayers, sounds, and so on, are the oscillations generated by cosmic plasmas across a band of the electromagnetic spectrum, which play a critical role in the process of cosmic creation. According to the Ṛgveda, Speeches or sounds are inherent in physical reality and are intimately associated with the creation, manifestation, and sustenance of physical reality. These Speech patterns are numerically coded as metres of the stanzas and hymns of the Ṛgveda. The metre of a stanza is associated with the cosmic phenomena explained in the stanza.

The Soma Pond is where discharged cosmic plasmas gather until they are matured and ready to be released. The shoreline of the Soma Pond is defined by the outer plasma sheath, the Vṛtra, which is symbolically represented as the serpent of the Heaven with its tail in its mouth. When the Soma Pond is addressed as one unit, it is called Samudra (समुद्रः), the Ocean of celestial Waters. When each plasma bloom in the Soma Pond is considered as an individual ocean, then it is addressed in the plural as oceans (समुद्राः samudrāḥ pl.). One pair of the Stone and the Soma Pond is located in the Upper Vault of the Heaven, the other pair is located in the Lower Vault of the Heaven (see Figure 6). In the Ṛgveda, the Soma Pond is called the abode or city of the Devas, the city of the gods.

8.1.11 The God Particles

According to the Ṛgveda, Agni, proton, and Soma, electron, are literally the god particles with no sub-constituents. They are 'fundamentals'. They are the Devas, the gods. And they are the two embryonic particles of celestial Waters, that is, cosmic plasmas, and of all beings. Agni is the fiery and dry (शुष्कः śuṣkaḥ) embryonic particle of cosmic plasmas; Soma, the cool and watery (आर्द्रः ārdraḥ) embryonic particle of cosmic plasmas. Both Agni and Soma know the intention of the Creator. They observe everything and know all beings by their births. Agni takes delight in drinking Soma. Soma is described as annaṃ (अन्नं), "food in a mystical sense". By drinking Soma, Agnis are exhilarated. The Ṛgveda explains that Somas enclose and restrain Agnis in the nucleus to form atomic structures.

In the magneto-plasma universe of the Ṛgveda, no neutrons, quarks, leptons, hardrons, gluons, photons, and so on are needed. Neither are force carrier particles, antimatter, antiproton, antielectron, antiquark, antilepton, nor anti-anything necessary. Interactions between Agnis and Somas and among themselves are determined by the patterns of their Speech and the directions of their spins. Agnis attract and combine with one another by the Speech they speak and modulate (RV vii.103.3 and 4). When Agnis are kindled, they become lights and fires. They carry forces and sounds.

Agni is the lord of the house. The house refers either to the atomic structure of an element or the individual plasma blooms, which grow in the Soma Pond. Agni is also called the lord of light. Plasma blooms maturing in the Soma Pond are called the eternal flames or the tongues of Agni. Agni has three forms or fires: Ekata (protium), Dvita (deuterium), and Trita (tritium). The Ṛgveda tells us that all forms are shaped by Tvaṣṭā, the magnetic field, according to the design of Agnis (RV viii.102.8). Soma is the lord of the mind and the lord of the Speeches. He is the Brahmā and the soul of the Devas. Soma spins a triple filament. Soma is the thunderbolt of Indra. While Agni is the lord of the light, Soma is the one who bestows that light. Agni and Soma dispatch their Watchers to observe everything. As noted, their Watchers carry nooses and bind to create ionic bonds and plasma sheaths. They are the keen-sighted eyes of the 'people' of cosmic plasmas.

8.1.12 By the Hymns of the Singers, Agni is Made into Many Forms

As stanzas RV x.114.5, RV i.164.46, RV ii.1.1, RV x.130.4-5, and RV vii.103.4-6 indicate, Agni is shaped into many forms by the Speeches, or hymns, of the singers. The Speeches, or hymns, of the singers are responsible for the manifestations of varying cosmic plasma phenomena we observe in the cosmic field and for the constitution of the atomic structures of elements. As noted, according to the Ṛgveda, Agnis and Somas capture one another and bind based on their Speech patterns.

8.1.13 The Milky Way Has Four Major Arms and Is Sustained by Cosmic Yajñas

The Milky Way is formed and sustained in the Innerfield, or the Midheaven, which is called the "regions of life". The Milky Way has four major Arms according to the Ṛgveda and the Bhuvanakośa of Siddhāntaśiromaṇi. According to the Bhuvanakośa, when the celestial river Gaṅgā falls upon Mt. Meru, the Galactic Bar, at the centre of the Milky Way, it divides itself into four parts and flows down along the four immovable buttresses of Mt. Meru. The four Arms of the Milky Way are called the four props or buttresses (विष्कम्भाः viṣkambhāḥ pl.) of Mt. Meru (stanza 37 Bhuvanakośa, Siddhāntaśiromaṇi).

The two Stones, one in the Upper Vault of the Heaven and the other in the Lower Vault of the Heaven, produce, discharge, and release cosmic plasmas, by which the four Arms of the Milky Way are sustained and remain alive (RV i.164.42). The Milky Way and the worlds and life within it are powered and sustained by the unbroken chain of cosmic Yajñas, that is, the events of cosmic ionisation and discharge and release of cosmic plasmas, day after day, three times a day.

8.2 COSMIC PLASMA DISCHARGE AND RELEASE MECHANISMS

Figure 13 shows the cosmic ionisation and plasma discharge and release mechanisms. The paths of cosmic plasma discharge and release shown in Figure 13 are based on what the Ṛgveda reveals.

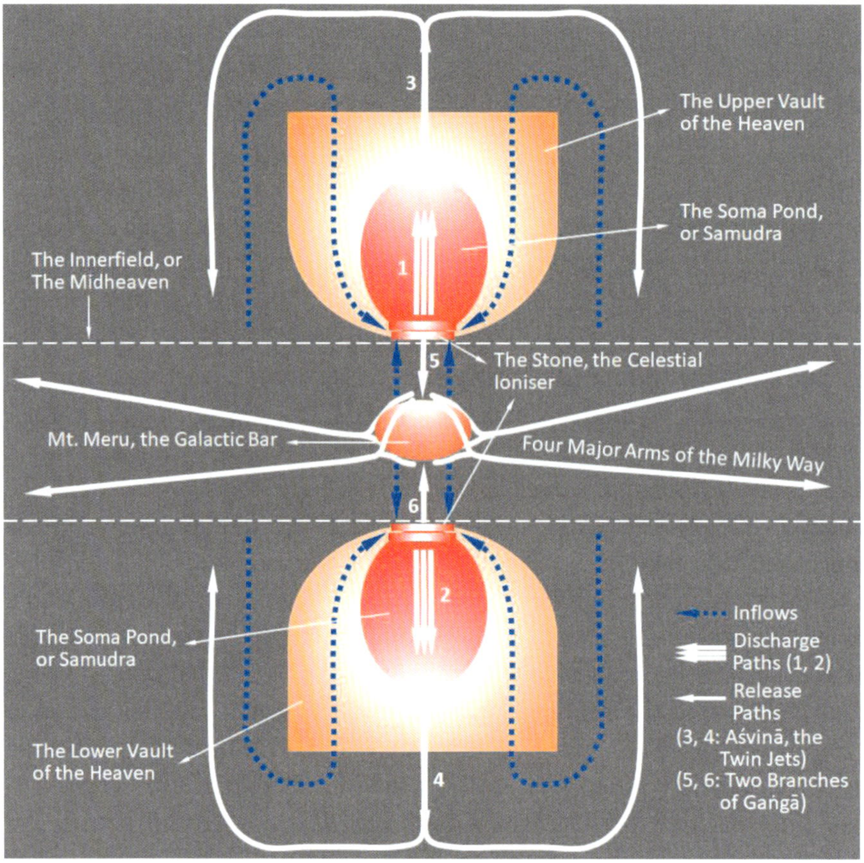

<Figure 13> The Creator's Marvellous Machine and Mechanisms of Cosmic Ionisation and Cosmic Plasma Discharge and Release. Diagram is NOT to scale. The shape and the path of arrows only show general directions of the flow and do not define the exact ranges of dispersal. The length and number of arrows do not indicate the volume, density, or speed of the flow.

When cosmic plasmas are produced by the Stone through the cosmic ionisation process, the produced cosmic plasmas are discharged from the Stone into the Soma Pond, or Samudra, the Ocean of celestial Waters. This discharge path is indicated by a threefold solid arrowed line.

When the cosmic plasmas, which are discharged into the Soma Pond, are fully matured, they are released from the Soma Pond through two different paths. These release paths are indicated by a solid arrowed line. The two release paths include: (i) the path that goes over the summit of the Vault of the Heaven and travels down towards the Innerfield where the Milky Way is formed and sustained, and (ii) the path that falls upon Mt. Meru, the Galactic Bar, through the base of the Soma Pond, and flows along the four major Arms of the Milky Way.

According to the Ṛgveda, the actual dispersal range of released cosmic plasmas will be much wider than what is shown in Figure 13 and far beyond the twofold Heavenly Vault and the Milky Way. The dispersal range of released cosmic plasmas defines the boundary of the Altar of cosmic Yajña, the field of the cosmic creation, the dominions of a galactic system.

The paths of the inflow of celestial hydrogen or other elements into the Stone are indicated by a dotted arrowed line in blue. The stanzas that explain the inflow paths to the Stone, the celestial ioniser, are not yet located in the extant text of the Ṛgveda. However, as can be seen in Figure 13, considering the layout of the Heavenly structures, one can see that there are only two possible paths into the Stone, the celestial ioniser.

Please note that the shapes and the paths of arrows only show general directions of the flows and do not define the exact ranges of dispersal. The length and number of arrows do not indicate the volume, density, or speed of the flow.

8.3 HYMNS OF BLESSING

8.3.1 The Devas Embody Their Wish and Adorn Their Creations

Though they are not the last stanzas written in the Ṛgveda, the Vedic poets could have concluded the Ṛgveda with stanzas RV x.66.6, 9, and 12 presented here. The sacrificers created the cosmic laws and the twofold Heaven and the Milky Way and sustain them. They produce the Waters, the cosmic plasmas, and the light-containing plants and the trees one after another. And they adorn their creations (RV x.66.9).

RV x.66.9 author वसुकर्णो वासुक्रः, to विश्वे देवाः, metre जगती छंदः
द्यावापृथिवी जनयन्नभि व्रताप ओषधीर्वनिनानि यज्ञिया ।
अंतरिक्षं स्व१ःरा पप्रुरूतये वशं देवासस्तन्वी३ नि मामृजुः ॥१०.६६.९॥

द्यावापृथिवी इति । जनयन् । अभि । व्रता । आपः । ओषधीः । वनिनानि । यज्ञिया ।
अंतरिक्षं । स्वः । आ । पप्रुः । ऊतये । वशं । देवासः । तन्वि । नि । ममृजुः ॥१०.६६.९॥
The sacrificers engendered the cosmic laws, begot the twofold Heaven and the Milky Way. They bring about the celestial Water, the light-containing plants, and the trees one after another, And nourish their own Innerfield to sustain it. The Devas embody their wish and adorn their creations. (RV x.66.9)

द्यावापृथिवी <m. acc. du of द्यावापृथिवी – Dyāvāpṛthivī, the twofold Heaven and the Milky Way>, जनयन् <inj. 3rd person pl. of √जन् 10P – they caused (acc.) to be born>, अभि <as a separate adverb or preposition – one after another>, व्रता <n. acc. pl. of व्रत – ordinances, laws>, आपः <n. acc. sg. of आपस् – Water (cosmic plasmas)>, ओषधीः <f. acc. pl. of ओषधि – light-containing plants (Soma Plants, plasma blooms, i.e. blooming cosmic plasmas)>, वनिनानि <n. acc. pl. of वनिन – woods, trees, forests (plasma blooms that grow in the field of Soma Pond are compared to woods or trees)>, यज्ञिया <(=यज्ञियानि) n. nom. pl. of यज्ञिय – the divines, "applied to gods", the sacrificers>;

अंतरिक्षं <n. acc. sg. of अंतरिक्ष – the Innerfield, the Midheaven>, स्वः <m. nom. sg/du/pl. of adj. स्व – one's own, their own>, आ <ind. prefix to verb पप्रुः (√पृ)>, आ-पप्रुः <prf. 3rd person pl. of आ-√पृ 3P – they fill, nourish>, ऊतये <f. dat. sg. of ऊति (dative infinitive) – to help, promote>, वशं <m. acc. sg. of वश – will, wish>, देवासः <m. nom. pl. of देव – the Devas>, तन्वी <m. nom. pl(?). of adj. तन्विन् (from √तन् 8P to put forth, manifest) – "possessed of a body", i.e. manifest, embody>, नि <ind. prefix to verb ममृजुः (√मृज्)>, नि-ममृजुः <prf. 3rd person pl. of नि-√मृज् – they embellish, adorn>.

8.3.2 Powerful Be the Cosmic Yajña

Powerful be the cosmic Yajña, strong be they who sing mighty hymns, mighty be the twofold Heavenly Vault and the Milky Way so that we may prosper (RV x.66.6). O Ādityas, O Rudras, O Vasus, quicken ye our prayer and the sacred hymns which we are singing now (RV x.66.12).

RV x.66.6 author वसुकर्णो वासुक्रः, to विश्वे देवाः, metre जगती छंदः
वृषा यज्ञो वृषणः संतु यज्ञिया वृषणो देवा वृषणो हविष्कृतः ।
वृषणा द्यावापृथिवी ऋतावरी वृषा पर्जन्यो वृषणो वृषस्तुभः ॥१०.६६.६॥
वृषा । यज्ञः । वृषणः । संतु । यज्ञियाः । वृषणः । देवाः । वृषणः । हविःऽकृतः ।
वृषणा । द्यावापृथिवी इति । ऋतऽवरी इत्यृतऽवरी । वृषा । पर्जन्यः । वृषणः । वृषऽस्तुभः ॥१०.६६.६॥
Powerful be the cosmic Yajña, strong be the sacrificers, strong be the Devas, strong be the preparers of oblation. Mighty be the twofold Heaven and the Milky Way, strong be the performer of the sacred work, strong be Parjanya, strong be they who utter mighty hymns. (RV x.66.6)

वृषा <m. nom. sg. of adj. वृषन् – strong, mighty>, यज्ञः <m. nom. sg. of यज्ञ – Yajña, sacrifice>, वृषणः <m. nom. pl. of adj. वृषन् – strong, mighty>, संतु <ipv. 3rd person pl. of √अस् – ye be>, यज्ञियाः <m. nom. pl. of यज्ञिय – sacrificers>, वृषणः <m. nom. pl. of adj. वृषन् – strong, mighty>, देवाः <m. nom. pl. of देव – the Devas>, वृषणः <m. nom. pl. of adj. वृषन् – strong, mighty>, हविष्कृतः <m. nom. pl. of हविष्कृत् – the preparers of oblation>;

वृषणा <m. nom. du. of adj. वृषन् – powerful, strong, mighty, great>, द्यावापृथिवी <mf. nom. of dual compound द्यावापृथिवी – Dyāvāpṛthivī, the twofold Heaven and the Milky Way>, ऋतावरी <f. nom. sg. of ऋतावन् (f. ऋतावरी) – the performer of the sacred work (refers to the celestial ioniser)>, वृषा <f. nom. sg. of adj. वृषन् (f. वृषा) – strong, mighty>, पर्जन्यः <m. nom. sg. of पर्जन्य – Parjanya, "the god of rain">, वृषणः <m. nom. pl. of adj. वृषन् – powerful, mighty>, वृषस्तुभः <mf. nom. pl. of वृषस्तुभ् (वृषन् powerful, mighty + √स्तुभ् to utter a sound) – they who utter mighty hymns>.

RV x.66.12 author वसुकर्णो वासुक्रः, to विश्वे देवाः, metre जगती छंदः
स्याम॑ वो॒ मन॑वो दे॒ववी॑तये॒ प्रांचं॑ नो य॒ज्ञं प्र ण॑यत साधु॒या ।
आ॒दि॑त्या॒ रु॒द्रा॒ वस॑वः॒ सुदा॑नव इ॒मा ब्रह्म॑ श॒स्यमा॑नानि जिन्वत ॥१०.६६.१२॥
स्याम॑ । वः॒ । मन॑वः । दे॒वऽवी॑तये । प्रांच॑ं । नः॒ । य॒ज्ञं । प्र । न॒य॒त । सा॒धु॒ऽया ।
आ॒दि॑त्याः । रु॒द्राः॑ । वस॑वः । सु॒ऽदा॑नवः । इ॒मा । ब्रह्म॑ । श॒स्यमा॑नानि । जि॒न्व॒त ॥१०.६६.१२॥
We men, may we be yours! For the divine feast, lead ye our Yajña duly forth to make it succeed. O Ādityas, O Rudras, O Vasus, bestowing abundantly, quicken ye our prayer and these sacred hymns which we are singing now. (RV x.66.12)

स्याम <opt. 1st person pl. of √अस् – may we be>, वः <m. gen. pl. of 2nd person pron. युष्मद् – of you, yours>, मनवः <m. nom. pl. of मनु – we men>, देववीतये <f. dat. sg. of देववीति – for feast for the gods, for the divine feast, enjoyment, draught>, प्रांचं <mf. acc. sg. of adj. प्रांच् (प्राञ्च्) – (with प्र-√नी) to advance, promote, further>, नः <m. gen. pl. of 1st person pron. अस्मद् – our>, यज्ञं <m. acc. sg. of यज्ञ – yajña, sacrifice>, प्र <ind. – forward, forth>, नयत <ipv. 2nd person pl. of √नी – ye lead>, साधुया <ind. – rightly, duly>;

आदित्याः <m. voc. pl. of आदित्य – O Ādityas>, रुद्राः <m. voc. pl. of रुद्र – O Rudras, the sons of Rudra (refers to the Maruts)>, वसवः <m. voc. pl. of वसु – O Vasus, a class of gods whose chief is Indra>, सुदानवः <m. voc. pl. of adj. सुदानु – pouring out or bestowing abundantly>, इमा <(=इमानि) n. acc. pl. of इदं – these>, ब्रह्म <n. acc. sg. of ब्रह्मन् – sacred utterance, prayer>, शस्यमानानि <n. acc. pl. of शस्यमान (शस्य adj. hymns to be recited + मान n. respect, honour) – honoured, sacred hymns recited>, जिन्वत <ipv. 2nd person pl. of √जिन्व् – ye quicken, impel, incite>.

Epilogue

In 1970, about five decades ago, Hannes Alfvén concluded his Nobel Lecture with the statement,

> It is possible that this *new era* [of plasma physics and space research] *also means a partial return to more understandable physics*. For the non-specialists four-dimensional relativity theory, and the indeterminism of atom structure have always been mystic and difficult to understand. I believe that it is easier to explain the 33 instabilities in plasma physics or the resonance structure of the solar system. The *increased emphasis on the new fields mean [sic] a certain demystification of physics*. In the spiral or trochoidal motion which science makes during the centuries, its guiding center has returned to those regions from where it started. It was the wonders of the night sky, observed by Indians, Sumerians or Egyptians, that started science several thousand years ago. It was the question why the wanderers – the planets – moved as they did that triggered off the scientific avalanche several hundred years ago. The same objects are now again in the center of science – only the questions we ask are different. *We now ask how to go there*, and *we also ask how these bodies once were formed*. And if the night sky on which we observe them is at a high latitude, outside this lecture hall – perhaps over a small island in the archipelago of Stockholm – we may also see in the sky an aurora, *which is a cosmic plasma, reminding us of the time when our world was born out of plasma. Because in the beginning was the plasma.*[89] (Italicised emphasis added)

Indeed, in the beginning was the plasma, produced and discharged by the Stone, the

---

[89] Hannes Alfvén, 'Plasma Physics, Space Research and the Origin of the Solar System', in *Nobel Lecture, Physics 1963-1970*, Amsterdam, Elsevier Publishing Company, 1972, pp. 315-316. Copyright © The Nobel Foundation 1970. Reprinted by permission of The Nobel Foundation.

celestial ioniser, and our world was born out of plasma. In the field of cosmic creation, what we observe are remarkable events and phenomena of cosmic plasmas, as Hannes Alfvén observed and as the Ṛgveda reveals, and not "clouds of hot gas". According to the Ṛgveda, the cosmic creation and life are magneto-plasma phenomena. These events and phenomena are self-powered and self-manifested by Nature.

For the first time, the mechanisms of cosmic ionisation and cosmic plasma discharge and release are presented in this book based on what the Ṛgveda reveals. The magneto-plasma cosmology revealed in the Ṛgveda is the most robust and complete cosmology and the cosmic creation model ever written or presented in the fields of cosmology, astrophysics, and cosmic plasma physics.

Knowledge of cosmic creation, recorded in the ancient Vedic language and concealed by veils of symbolism and beautiful poetic language, is deciphered and presented in this book. The knowledge of magneto-plasma cosmology revealed in the Ṛgveda is the knowledge of cosmic creation, the great work of Nature, the celestial Alchemy, which eluded the enlightened minds of priests, mystics, scholars, oligarchs, and royals for millennia.

The implications of the knowledge revealed in the Ṛgveda are clear and extensive. It, foremost, presents how the cosmic creation was initiated in the beginning and how our galactic system is created and sustained through the events of cosmic ionisation and cosmic plasma discharge and release. And the Ṛgveda reveals that physical law, the law of Nature, and the spiritual law are one and the same.

The magneto-plasma cosmology revealed in the Ṛgveda provide an alternative frame for coherent analyses and interpretations of the celestial objects and cosmic plasma phenomena observed by NASA space probes and telescopes and a better understanding of the history of the universe and the galactic systems that have evolved within the universe.

Natural philosophy and the knowledge of cosmic creation and the associated symbolism revealed in the Ṛgveda provide a key to understanding the origin and philosophical heritage of alchemy, Gnosticism, and Hermeticism and to unwrapping the occult symbolism that permeates in these fields as well as in occult societies. The Avesta, the sacred text of Zoroastrianism, which has special linguistic and philosophic affinities to the Ṛgveda, may be read anew in light of the knowledge revealed in the Ṛgveda. And it may provide the foundation for the renewed revival of natural theology of ancient Greek.

In the land of its birth, the Upaniṣads (उपनिषदः pl.), Purāṇas (पुराणानि pl.), Smṛtis (स्मृतयः pl.), and Itihāsas (इतिहासाः pl.) can be read anew in light of the knowledge revealed in the Ṛgveda. Smṛtis, the works derived from the Vedas or the auxiliary arms of the Vedas, include the Vedāṅgas, Sūtras (सूत्रा pl.), and the Law Book of Manu (मनुः). Vedāṅgas are the six limbs, or auxiliary sciences, of the Veda. Sūtras consist of Śrautasūtras (श्रौतसूत्रा pl.) based on the Vedas and Gṛhyasūtras (गृह्यसूत्रा pl.) that contain directions for domestic rites and ceremonies. The Law Book of Manu (मानवं धर्मशास्त्रं) is regarded as the ancient law book of Hindus. Itihāsas, the history, include Mahābhāratam and Rāmāyaṇam.

We have to rethink the heritage and history of mankind. We are in need of rethinking the field of physics, especially the field of plasma universe and cosmic plasma physics.

As interpretation and translation of each of the ten maṇḍalas of the Ṛgveda continue, they will be published in time. The translation of maṇḍalas II-IX of the Ṛgveda will be done first and maṇḍalas I and X will follow.

<Figure 14> "*Shiva as the Lord of Dance*, India (Tamil Nadu), c. 950–1000, Los Angeles County Museum of Art, anonymous gift". Image source: https://collections.lacma.org/node/240893 (View No. 3 of 5 Views available). In public domain. (accessed 28 January 2020). Śiva is the Creator and Sustainer of the World, not the Destroyer. Śiva destroys the ionic bonds of hydrogen to produce the sacred cosmic lights and fires, the cosmic plasmas. The infant-like demon figure trampled under his right foot is the symbolically represented body of hydrogen. With his drum, the Ḍamaru (डमरुः), the marvellous Creator's machine, in his upper right hand and the flames of the Stone, the celestial ioniser, in his upper left hand, Śiva (शिवः) dances to produce and discharge the sacred cosmic lights and fires, the secret fires of cosmic creation. It is the dance of eternity, for Aśvattha (अश्वत्थः), the Tree of Life that represents our galactic system, is imperishable.

Appendix:
Stanzas Translated

1. STANZAS LISTED BY SECTIONS IN THIS BOOK

SECTION	STANZAS	ADDRESSED TO	METRE

I. The Ṛgveda and Its Interpretation

SECTION	STANZAS	ADDRESSED TO	METRE
1.1 THE ṚGVEDA AND THE VEDIC LITERATURE	–	–	–
1.2 THE VEDĀṄGAS, THE SIX LIMBS OF THE VEDA	–	–	–
1.3 THE DATE OF THE ṚGVEDA	–	–	–
1.4 PROBLEMS ASSOCIATED WITH TRADITIONAL INTERPRETATIONS OF THE ṚGVEDA	–	–	–
1.5 A NEW INTERPRETATION OF THE ṚGVEDA	–	–	–
1.6 THE TEXT OF THE ṚGVEDA	–	–	–

II. The Heavenly Structures Revealed in the Ṛgveda

SECTION	STANZAS	ADDRESSED TO	METRE
2.1 THE SHINING ONES, THE DEVAS AND THE ṚṢIS			
2.2 THE THREEFOLD HEAVENLY SPHERE, THE ALTAR OF COSMIC YAJÑA			
2.2.1 The Upper Heaven, the Innerfield, and the Lower Heaven	RV v.60.6	मरुतो मरुतोवाग्निश्च	त्रिष्टुप्
2.2.2 The Cosmic Yajña, the Origin of the World and Life	RV i.164.35	विश्वे देवाः	त्रिष्टुप्
2.2.3 Make Me Immortal in That Luminous Threefold Heaven	RV ix.113.9	पवमानः सोमः	पंक्तिः
2.3 THE HEIGHT OF THE HEAVENLY VAULT			
2.3.1 The Height of the Heavenly Vault Revealed in the Sūryasiddhānta	90 Bhūgolādhyāya, Sūryasiddhānta	–	–
2.3.2 The Height of the Heavenly Vault by the Siddhāntaśiromaṇi	67 Bhuvanakośa, Siddhāntaśiromaṇi	–	–
2.4 THE TWOFOLD HEAVENLY VAULT AND ITS SUBSTRUCTURES			
2.4.1 The Great Heavens of Two, the Upper and the Lower Heavens	RV iv.56.5	द्यावापृथिव्यौ	गायत्री
	RV x.88.17	सूर्यवैश्वानरौ	त्रिष्टुप्
2.4.2 The twofold Heavenly Vault Observed by Chandra and Hubble	–	–	–
2.4.3 The Soma Pond in the Vault of the Heaven	RV i.164.33	दीर्घतमा औचथ्यः	त्रिष्टुप्
2.4.4 Vimāna, the Chariot of the Devas	RV iii.26.7	अग्निरात्मा वा	त्रिष्टुप्
2.4.5 The Upper Heaven Revolves Westward, the Lower Heaven Eastward	RV i.164.38	विश्वे देवाः	त्रिष्टुप्
	55 Bhūgolādhyāya, Sūryasiddhānta	–	–
2.4.6 The Heaven, the Red Bull	RV v.58.6	मरुतः	त्रिष्टुप्
2.4.7 The Structures of the Heaven, the Complete Picture	RV i.20.6	ऋभवः	गायत्री
2.5 THE SOMA POND, THE WOMB OF COSMIC CREATION			
2.5.1 The Soma Pond Observed by Chandra	–	–	–
2.5.2 The Soma Pond, the Golden Womb	RV x.121.1-4	कः	त्रिष्टुप्
2.5.3 Samudra, the Ocean of Celestial Waters	RV x.142.7-8	अग्निः	अनुष्टुप्
2.5.4 Rodasī, the Pair of Mātā and Duhitā, the Pair of the Stone and the Soma Pond	RV iii.55.12-13	रोदसी	त्रिष्टुप्
2.5.5 The Radiant Udder of Pṛśni	RV ii.34.2	मरुतः	जगती
2.5.6 The Moon with Beauteous Wings Glides in the Heaven	RV i.105.1-2	विश्वे देवाः	पङ्क्तिः

SECTION	STANZAS	ADDRESSED TO	METRE
2.5.7 Sūrya, the Best of Lights	RV x.170.1-4	सूर्यः	जगती
2.5.8 Sūrya, the Single Eye, the All-Seeing Eye	RV ix.9.4	पवमानः सोमः	गायत्री
	RV i.50.2,8	सूर्यः	गायत्री
2.5.9 Sūrya, the Eye of Mitra and Varuṇa	RV x.37.1	सूर्यः	जगती
2.5.10 Sūrya Takes Hold of the Wondrous Weapon	RV v.63.4	मित्रावरुणौ	जगती
2.5.11 Sūrya, the Refulgent Chariot	RV v.63.7	मित्रावरुणौ	जगती
2.5.12 Uṣā, the Dawn	RV i.92.1-4	उषाः	जगती
2.5.13 The Celestial Boat Equipped with Oars	RV x.101.2	विश्वे देवा ऋत्विजो वा	त्रिष्टुप्
2.5.14 The Heads of Sorceresses	RV i.133.2	इंद्रः	अनुष्टुप्
2.5.15 An Owl, an Owlet, and the Sorcerer	RV vii.104.22	इंद्रः	त्रिष्टुप्
2.5.16 The Mighty Thunderbolt of Indra	RV i.52.7	इंद्रः	जगती
2.5.17 Rudra	RV vii.46.1-4	रुद्रः	जगती
2.5.18 The Maruts Arrive in Lightning Laden Chariots with Spears	RV i.88.1	मरुतः	प्रस्तारपंक्तिः
	RV ii.34.5	मरुतः	जगती
2.5.19 The Maruts Make Streaming Currents and Flashing Thunderbolts	RV i.64.5,7	मरुतः	जगती
2.5.20 The Maruts Move Swiftly Like Vātas and Are Effulgent Like Tongues of Fires	RV x.78.3	मरुतः	त्रिष्टुप्
2.5.21 The Maruts Are Spirited Like Serpents	RV i.64.8	मरुतः	जगती
2.5.22 The Maruts, the Eternal Flames	RV i.64.10	मरुतः	जगती
2.5.23 Triṣṭup, a Libation for the Maruts	RV viii.7.1	मरुतः	गायत्री
2.5.24 Vāyu and Vāta	RV x.168.1-4	वायुः	त्रिष्टुप्
2.5.25 Indra Slays the Serpent of the Heaven	RV viii.93.1-3	इंद्रः	गायत्री
2.5.26 Indra Delights in Soma Drinking	RV viii.32.1,5	इंद्रः	गायत्री
	RV iii.36.8	इंद्रः	त्रिष्टुप्
2.5.27 Indra, the Maker of Light and Form	RV i.6.3	इंद्रः	गायत्री
	RV i.6.5	मरुत इंद्रश्च	गायत्री
2.5.28 Saṃvatsarakālacakraṃ, the Soma Pond and the Wheel of Cosmic Ionisation	RV i.164.48	संवत्सरसंस्थं कालचक्रवर्णनं	त्रिष्टुप्
2.5.29 The Thousand-Pillared Abode of Kings Varuṇa and Mitra	RV ii.41.5	मित्रावरुणौ	गायत्री
2.6 THE SACRIFICIAL POSTS			
2.6.1 The Sacrificial Post, the Luminous Pillar of the Heaven	RV iii.8.1-5	यूपः (sg.)	त्रिष्टुप्
2.6.2 A Hundred or a Thousand Sacrificial Posts	RV iii.8.6-10	यूपाः (pl.)	त्रिष्टुप्
2.6.3 The Remaining Stump of the Destroyed Sacrificial Post	RV iii.8.11	छिन्नयूपस्य मूलभूतः स्थाणुः	त्रिष्टुप्
2.7 THE STONE, THE CELESTIAL MACHINE OF IONISATION AND COSMIC PLASMA DISCHARGE AND RELEASE			
2.7.1 The Stone and the Keystone	–	–	–
2.7.2 The Two Stones of Trita	RV ix.102.2	पवमानः सोमः	उष्णिक्
2.7.3 The Two Stones at the Base of the Soma Pond	RV i.109.3-4,7	इंद्राग्नी	त्रिष्टुप्
2.7.4 The Stones Are Impelled by Savitā	RV x.175.1-4	ग्रावाणः	गायत्री
2.7.5 The Stones Sing and Make Rumbling Sounds	RV x.94.1-4	ग्रावाणः	जगती
	RV x.94.5	ग्रावाणः	त्रिष्टुप्

SECTION	STANZAS	ADDRESSED TO	METRE
2.7.6 The Two Arms and Ten Fingers of the Stone	RV v.43.4	विश्वे देवाः	त्रिष्टुप्
2.7.7 Tvaṣṭā's Ten Maidens	RV i.95.2	अग्निरग्निरौषसो वा	त्रिष्टुप्
2.7.8 The Stone Wheel	RV x.101.7	विश्वे देवा ऋत्विजो वा	त्रिष्टुप्
2.7.9 The Two Stones, Mortar and Pestle	RV i.28.1-4	इंद्रः	अनुष्टुप्
	RV i.28.5-6	उलूखलं	अनुष्टुप्
	RV i.28.7-8	उलुखलमुसलौ	गायत्री
	RV i.28.9	प्रजापतिर्हरिश्चंद्रोऽधिषवणचर्म सोमो वा	गायत्री
2.7.10 Like a Real Comb	RV i.191.1-5	अतृणसूर्याः	अनुष्टुप्
2.7.11 Kṣetrapati, the Lord of the Field	RV iv.57.1	क्षेत्रपतिः	अनुष्टुप्
	RV iv.57.2-3	क्षेत्रपतिः	त्रिष्टुप्
	RV iv.57.4	शुनः	अनुष्टुप्
	RV iv.57.5	शुनासीरौ	पुर उष्णिक्
	RV iv.57.8	शुनासीरौ	त्रिष्टुप्
2.7.12 The Adhvaryus Make Sweet Soma Mead	RV v.43.3	विश्वे देवाः	त्रिष्टुप्
2.8 PṚTHIVĪ, THE MILKY WAY			
2.8.1 The Milky Way Has Four Major Spiral Arms	RV i.164.41	विश्वे देवाः	जगती
	RV i.164.42	वाक् and आपः	प्रस्तारपंक्तिः
2.8.2 Different Regions of the Milky Way	–	–	–
2.8.3 Five Regions	RV ix.86.29	पवमानः सोमः	जगती
2.8.4 Six Divine Expanses	RV x.128.5	विश्वे देवाः	त्रिष्टुप्
2.8.5 Seven Domains	RV i.22.16	विष्णुर्देवा वा	गयत्री
2.8.6 Eight Regions	RV i.35.8	सविता	त्रिष्टुप्
2.8.7 A Hymn to Pṛthivī	RV v.84.1-3	पृथिवी	अनुष्टुप्
2.9 AŚVATTHA, THE INVERTED TREE, THE VEDIC TREE OF LIFE	RV i.24.7	वरुणः	त्रिष्टुप्
	Kaṭha Upaniṣad ii.3.1	–	–

III. The Divine Architects

SECTION	STANZAS	ADDRESSED TO	METRE
3.1 TVAṢṬĀ, MAGNETISM IN GENERAL AND THE MAGNETIC FIELD IN PARTICULAR			
3.1.1 Tvaṣṭā, the First Born	RV i.13.10	त्वष्टा	गायत्री
3.1.2 The Shaper of All Forms	RV x.110.9	त्वष्टा	त्रिष्टुप्
	RV viii.102.8	अग्निः	गायत्री
3.1.3 The Possessor of the World	RV ii.31.4	विश्वे देवाः	जगती
3.1.4 Tvaṣṭā Holds the Goblets That Serve the Drink of the Devas	RV x.53.9	अग्निः सौवीकः	जगती
3.1.5 Tvaṣṭā, the Divine Architect, the All-Seeing One with Two Winged Arms	RV x.81.1-7	विश्वकर्मा	त्रिष्टुप्
3.2 THE ṚBHUS, THE MAGNETIC FORCES			
3.2.1 The Ṛbhus Revive Their Old Parents and Impel the Yajña	RV iv.33.1-3	ऋभवः	त्रिष्टुप्
3.2.2 The Ṛbhus Guard the Cow and Make the Four Celestial Bowls	RV iv.33.4-6	ऋभवः	त्रिष्टुप्
3.2.3 The Skilful Architects Repose at Agni's Rite of Soma Reception	RV iv.33.7-9	ऋभवः	त्रिष्टुप्
3.2.4 The Ṛbhus Make Two Bay Steeds for Indra	RV iv.33.10-11	ऋभवः	त्रिष्टुप्

SECTION	STANZAS	ADDRESSED TO	METRE
IV. Celestial Ionisation and Other Plasma Phenomena			
4.1 CELESTIAL IONISATION			
4.1.1 Praise of the Ploughing Wheel	RV x.34.1,9	अक्षकृषिप्रशंसा	त्रिष्टुप्
	RV x.34.7	अक्षकृषिप्रशंसा	जगती
4.1.2 The Gamester's Reproach of the Ploughing Wheel That Shatters the Ionic Bond	RV x.34.2-6	अक्षकितवनिंदा	त्रिष्टुप्
4.1.3 Cosmic Plasma Discharge	RV x.135.1-7	यमः	अनुष्टुप्
4.2 CELESTIAL WATERS, THE COSMIC PLASMAS			
4.2.1 The Adhvaryus Press Out the Celestial Waters	RV x.30.2,3,15	आप अपंनपाद्वा	त्रिष्टुप्
4.2.2 Celestial Waters Rise Up and Are Released Day After Day	RV i.164.51	सूर्यः पर्जन्याग्नी वा	अनुष्टुप्
	RV i.164.52	सरस्वान् सूर्यो वा	त्रिष्टुप्
4.2.3 Sūrya Spins the Celestial Waters with His Beams of Light	RV vii.47.1-4	आपः	त्रिष्टुप्
4.2.4 Varuṇa, Mitra, Aryamā: Cosmic Plasma Discharge Modes	RV v.3.1-2	अग्निः	त्रिष्टुप्
4.2.5 Varuṇa, the Supreme Monarch	RV viii.42.1-2	वरुणः	त्रिष्टुप्
4.2.6 Within Varuṇa the Three Heavens Are Placed	RV vii.87.5	वरुणः	त्रिष्टुप्
4.2.7 Mitra and Varuṇa Sustain the Cosmic Sacrifice Through the Cosmic Sacrifice	RV i.23.5	मित्रावरुणौ	गायत्री
4.2.8 Mitra Beholds the Tillers of the Field with an Unwinking Eye	RV iii.59.1	मित्रः	त्रिष्टुप्
	RV iii.59.8-9	मित्रः	गायत्री
4.3 ADITI, THE DOUBLE HEADED			
4.3.1 Aditi, Mother of Kings Mitra and Varuṇa	RV ii.27.7	आदित्याः	त्रिष्टुप्
	RV viii.25.3	मित्रावरुणौ	उष्णिक्
4.3.2 Aditi, the Milch Cow	RV i.153.3	मित्रावरुणौ	त्रिष्टुप्
4.3.3 Aditi Unbinds the Sin of the Ionic Bond	RV vii.93.7	इंद्राग्नी	त्रिष्टुप्
4.3.4 Aditi, Daughter of Dakṣa	RV x.72.1-5	देवाः	अनुष्टुप्
4.3.5 Aditi and Diti	RV v.62.8	मित्रावरुणौ	त्रिष्टुप्
	RV iv.2.11	अग्निः	त्रिष्टुप्
4.3.6 The Ādityas, the Sons of Aditi	RV ii.27.1	आदित्याः	त्रिष्टुप्
4.4 SAVITĀ			
4.4.1 Savitā, Tvaṣṭā, Viśvarūpa	RV x.10.5	यमो वैवस्वतः	त्रिष्टुप्
4.4.2 By Savitā's Propulsion Celestial Waters Flow	RV iii.33.6	इंद्रः	त्रिष्टुप्
4.4.3 Savitā Impels Our Hymns	RV iii.62.10	सविता	गायत्री
4.4.4 Three Times a Day Savitā Brings Celestial Waters	RV iii.56.6	विश्वे देवाः	त्रिष्टुप्
4.5 VṚTRA, THE SERPENT OF THE HEAVEN, THE PLASMA SHEATH			
4.5.1 Dāsa and Ārya, the Two Opposite Charges of the Double Layer	RV vi.33.3	इंद्रः	त्रिष्टुप्
4.5.2 Let Not My Thread Be Severed Before Time	RV ii.28.5	वरुणः	त्रिष्टुप्
	RV viii.67.20	आदित्याः	गायत्री
4.6 VARUṆA'S NOOSE, THE SIN OF THE IONIC BOND			
4.6.1 The Bonds Not Wrought of Ropes	RV vii.84.2	इंद्रावरुणौ	त्रिष्टुप्
4.6.2 Remove the Noose from Us O Varuṇa	RV i.24.14-15	वरुणः	त्रिष्टुप्

SECTION	STANZAS	ADDRESSED TO	METRE
4.7 VĀK, THE SPEECH OF COSMIC PLASMAS			
4.7.1 Vāk, the Queen of the Devas	RV viii.100.10-11	वाक्	त्रिष्टुप्
4.7.2 A Hymn of Vāk	RV x.125.1,3-8	वागांभृणी (वाक् आम्भृणी)	त्रिष्टुप्
	RV x.125.2	वागांभृणी (वाक् आम्भृणी)	जगती
4.8 THE WATCHERS			
4.8.1 Varuṇa's Watchers	RV vii.61.3	मित्रावरुणौ	त्रिष्टुप्
	RV vii.87.3	वरुणः	त्रिष्टुप्
4.8.2 Agni's Watchers	RV iv.4.3	अग्नी रक्षोहा	त्रिष्टुप्
4.8.3 Soma's Watchers	RV ix.73.4,7	पवमानः सोमः	जगती
4.9 MĀYĀ			
4.9.1 Māyā, the Magic Force of Varuṇa	RV v.85.5	वरुणः	त्रिष्टुप्
4.9.2 Māyā Drives the Flying Horse	RV x.177.1	मायाभेदः	जगती
	RV x.177.2-3	मायाभेदः	त्रिष्टुप्
4.10 ĀTMĀ			
4.10.1 Ātmā, the Soul of Cosmic Yajña	RV ix.6.8	पवमानः सोमः	गायत्री
4.10.2 Have I Indeed Drunk Soma?	RV x.119.1-13	आत्मस्तुतिः	गायत्री
4.11 MANA, THE COSMIC MIND	RV x.58.1-12	मनआवर्तनं	अनुष्टुप्

V. In the Beginning

5.1 A HYMN OF THE CREATION	RV x.129.1-7	भाववृत्तं	त्रिष्टुप्
5.2 WEAVING THE NEBULOUS FABRIC WITH SĀMA HYMNS	RV x.130.1-7	भाववृत्तं	त्रिष्टुप्

VI. The Events of Cosmic Plasma Release

6.1 AŚVINĀ, THE TWIN JETS			
6.1.1 Aśvinā Transport the Plasma Sheaths Filled with Soma Mead	RV iv.45.1-3	अश्विनौ	जगती
6.1.2 The Spectacle of Cosmic Eruption, the Dappled Cows Are Released	RV x.61.4-9	विश्वे देवाः	त्रिष्टुप्
6.1.3 The Mighty Aśvinā Perform Wondrous Deeds	RV i.46.1-2	अश्विनौ	गायत्री
6.1.4 The Coursers of Aśvinā Rush Over the Blazing Summit of the Heaven	RV i.46.3-5	अश्विनौ	गायत्री
6.1.5 Aśvinā Bring the Two Streams of Light	RV i.46.6-7	अश्विनौ	गायत्री
6.1.6 The Two Circumambulatory Paths of Aśvinā	RV i.46.14	अश्विनौ	गायत्री
6.2 THE CELESTIAL RIVERS			
6.2.1 Sindhu Flows Upwards Roaring Like a Charging Bull	RV x.75.1-3	नद्यः	जगती
6.2.2 Sindhu Leads the Two Wings of Army Like a Warrior King	RV x.75.4	नद्यः	जगती
6.2.3 Seek My Hymn O Gaṅgā, Yamunā, Sarasvatī, Śutudrī, and Paruṣṇī	RV x.75.5-6	नद्यः	जगती
6.2.4 Sindhu, Most Rapid of the Rapids, Traverses the Heavenly Spheres	RV x.75.7,9	नद्यः	जगती
6.2.5 Exalted by Soma Mead Sindhu Grows Bright	RV x.75.8	नद्यः	जगती
6.2.6 Mighty Gaṅgā Falls Upon Mt. Meru	37,38 Bhuvanakośa, Siddhāntaśiromaṇi	–	–

SECTION	STANZAS	ADDRESSED TO	METRE
VII. Agni and Soma			
7.1 AGNI, PROTON			
7.1.1 Agni, an Embryonic Particle of Celestial Waters and All Beings	RV i.70.3/4	अग्निः	द्विपदा विराट्
7.1.2 Agni Comes Out of the Stone as Celestial Waters	RV ii.1.1	अग्निः	जगती
7.1.3 The Three Forms of Agni	RV iii.20.2	अग्निः	त्रिष्टुप्
7.1.4 Ekata, Dvita, Trita: Protium, Deuterium, Tritium			
7.1.4.1 Ekata	–	–	–
7.1.4.2 Dvita Brings Slayed Offerings	RV v.18.1-3	अग्निः	अनुष्टुप्
7.1.4.3 Trita Blasts and Whets Like a Blower in the Smelting Furnace	RV v.9.5	अग्निः	अनुष्टुप्
	RV i.52.5	इंद्रः	जगती
	RV viii.47.13	आदित्याः	महापंक्तिः
	RV viii.47.14-16	आदित्या उषाश्च	महापंक्तिः
7.1.5 Agni's Path: Black, White, Yellow, Red, Purple, and Glorious	RV x.20.9	अग्निः	विराट्
7.1.6 Frogs Capture One Another Based on Their Speech Patterns	RV vii.103.1	मंडूकाः	अनुष्टुप्
	RV vii.103.2-10	मंडूकाः	त्रिष्टुप्
7.1.7 The Bird That Is but One, Singers with Hymns Shape It into Many Forms	RV x.114.5	विश्वे देवाः	त्रिष्टुप्
	RV i.164.46	सूर्यः	त्रिष्टुप्
7.2 SOMA, ELECTRON			
7.2.1 Soma, an Embryonic Particle of Celestial Waters	RV ix.97.41	पवमानः सोमः	त्रिष्टुप्
7.2.2 The Thunderbolt of Indra	RV ix.72.7	पवमानः सोमः	जगती
7.2.3 Soma Spins a Triple Filament	RV ix.86.32	पवमानः सोमः	जगती
7.2.4 Soma, the Brahmā of the Devas	RV ix.96.5-6	पवमानः सोमः	त्रिष्टुप्
7.2.5 The Bestower of Light Skilled in a Thousand Arts	RV ix.96.18-19	पवमानः सोमः	त्रिष्टुप्
7.2.6 Soma, the Lord of the Mind	RV ix.28.1	पवमानः सोमः	गायत्री
7.2.7 Soma, the Lord of Speech	RV ix.26.1,4	पवमानः सोमः	गायत्री
7.2.8 Flow Indu, the Frog is Eager for Water	RV ix.112.1-4	पवमानः सोमः	पंक्तिः
VIII. The Cosmology of the Ṛgveda: Summary			
8.1 WHAT THE ṚGVEDA REVEALS			
8.1.1 The Universe, the Infinite Ocean of Celestial Hydrogen	–	–	–
8.1.2 A Single Source and a Single Cause	–	–	–
8.1.3 The Creation and Life are the Magneto-Plasma Phenomena	–	–	–
8.1.4 The Twofold Heavenly Vault and Its Substructures are the Magneto-Plasma Structures	–	–	–
8.1.5 The Stone, the Celestial Ioniser, the Cold Fire	–	–	–
8.1.6 Magnetic Field and Magnetic Forces	–	–	–
8.1.7 Māyā	–	–	–
8.1.8 Cosmic Yajñas, the Sacred Cosmic Sacrifices	–	–	–
8.1.9 The Rains of Cosmic Plasmas and the Watchers	–	–	–
8.1.10 The Soma Pond and Plasma Blooms	–	–	–

SECTION	STANZAS	ADDRESSED TO	METRE
8.1.11 The God Particles	–	–	–
8.1.12 By the Hymns of the Singers, Agni is Made into Many Forms	–	–	–
8.1.13 The Milky Way Has Four Major Arms and Is Sustained by Cosmic Yajñas	–	–	–
8.2 COSMIC PLASMA DISCHARGE AND RELEASE MECHANISMS	–	–	–
8.3 HYMNS OF BLESSING			
8.3.1 The Devas Embody Their Wish and Adorn Their Creations	RV x.66.9	विश्वे देवाः	जगती
8.3.2 Powerful Be the Cosmic Yajña	RV x.66.6, 12	विश्वे देवाः	जगती

2. Stanzas Listed by the Ṛgveda Maṇḍalas

Ṛgveda (328 stanzas)

Maṇḍalaṃ I (63 stanzas)					
Stanza	Section	Page	Stanza	Section	Page
RV i.6.3,5	2.5.27	74	RV i.64.8	2.5.21	66-67
RV i.13.10	3.1.1	119	RV i.64.10	2.5.22	67
RV i.20.6	2.4.7	37	RV i.70.3/4	7.1.1	218
RV i.22.16	2.8.5	112	RV i.88.1	2.5.18	64
RV i.23.5	4.2.7	152	RV i.92.1-4	2.5.12	56-58
RV i.24.14-15	4.6.2	167-168	RV i.95.2	2.7.7	96-97
RV i.24.7	2.9	116	RV i.105.1-2	2.5.6	47-48
RV i.28.1-4	2.7.9	98-99	RV i.109.3-4,7	2.7.3	87-89
RV i.28.5-6	2.7.9	99-100	RV i.133.2	2.5.14	59
RV i.28.7-8	2.7.9	100-101	RV i.153.3	4.3.2	156
RV i.28.9	2.7.9	101	RV i.164.33	2.4.3	32
RV i.35.8	2.8.6	113	RV i.164.35	2.2.2	20-21
RV i.46.1-2	6.1.3	205	RV i.164.38	2.4.5	34-35
RV i.46.3-5	6.1.4	205-206	RV i.164.41	2.8.1	109
RV i.46.6-7	6.1.5	207	RV i.164.42	2.8.1	109
RV i.46.14	6.1.6	207-208	RV i.164.46	7.1.7	231-232
RV i.50.2,8	2.5.8	52	RV i.164.48	2.5.28	76
RV i.52.5	7.1.4.3	223	RV i.164.51	4.2.2	146
RV i.52.7	2.5.16	60	RV i.164.52	4.2.2	146-147
RV i.64.5,7	2.5.19	65-66	RV i.191.1-5	2.7.10	102-104

Maṇḍalaṃ II (8 stanzas)			Maṇḍalaṃ III (22 stanzas)		
Stanza	Section	Page	Stanza	Section	Page
RV ii.1.1	7.1.2	218-219	RV iii.8.1-5	2.6.1	78-80
RV ii.27.1	4.3.6	161	RV iii.8.6-10	2.6.2	81-83
RV ii.27.7	4.3.1	155	RV iii.8.11	2.6.3	83
RV ii.28.5	4.5.2	165	RV iii.20.2	7.1.3	219
RV ii.31.4	3.1.3	121	RV iii.26.7	2.4.4	33-34
RV ii.34.2	2.5.5	47	RV iii.33.6	4.4.2	162-163
RV ii.34.5	2.5.18	64-65	RV iii.36.8	2.5.26	73-74
RV ii.41.5	2.5.29	77	RV iii.55.12-13	2.5.4	46
–	–	–	RV iii.56.6	4.4.4	163-164
–	–	–	RV iii.59.1	4.2.8	153
–	–	–	RV iii.59.8-9	4.2.8	153-154
–	–	–	RV iii.62.10	4.4.3	163

Maṇḍalaṃ IV (23 stanzas)			Maṇḍalaṃ V (17 stanzas)		
Stanza	Section	Page	Stanza	Section	Page
RV iv.2.11	4.3.5	160	RV v.3.1-2	4.2.4	149-150
RV iv.4.3	4.8.2	174	RV v.9.5	7.1.4.3	222
RV iv.33.1-3	3.2.1	127-128	RV v.18.1-3	7.1.4.2	220-221
RV iv.33.4-6	3.2.2	128-130	RV v.43.3	2.7.12	107-108
RV iv.33.7-9	3.2.3	130-132	RV v.43.4	2.7.6	96
RV iv.33.10-11	3.2.4	132-133	RV v.58.6	2.4.6	36-37
RV iv.45.1-3	6.1.1	199-200	RV v.60.6	2.2.1	19
RV iv.56.5	2.4.1	28	RV v.62.8	4.3.5	159-160
RV iv.57.1	2.7.11	104-105	RV v.63.4	2.5.10	54
RV iv.57.2-3	2.7.11	105-106	RV v.63.7	2.5.11	55
RV iv.57.4	2.7.11	106	RV v.84.1-3	2.8.7	114-115
RV iv.57.5	2.7.11	106	RV v.85.5	4.9.1	176
RV iv.57.8	2.7.11	107	–	–	–

Maṇḍalaṃ VI (1 stanza)			Maṇḍalaṃ VII (24 stanzas)		
Stanza	Section	Page	Stanza	Section	Page
RV vi.33.3	4.5.1	164-165	RV vii.46.1-4	2.5.17	61-63
–	–	–	RV vii.47.1-4	4.2.3	147-149
–	–	–	RV vii.61.3	4.8.1	173
–	–	–	RV vii.84.2	4.6.1	166-167
–	–	–	RV vii.87.3	4.8.1	173-174
–	–	–	RV vii.87.5	4.2.6	152
–	–	–	RV vii.93.7	4.3.3	157
–	–	–	RV vii.103.1	7.1.6	227
–	–	–	RV vii.103.2-10	7.1.6	227-231
–	–	–	RV vii.104.22	2.5.15	59-60

Maṇḍalaṃ VIII (17 stanzas)			Maṇḍalaṃ IX (21 stanzas)		
Stanza	Section	Page	Stanza	Section	Page
RV viii.7.1	2.5.23	68	RV ix.6.8	4.10.1	178-179
RV viii.25.3	4.3.1	155-156	RV ix.9.4	2.5.8	51
RV viii.32.1,5	2.5.26	73	RV ix.26.1,4	7.2.7	238
RV viii.42.1-2	4.2.5	150-151	RV ix.28.1	7.2.6	237
RV viii.47.13	7.1.4.3	224	RV ix.72.7	7.2.2	233
RV viii.47.14-16	7.1.4.3	224-225	RV ix.73.4,7	4.8.3	175
RV viii.67.20	4.5.2	166	RV ix.86.29	2.8.3	111
RV viii.93.1-3	2.5.25	71-72	RV ix.86.32	7.2.3	234
RV viii.100.10-11	4.7.1	168-169	RV ix.96.5-6	7.2.4	235
RV viii.102.8	3.1.2	120-121	RV ix.96.18-19	7.2.5	236-237
–	–	–	RV ix.97.41	7.2.1	232
–	–	–	RV ix.102.2	2.7.2	86-87
–	–	–	RV ix.112.1-4	7.2.8	239-240
–	–	–	RV ix.113.9	2.2.3	23

Maṇḍalaṃ X (132 stanzas)					
Stanza	Section	Page	Stanza	Section	Page
RV x.10.5	4.4.1	162	RV x.101.2	2.5.13	58
RV x.20.9	7.1.5	226	RV x.101.7	2.7.8	97
RV x.30.2,3,15	4.2.1	144-146	RV x.110.9	3.1.2	120
RV x.34.1,9	4.1.1	136-137	RV x.114.5	7.1.7	231
RV x.34.2-6	4.1.2	138-140	RV x.119.1-13	4.10.2	179-182
RV x.34.7	4.1.1	136-137	RV x.121.1-4	2.5.2	42-44
RV x.37.1	2.5.9	53-54	RV x.125.1,3-8	4.7.2	169-173
RV x.53.9	3.1.4	122	RV x.125.2	4.7.2	170
RV x.58.1-12	4.11	183-186	RV x.128.5	2.8.4	112
RV x.61.4-9	6.1.2	201-204	RV x.129.1-7	5.1	188-191
RV x.66.6	8.3.2	252-253	RV x.130.1-7	5.2	193-195
RV x.66.9	8.3.1	252	RV x.135.1-7	4.1.3	141-144
RV x.66.12	8.3.2	253	RV x.142.7-8	2.5.3	44-45
RV x.72.1-5	4.3.4	157-159	RV x.168.1-4	2.5.24	69-71
RV x.75.1-3	6.2.1	209-210	RV x.170.1-4	2.5.7	49-51
RV x.75.4	6.2.2	210-211	RV x.175.1-4	2.7.4	89-90
RV x.75.5-6	6.2.3	211-212	RV x.177.1	4.9.2	177
RV x.75.7,9	6.2.4	213-214	RV x.177.2-3	4.9.2	177-178
RV x.75.8	6.2.5	214	–	–	–
RV x.78.3	2.5.20	66	–	–	–
RV x.81.1-7	3.1.5	123-125	–	–	–
RV x.88.17	2.4.1	29	–	–	–
RV x.94.1-4	2.7.5	91-93	–	–	–
RV x.94.5	2.7.5	94	–	–	–

Siddhāntaśiromaṇi, Sūryasiddhānta, Kaṭha Upaniṣad (6 stanzas)

Book	Stanza	Section	Page
Siddhāntaśiromaṇi (3 stanzas)	37, 38 Bhuvanakośa (Chapter 3)	6.2.6	215-216
	67, Bhuvanakośa (Chapter 3)	2.3.2	26-27
Sūryasiddhānta (2 stanzas)	55, Bhūgolādhyāya (Chapter 12)	2.4.5	35
	90, Bhūgolādhyāya (Chapter 12)	2.3.1	25
Kaṭha Upaniṣad (1 stanza)	Kaṭhopaniṣad ii.3.1	2.9	116-117

Bibliography

Advaita Ashrama, *The Life of Swami Vivekananda by his Eastern and Western Disciples*, Vol. 2, 6th edition, 3rd reprint, Mayavati, India, 1998.

Advaita Ashrama, *The Complete Works of Vivekananda*, 20th print, Mayavati, India, Vol. 3, 2007.

Agrawala, Vasudeva S., *Sparks from the Vedic Fire: A New Approach to Vedic Symbolism*, Varanasi, Director, School of Vedic Studies, Banaras Hindu University, 1962.

Alfvén, Hannes, 'Plasma Physics, Space Research and the Origin of the Solar System', in *Nobel Lecture, Physics 1963-1970*, Amsterdam, Elsevier Publishing Company, 1972.

Alfvén, Hannes, *Cosmic Plasma*, Dordrecht, Holland, D. Reidel Publishing Company, 1981. Astrophysics and Space Science Library Series 82.

Arkasomayaji, D., *Siddhāntaśiromaṇi of Bhāskarācārya*, 2nd ed., Tirupati, Rashtriya Sanskrit Vidyapeetha, 2000, p. 65 (६५). Rashtriya Sanskrit Vidyapeetha Series, 67.

Arnold, E. Vernon, *Vedic Metre in Its Historical Development*, Cambridge, Cambridge University Press, 1905.

Barnstone, Willis and Marvin Meyer (eds.), *The Gnostic Bible: Gnostic Texts of Mystical Wisdom from the Ancient and Medieval Worlds*, Boston, Shambhala, 2009.

Becker, Robert O. and Gary Selden, *The Body Electric: Electromagnetism and the Foundation of Life*, reprint, New York, William Morrow, 1985.

Benfey, Theodor, *Göttingische gelehrte Anzeigen unter der Aufsicht der Königl. Gesellschaft der Wissenschaften*, 1858, cited in Karen Thomson, 'A Still Undeciphered Text: How the Scientific Approach to the Rigveda Would Open Up Indo-European Studies', *The Journal of Indo-European Studies*, Vol. 37, No. 1 & 2, Spring/Summer 2009a.

Bloomfield, Maurice, *A Vedic Concordance*, Charles Rockwell Lanman (ed.), reprint, Cambridge, Massachusetts, Harvard University, 1906. Harvard Oriental Series Volume 10.

Böhtlingk, Otto and Rudolf Roth, *Sanskrit-Wörterbuch*, St. Petersburg, Kaiserliche Akademie der Wissenschaften, 1855-75, cited in Karen Thomson, 'A Still Undeciphered Text: How the Scientific Approach to the Rigveda Would Open up Indo-European Studies', *The Journal of Indo-European Studies*, Vol. 37, No. 1 & 2, Spring/Summer, 2009a.

Carretti, Ettore et al., 'Giant Magnetized Outflows from the Centre of the Milky Way', *Nature* 493, 03 January 2013, pp. 66-69. doi:10.1038/nature11734.

CERN, *Physics*, [website] https://home.cern/science/physics (accessed 30 January 2020).

Colebrooke, H.T., *Kosha or Dictionary of the Sanskrit Language by Umura Singha with an English Interpretation and Annotations*, Calcutta, Haragobinda Rakshit, 1891.

Copenhaver, Brian P., *Hermetica: The Greek Corpus Hermeticum and the Latin Asclepius in a New English Translation with Notes and Introduction*, Cambridge, Cambridge University Press, 1992.

Cremo, Michael A. and Richard L. Thompson, *Forbidden Archeology: The Hidden History of the Human Race*,1st edition, revised, Los Angeles, Bhaktivedanta Book Publishing, Inc., 1998.

Dasgupta, Surendranath, *A History of Indian Philosophy*, reprint, Vol. I, London, Cambridge University Press,1957.

Davis, Albert Roy and Walter C. Rawls, Jr., *The Magnetic Blueprint of Life*, New York, Exposition Press, 1979.

Davis, Albert Roy and Walter C. Rawls, Jr., *Magnetism and its Effects on the Living System*, 12th printing, Metairie, Louisiana: Acres U.S.A., 1996.

Deshpande, Madhav M., *A Sanskrit Primer*, Ann Arbor, Michigan, The University of Michigan Center for South and Southeast Asian Studies, 2003. Michigan Papers on South and Southeast Asia No. 47.

Emiliani, Cesare, *Planet Earth: Cosmology, Geology, and Evolution of Life and Environment*, reprint, New York, Cambridge University Press, 1992.

Filliozat, Pierre-Sylvain, 'Ancient Sanskrit Mathematics: An Oral Tradition and a Written Literature', in Karine Chemla (ed.), *History of Science, History of Text*, Dordrecht, Netherlands, Springer, 2004. Boston Studies in the Philosophy of Science Vol. 238.

Fishman, Gerald J. and Dieter H. Hartmann, 'Gamma-Ray Bursts: New Observations Illuminate the Most Powerful Explosions in the Universe', *Scientific American*, July 1997.

From Colombo to Almora: Being a Record of Swami Vivekānanda's Return to India after His Mission to the West, 2nd ed., Madras, Brahmavadin Press, 1904.

Gangooly, P. (ed.), *Translation of the Sūrya-siddhānta: a Textbook of Hindu Astronomy with Notes and an Appendix by Rev. Ebenezer Burgess*, reprint, Calcutta: The University of Calcutta, 1935.

Griffith, Ralph T.H. (trans.), *The Hymns of the Rigveda: Translated with a Popular Commentary*, 2nd ed. (complete in two volumes), Vol. I and Vol. II, Benares, E.J. Lazarus and Co., 1896 and 1897.

Hauck, Dennis William, *The Complete Idiot's Guide to Alchemy*, New York, Alpha Books, Penguin Group (USA) Inc., 2008.

Holland, Gary B., Professor, University of California, Berkeley. Email response to the author, 10 May 2013. Personal communication.

Jamison, Stephanie, *The Ravenous Hyenas and the Wounded Sun: Myth and Ritual in Ancient India*, Ithaca, Cornell University Press, 1991, cited in Karen Thomson, 'A Still Undeciphered Text: How the Scientific Approach to the Rigveda Would Open Up Indo-European Studies', *The Journal of Indo-European Studies*, Vol. 37, No. 1 & 2, Spring/Summer, 2009a.

Jamison, Stephanie W. and Joel P. Brereton (trans.), *The Rigveda: The Earliest Religious Poetry of India*, Vols. I-III, New York, Oxford University Press, 2014.

Linden, Stanton J. (ed.), 'Zosimos of Panopolis (fl. c. 300AD): Of Virtue, Lessons 1-3', *The Alchemy Reader: From Hermes Trismegistus to Isaac Newton*, 6th printing, New York: Cambridge University Press, 2014.

Macdonell, A. A., *Vedic Mythology*, Strassburg, Verlag von Karl J. Trübner, 1897.

Macdonell, A. A., *Vedic Grammar*, Strassburg, Verlag von Karl J. Trübner, 1910.

Macdonell, A. A., *A Vedic Grammar for Students*, Oxford, The Clarendon Press, 1916.

Macdonell, A. A., *A Vedic Reader for Students*, Oxford, The Clarendon Press, 1917

Martin, Thomas C., *The Inventions Researches and Writings of Nikola Tesla*, 2nd ed., New York, The Electrical Engineer, 1894.

McEnery, Julie, Fermi Project Scientist, 9 November 2010 [website] https://www.nasa.gov/mission_pages/GLAST/news/new-structure-briefing.html (accessed 5 February 2020).

Meyer, Marvin (ed.), *The Nag Hammadi Scriptures: The Revised and Updated Translation of Sacred Gnostic Texts Complete in One Volume*, New York, HarperOne, 2007.

Monier-Williams, Monier, *A Sanskrit-English Dictionary: Etymologically and Philologically Arranged with Special Reference to Cognate Indo-European Languages*, Oxford, The Clarendon Press, 1899, pp. ix-x. [online] https://www.sanskrit-lexicon.uni-koeln.de/scans/MWScan/2014/web/webtc/indexcaller.php.

Müller, F. Max, *A History of Ancient Sanskrit Literature*, 2nd ed., revised, London, Williams and Norgate, 1860.

Müller, F. Max (ed.), *Hymns of the Rig-Veda in the Samhita & Pada Texts*, 2nd ed., Vol. I and Vol. II, London, Trübner and Co., 1877.

Müller, F. Max (ed.), *Rig-Veda-Saṃhitā: The Sacred Hymns of the Brāhmans together with the Commentary of Sāyaṇācārya*, 2nd edition, Vols. I-IV, Oxford, The Clarendon Press. 1890-1892.

Müller, F. Max (trans.), *Vedic Hymns Part I: Hymns to the Maruts, Rudra, Vāyu, and Vāta*, Oxford, The Clarendon Press, 1891. The Sacred Books of the East Series, Vol. XXXII.

Müller, F. Max, *India: What Can It Teach Us? A Course of Lectures Delivered before the University of Cambridge*, new edition, London, Longmans, Green, And Co., 1892.

Müller, F. Max, *The Six Systems of Indian Philosophy*, new impression, London, Longmans, Green and Co., 1919.

NASA Chandra X-Ray Observatory, *Discovering the X-Ray Universe: Colossal Clouds of Hot Gas*, http://chandra.harvard.edu/xray_astro/discover.html (accessed 25 June 2019).

NASA Chandra X-Ray Observatory, *Pictor A: Blast from Black Hole in a Galaxy Far, Far Away*, http://chandra.si.edu/photo/2016/pictora/ (accessed 26 June 2019).

NASA Chandra X-Ray Observatory, *X-Ray & Optical Images of Planetary Nebulas*, http://chandra.harvard.edu/photo/2012/pne/pne.jpg (accessed 25 June 2019).

NASA/ESA Hubble Space Telescope, in the image description of *A Cosmic Garden Sprinkler*, https://www.spacetelescope.org/images/potw1241a/ (accessed 25 June 2019).

NASA/ESA Hubble Space Telescope, *The Hourglass Nebula*, http://www.spacetelescope.org/images/opo9607a/ (accessed 26 June 2019).

NASA/ESA Hubble Space Telescope, *Open Cluster NGC 290: Gemstones in the Southern Sky*, https://www.nasa.gov/multimedia/imagegallery/image_feature_550.html (accessed 25 June 2019).

NASA/ESA Hubble Space Telescope, *Southern Crab Nebula*, http://www.spacetelescope.org/images/opo9932b/ (accessed 26 June 2016).

NASA/Jet Propulsion Laboratory, *A Road Map to the Milky Way (Annotated)*, http://www.spitzer.caltech.edu/images/1925-ssc2008-10b-A-Roadmap-to-the-Milky-Way-Annotated (accessed 26 June 2019).

Nikola Tesla Museum, *Nikola Tesla: Colorado Springs Notes: 1899-1900*, Beograd, Yugoslavia, Nolit, 1978.

O'Flaherty, Wendy Doniger, *The Rig Veda: An Anthology*, Harmondsworth, Penguin Books, 1981.

Particle Data Group of Lawrence Berkeley National Laboratory, 'The Standard Model: The Theory of Fundamental Particles and Forces', *The Particle Adventure*, 2014. [website] https://www.particleadventure.org/standard-model.html (accessed on 31 January 2020).

Peratt, Anthony L., *Physics of the Plasma Universe*, 2nd ed., New York, Springer, 2015.

Sachau, Edward C. (ed.), *Alberuni's India: An Account of the Religion, Philosophy, Literature, Geography, Chronology, Astronomy, Customs, Laws and Astrology of India about A.D. 1030*, Volumes I & II (Bound in One), reprint, Delhi, Low Price Publications, Vol. I, 2011.

Sarup, Lakshman, *The Nighaṇṭu and the Nirukta of Śrī Yāskācārya: The Oldest Indian Treatise on Etymology, Philology, and Semantics*, reprint, Delhi, Motilal Banarsidass Publishers Private Limited, 2009.

Sāstri, Bāpū Deva (ed., trans.), *Translation of the Sūryasiddhānta by Bāpū Deva Sāstri, and of the Siddhānta Śiromani by the late Lancelot Wilkinson*, Calcutta, Baptist Mission Press, 1861.

Speeches and Writings of Swami Vivekananda: A Comprehensive Collection with Four Portraits, 3rd ed., Madras, G.A. Natesan & Co., 1899.

Sri Aurobindo Ashram Trust, *Sri Aurobindo Secret of the Veda*, Twin Lakes, WI, Lotus Press, 1995.

Sri Aurobindo Ashram Trust, *Sri Aurobindo Hymns to the Mystic Fire with Sanskrit Text, English Translation and Commentary*, Twin Lakes, WI, Lotus Light Publications, 1996.

Sri Yukteswar Giri, *The Holy Science*, 8th ed., Los Angeles, Self-Realization Fellowship, 1990.

Su, Meng and Douglas P. Finkbeiner, 'Evidence for Gama-Ray Jets in the Milky Way', *The Astrophysical Journal* 753:61(13pp), 01 July 2012. doi:10.1088/0004-637X/753/1/61.

Thompson, Richard L., *Vedic Cosmography and Astronomy*, 3rd printing, Delhi, Motilal Banarsidass, 1996.

Thomson, Karen, 'Sacred Mysteries: Why the *Rigveda* Has Resisted Decipherment', *Times Literary Supplement*, 26 March 2004, pp. 14-15.

Thomson, Karen and Jonathan Slocum, *Ancient Sanskrit online*, [A ten-class course on the language of the Rigveda], Linguistic Research Center, The University of Texas at Austin, 2006. [website] https://lrc.la.utexas.edu/eieol/vedol .

Thomson, Karen, 'A Still Undeciphered Text: How the Scientific Approach to the Rigveda Would Open Up Indo-European Studies', *The Journal of Indo-European Studies*, Vol. 37, No. 1 & 2, Spring/Summer 2009a, pp. 4, 11.

Thomson, Karen, 'A Still Undeciphered Text, continued: the reply to my critics', *The Journal of Indo-European Studies*, Vol. 37, No. 1 & 2, Spring/Summer 2009b, pp. 78-79.

Thomson, Karen, 'The Plight of the Rigveda in the Twenty-First Century', *The Journal of Indo-European Studies*, Vol. 38, No. 3 & 4, Fall/Winter 2010, p. 423.

Thomson, Karen, 'Speak for itself: How the long history of guesswork and commentary on a unique corpus of poetry has rendered it incomprehensible', *Times Literary Supplement*, 8 January 2016, pp. 3, 4.

Van Nooten, Barend A. and Gary B. Holland (ed.), *Rig Veda: A Metrically Restored Text with an Introduction and Notes*, Department of Sanskrit and Indian Studies, Harvard University, 1994. Harvard Oriental Series, 50.

Wallis, H. W., *The Cosmology of the Rigveda: An Essay*, London, Williams and Norgate, 1887.

Whitney, William D., *The Roots, Verb-Forms and Primary Derivatives of the Sanskrit Language*, Leipzig, Breitkopf and Härtel, 1885.

Whitney, William D., *Sanskrit Grammar: Including Both the Classical Language and the Older Dialects of Veda and Brahmana*, Delhi, Motilal Banarsidass, 1888.

Wikimedia, Bradford Industrial Museum, *Spinning Machine at Bradford Industrial Museum, Bradford, West Yorkshire, England. Understood to be a Noble Comb*, [website] https://commons.wikimedia.org/wiki/Category:Bradford_Industrial_Museum#/media/File:Bradford_Industrial_Museum_073.jpg (Accessed 2 November 2018).

Wilson, H. H. (trans.), *Ṛig-Veda Sanhitā: A Collection of Ancient Hindu Hymns Constituting the First Ashṭaka or Book of the Ṛig-Veda; the Oldest Authority for the Religious and Social Institutions of the Hindus*, 2nd ed., London, N. Trübner and Co., 1866.

Wilson, H. H. (trans.), *Ṛig-Veda Sanhitā: A Collection of Ancient Hindu Hymns Constituting the Second Ashṭaka or Book of the Ṛig-Veda; the Oldest Authority for the Religious and Social Institutions of the Hindus*, Wm. H. Allen and Co., 1854.

Wilson, H. H. (trans.), *Rig-Veda Sanhitā: A Collection of Ancient Hindu Hymns Constituting the Third and Fourth Ashṭakas or Books of the Rig-Veda; the Oldest Authority for the Religious and Social Institutions of the Hindus*, London, Wm. H. Allen and Co., 1857.

Wilson, H.H. (trans.), *Rig-Veda Sanhitā: A Collection of Ancient Hindu Hymns Constituting the Fifth Ashṭaka or Book of the Rig-Veda; the Oldest Authority for the Religious and Social Institutions of the Hindus,* E.B. Cowell (ed.), London, N. Trübner and Co., 1866.

Wilson, H. H. (trans.), *Ṛig-Veda Sanhitā: A Collection of Ancient Hindu Hymns Constituting the Sixth and Part of the Seventh Ashṭaka of the Ṛig-Veda,* E.B. Cowell and W.F. Webster (ed.), London, Trübner & Co., 1888.

Wilson, H. H. (trans.), *Ṛig-Veda Sanhitā: A Collection of Ancient Hindu Hymns Constituting Part of the Seventh and the Eighth Ashṭaka of the Ṛig-Veda*, edited by W.F. Webster, London, Trübner & Co., 1888.

Winternitz, M., *A History of Indian Literature: Introduction, Veda, Epics, Purāṇas, and Tantras*, S. Ketkar (trans.), revised by the author, Calcutta, The University of Calcutta, Vol. I, 1927.

Witzel, Michael, 'The Development of the Vedic Canon and its Schools: The Social and Political Milieu', in Michael Witzel (ed.), *Inside the Texts, Beyond the Texts*, Cambridge, MA, Harvard University, 1997. Harvard Oriental Series, Opera Minora 2.

Index

Page numbers provided for each subject are not exhaustive. Pages listed are mostly the ones that provide substantial information for the subject. Transliterated Sanskrit words are placed roughly following the alphabetical order of English. The letter 'f' that accompanies the page number denotes that the word is included in the footnote on that page; the letter 'g' denotes that the word is in the grammatical analyses on that page.

A

abhakta (abhaktaḥ) the freed One 87

Adhvaryu(s) 2, 92, 107, 144-145, 229; Adhvaryus make the sweet Soma mead 107; Adhvaryus press out celestial Waters 144-145; Adhvaryu's three assistants are Pratiprasthātā, Neṣṭā, Unnetā 107

Ādi Brahma Samāj xvi; congratulatory note to Müller xvi

Aditi 81, 154-160, 238; Aditi milch Cow 154, 156; daughter of Dakṣa 157-159; double headed 154; mother of all created beings 154; mother of Kings Mitra and Varuṇa 155; produces inexhaustible cosmic plasmas and discharge them 154; unbinds the sin of the ionic bond 156-157

Ādityas 81-82, 154-155, 159-161, 223; the chief of the Ādityas is Varuṇa 160; sons of Aditi 81, 160; *seven Ādityas: Varuṇa, Mitra, Aryamā, Bhaga, Dakṣa, and Aṃśa, Savitā 160

Adri (Soma pressing Stone) 95

Adṛśyadṛśyakagiriṃ 27

aftergrowth(s) 49, 62-63, 112g, 204g, 247

Agni (proton) 33, 44, 92, 204, 217-219, 226; Agni's paths (black, white, yellow, red, purple, glorious) 226; born of the fire sacrifice 219; comes out of the Stone 218; dry embryonic particle 12, 217; forms are based on the design of Agnis 120; magnificent One 120; gamester 135-137; god particle(s) 217, 248; lord of the house 220; lord of light 217; lord of men 218; shaped into many forms by the Speeches of the wise singers 231, 249; Trita Āptya 220, 223-225; *three forms or bodies or fires of Agni 46, 84, 238; Ekata (protium) 34, 46, 85, 95, 219, 249; Dvita (deuterium) 34, 46, 85, 95, 219-223, 225, 249; Trita (tritium) 34, 46, 74, 85-87, 92, 204, 219-225, 238, 246, 249

Agni Yajña 120, 231g, 246; see also 'fire sacrifice'

Agrawala, Vasudeva S. xvii

Ahi (serpent of the Heaven) 40, 71-72

Ajā 111

Alberuni xv, 6

alchemical text(s) xxix, 52, 137

alchemy xxv-xxvi, 53f, 137f, 256; celestial Alchemy xxvi, 256

Alfvén, Hannes xix, 255-256

All-Seeing Eye 51-52

Altar of creation 40, 243, 251

Altar of cosmic Yajña 18-21, 40, 251

Amarakoṣa (a Sanskrit dictionary) xviii

aṃbha (hydrogen, prima materia) xxiii-xxiv 35, 182, 187-188, 242

Aṃśa (cosmic force associated with Agni) 160-161
ancestors 66; see also 'forefathers'
Ancient High Indian (language) 8
androgyne 102
angel(s) see 'Aṅgirā(s)'
Aṅgirā(s) 21, 55, 104, 127, 223; angel(s) 21, 41; warring angels 21; see also 'Messenger(s) of the Heaven' and 'Messenger(s) of celestial lights'
annaṃ (mystical food) see 'Soma (electron)'
anthropomorphic language 10
anudātta see 'Vedic accents'
Anuṣṭup see 'metres, metrical structures'
Āpaḥ xxiii, 36, 144, 147, 247; see also 'celestial Water(s)', 'cosmic plasma(s)'
Āraṇyaka(s) xxv, 1
archers 64, 67
Arjunī 114
arms two arms 44, 95-96; two winged arms124
Arm(s) of the Milky Way see 'Milky Way, the'
Arnold, Vernon xvii, 192
arrow(s) 53-54, 67, 76, 143, 215, 233; Agni's arrows 54, 77-78, 143, 215, 247; archers hold arrows 64, 67; bright arrows 21, 54, 80; launched arrows 62; Rudra's arrows 40, 171, 215; swiftly flying arrows 61; three hundred and sixty arrows or spears 76; Vivasvān's arrows 165-166
Āryan(s) xxii, 4
Ārya and Dāsa 4, 164, 176
Aryamā 12, 149-150, 155-157, 160-161, 196, 245-246
as above, so below xxv, 18
Asat (the state of non-Real, non-Existence,) 157, 187-188
Aśvattha (inverted tree, Tree of Life) 115-117, 116 (Fig. 11)
Aśvinā (du.) 36, 71, 76, 88, 169, 197-201, 205-207; Aśvinā's chariot flies with winged coursers 205; Aśvinā's journey thrice a day 76; circumambulatory paths of Aśvinā 207-208; coursers of Aśvinā rush over the blazing summit of the Heaven 205; perform wondrous deeds 198, 205; two black sons of the Heaven 201
Atharvaveda xxii, xxv, 1-2, 8; Atharvaveda Saṃhitā 1; the Veda of Atharvan 2
Atithi 220
Ātmā (soul) xxiv, 33-34, 70, 178-179, 182, 242; Ātmā the soul of Yajña 178; ātmā individuated soul 34, 178; soliloquy of Ātmā 179-182
atomic structure(s) 161, 217, 222; a stronghold and a cave 74; bound in 138; breaks, destroys, burn down 94-95, 136-137, 220, 222, 244-245; cut asunder 106, 122; forming the shell of 232; freed from, not bound 69, 130g, 155; pressing out from 89; the constitution of 231, 249
atoned for their sin 93
atonement 93, 140; atonement for the sin 93; hall of atonement 140; Place of the Atoned 93; Place of Atonement 93
Atri (devourer) 18, 36, 114, 196, 220, 222
Aurobindo, Sri xvii
Avantī (and its inhabitants) 61-62
Avesta 257
axe 78; double-headed, single-headed axe 78; axe wrought of gold 78; see also 'hatchet (heavenly)'

B
battle see 'cosmic battle'
bearers of seeds 190; see also 'aṃbha (hydrogen, prima materia)' and 'water-bearer'
Becker, Robert O. xix
Benfey, Theodor 6
Bennu 93
Bhaga 121, 160-161, 170; dispenser, name of Rudra 160
Bharadvāja (the bearer of speed) 195-196
Bhāskara 23, 27
Bhūmi 65, 114
bhūtasaṃkhyā system 25, 27
Bhuvanaṃ xxv, 20, 41, 122
big bang theory xxiii, xxv, xxix
Bird, ruddy 145
Birkeland currents 233; see also 'filament(s)'
Birkeland, Kristian xix
Bloomfield, Maurice xvii
blooming plasma(s) 44, 51, 96, 102, 175, 205, 234, 245
boat, celestial 39, 58, 142, 207; boat equipped with oars 39, 58; boat of hymns 207; boat of sacrifice 58; hundred-oared boat 58; winged-boat 58

brahma (n. nom. sg. of ब्रह्मन् brahman) 117; prayer 79, 229; exhilarating hymn 60; sacred Speech 203
Brahmā (m. nom. sg. of ब्रह्मन्) 21, 117, 171, 239, 243, 249; Brahmā Creator 21; Brahmā's Egg, Egg of Brahmā 24-27; Brahmā of the Devas 70, 234-235; the Speech and the supreme Vault of the Heaven 21
Brāhmaṇaṃ (Brāhmaṇas pl.), the Vedic literature xxiii, 1
Brāhmaṇa(s), priest xv-xvi, 9, 26, 27g, 92, 157-158, 226-227, 229
Brahmaṇaspati blasts and smelts hydrogen 157-158; cuts the ionic bond asunder 78; lord of the prayer 122, 157; the name of the Stone deified 122
Brahmāṇḍa (Egg of Brahmā) 24, 26, 52
Brahmāṇḍasampuṭa (the hemispherical bowl-shaped Egg of Brahmā) 25
Brereton, Joel P. 9
Bṛhaspati (lord of the prayer, the Stone) 194
Bṛhatī see 'metres, metrical structures'
burning bush (on the mountain of God) 41

C

camvā (du. two celestial Bowls) 18
Carretti, Ettore 17f, 30
cattle 43, 72-73, 169, 201, 236, 238; abode for cattle 39; herd(s) of cattle 169, 211, 224, 238; herd of ruddy cattle 201
cauldron(s) 26-27, 33, 117, 229; Hemispherical Cauldron Bowl 117; inexhaustible cauldron 33; red-hot cauldrons 230
celestial being(s) xxiv, 1, 11-12, 15-16, 191, 242, 245
celestial boat see 'boat, celestial'
celestial (cosmic) ionisation xxiii-xxiv, 3, 30, 39, 95, 102, 135, 150, 152, 242; cosmic ionisation violent process 137; cycle of celestial ionisation 128
celestial ioniser see 'Stone, the'
celestial jewel(s) or gemstone(s) 15-16, 113; aquatic jewels 113
celestial people 93; people of cosmic plasmas 77, 153, 175, 233-234, 247, 249
celestial river(s) 24, 36, 56, 130, 144, 197, 208-209, 211; Ārjīkīyā 212; Asiknī 208, 212; flow upwards into, flow out 208; Gaṅgā 208, 212; Kubhā 212; Marudvṛdhā 208, 212; Mehatnū 212; Paruṣṇī 208, 211-212; Rasā 208, 212; Sarasvatī 208, 212; set out triply 209; Sindhu 209, 212; Susartu 212; Suṣomā 208, 212; Śutudrī 211-212; Śvetī 212; Tṛṣṭāmā 212; Vitastā 212; Yamunā 208, 212; see also 'Sindhu'
celestial sphere (of astronomy) 20
celestial Water(s) xxiii, 22, 36, 44, 47-48, 75, 109, 130, 144-148, 162-163, 184-185, 205-209, 214, 217-218, 225-228, 232-233, 247; aquatic jewels 113; celestial Waters Devīs 147-148; Ocean of 44-46; living Waters 209; refreshing draught of 207; waves of 109; see also 'cosmic plasma(s)'
Chanda 3, 194-195; study of the patterns of Speech 3; mighty Chanda 195
chandas (pl.) metres, Speeches 191-192
Chandramā 39; see also 'Moon'
Chandra X-Ray Observatory xix, 15, 29, 38, 39; Chandra images 16 (Fig. 1), 31 (Fig. 5), 41 (Fig. 7), 85 (Fig. 8)
charcoals (heavenly) cold charcoals 137, 244; shatter the ionic bonds and burn down the atomic structures 137
charioteer(s) 163, 195-196, 198; charioteer of one's soul 163; the two charioteers 198
chariot(s) 51-52, 55, 69, 121, 142, 159, 180; beautifully running chariot 131; chariot of Aśvinā 198-199; chariot laden with riches 97; chariots of fire 37; chariot of the Devas xxx, 33; majestic chariot of Vāta 69; lightning laden chariots 63, 64; Maruts' chariot 36-37; one-poled chariot 142; refulgent chariot 55; sound of the majestic chariot reverberates 69; Sūrya's chariot 51-52; wheelless chariot 142
Christ identified with the Greek word for fish 165-166; the anointed one 165
circumambulatory paths of Aśvinā 198, 207-208
classical Sanskrit 8-9
coals see 'charcoals (heavenly)'
cold fire 137, 244
Colebrooke, H.T. xviii(f)
colour(s) 75, 94, 201, 226; dark-blue 75, 94, 201; royal blue 94; black 75, 94; black teeth 196; black, white, yellow, red, purple, and glorious 226; two black sons of the Heaven 201
comb(s) 84, 101-103; real comb 102; see also 'Stone, the'
commentaries on the Ṛgveda xvi, 7

constellations see 'cosmic plasma blooms' and 'Nakṣatras of the Heaven'
cosmic battle(s) 21, 82
cosmic creation xxi-xxvi, 150, 187, 241-243; field of cosmic creation 18-19, 254; knowledge of cosmic creation xxi-xxii, xxv; robust and complete creation model xxvi, 256; Whence comes this creation? 190; see also 'wheel of cosmic creation' and 'Womb of cosmic creation'
cosmic field 89, 231, 245; see also 'Altar of cosmic Yajña' and 'field of cosmic creation'
cosmic force(s) xxiii-xxiv, 68, 160, 182, 242
cosmic ionisation see 'celestial ionisation'
cosmic law xxv, 251-252; divine law 55; law of Nature xxv, 256
cosmic Mind xxiv, 89, 182-183, 188-189, 242
cosmic plasma(s) xxii-xxvi, 15, 21, 25, 39-42, 45-47, 60-61, 71-72, 93, 144, 146; discharged and released three times a day 19, 72, 212, 246; make Speeches 60, 90, 191, 237; noisy 237; most rapid of the rapids 212-213; pathways of cosmic plasma flow 56; sing hymns 237; streaming cosmic plasmas 54, 127, 213; see also 'celestial Waters'
cosmic plasma blooms 18, 22, 31, 34, 36-37, 40-41, 43-44, 49, 54, 62-63, 122, 199, 201-202, 233-234, 239-240, 247; constellations of cosmic plasma blooms 51
cosmic plasma discharge plasma dark discharge 12, 22, 75, 94, 149, 160, 196, 201, 208, see 'Varuṇa'; plasma light discharge 12, 44, 149, 153, 160, 196, 245-246, see 'Mitra'; plasma fire discharge 12, 149, 155, 160, 196, 247, see also 'Aryamā'
cosmic plasma discharge and release mechanisms 250-251
cosmic plasma(s) discharged into the Soma Pond (or the Samudra) 32, 44, 102, 111-112, 160, 182, 238, 246-247
cosmic plasma(s) released from the Soma Pond 70, 112, 198, 251; downward release 36, 93, 214, 246; shoot upwards 36; upward release 246;
cosmic sacrifice see 'cosmic Yajña'
cosmic Yajña(s) xxiv-xxv, 18-21, 39, 55, 121 152, 188, 202, 237, 242-243, 249; Altar of cosmic Yajña 18-21, 251; cosmic Yajña Origin of the World and Life 20, 21; field of cosmic Yajña 19; unbroken chain of cosmic Yajñas xxii, 30, 42, 55, 152; see also 'Agni Yajña' and 'Soma Yajña'
cosmology of the Ṛgveda, magneto-plasma xxii, xxv-xxvi, 241; magneto-plasma universe 217, 248
Cowell, E.B. 7
cows 56, 74, 103, 230; cowshed 57, 103 dappled cows 36-37, 111, 201, 203-204; red cows 56; ruddy cows medicine 89-90; ruddy cows made Speech 138,140; stalls of cattle 40
creation see 'cosmic creation'
Creator's marvellous machine xxiii, 30, 38 (Fig. 6), 242, 250 (Fig. 13)
cultivators of the field 81

D

Dakṣa (cosmic force associated with Soma) 157-161
Ḍamaru see 'Śiva'
Darbha grass see 'grass(es)'
Dāsa and Ārya 4, 164, 176
Dasgupta, Surendranath 8
daughter milch Cow see 'milch cow'
Davis, Albert Roy xix, 30
Dawn (uṣāḥ, sg.) 55, 57, 185, 205, 208, 225
Dawns (uṣasaḥ, pl.) 55-57, 74, 223
Day (Mitra) 189
deer 94, 136; tawny deer 67, 136-137
deuterium see 'Agni (proton)'
Deva(s) xxiv, 1, 15, 33, 35-37, 45, 153, 190; even the Devas came after the emergence of the creation 190; 'Of the Devas' and 'Of the Opposite of the Devas' 35-36; origins of the Devas 157-158
Devī(s) xxiv, 45, 94, 122, 147, 168-169; Devī Iḷā 45,121-122; Devī Kālī 94; Devī Vāk 169; mother milch Cow 45; see also 'Sarasvatī' and 'celestial river(s)'
Dhenu see milch cow
Dhiṣaṇe (du.) 18, 28; sujanmanī dhiṣaṇe (two bowls of noble birth) 28
Dīrghatamā (the longest One) 21-22, 156
Diti 159-160
divine architect(s) xxiii, 38, 48, 51, 119-133, 242, see also 'Viśvakarmā'
dominions of a galactic system 251; see also 'altar of cosmic Yajña' and 'field of cosmic creation'
double layer 164, see 'plasma sheath(s)'
dragons see 'serpent(s) of the Heaven'

droṇāni (vats) 40
Duhitā see 'milch Cow'
Dura another name for the Stone 61-62
Dūrvā grass see 'grass(es)'
Dvāpara Yugaṃ see 'Yuga cycle'
Dvīpas 108, 113; seven Dvīpas 113
Dvita (deuterium) see 'Agni (proton)'

E
Egg of Brahmā see 'Brahmāṇḍa (Egg of Brahmā)'
Ekata (protium) see 'Agni (proton)'
electricity, electric xix(f), 89, 113, 245; are subsequent to the magnetic field 89
electron see 'Soma (electron)'
embryonic particle see 'Agni (proton)', 'Soma (electron)'
eternal flames 41, 60, 64, 67, 81, 249
exerting force 188, 190, 245
Existence (Real) see 'Sat (Real, the state of Existence)'
Exodus (3:1-3) 41

F
fasces see 'Roman fasces'
Fermi Bubbles 16-17, 30, 35
field of cosmic creation 18-19, 251, 256; see also 'Altar of cosmic Yajña'
filament(s) filament(s) of Soma Plant(s) 40, 95-96, 166, 233; a hundred purifying filaments 148; magnetic field-aligned filamentary structures 39, 130; magnetic field-aligned triple filament 199; triply-spun (braided) filament 56, 95-96, 96, 166, 199, 233-234, 247
Filliozat, Pierre-Sylvain xvi(f)
fingers see 'ten fingers'
Finkbeiner, Douglas P. 17f, 24, 26-27, 199, 243
firebird 59, 93, 144
fire sacrifice 2, 120, 138, 156, 219, 246
firestones see 'Stone, the'
fish fish caught in the net 165; king of fish 165
Fishman, Gerald J. 76
flaming locks 51-53
flaming swords 233
fleece(s) golden fleece 21, 40, 214, 234; sheep's fleece(s) 40, 214
flying horse see 'horse(s)'
forest(s) 1, 40, 65, 211, 218, 247; forests-hermits 1; forest texts 1; forests of Soma plants 78, 128, 218;
Forest, the 39, 78, 124; Goddess of the Forest (the Soma Pond) 39; Lord of the Forest (the Soma Pond) 39, 78-79, 83; Lord of the Forest (the Stone) 84, 100
forefathers xxiv, 10, 66, 87-88, 183, 193, 195
fountain of sacrifice 165
freed One (abhaktaḥ) 87
Freund, Peter xviii-xix, 12
Frog(s) 226-230, 238, 240; capture one another based on their Speech Patterns 226, 228; frogs bestow riches, cows 230; modulate their Speech 229
Funderburk, Jim xix
fundamental building blocks (of all beings) 122, 123g
funeral gift 141, 143

G
Galactic Bar 36, 93, 108, 110, 113, 197, 208, 215, 246, 249; Place of the Atoned 93; see also 'Mt. Meru'
galactic structure, system(s) xxiii-xxv, 17, 19-20, 24, 38, 41, 55, 86, 150, 116, 241-244, 247, 250 (Fig. 13), 251
Gaṅgā 24, 36, 109-110, 197, 208, 212, 214-215; divides itself into four parts 215; falls upon Mt. Meru 214; two branches of Gaṅgā 197; see also 'celestial river(s)'
Gangooly, P. 23, 25f, 108f
Garuḍa beautifully winged bird 231; great Bird 75; mythical firebird 59, 93, 144
Gāyatrī see 'metres, metrical structures'
gāyatrī stanza 163
Genesis 2:9 136
gnostic texts, fragmentary xxvi, xxvi(f)
Gnosticism xxv
god particles 217, 248
golden fleece see 'fleece(s)'
golden reed 40
golden womb 39, 42
Gomatī and Krumu 211-212; see also 'Sindhu'
goose, ruddy 59
grain 72, 240; hydrogen atoms 72, 240g
grass(es) 40, 44-45, 102; Darbha 40,102-103; Dūrvā 44-45; Kuśara 40, 102-103; Muñja 40, 102-103; sacrificial grass(es) 60, 63; Śara 22, 40, 102-103; Sairya 102-103; Vīraṇa 102-103

gravity 17, 126, 245; interactive exertion of magnetic forces ... 245
great work of Nature xxvi, 256
Griffith, Ralph xvii, 4, 12, 198

H
Hartmann, Dieter H. 76
hatchet (heavenly) 78, 80-81, 122; fiery hatchet 83; hatchet wrought of gold 78, 122
hearth 61, 160
Heaven, the xxiv-xxv, xxx, 18-21, 31-32, 36-38, 77-78, 198-199, 201-203, 205, 207; complete structural component of the Heaven 37-38 (Fig. 6); Heaven the Red Bull 36; Lower Heaven 18-20, 28, 37; Lower Vault of the Heaven revolve eastward 34-36; Midheaven (the Innerfield) 17-20, 37; threefold Heaven 18-23; Upper Heaven 18-20, 28, 37; Upper Vault of the Heaven revolves westward 34-36;
Heavenly Sphere(s) 18-20, 28, 33-34, 40, 56 114, 199-200, 212-213; different from the celestial sphere of astronomy 20
Heavenly Vault, the 20, 21, 23-27, 30, 35, 38 (Fig. 6), 52, 56, 71, 75, 86, 198, 212-214, 234, see also 'Vault of the Heaven'; bowl-shaped (Heaven, Heavenly Vault) 18, 30; different stages of structural development 29; eastern region of the Heavenly Vault 56; Heavenly Vault Brahmā 21; threefold Heavenly Vault 19, 22-23; twofold Heavenly Vault xxiii-xxiv, 28-30, 34-38, 42, 94, 101, 241-242
height of the Heavenly Vault 23-27, 243
hermaphrodite 102
hermetic texts, fragmentary xxvi, xxvi(f)
Hermetica (a book title) xxvi(f)
hermetically sealed mind xxix
Hermeticism xxv, 256
Holland, Gary B. xvii, xix, 13(f), 192, 192f
horns exalted horns of oxen 63; horns of Altar 40; horns of bulls 40, 82, 247
horse(s) 64, 72-73, 97, 100, 126, 213-214; dark horse 36; flying horse pataṃga 177; swift horse 126; winged horse 64; see also 'steed(s)'
hot gas xix, xxix-xxx, 17(f), 256
Hotā (sacrificer) 91-92, 107, 120, 122-123, 156
hound sorcerer like a hound 59
house of death and rebirth 183
Hubble Space Telescope xix, 16, 29, 38; Hubble images 16 (Fig. 1, Fig. 2), 31 (Fig. 5), 85 (Fig. 8), 198 (Fig. 12)
hydrogen xxiii-xxiv, 2, 24, 30, 35, 69, 107, 141, 150, 155, 207, 242, see also 'aṃbha (hydrogen, prima materia)'; body(ies) of hydrogen 42, 94-95, 141, 143; androgyne or hermaphrodite 102; Brahmaṇaspati blasts and smelts hydrogen 157-158; death of hydrogen 183, 223; destroyed by the fire of the Stone 42; giving forth sound 187; hydrogen atoms throw their bodies off 135; hydrogen ions 48; Invisibles 102-103; ionic bonds of hydrogen 42, 45, 61, 74, 102, 155; magnificent hydrogen ions 120; 'sin' ionic bonds of hydrogen 138, 156; three isotopes of hydrogen 219; water-bearer 187
hymn, stanza, verse the units of Vedic metres 191; Vedic metres 191-192; seven most common metres of the Ṛgveda 192

I
Iḷā Goddess Iḷā 84; mother milch cow, the Stone 45, 121; overflowing with the milk of sacrifice 46
Indian systems of thoughts xviii-xix, xxi, xxv, 8
Indra xxx, 23g, 59-60, 71-75, 97-100, 131-132, 147-149, 162, 164, 233; bull 71; cuts channel for celestial Waters 144, 147; delights in Soma drinking 71-72; different aspect of Trita 75; has beautiful jaws 71; Indra's belly 39, 71, 73; slays the serpent of the Heaven 71; thunderbolt a hundred-limbed or thousand-spiked 71; maker of light, shaper of all forms 74; two jaws 71; vajra 233
indu (bright drop of Soma) 22-23, 64, 71, 232, 238-240
Induja (name of Mecury) 53
infant(s) xxiv, 95, 238-239, 257 (Fig. 14)
inherent force 188, 190
Innerfield (antarikṣaṃ) 17-20, 30, 37, 150-151, 243, 251, see also 'Heaven, the'
Invisibles see under 'hydrogen'
ionic bond(s) xxiv, 34, 42, 66, 74, 86-87, 89, 93, 135, 154-156, 166-168, 222-223, 238, 244; demons, enemies of the Devas 89; sin of the ionic bond(s) 138,156, 166-167; wicked, wicked demon 89, 95; wicked evil-doer 224
ionisation see 'celestial ionisation'
ionised and unbound state 22
ioniser see 'celestial ioniser'

isotopes of hydrogen 219
Itihāsas (pl.) 257

J
Jagatī see 'metres, metrical structures'
jaghana (sg.), jaghanā (du.) 98
Jamadagni, Agni the blazing fire 120, 196
Jamison, Stephanie W. 9
Jupiter (the Vault of the Heaven) 52

K
Ka (who represents the Soma Pond) 42-44
kalaśā (water-pots) 40
Kālī (goddess) 94
Kali Yugaṃ see 'Yuga cycle'
kāma (desire, intent) xxiv, 188-189, 190g, 243
Kaśyapa 22, 196; has black teeth 22, 196; his wont of Soma drinking 22
Kaṭha Upaniṣad (ii.3.1) 116-117
Keystone 85-86, 138
King(s) King Soma 112; King of fish 165; see 'Mitra' and 'Varuṇa'
Kṛṣṇa(s) 94, 177
Kṛta (or Satya) Yugaṃ see 'Yuga cycle'
Krumu (woodland or forest of kṛmuka) 211-212
Kṣetrapati (lord of the field) 104
Kumāra (child, youth, Agni) 140-141
Kuśara grass see 'grass(es)'

L
lances see 'spears, lances'
Langmuir, Irving xix
law of Nature xxv, 256
light-containing plant 22, 31, 40-41, 53, 105 130, 185, 197, 218, 222, 239, 247, 252; Agni's smoking flames 222; stupendously long 147
Linden, Stanton J. 137f
liṅgaṃ 86
Lokāloka 27
Lord of the Forest (the Soma Pond) 39, 78-79, 83
Lord of the Forest (the Stone) 84, 100; two lords of the Forest 84, 100; lords of the Forest 84, 115
luna 52-53

M
Macdonell, A.A. xvii-xviii, 6
magnetic field(s) xxiii, 17-18, 20, 30, 37-38, 41, 68, 89, 119-121, 129-130, 174, 213, 243-244; bowl-shaped 20, 30; electric phenomena subsequent to and born of magnetism 89, 245; innate topology of the magnetic field structure 30; in the shape of Ḍamaru 30; stupendously large magnetic field structures cause the speed of streaming cosmic plasmas (1,794 light-years per hour) 213; see also 'Tvaṣṭā'
magnetic forces xxiii, 18, 37-38, 89, 119, 126, 128-130, 244-245; three divisions of magnetic forces 126; see also 'Ṛbhus (pl.)'
magnetic ropes 233, see also 'filament(s)'
magneto-plasma cosmology see 'cosmology of the Ṛgveda, magneto-plasma'
magneto-plasma structure(s) 30, 243
magnum opus xxvi
maiden(s) 57, 76, 84, 92; maidens of Trita 84, 150; maidens of Tvaṣṭā 57; ten maidens 76; Tvaṣṭā's ten maidens 84, 96;
Makino, Karen xix
man, men, heroes deceased man 149g; man xxiv, 153, 156, 193; men xxiv, 43, 64, 81, 87, 96; heroes xxiv, 64, 67, 96, 206; strong man 64, 67, 67g; wretched man 139
Mana (manaḥ, cosmic Mind) 182-183; cyclical journey, revolving Mana 183-186
maṇḍalas ten maṇḍalas of the Ṛgveda xvii, xxvi, 10, 12; a total of 1028 hymns 2;
Manu's Law Book 257
Manuṣyas (a class of deceased (ionised) ancestors) 194-195
Marīci (a particle of light) 22
Mars 52-53; celestial warrior 53
Martin, Thomas C. xix(f)
Maruts (pl.) 36-37, 40, 46-47, 63-68, 215, 247; adorned with bright lances 64; eternal flame 67; impel darkness 66-67; lances of the Maruts 54, 247; lightning laden chariots of 63-64; Maruts' chariots 37; produce streaming currents 63; sing delightful songs 64; Soma plants, or the plasma blooms, personified 63; tongues of Agni 63; Triṣṭup a libation for the Maruts 68; wear strong man's ring 64, 67
Mātā, Duhitā 33, 45-46
māyā xxx, 54, 126, 176-177, 245; magic force of Varuṇa 54, 176; revolves fast within all beings 177, 245; Suparṇī and Kadru 176; twofold magic force 177
McEnery, Julie 17

melody is born of the stanza 2
Mercury 52-53; father of the bright drop of Soma 53; Soma Press, the Stone 53
messenger(s) of celestial lights 104, 112, 127, 150
messenger(s) of the Heaven 21, 55, 156, 223
metres, metrical structures 191-192; Anuṣṭup 192, 194; Bṛhatī 192, 194; Gāyatrī 163, 192, 194; Jagatī 192, 194; Triṣṭup 192, 194; Triṣṭup a libation for the Maruts 68; Uṣṇihā 192, 194; Virāṭ 192, 194, 236; units of Vedic metres 191
Midheaven see 'Heaven, the'
milch cow daughter milch cow (the Soma Pond) 31, 33, 45, 203, 204, 224; mother milch cow (the Stone) 33, 45, 84, 108, 121, 154, 156, 168, 224; Aditi milch cow 154, 156; brilliant milch cow 127; Vāk milch cow 168-169
Milky Way, the xxii-xxv, 5, 17-21, 26-28, 32, 108-117; eight regions 113; five regions 111; four major spiral Arms 108-110, 249; Heaven above, Heaven below 18; name of four arms (Bhadrāśva Varṣa, Bhārata Varṣa, Ketumāla Varṣa, Uttara Kurus) 215; seven domains 112; six divine expanses 111-112; sustained by cosmic Yajñas 249; sustained by imperishable Waters (cosmic plasmas) 109
Mind see 'Mana (manaḥ, cosmic Mind)'
Mitra 53-55, 77, 149-150, 152-157; cosmic plasma light discharge 149; Kings Mitra and Varuṇa 77, 155; Mitra and Varuṇa sustain cosmic sacrifice through cosmic sacrifice 152; a thousand-pillared abode of Kings Varuṇa and Mitra 77
modes of Soma the third, fourth modes of Soma 236
Monarch Empress Aditi 154; of the Heaven and the Milky Way 94; of the whole World 69; sole Monarch 43; supreme Monarch 150-151
Monier-Williams, Monier xvii-xviii, xxxiii, 11; his Sanskrit-English Dictionary xvii-xviii, 11; when English definitions of a Vedic word are out of context 11
mons Veneris 86
Moon moon (Soma, electron) 24, 52-53; moon (the Soma Pond) 39, 47, 61, ruddy moon 97-98
Mortar and Pestle 84, 97
mother milch Cow see 'milch cow'
Mt. Meru 108, 113, 197, 214-215, 246, 249, 250 (Fig. 13), 251, see also 'Galactic Bar'
Mt. Mūjavān 22, 39, 135-136
Mukherjee, Krishna xix
Müller, Max F. xvi-xvii, xxi, 4, 7, 12; the Ṛgveda text compiled by Müller 12-13
Muñja grass see 'grass(es)'
mythical bird 59g, 144, 232g, see also 'Bennu', 'Firebird', 'Garuḍa', 'Phoenix'

N
Nag Hammadi Scripture, the (book) xxvi(f)
Nāgas (the serpent people) 247
Nāka (the Heavenly Vault) 21
Nakṣatras (of the Heaven) 40, 51-52, 86; constellations of plasma blooms 51-52; Nakṣatras revolve within the Vault of the Heaven 52
Nāsatyā 198, 206
Naṭarāja (lord of the dance) 95, 258 (Fig. 14)
Nature xxv-xxvi, 30; does not operate in a convoluted manner xxv; great work of Nature xxvi, 256; law of Nature xxv; magnetic field structure of Nature 30
nebula(s) 15,16 (Fig. 1), 29-30, 31 (Fig.5)
net net of the instigator 63; fish caught in the net 165-166
Nighaṇṭu xvii, 6f
Night (Varuṇa) 189, 208
Nikola Tesla Museum xix(f)
Niruktaṃ xvii, 3, 6
niṣkṛtaṃ see 'Place of the Atoned' and 'Place of Atonement'
non-Existence (non-Real) see 'Asat (non-Real, the state of non-Existence)'
noose see 'Varuṇa'
nucleus Agnis intimately joined within 161; bond between the nucleus and the outer shell 167; protons in the nucleus 34, 217; Soma encloses Agnis in the nucleus 232; a stronghold and a cave 74

O
oligarchs xxx, 138, 256
order of six 151-152
original sin see 'sin'
oṣadhi see' light-containing plant'
ouroboros or ophis xxx, 40, see 'Vṛtra(s)'
outposts of advanced civilisations xxvii
owl, owlet 59

P

padapāṭha xvi, xxii, xxxiv, 12-13
Pāṇini Pāṇini's grammar xxxi, 8
Parjanya streaming plasma currents personified 54; the rain of cosmic plasma personified 107, 252, 253g
particle of light see 'Marīci (a particle of light)'
Pataṃga (flying horse) see 'horse(s)'
Pavamāna Soma 22, 111, 232-235
peacock feathers of a peacock 94; peacock's plumes 40; peacock-tailed 40;
penance (tapaḥ) 188, 244
Peratt, Anthony L. xix, 173f
Phoenix 93, 144
Pillar(s) of the Heaven a hundred or a thousand sacrificial posts 77, 80-82; liṅgaṃ 86; luminous and lofty Heaven's prop 50; luminous Pillar of the Heaven 77; obelisk 86; Pillar the Soma Pond 23, 39, 42; remaining stump of the destroyed sacrificial Post 83; the Sacrificial Post 77-80
Place of Atonement (the Stone) 93, 138, 140
Place of the Atoned Mt. Meru (the Galactic Bar) 93-94; the Soma Pond 93, 234
planetary nebulas see 'nebula(s)'
plasma(s) see 'cosmic plasma(s)'
plasma bloom(s) see 'cosmic plasma bloom(s)'
plasma discharges see 'cosmic plasma discharge(s)'
plasma sheath(s) xxx, 40-41, 50, 55, 67, 128, 164, 173, 199, 201; see also 'Vṛtra(s)'; boat of hymns 207; Dāsa and Ārya, the two opposite charges of the double layer 164; demons, enemies of the Devas 89; fetters 128; a fleet of plasma sheaths 40, 70; how it is fashioned 203; new sheath is fashioned by their Speeches 203; protector of the house 203; a set of three swift steeds and a sheath filled full with sweet mead 199; tubular stalks 143; wicked sorcerer 59; see also 'serpent(s) of the Heaven'
ploughed field 104
ploughing wheel 135, 138
political correctness 11
prayer(s) see 'Speech(es) of cosmic plasmas'
prima materia xxiii, 242
protector of the hearth, or house 61, 160, 203
protium see 'Agni (proton)'
proton see 'Agni (proton)'
Pṛthivī (the Milky Way) 18, 28, 34, 52, 108-115; see 'Milky Way, the'
Pṛśni 47, 84; udder of Pṛśni 39, 46-47
Psalm (140:10) 137
Purāṇa(s) xxv, 26-27, 257

R

rain-bearer the Soma Pond, the rain-bearer 79
rainy season 128, 226-227, 229-230
Rajasī (du. the two Heavenly Spheres of light) 28
Ram a male ship, refers to the Soma Pond 39
Rāma 120
Rawls, Walter C. Jr. xix, 30
Ṛbhukṣā see 'Ṛbhus (pl.)'
Ṛbhus (pl.) xxiii, 18-19, 37, 126-133, 234 242-244; a triad of the Ṛbhus 126; divine architect(s) 38, 119; guard the cow 129; make the four celestial Bowls 129; Ṛbhu or Ṛbhukṣā 126, 244; repose at Agni's rite of receiving Soma 130; revive their old parents 127; skilful architects 131; Vāja 126, 244; Vibhvā 126, 244
Real see Sat (Real, the state of Existence)
reciprocal relationship 169, 234
remnant(s) of the sacrifice 83, 247
reṇu (charged particle) 69
Ṛgveda Ṛgveda Saṃhitā 1, 2; Ṛgveda systematic expositions of the Vedic cosmology 10; Ṛgveda the knowledge of cosmic creation xxi-xxii, xxv, xxix; What are revealed in the Ṛgveda xxiv, 38 (Fig. 6), 256
Ring of Fire 57, 84-85
river(s) see 'celestial river(s)'
Rodasī (du.) the pair of milch Cow Mātā (माता) and Duhitā 45; the pair of the Stone and the Soma Pond 28, 121, 174, 181
Roman fasces 77
Ṛṣi(s) xxiv, 15, 123, 171, 194-195, 234-236, see also 'Seven Ṛṣis'
Ṛtvijas (priests) 92, 97; four Ṛtvijas: Hotā, Adhvaryu, Udgātā, and Brāhmaṇa 92
ruddy Bird see 'Bird, ruddy'
Rudra 39-41, 46-47, 60-63; arrows of Rudra 40, 54; father of the Maruts 46-47, 61; has a strong bow and swiftly flying arrows 61; owner of his own abode 61; protector of the hearth 61; Rudra not Śiva 61; Rudra's launched arrow 62
Rudras (pl. sons of Rudra) 81

S

sacrifice (celestial, cosmic) see 'cosmic Yajña(s)'
sacrifice (Vedic) see 'Vedic yajña(s)'
sacrificial fire (Trita) 194, see also under 'Agni (proton)'
sacrificial food 100
sacrificial post(s) see 'Pillar(s) of the Heaven'
Sairya grass see 'grass(es)'
Sāma hymn(s) 142, 191, 193; Sāma hymns weaving shuttles 193
Sāmaveda xxv, 1, 2; Sāmaveda Saṃhitā 1; songs for the Udgātā 2
Saṃhitā(s) 1-2
Saṃhitāpāṭha xvi, 12-13
Samudra (the Ocean of celestial Waters) 18, 22, 37, 39, 44-45, 177, 197, 208, 236, 248
Saṃvatsara 39, 75
Saṃvatsarakālacakraṃ 75
Śara grass see 'grass(es)'
Sarasvān 11, 146
Sarasvatī celestial river 24, 208; the Soma Pond 11, 39; see also 'celestial river(s)'
Sat (Real, the state of Existence) 157, 187-188
Śatapathabrāhmaṇaṃ xxii, 75, 176
Saturn 52, 53; name of Saturn mandaḥ derived from √mand (to rejoice, exhilarated, be splendid, beautiful) 53; physician dispenses the salubrious medicines 53
Satya Yugaṃ see 'Yuga cycle'
satyaṃ (reality, truth) 49-50, 206
Savitā 89-90, 113, 161-163; Savitā carries the thoughts and prayers 163; illumines the Milky Way's eight regions 113; impels the firestones of the Stone 89; vivifier 161
sāvitrī stanza see 'gāyatrī stanza'
Sāyaṇa xvii, 6-7, 12
Selden, Gary xix
self-powered One see 'svadhāvān (self-powered One)'
Sengupta, Prabodhchandra 23
Serpent of the Heaven 40, see also 'Vṛtra(s)'
serpent(s) of the Heaven xxx, 21, 40, 59, 71-72, 201, 233, 246; see also 'Vṛtra(s)' and 'plasma sheath(s)'
serpent people (Nāgas) 40, 247
Seven Ṛṣis 196; names of seven Ṛṣis: Atri (devourer), Bharadvāja (the bearer of speed), Gotama (the best of oxen), Jamadagni (Agni the blazing fire), Kaśyapa (one who has black teeth), Vasiṣṭha (the owner of the cow of plenty), Viśvā-Mitra (all-containing Mitra)
sheep's fleeces, golden fleece see 'fleece(s)'
Siddhāntaśiromaṇi 23-24, 26-27, 113, 214
sin 93, 138, 154, 156-157, 166-167; original sin that must be expiated by scaricfice 138
Sindhu 205, 208-214; beautiful to behold 212-213; advances upwards roaring like a charging bull 210; grows bright 214; leads two wings of army 210-211; most rapid of the rapids 212-213; reaches Gomatī and krumu 211-212; see also 'celestial river(s)'
single cause and a single source xxiv, 242
Single Eye 51
Śiśu (infant, child) xxiv, 238-239
Śiva xxiv, xxvi, 28, 38, 75, 86; Ḍamaru (the drum of Śiva) xxiv, xxvi, 18, 28, 38, 94; Rudra not Śiva 61; trident of Śiva 95; see also 'Naṭarāja (lord of the dance)'
Śiva liṅgaṃ 86
Slocum, Jonathan xvii
Smṛtis (pl.) xxv, 257
sol 52-53
Soma (electron) 232-238; annaṃ (mystical food) 248; begetter of sacred hymns 235; bestower of light 236; Brahmā of the Devas 234-235; bright drops of Soma 53, 144; bull midst the wild beasts 234-235; daddy is a physician 240; embryonic particle 12, 247-248; exhilarating 233; falcon among the vultures 234-235; guides newly atoned people to the Place of the Atoned 233-234; King Soma 112; leader of the bards 235; lord of Speech 237-238; lord of the mind 237; lord of the sacrifice 233-234; lord of the maidens 233-234; mummy grinds grain 240; orbits the nucleus 161, 167; skilled in a thousand arts 236; leader of the bards 236; spins a triple filament 233-234; Soma of enlightened mind 236; steed of a war-chariot 237-238; third press of Soma 72-73; third Soma sacrifice 132; water buffalo 232
somadhanā (Soma receptacles) 40
Soma draught(s) 73, 138, 223-225, 230, 235
Soma plants 18, 21, 39-40, 143-144, 158, 208-209, 211, 217-219, 247; continuously spring up 62; cosmic plasmas blooming into ridge-like structures are Soma plants 31; hundred filaments of Soma plants 148; Maruts are Soma plants 36; medicines 62; new shoots of

Soma plants 239; regrowing Soma plants 49; ridge-like structures 31; Soma plants sacrificial posts 77; tubular stalk of Soma plant blown 143
Soma Pond xxx, 18, 28-34, 39-40, 41 (Fig. 7), 44-48, 75-78, 86-88, 226-227, 246-249; ancient Ocean of splendour 203-204; armoured with a plasma sheath, furnished with the Stone Wheel 97; one in the Upper Vault the other in the Lower Vault of the Heaven 28; Reedy Soma Pond 37, 39; Saturn Soma Pond 52-53; seat of virtuous deed 202; sits upon the Stone 37; Two Soma Ponds coordinate to perform Soma Yajña 29; where discharged cosmic plasmas gather, grow, and bloom 52, 93; water-bearer 115, 210; womb of cosmic creation 39-40, 42, 52, 94
Soma Press 53, 84, 97-98, see also 'Stone, the'
Soma Yajña, Soma sacrifice 29, 69, 114, 132, 201, 229, 246
sorcerer 59
sorceresses 40, 59
spears, lances glittering spears of Maruts 40; spears 21, 53, 63-64, 143, 247; lances 54, 64, 66-67, 233, 247
spears of destiny 247
Speech(es) of cosmic plasmas 21, 60, 90, 106, 108, 121, 169, 191-192, 203, 226-232, 244, 248-248; numerically coded as metres of the stanzas 68, 191; Speeches are Brahmā, the Creator 21; Speeches of the singers are responsible for the manifestations of varying cosmic plasma phenomena we observe in the cosmic field 249; Speeches, or sounds, are inherent in physical reality and are intimately associated with the creation, manifestation, and sustenance of physical reality 192, 248
Sphere(s) (Heavenly) the Heavenly Spheres 40, 114; threefold Heavenly Sphere 18-20
steed(s) 43, 84, 105, 179, 213; neighing steed 114; of a war chariot 237-238; Soma pressing Etaśa 84; swift steeds 126-127, 131; winged steed 197-198
St. Peter's Square see 'Vatican'
standard model (of physics) xxii, xxx, 126
Stone, the xxiv-xxv, 2, 20, 28-33, 37, 44-48, 77-78, 84-87, 89-107, 204, 218-220, 222, 237-240, 243-244; at the base of the Heavenly Vault 69; comb 101-103; charcoals (Heavenly) 137; firestones 2, 30, 56-57, 75-76, 84-85, 89, 92, 136-137, 149, 233, 244; firestones Tvaṣṭā's maidens 92, 96; Lord of the Forest 78-79, 81, 83-84, 100; lords of the Forest 84, 100, 115; many-footed 98; Mother cows 56, 211; Philosopher's Stone 84; sing and make rumbling sounds 90; ten fingers 76, 84, 95-96; Trita fire(s) of the Stone 74, 85, 155; two arms 43-44, 76, 95-96, 238; two discs of the Stone 2, 37, 71, 84, 87, 98, 240; two lords of the Forest 84,100; two (press) stones 37, 76, 84, 86-87, 97; two stones of Trita 84, 86-87; Vivasvān 161,165-166, 183, 238
Stone Wheel 97; stone wheels 86
strong man's ring 64, 67
Su, Meng 17f, 24, 26-27, 199f
sun (of our solar system) 5, 24, 48, 52
Śuna (the auspicious One, the Stone) 106
Śuna and Sīra (ploughshare and plough) 106-107
Śunaḥśepa (tail of the auspicious One) 98, 115
suparṇa 144; mahāsuparṇa (great Bird) 75; see also 'mythical bird' and 'Garuḍa'
superstition(s) xxvi-xxvi, 10
supreme monarch xxiv, 150-151; see also 'King(s)'
Sūra (another name for Sūrya) 51-52
Sūrya 24, 39, 48-55, 94-95, 102, 147, 232, 234-235; gold-hued pillar of Sūrya 159; Sūrya, satyaṃ cast into the receptacle 49; sees far and wide 49
Sūryasiddhānta xxiii, 23-27, 35, 52-53, 108, 110, 113f
Sutton, Nicholas xviii
svadhāvān (self-powered One) 150
Svarita see 'Vedic accents'
symbol X 233
symbolism occult symbolism xxv, 256; Vedic symbolism xvii, xxvi; practicing symbolism 138

T
tail of the auspicious One see 'Śunaḥśepa' and 'peacock'
tapa (tapaḥ, penance) 188, 244
ten fingers see 'Stone, the'
Tesla, Nikola xix, xix(f)
third press of Soma see 'Soma (electron)'
Thomson, Karen xvii, xix, 8-10
three isotopes of hydrogen 219
three-tongued flame(s) 95

thunderbolt(s) see 'Indra'
tongues of fires 66; tongues of Agni's flame 143; tongues of flame 219
Tree (the Soma Pond) radiant red Tree 92; Tree clothed with beautiful leaves 141
Tree of Life (Aśvattha, inverted tree) 115-117, 116 (Fig. 11); imperishable, resplendent Tree 117; the maker and discharger of celestial lights 117
tree(s) Soma plants, or plasma blooms, or light-containing plants 22, 37, 20, 211, 218, 247; kṛmuka trees 211; trees are shattered 36; Vibhīdaka tree 135-136
Tresman, Ian xix
Tretā Yugaṃ see 'Yuga cycle'
Tridhātu Rajasa (the threefold stratum of the Heavenly Sphere) 19
Trident of Śiva 95
Tridiva (the threefold Heaven) 19
Trimūrti, the first embodied three 19
Trināka (the threefold Heavenly vault) 19
Trīṇirajāṃsi (the three cultivated Spheres) 19
triple filament 233-234, see also 'filament(s)' and 'triply-spun filament(s)'
triply-spun filament(s) 56, 69, 87, 95-96, 166, 199, 209, 233, 245, 247
Trirocanā (threefold luminous sphere) 19
Triṣṭup see 'metres, metrical structures'
Trita (tritium) see 'Agni (proton)'
Tvaṣṭā xxiii, 18, 37, 41, 60-61, 96, 119-125, 161-162, 242, 244-245; the first born 89, 119, 188, 244; divine architect 48, 51, 122-123, 161; self-powered One 124; shaper of all forms 74, 119-120; origin of cosmic mind 89; possessor of the World 119, 121; proficient in the art of magic 121; wearer of all forms 119
twin Jets 29-30, 36-37, 88, 121, 197-200, 245-246, see also 'Aśvinā'
two arms see 'Stone, the'
two discs of the Stone see 'Stone, the'
two opposite charges (of the plasma sheath) 4,164, 173; see also 'plasma sheath(s)'
two winged arms 122; two arms and wings 124

U

Udātta see 'Vedic accents'
udder of Pṛśni see 'Pṛśni'
udders of flaming milch cows 64
Udgātā (the singer of hymns) 2, 92
universe xxiii-xxiv, 217, 242; the infinite ocean of celestial aṃbha, hydrogen xxi, 242
Upaniṣads xxv, 1
uṣā (sg.) see 'Dawn (uṣāḥ, sg.)'
uṣasaḥ (pl.) see 'Dawns (uṣasaḥ, pl.)'
Uṣṇihā see 'metres, metrical structures'

V

vahni (vahniḥ, fire, the god of fire) 74, 84, 204
Vāja see 'Ṛbhus (pl.)'
vajra (thunderbolt) see 'Indra'
Vāk 168-172, 237; bow for Rudra 171; hymn of Vāk 168-172; Speeches, the sounds personified 169; Speeches of cosmic plasmas 168-169; Vāk Queen of the Devas 168
Van Nooten, Barend A. xvii, 192
Varna, Avinash L. xix
Varṣas (pl.) Arms of the Milky Way 108, 215
Varuṇa 53-55, 77, 149-153, 155-157, 165-168, 173-174, 176, 201, 208-209; crown of King Varuṇa 37; cuts the channels for Sindhu 209; King of hallowed might 116; King Varuṇa 37, 86, 94; magic force of Varuṇa 126, 176; measures out the Milky Way with Sūrya 176; supreme Monarch 150-151; thousand-pillared Abode of Varuṇa 39, 77; Varuṇa's noose 166-168; Varuṇa's sheath 166-167
Vasukra recorded the Veda in writing xv-xvi, 5
Vasus (pl.) 81-82, 147, 169; a class of the Devas whose chief is Indra 81
Vāta(s) 49, 66, 68-70, 100, 127; germ of the World 70; soul of the Devas 70; Vāta drives celestial Waters 127; Vāta blew forth the pressed Soma 100
Vatican flag of the Vatican City 86; St. Peter's Square 86, 138; Vatican City 138
Vault of the Heaven xxiv, 21, 24-30, 38 (Fig. 6); bowl-shaped 17 (Fig. 3), 18, 20, 25-26, 28; height of the Heavenly Vault 23-27; twofold Vault of the Heaven 28, 30, 34-35, 171; Lower Vault, Upper Vault of the Heaven, xxiv, 28-30, 34-36, 77, 88, 93; Lower Vault of the Heaven revolves eastward 34-36; Upper Vault of the Heaven revolves westward 34-36; Vault of the Heaven Brahmā 21
Vāyu 68-70, 107; companion of celestial Waters 70; majestic chariot of Vāta 69; Monarch of the

whole World 69
Vedāṅga(s) auxiliary sciences of the Veda 3; six Vedāṅgas (Śikṣā, Chanda, Vyākaraṇaṃ, Niruktaṃ, Jyotiṣaṃ, and Kalpa) 3
Veda(s) xxi-xxii, 1, 2; recitation methods xvi; four branches of the Veda (Ṛgveda, Yajurveda, Sāmaveda, Atharvaveda) 1; loss of the original meaning xvi, 6, 24; Müller recompiled and published the Ṛgveda xvi; study of the Veda will kill out superstitions 10; Veda first recorded in writing xv-xvi; Veda lost, obliterated xxii, xv-xvi, 4-6
Vedi (Altar of cosmic Yajña) 20, see also 'Altar of cosmic Yajña'
Vedic accents Anudātta, Svarita, Udātta xxxii
Vedic language xviii, xxi-xxii, xxvi, 8; Ancient High Indian 8; classical Sanskrit 8
Vedic literature xxv, 1-2
Vedic yajña(s) xxiv-xxv, 242
Venus 52-53, 86
Vibhīdaka tree 135-136
Vibhvā see 'Ṛbhus (pl.)'
Vimāna xxx, 33
Vīraṇa grass see 'grass(es)'
Virāṭ see 'metres, metrical structures'
Viṣṇu 109, 112, 215, 235
Viśvakarmā 50-51, 119, 122-125, see also 'divine architect(s)'
Viśvā-Mitra 196
Viśvarūpa 22, 161-162; infinite forms contained in him 22; shaper of all form 162
Viśve Devas 22, 50-51, 111-112; ; born and grow in the Soma Pond 51, 111
Vivasvān see 'Stone, the'
Vṛtra(s) xxx, 21, 40-41, 41 (Fig. 7), 50, 71-72, 93; Dāsa and Ārya, the two opposite charges of the double layer of the plasma sheath 164; ouroboros or ophis xxx; protector of the house 203; serpents of the Heaven 40; the Serpent of the Heaven xxx, 40; the serpent race 93;
Vivekānanda xix, xxi, xxi(f), xxii, 6f, 10
Vyāsa (Vedavyāsa) xv-xvi, xxii, 5, 23-24
Vyoma (n.) xxiv, 21
Vyomā (m.) 109

W
Wallis, H.W. xvii
warring angels 21, see also 'Aṅgirā(s)'
Watchers 173-175, 246-247; Agni's Watchers 174; Soma's Watchers 175; Varuṇa's Watchers 173
Water(s) see 'celestial Water(s)' and 'cosmic plasma(s)'
water-bearer 115, 187, 188g, 210, 242; aṃbha, hydrogen, the original water-bearer 187, 242; the Soma Pond, the Heaven's water-bearer 115, 210
What are revealed in the Ṛgveda, what the Ṛgveda reveals xxiv, 241
wheel of cosmic ionisation see 'Saṃvatsarakālacakraṃ'
Whitney, William D. xvii
wild beast(s) 65, 103, 234-235
Wilson, H.H. xvii, 7, 198; Wilson's translation of the Ṛgveda xvii
winged arms see 'two-winged arms'
winged horses see 'horse(s)
Winternitz, M. xxxi, 2, 3f, 4, 6, 8
Womb of cosmic creation 39-40, 42

Y
Yajña(s) see 'cosmic Yajña(s)' and 'Vedic yajña(s)'
Yajurveda xxv, 1-2
Black Yajurveda (Kṛṣṇa Yajurveda) 1-2; White Yajurveda (Śukla Yajurveda) 1-2; Yajurveda Saṃhitās 1-2
Yama 141, 143, 161-162, 183-186; house of death and rebirth 183; house of Yama 183; seat of Yama 143
Yamī 161
Yamunā 24, 208, 211-212, see also 'celestial river(s)'
Yāska xvii, 6-7
yojana(s) 25-27
yoni (the womb, source, seat, fountain) 2, 32
Yuga cycle 4-5; ascending and descending Yuga cycle 4-5; Kṛta or Satya Yugaṃ 4-5; Tretā Yugaṃ 4-5; Dvāpara Yugaṃ 4-7; Kali Yugaṃ 4-5; Solar Yuga cycle 4

About the author

The author has a doctorate in remote sensing from the University of Utah, taught at several universities, and pursued research in multispectral image analysis algorithms for environmental mapping. After retirement, she has been pursuing research in the Vedic and the classical Sanskrit literatures, focusing on the Ṛgveda (ऋग्वेदः), the Atharvaveda (अथर्ववेदः), the Upaniṣads (उपनिषदः pl.), the Bhagavadgītā (भगवद्गीता), and the Mahābhāratam (महाभारतं).